Life Cycle of Sustainable Packaging

Life Cycle of Sustainable Packaging

From Design to End of Life

Rafael A. Auras
Michigan State University, MI, USA

and

Susan E. M. Selke
Michigan State University, MI, USA

WILEY

Published by John Wiley & Sons, Inc., Hoboken, New Jersey.
Published simultaneously in Canada.

Library of Congress Cataloging-in-Publication Data
Names: Auras, Rafael Antonio, author. | Selke, Susan E. M., author.
Title: Life cycle of sustainable packaging : from design to end of life / Rafael A. Auras, Susan E.M. Selke.
Description: Hoboken, New Jersey : John Wiley & Sons, Inc., [2023] | Includes bibliographical references and index.
Identifiers: LCCN 2022012690 (print) | LCCN 2022012691 (ebook) | ISBN 9781119878100 (hardback) | ISBN 9781119878117 (pdf) | ISBN 9781119878124 (epub)
Subjects: LCSH: Packaging--Environmental aspects. | Packaging waste--Environmental aspects. | Packaging--Design.
Classification: LCC TD195.P26 A97 2023 (print) | LCC TD195.P26 (ebook) | DDC 688.8--dc23/eng/20220812
LC record available at https://lccn.loc.gov/2022012690
LC ebook record available at https://lccn.loc.gov/2022012691

Cover image: Courtesy of Rafael A. Auras
Cover design by Wiley

Set in 9.5/12.5pt STIXTwoText by Integra Software Services Pvt. Ltd, Pondicherry, India

To our families

Contents

List of Abbreviations

Abbreviation	Definition
AAGI	Annual Greenhouse Gas Index
AD	Average Dose
AFO	Animal Feeding Operation
AFPR	Alliance of Foam Packaging Recyclers
ALCA	Attributional Life Cycle Assessment
APC	Air Pollution Control
APR	Association of Packaging Recycling
AQI	Air Quality Index
ASTM	ASTM International, formerly known as American Society for Testing and Materials
ATSDR	Agency for Toxic Substances and Disease Registry
BACT	Best Available Control Technology
BF	Bioreactor Landfill
BMP	Best Management Practices
BOD	Biological Oxygen Demand
BOF	Basic Oxygen Furnace
BOPP	Biaxially Oriented Polypropylene
BPI	Biodegradable Products Institute
CAA	Clean Air Act
CAFO	Concentrated Animal Feedlot Operation
CAGR	Compound Average Growth Rate
CBW	Corrugated Board Basis Weight
CDC	Centers for Disease Control
CE	Circular Economy
CERCLA	Comprehensive Environmental Response, Compensation, and Liability Act
CFC	Chlorofluorocarbon

CFR	Code of Federal Regulations
CHP	Combined Heat and Power
CISWI	Commercial and/or Industrial Solid Waste Incineration
CLCA	Consequential Life Cycle Assessment
COD	Chemical Oxygen Demand
CWA	Clean Water Act
DALY	Disability-Adjusted Life Years
DDT	Dichlorodiphenyltrichloroethane
DEQ	Department of Environmental Quality
DRC	Display Ready Corrugated
EDF	Environment Defense Fund
EF	Ecological Footprint
EFP	Environmental Footprint
EG	Emission Guidelines
EGLE	Department of Environment, Great Lakes, and Energy
EKC	Environmental Kuznets Curve
EPA	Environmental Protection Agency
EPCA	Environment Pollution (Prevention and Control) Authority
EPD	Environmental Product Declaration
EPR	Extended Producer Responsibility
EUROPEN	The European Organization for Packaging and the Environment
EVOH	Ethylene Vinyl Alcohol
FA	Fly Ash
FCM	Food Contact Material
FDA	Food and Drug Administration
FFDCA	Federal Food, Drug, and Cosmetic Act
FIFRA	Federal Insecticide, Fungicide, and Rodenticide Act
FLW	Food Loss and Waste
FTIR	Fourier Transform Infrared Spectroscopy
FWG	Food Waste Generation
GDP	Gross Domestic Product
GHG	Greenhouse Gas Emissions
GMA	Grocery Manufacturers Association
HAP	Hazardous Air Pollutant
HC	Hydrocarbon
HDPE	High-Density Polyethylene
HFC	Hydrofluorocarbon

HMIWI	Hospital/Medical/Infectious Waste Incineration
IPCC	Intergovernmental Panel on Climate Change
ISBM	Injection Stretch Blow Molding
ISO	International Organization for Standardization
IWLF	Industrial Waste Landfill
LCA	Life Cycle Assessment
LCI	Life Cycle Inventory
LCIA	Life Cycle Impact Assessment
LCM	Life Cycle Management
LCRS	Leachate Collection and Removal System
LCT	Life Cycle Thinking
LCV	Lower Calorific Value
LDPE	Low Density Polyethylene
LF	Landfill
LFG	Landfill Gas
LLDPE	Linear Low Density Polyethylene
LMOP	Landfill Methane Outreach Program
MACT	Maximum Achievable Control Technology
MCL	Maximum Contaminant Level
MCLG	Maximum Contaminant Level Goals
MRDL	Maximum Residual Disinfectant Level
MRDLG	Maximum Residual Disinfectant Level Goal
MRF	Material Recovery Facility
MSU	Michigan State University
MSW	Municipal Solid Waste
MSWI	Municipal Solid Waste Incineration
MSWLF	Municipal Solid Waste Landfill
MWI	Manufacturing Waste Incinerators
NAAQS	National Ambient Air Quality Standards
NAPCOR	National Association for PET Container Resources
NDIR	Nondispersive Infrared Spectroscopy
NFP	Nutrition Facts Panel
NGO	Nongovernmental Organization
NIAS	Nonintentionally Added Substances
NIMBY	Not in My Backyard
NMOC	Nonmethane Organic Compounds
NOAA	National Oceanic and Atmospheric Administration

NOL	No Objection Letter
NPDES	National Pollution Discharge Elimination System
NSPS	New Source Performance Standards
OCC	Old Corrugated Containers
OECD	Organization for Economic Cooperation and Development
OLR	Open-Loop Recycling
ONP	Old Newspaper
OPP	Oriented polypropylene
OSWI	Other Solid Waste Incineration
PAH	Polycyclic Aromatic Hydrocarbon
PBAT	Poly(butylene adipate-*co*-terephthalate)
PBB	Polybrominated Biphenyls
PBT	Persistent, Bio-accumulative, and Toxic
PCB	Polychlorinated Biphenyl
PCL	Polycaprolactone
PCR	Post Consumer Recyclate
PDO	1,3-Propanediol
PE	Poly(ethylene)
PERC	Tetrachloroethene
PET	Poly(ethylene terephthalate)
PETG	Glycol-modified PET
PFAS	Per- and Polyfluoroalkyl Substances
PHA	Poly(hydroxyalkanoate)
PHB	Poly(hydroxybutyrate)
PHO	Poly(hydroxyoctanoate)
PKG	Packaging
PLA	Poly(lactic acid)
PM	Particulate Matter
PP	Poly(propylene)
PRF	Plastic Recycling Facility
PS	Poly(styrene)
PTMAT	Poly(tetramethylene adipate-*co*-terephthalate)
PVC	Poly(vinyl chloride)
PVDC	Poly(vinylidene chloride)
PVOH	Poly(vinyl alcohol)
RCRA	Resource Conservation and Recovery Act
RDF	Refuse Derived Fuel

RPC	Reusable Plastic Containers
RPP	Reusable Plastic Pallets
SDG	Sustainable Development Goal
SDWA	Safe Drinking Water Act
SI	Sustainability Indicators
SoP	School of Packaging
SPA	Sustainable Packaging Alliance
SPC	Sustainable Packaging Coalition
SPI	Society of the Plastics Industry
SRI	Steel Recycling Institute
TCE	Trichloroethylene
TFI	The Fertilizer Institute
TMDL	Total Maximum Daily Load
TPS	Thermoplastic Starch
TSCA	Toxic Substances Control Act
TSS	Total Suspended Solids
TVA	Tennessee Valley Authority
UFP	Ultrafine Particle
UN	United Nations
USAID	United States Agency for International Development
USCC	US Composting Council
USDA	United States Department of Agriculture
USP	US Pharmacopeia
VOC	Volatile Organic Compound
WHO	Word Health Organization
WRAP	Waste & Resource Action Programme
WtE	Waste to Energy

Preface

Packaging has become an essential instrument for sustainable development since it can guarantee protection, distribution, and safe consumption of food, medicine, and general goods. Civilization's history and development are highly intertwined with the materials used to produce packages and their construction, transportation, and commerce. From the clay amphoras used to deliver olive oil and wine across the Mediterranean during Roman and Egyptian times, to the local delivery of food in modern societies, to the freight transportation of goods from China to the USA, to the delivery of goods to the current space mission, and the future Mars exploration program, packaging has played a fundamental role in making all these enterprises possible.

Yet, the other side of packaging that has become an issue for modern society as population numbers expand is its disposal and the contamination of the environment by single-use packaging, mainly due to the lack of formal collection and waste management systems in low-income economies worldwide. As societies have become more dependent on packaging systems to guarantee their well-being and development, our relationship with packaging has become more complex, requiring a deep and all-inclusive understanding of the benefits and responsibility of implementing packaging systems.

On the journey of creating, teaching, and researching packaging systems, the soul of this book was born through a long relationship by the authors with packaging systems and their challenges and opportunities for creating packaging for sustainable development. Although it took a long time to complete this first edition, several different pieces of the work presented here are an evolution of material previously developed for several audiences in conferences and publications. In this journey, we learned that creating packaging systems for sustainable development requires a transdisciplinary approach, extensive and meaningful collaborations, and an examination of the entire life cycle of the packaging materials, processing, and end of life so that the future packaging systems can be conceived. So, to reflect on this journey and facilitate learning, we decided to organize this book so that students and teachers can learn and develop future packaging systems considering the holistic benefits and impacts they produce for society.

We start the book by providing a general framework of sustainability and circular economy related to packaging – Chapter 1. This chapter provides an available description of sustainability and the evolution of this concept, and implications for packaging. We introduce the United Nations Sustainable Development Goals (UN-SDGs) and how to create sustainable indicators for the evaluation of systems in general and in packaging. As any new packaging begins with an idea and a desire to solve a need, we continue with an invited chapter about design thinking and how to use this methodology in packaging – Chapter 2. This chapter also provides a general discussion of the tools used to create packaging systems. Since not all the readers will have training on the primary packaging materials and the types of energy required to produce them, Chapter 3 provides a general overview of packaging materials, processing, and the energy used to make them as well as a general description of the main components of packaging systems.

In this book, we gave much attention to the concept of life cycle assessment (LCA) as the primary tool to evaluate the environmental footprint (EFP) of packaging systems; therefore, we spend several chapters building the necessary background knowledge about pollution and how to create EFP indicators. Chapter 4 defines the main criteria of pollution created by natural and anthropogenic factors and explains risk assessment and management. Chapter 5 further explores the main reasons and culprits causing soil pollution. Chapter 6 introduces the concepts of groundwater and surface water pollution. It describes the main vectors responsible for water pollution and dives into the primary US legislation regulating it. Chapter 7 discusses air pollution and examines the primary or criteria air pollutants responsible for modern air pollution. In this chapter, we also introduce the main framework of the Clean Air Act. Although most of the chapters in this section are US centered due to the regulatory bodies for soil, water, and air pollution, they apply to almost any region worldwide with some considerations. We close the section on pollution with Chapter 8 – global climate change that has become one of modern society's most significant and most concerning issues. This chapter elaborates on the primary sources of greenhouse gases and their impact on global climate. We also provide a discussion of the leading climate change agreements. After the background for dealing with pollution and creating EFP indicators are presented, we concentrate in Chapter 9 on providing a general review of LCA and the main steps to conduct a streamlined LCA when considering cradle to gate, grave, or cradle boundary conditions. This chapter also provides a general discussion of the main steps to complete an LCA according to the International Organization for Standardization (ISO) standards.

Chapter 10 introduces the concept of municipal solid waste (MSW) and the leading waste management systems for MSW – reduction, reuse, recycling, incineration with energy recovery, and residuals management. Chapter 11 expands on the reduction method of MSW and combines streamlined LCA to quantify the EFP of the reduction strategies, further focusing on the primary packaging materials (i.e., glass, metal, paper, and plastic). Chapter 12 focuses on reusing and the main scenarios and systems where reusing is beneficial, discussing how to improve them. Chapter 13 deals with recycling from the consumer, collection, and separation perspectives. Chapter 13 also includes issues about the economy and the main systems to recycle packaging materials. Chapter 14, although also considered a recycling method, deals with aerobic and anaerobic biodegradation of packaging and explores the requirements to certify a package as compostable and derived from bio-based resources. Incineration with energy recovery is an essential and growing method to deal with MSW, more in other countries than in the US, and it is discussed in Chapter 15. This chapter also discusses the main types of waste combusting units. Finally, we finish this section with Chapter 16 discussing sanitary landfills and the methods to control and deal with the pollution created by these units. The principal regulations for building and operating landfills according to the US framework are discussed.

Although not a waste management system but an important consideration when discussing packaging and its impact on current society, Chapter 17 introduces the impact of littering in the terrestrial and marine environments. It provides some relevant data mainly on the effect of littering plastics on the environment and the impact of global climate change on increasing littering in the environment. We finalize the book with Chapter 18, which we called "keeping in perspective," where we evaluate the entire supply chain and the impact of packaging systems on the whole EFP of packaging systems and their products. We also introduce some notions of the impact between consumer perceptions and the creation of packaging for sustainable development.

As a reader, learner, or instructor of or for this book, please, feel free to combine, remove, and mix and match the chapters as they fit you and your learners since the book was written principally keeping in mind students with packaging, material science, and supply chain backgrounds. But it should be easily implemented for an array of learners with different experiences by complementing or reducing the material covered in selected chapters. We provided case studies for each chapter to help illustrate the main concepts. Still, since pollution as well as packaging are transdisciplinary sciences, the presentation of each case study can be seen and expanded using different perspectives, and we encourage that. We hope that you enjoy your time spent with this book, and please provide any feedback on how to improve it.

We want to express our gratitude to our students, who allowed us to develop and try out this material through the years. Our colleagues helped us discuss and kept us accurate to be accountable about general claims and rigorous about the calculations. Special thanks to our editorial team at Wiley for helping bring this book to life, and not least to our personal and extended packaging families that are part of this exciting journey to create a better future for society through the packaging lens. Although any book edition is imperfect, we reviewed the content several times; however, there are always fairies and goblins playing with our writing and editing process, introducing typos and mistakes, mixing with our best intentions. So, any typos and mistakes are the responsibility of the authors and the authors only.

Rafael A. Auras and Susan E. M. Selke
February 2022
East Lansing, MI, USA

About the Companion Website

This book is accompanied by companion website:

www.wiley.com/go/Auras/lifecycleofsustainablepackaging

This website includes Solution Manual and Teacher Materials

1

The Role of Packaging in Sustainable Development

1.1 Learning Objectives

After learning from this chapter, you are expected to:

- Understand the role of packaging in social, environmental, and economic development contexts.
- Describe the main concepts of sustainability and its history.
- Understand the main areas of sustainability.
- Explain the 17 United Nations Sustainable Development Goals (UN-SDGs).
- Describe the main organization working on creating packaging for sustainable development.
- Understand the main concepts of life cycle thinking and circular economy.
- Describe the main tools designed to evaluate the environmental footprint of packaging systems.

1.2 Introduction

Along with the world's dramatic population growth over the past 50 years, the production of goods and services and the size of the global economy have massively increased. This exponential growth has strained our environment by increasing the levels of contaminants and the depletion of natural resources. Packaging is an essential part of valuable goods, but packaging production has environmental impacts, and packages are usually disposed of after use. This disposal contributes to environmental distress if proper recovery methods are not in place. Significant efforts have been made in optimizing packaging along the packaging value chain to reduce waste and reduce its environmental footprint. Even though continuous technology development and attitude changes have helped to reduce packaging waste, packaging and containers remain crucial to achieving sustainable systems.

1.3 Packaging and Sustainable Development

Packaging is a crucial component of the economy, environment, and society of any country. Packaging is highly intertwined in individuals' daily lives, helping societies transport, protect, and contain materials for producing and consuming goods. Packaging enables consumers to access products such as electronics and clean and safe medicine and food. Different packaging categories,

Life Cycle of Sustainable Packaging: From Design to End of Life, First Edition. Rafael A. Auras and Susan E. M. Selke.

Companion Website: www.wiley.com/go/Auras/lifecycleofsustainablepackaging

such as primary, secondary, and distribution packaging, have different attributes and deliver various functions. Primary and secondary packaging help the consumer to decide which product to buy, how to use it, and assist in product consumption. Distribution packaging helps to transport raw and processed materials to manufacturers for making additional goods, as well as to retailers and consumers. Although the importance of packaging and its ancillary processes and systems is not always fully understood, packaging is a crucial part of our global and circular economy. Packaging is an intrinsic component of any company, culture, and society, so it plays a vital role in any organization's triple bottom line. Packaging created for achieving sustainable development impacts the economy, environment, and social well-being of any society.

Economy: Packaging not only creates markets for commercializing products, but it also is the birth of industries to produce materials and their commercialization. The primary packaging materials, glass, metal, paper, and plastics, are crucial commodity materials in any economy. Packaging not only protects a product during distribution and until reaching the consumer, but it also communicates about each product. With a global packaging materials market estimated as U$D 1 trillion in 2020, packaging is one of the leading economic drivers for many countries [1]. The packaging market is split evenly between food, beverage, and others, and it is composed of different materials, as shown in Figure 1.1. The North American packaging market alone in 2020

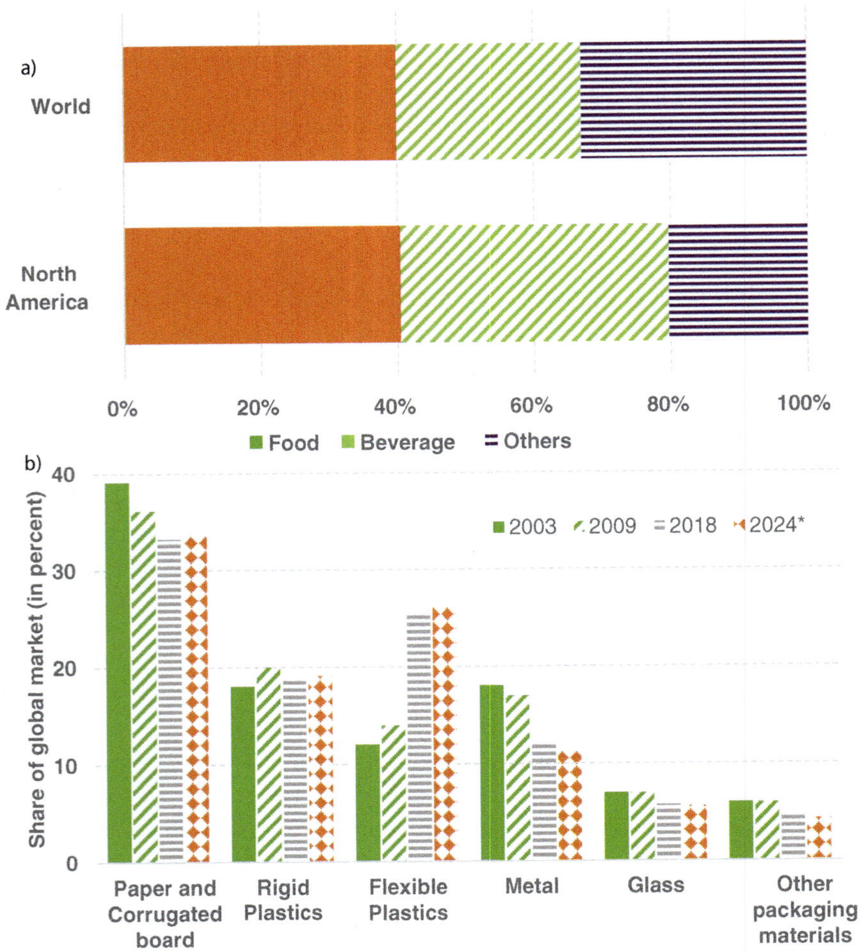

Figure 1.1 (a) World (US$ 1 Trillion in 2020) and North American (US$ 180 billion) packaging market in 2020 and (b) market share of different packaging materials in the world packaging market in 2003, 2009, 2018, and *expected in 2024, data obtained from [1].

was valued at more than $180 billion, and it is expected to grow to more than $220 billion by 2025 (Figure 1.2) [1]. Figure 1.3 shows that the global packaging consumption is expected to increase at a 3.0% CAGR (compound average growth rate) between 2019 and 2024, with the most significant CAGR increase in Eastern Europe, Asia, Africa, and the Middle East [1].

Environment: Packaging can positively and negatively impact the environment across its entire life cycle. First, during its production, it consumes resources such as land, water, and energy and creates waste. Second, it protects and contains the product during its life and communicates dietary and nutritional information. Packaging extends product shelf life. After its useful life, it can create waste and pollution if not adequately managed. The average waste generation per capita around the world was around 1.2 kg/person/day (1.3 billion tons of solid waste per year) in 2012,

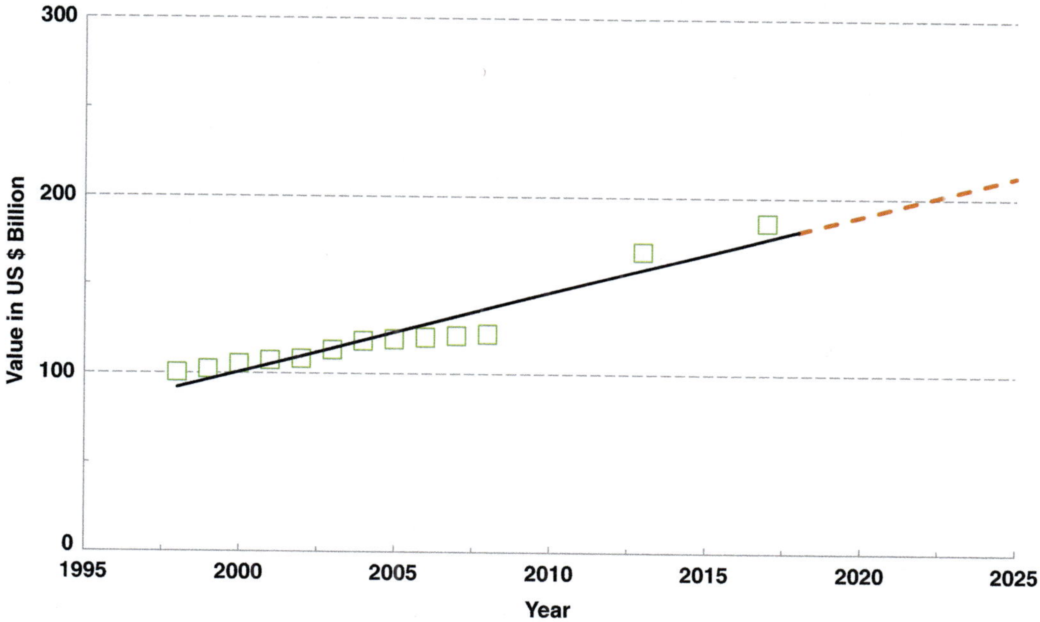

Figure 1.2 North American packaging market, values from 1995 to 2008, 2013, and 2017, and extrapolation to 2025, data obtained from [2].

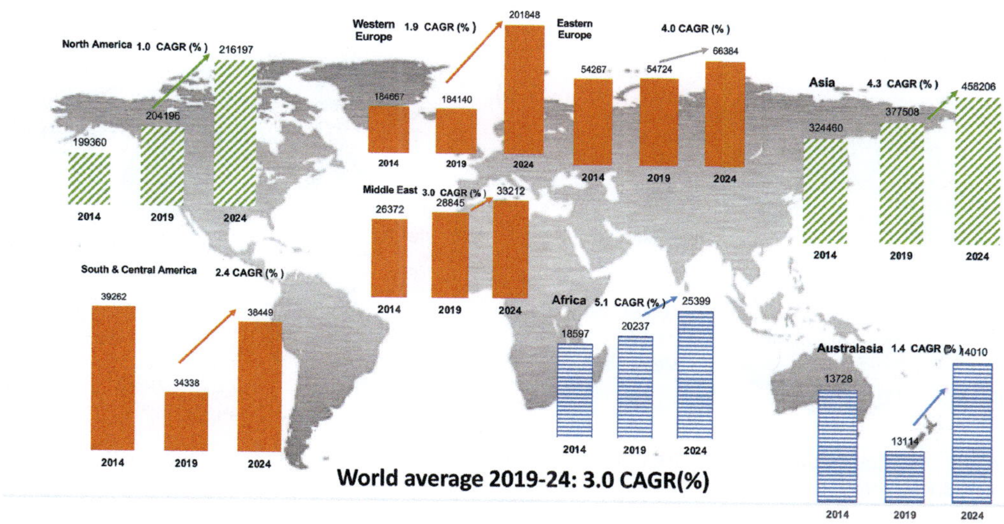

Figure 1.3 Forecast packaging consumption by world region, 2019–2024 in $ million, data obtained from [1].

and it is expected to grow to 1.42 kg/person/day by 2025 (2.2 billion tons of solid waste per year.) High-income-level countries generate waste even as high as 14 kg/person/day [3]. Packaging is responsible for a large part of this waste, aggravated by single-use packaging. Figure 1.4 shows the global and the United States composition of municipal solid waste (MSW). In the US, packaging and containers represent around 30% of the MSW (i.e., "commonly known as trash or garbage discarded by consumers – consists of everyday items such as product packaging, grass clippings, furniture, clothing, bottles, food scraps, newspapers, appliances, paint, and batteries") collected in the US [4]; further discussed in Chapter 10. In addition to the amount of packaging material ending in the MSW, packaging can generate litter – white pollution – and enter the ocean creating marine pollution. Estimates for plastic waste entering the ocean range between 5 and 18 million metric tons annually, and scientists estimate 2% to 5% of the total plastic waste generated by coastal countries annually ends up in the ocean [5, 6]; further discussed in Chapter 17.

Social. Packaging industries can create healthy and vibrant communities, providing labor and economic growth, but they can also generate pollution, damaging humans and wildlife. It is estimated that several million people work in the packaging industry worldwide. In the US, in 2020, according to the Bureau of Labor Statistics, 49,755 people were working in Packaging and Labeling Services, 252,000 people were employed in Packaging and Filling machine operation and tenders, and around 100,000 people were employed in plastics packaging materials, film, and sheet [7]. Since the conversion of packaging raw materials and production of goods requires modest industrialization, medium and high-income countries tend to host these industries, creating benefits for their society. Scandinavian countries, the home of companies such as Tetrapack® and Ikea®, show better social performance indicators such as wealth distribution, health, and education access than many other nations, indicating that social justice and economic development are not mutually exclusive.

Furthermore, new company successes such as Apple® and Patagonia® suggest that economic performance and social responsibility are not only mutually inclusive, but they may also be correlated. Figure 1.5 shows the environmental performance index (EPI) of different countries. The EPI includes indicators for environmental health such as environmental risk exposure, air and water quality, sanitation, and ecosystem vitality such as climate and energy, biodiversity, habitat, fisheries, forests, agriculture, and water resources [8]. We can observe that high-income countries tend to have high EPI.

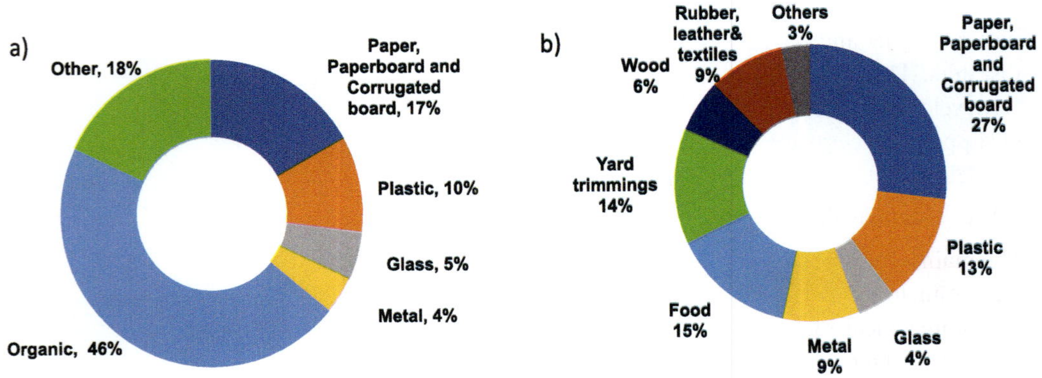

Figure 1.4 (a) Global solid waste composition; adapted from [3] and (b) US municipal solid waste (MSW) composition; adapted from [4].

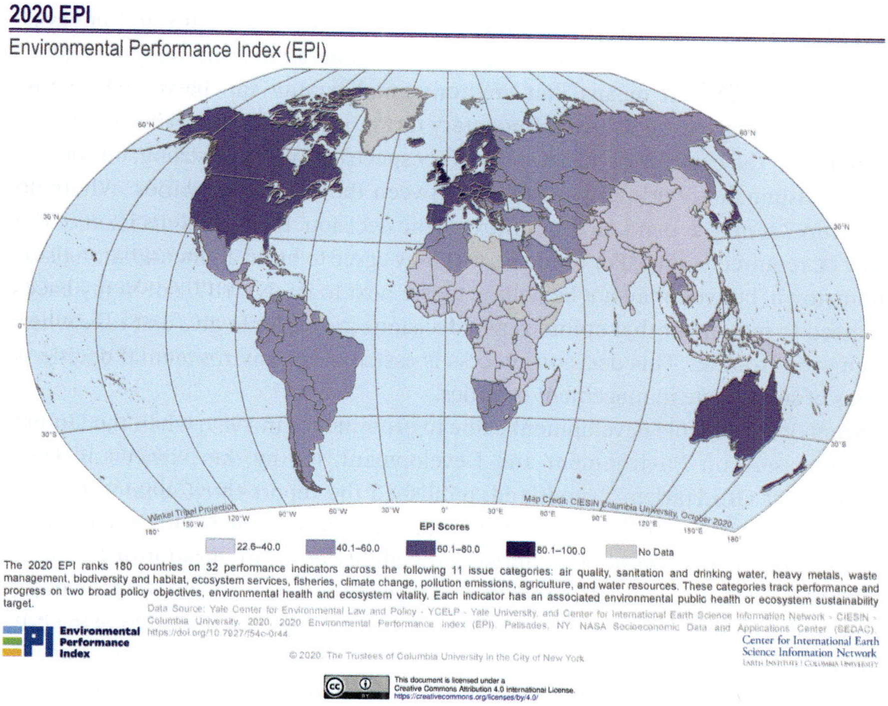

Figure 1.5 Environmental performance index (EPI) by countries. A value of 100 indicates the highest score; reproduced from [8] licensed under CC BY 4.0.

1.4 Sustainability

The precise meaning of sustainability and what it embraces varies depending on who uses the term and in what context. Several definitions have been provided during the years. Some of them define sustainability as:

- "...the capacity of a system to maintain output at a level approximately equal to or greater than its historical average, with the approximation determined by the historical level of variability." [9]
- "The sustainability of natural ecosystems can be defined as the dynamic equilibrium between natural inputs and outputs, modified by external events such as climatic change and natural disasters." [10]
- "... development that improves the quality of human life while living within the carrying capacity of supporting ecosystems." [11]
- "Development that meets the needs of the present without compromising the ability of future generations to meet their own needs." [12]

The main question is, how can something so vague be so important?

According to Kidd (1992), "there is not, and should not be, any single definition of sustainability that is more logical and productive than other definitions" [13]. According to Schaller (1993), "sustainability is like truth and justice – concepts do not readily capture in concise definitions" [14]. Although we cannot fully agree on the meaning of sustainability, we can distinguish two main extremes of sustainability, strong and weak sustainability.

The *strong sustainability* approach means no *trade-off* between economic gain and environmental quality. Our environment is a critical resource for future generations' survival.

The *weak sustainability* approach means that environmental quality can be traded against economic gain since everything can be valued in monetary terms, including the environment.

Most of us may not be at the extreme of this sustainability spectrum, weak or strong; our beliefs and behavior toward sustainability primarily reside in between these two boundaries. Where do you fall on the spectrum? It may depend on the context of the decision, the type of decision or the action that you may be required to take. For example, you may agree to build an incineration plant in your state or hometown, but you may not be willing to live next to it – NIMBY, not in my backyard, concept. We need to remember that opinions and decisions are easier to give and take when they are farther removed from us. This decision process is essential for environmental decisions and the implications of solid waste management facilities.

The modern concept of sustainable development came to prominence in 1987, when the United Nations' World Commission on Environment and Development, led by the former and later Norwegian prime minister Gro Harlem Brundtland, published the report *Our Common Future*, often called the Brundtland report [12], and defined sustainable development as "development that meets the needs of the present without comprising the ability of future generations to meet their own needs." This concept is commonly illustrated in Figure 1.6, where the three main areas, social, environment, and economic, intersect. The same concept is frequently referred to as the 3-P, People, Planet, and Profit.

In the case of the US Environmental Protection Agency (US-EPA), sustainability is defined as "to create and maintain conditions, under which humans and nature can exist in productive harmony, that permit fulfilling the social, economic, and other requirements of present and future generations" [15]. The increasing interest in sustainable development advances the discussion about equity and the moral responsibility of the present society to future ones. The Brundtland report elevated the debate from the seventies from environmental issues to a more significant discussion about social equality and equity within and between generations. So, the subject of inequality found a place in policy agendas worldwide and became the subject of a much larger debate in the progress of cities, states, and countries [16].

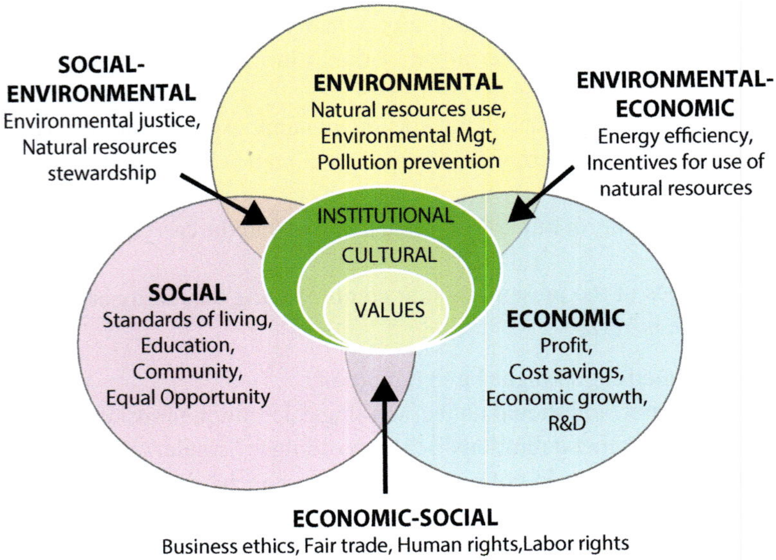

Figure 1.6 Schematic view of sustainable development.

1.5 Sustainability Timeline

This section provides an essential timeline of sustainable development. To understand how we reached the current sustainability concepts, a short tour of the main milestones in the history of sustainability helps. Table 1.1 shows a timeline of the main ideas related to the concept of sustainability since the early seventeenth century; mostly adapted from references [17, 18].

Table 1.1 Development of the sustainability concept over time.

Century	Year	Authority	Major Ideas and/or Events	Era
17th	1627	Francis Bacon	Described the creation of a utopian land [20]	Scientific revolution
		Rene Descartes	Discourse of a Method	
		John Locke	Theory of life, liberty, and property	
		Adam Smith	Labor division	
	Early years	Jean-Jacques Rousseau	Society corrupts man	
18th	1713	Hans Carl von Carlowitz	Outlined methods for sustainable use of forests [21]	Industrial revolution, French revolution
	1798	Thomas Robert Malthus	*An Essay on the Principle of Population* [22]	
19th	1800s	B. Alcott, M. Fuller, Ralph W. Emerson	Described the natural world as a source of guidance and a mirror [17]	Transcendentalist movement
	1818	Marry Shelley	*Frankenstein*	
	1844	Friedrich Engels	Outlines of a critique of political economy	
	1848	John Stuar Mill	Principles of Political Economy	
	1875	Karl Max	Critique of the Gotha Programme	
	Late 1800s	Gifford Pinchot	Developed a conservation ethic – Conservationism	
20th	1901	John Muir	*Our National Parks* [23]	Environmental movement
	1912		*The Yosemite* [19]	
	1949	Aldo Leopold	Land ethic [24]	
	1962	Rachel Carson	*Silent Spring*: investigated the impact of pesticides and other industrial chemicals [25]	
	1966	Barbara Ward	*Spaceship Earth;* discussing the impact of fine resources on Earth	
	1967	Edward Mishan	The cost of Economic Growth – GNP	
	1968	Paul and Anne Ehrlich	The population bomb	

(*Continued*)

Table 1.1 (Continued)

Century	Year	Authority	Major Ideas and/or Events	Era
	1972	Donella Meadows	The limits to growth – described impact of populations, pollution, and resource consumption [26]	Contemporary Environmentalism
		United Nations	First UN Conference on the Human Environment in Stockholm, Sweden	
	1973	Cole et al.	Thinking about the future: A Critique of "The limits of Growth"	
	1973	E. Schumacher	Small is beautiful	
	1975	World Council of Churches	Concept of "Sustainable Society"	
	1977	Herman Daly	*Steady-State Economics*	
	1977	Denis Pirages	The sustainable society	
	1977–1981	Carter Administration	Renewable energy sources comprehensive Environmental Response, Compensation, and Liability Act (CERCLA) – 1980 Superfund	
	1980	Robert Allen	How to save the world	
	1980	IUCN[1]	*World Conservation Strategy* [27]	Sustainability revolution
	1981	Lester Brown	*Building a sustainable society*	
	1982	United Nations	Environmental conference in Stockholm, Sweden – Sustainable development	
	1987	Gro Harlem Brundtland	*Our Common Future* [12], WCED[2] – meeting in 1983	
	1990	IPCC	Creation of the Intergovernmental Panel on Climate Change (IPCC) – First climate report – FAR Climate Change 1990	
	1992	United Nations	Environmental conference in Rio de Janeiro – Earth Summit 1992	
	1995	IPCC	Second climate change report – Tar Climate Change 1995	
	1997	United Nations	Earth Summit II – New York – Agenda 21	
	1997		Kyoto Protocol – Conference of the partners (COP) climate change	
	2000		The Hague Conference – COP 6	

Table 1.1 (Continued)

Century	Year	Authority	Major Ideas and/or Events	Era
21st	2001	IPCC	Third climate change report – TAR	
	2002	United Nations	Environmental conference in Johannesburg – World Summit	
	2007	IPCC	Fourth climate change report – AR4 Climate Change 2007	
	2012	United Nations	Environmental conference in Rio de Janeiro – "Rio + 20" Earth Summit	
	2012	Ellen MacArthur Foundation	Circular Economy	
	2013	IPCC	Fifth climate change report – AR5 Climate change	
	2015	United Nations	Environmental conference in Paris – The Paris Agreement	
	2015	United Nations	Global Sustainable Development Goals – The 2030 Agenda for Sustainable Development	
	2017	US	US withdrawal from the COP agreement	
	2021	IPCC	Sixth climate change report – AR6 Climate Change 2021: The Physical Science Basis	
	2021	UN	Glasgow Climate Change Conference	

Note: [1]The International Union for the Conservation of Nature and Natural Resources. [2]The World Commission on Environment and Development.

We can divide this timeline into different eras of sustainability evolution. During the early period of the scientific revolution, the 1627 book by Francis Bacon – *The New Atlantis* – provides a description and an indication of the time to come when humans believed that *science would enable the domination of nature.* In the early eighteenth century, the Age of Enlightenment, also known as the Age of Reason, was an intellectual movement based on science and technology disseminated in Europe and North America, bringing population growth and resources to the center of the discussion. During this period, Jean-Jacques Rousseau expressed that man was born good but corrupted by society. By 1798, Malthus published his *Essay on Population* growth, where he stated that the geometric growth of population will always outgrow food supplies, creating a general concern about resources and population. In his second edition of the essay (1803), he expressed that poverty could be improved by education and delayed marriage. In the meantime, the expression of human domination of science and power was reflected in popular books such as the 1818 *Frankenstein* by Mary Shelly, which expressed humans' obsessive quest for power. It was not until ten years after Malthus died that Friedrich Engels and Karl Marx refuted his opinion. In 1844, Engels wrote that humans could conquer nature if we used resources wisely, through science and technology, and there was no limit to growth [17]. He even went further to say that overpopulation was an absurd concept when only one-third of Earth's land was cultivated. Marx expressed concern that poverty and misery were not due to overpopulation but oppressive economic and political structures. By 1848, John Stuart Mill, the English philosopher, published *Principles of Political Economy*, stating that profit and economic growth would decline over time as productivity limits were reached, so population growth must be restrained.

By the beginning of the twentieth century, population growth, uncontrolled use of natural resources, and weak environmental policies gave birth to the environmental movement. John Muir, a preservationist, organized the environmental movement and published his first book, *The Mountains of California*, describing the beauty of California nature and wilderness. His more significant achievements were the establishment of Yosemite National Park (1890) and leadership in the foundation of the Sierra Club. By 1910, President Teddy Roosevelt appointed the first director of the US Forest Service, Gifford Pinchot. He was known as a conservationist with a utilitarian point of view, "the greatest good for the greatest number for the longest time." Some historians and environmentalists considered this the first attempt at a sustainability definition. With Pinchot, the forestry profession was born under the precept of conservationism, and America's National Forest Service set areas of public land aside, mostly for timber production. However, forests were cultivated with one tree species to obtain the maximum amount of wood, creating biodiversity issues. The preservationists claimed that the problem with conservationism was the lack of ecologic knowledge. So, the science of ecology was born.

In the mid-1950s, reaction to overexploitation of natural resources was reflected by the work of Aldo Leopold, Rachel Carson, and Barbara Ward. Aldo Leopold (1949), in his essay Land of Ethics in *A Sand County Almanac*, wrote, "A thing is right when it tends to preserve the integrity, stability, and beauty of the biotic community. It is wrong when it tends otherwise." Rachel Carson (1962) published her famous book *Silent Spring* describing the unintended and unpredicted consequences of the production of dichlorodiphenyltrichloroethane (DDT) by the Michigan Chemical Company (by 1962 already owned by the Velsicol Corporation). The production and release of DDT and PBB (polybrominated biphenyls) have affected the Pine River, the St. Louis community in Michigan, and the state of Michigan even until today. A recount of this tragic story will be discussed in Chapter 6 – Case Study 6.1. Barbara Ward (1966) published *Spaceship Earth*, stating that human activity had grown to a scale where further expansion was impossible. In 1967, Edward Mishan published *The Cost of Economic Growth*. The central premise was that the calculations of Gross National Product (GNP) and Gross Domestic Product (GDP) were misleading measurements of human welfare. With the concern of global environmental limits, additional work about environmental pollution started to get attention. Paul Ehrlich (1968) published *The Population Bomb* arguing that population growth would lead to hunger, especially in Asia and overcrowded India. *The Limits to Growth* (1972), the book reporting the first computer model of population and resources, created, published, and translated into 28 languages, sold more than 9 million copies, and declared that the limit to growth on Earth would be reached in the next hundred years, creating shortages of food. Much criticism was directed at this work, mostly concentrating on its gloomy Malthusian point of view.

A new era, the sustainability revolution, started with the involvement of the United Nations. In 1972, the first UN Conference on the Human Environment took place in Stockholm, Sweden. While only two prime ministers attended the meeting, the Swedish and Indian prime ministers, it led to the creation of the United Nations Environment Programme (UNEP). The involvement of the World Council of Churches by 1975 helped to formulate the concept of a "sustainable society." However, it was not until 1977 that the first book titled *The Sustainable Society* by Denis Pirages was published, giving birth to the actual word "sustainable." In the US, new regulations such as the Clean Air Act (1963) and Clean Water Act – created in 1948 as the Federal Water Pollution Control

Act and later extensively amended in 1972, established frameworks for environmental regulations. Between 1977 and 1981, the administration of President Jimmy Carter in the US created legislation such as the Superfund act (1980), which was intended to mitigate soil, water, and land pollution.

The establishment of the World Commission on the Environment and Development (WCED) in 1983, by the UN, created the most significant statement about sustainable development when the *Our Common Future* report was published in 1987 with Gro Harlem Brundtland, former prime minister of Norway, as the chairperson. By 1990, the first Intergovernmental Panel on Climate Change (IPCC) report was published; subsequent reports were published in 1995, 2001, 2007, 2013/14, and 2021, with the IPCC winning the 2007 Nobel Peace Prize for their foundational work on quantifying the impacts of humans on climate change.

Since 1990, several UN and world summits have taken place to create a common agenda for global climate change and sustainability; among the most notable were the 1992 Earth Summit in Rio de Janeiro, Brazil; the 1997 Kyoto meeting in Kyoto, Japan, which created the Kyoto protocol, committing industrialized nations to an overall reduction of 5.2% in greenhouse gas emissions (GHG); the 2002 World Summit of sustainable development in Johannesburg, South Africa; and the 2015 Paris Agreement. During the 2015 Paris meeting, 197 countries met in Paris and acknowledged that climate change is a common concern for humankind. Hence, they agreed to reduce emissions and limit temperature rise, as stated in Article 2:

a) Holding the increase in the global average temperature to well below 2°C above pre-industrial levels and pursuing efforts to limit the temperature increase to 1.5°C above pre-industrial levels, recognizing that this would significantly reduce the risks and impacts of climate change;
b) Increasing the ability to adapt to the adverse impacts of climate change and foster climate resilience and low greenhouse gas emissions development, in a manner that does not threaten food production; and
c) Making finance flows consistent with a pathway towards low greenhouse gas emissions and climate-resilient development. [19]

By September 2017, 166 of the 197 parties entered the agreement. However, on June 1, 2017, the US, one of the two largest GHG emitting countries besides China, announced it would leave the agreement. In 2021, under the Biden administration, the US announced it was rejoining the agreement. A detailed historical and chronological description of the development of the modern sustainability concept can be found elsewhere [17, 18].

1.6 United Nations Sustainable Development Goals (UN-SDGs)

All UN Member States adopted the United Nations (UN) sustainability goals stated in the 2030 Agenda for Sustainable Development in 2015. The UN-SDGs – seventeen – provide a common plan for peace and prosperity for people and the planet, now and into the future. The 17 SDGs are a call for action from low- to high-income countries in a global partnership. The SDGs recognize "that ending poverty and other deprivations must go hand-in-hand with strategies that improve health and education, reduce inequality, and spur economic growth – all while tackling climate change and working to preserve our oceans and forests." Table 1.2 and Figure 1.7 show the seventeen SDGs. A detailed description of the SDGs can be found elsewhere [28]. The Division for Sustainable Development Goals (DSDG) in the UN Department of Economic and Social Affairs provides support and capacity-building for the SDGs and their related thematic issues, including water,

Table 1.2 United Nations 17 Goals to transform our world, targets' information; reproduced from [28].

Goal	Focus	Targets
1	No Poverty	• "By 2030, *eradicate extreme poverty* for all people everywhere, currently measured as people living on less than $1.25 a day. • By 2030, *reduce at least by half the proportion of men, women and children of all ages living in poverty* in all its dimensions according to national definitions. • Implement *nationally appropriate social protection systems* and measures for all, including floors, and by 2030 achieve substantial coverage of the poor and the vulnerable. • By 2030, ensure that all men and women, in particular the poor and the vulnerable, have *equal rights to economic resources*, as well as access to basic services, ownership and control over land and other forms of 13 property, inheritance, natural resources, appropriate new technology and financial services, including micro-finance. • By 2030, *build the resilience of the poor and those in vulnerable situations* and reduce their exposure and vulnerability to climate-related extreme events and other economic, social and environmental shocks and disasters. • Ensure significant *mobilization of resources* from a variety of sources, including through enhanced development cooperation, in order to provide adequate and predictable means for developing countries, in particular least developed countries, to implement programmes and policies to end poverty in all its dimensions. • Create *sound policy frameworks* at the national, regional and international levels, based on pro-poor and gender-sensitive development strategies, to support accelerated investment in poverty eradication actions."
2	Zero Hunger	• "By 2030, *end hunger and ensure access* by all people, in particular the poor and people in vulnerable situations, including infants, *to safe, nutritious and sufficient food all year round.* • By 2030, *end all forms of malnutrition*, including achieving, by 2025, the internationally agreed targets on stunting and wasting in children under 5 years of age, and address the nutritional needs of adolescent girls, pregnant and lactating women and older persons. • By 2030, *double the agricultural productivity* and incomes of small-scale food producers, in particular women, indigenous peoples, family farmers, pastoralists and fishers, including through secure and equal access to land, other productive resources and inputs, knowledge, financial services, markets and opportunities for value addition and non-farm employment. • By 2030, *ensure sustainable food production systems and implement resilient agricultural practices* that increase productivity and production, that help maintain ecosystems, that strengthen capacity for adaptation to climate change, extreme weather, drought, flooding and other disasters and that progressively improve land and soil quality. • By 2020, *maintain the genetic diversity of seeds, cultivated plants and farmed and domesticated animals* and their related wild species, including through soundly managed and diversified seed and plant banks at the national, regional and international levels, and promote access to and fair and equitable sharing of benefits arising from the utilization of genetic resources and associated traditional knowledge, as internationally agreed. • *Increase investment*, including through enhanced international cooperation, in rural infrastructure, agricultural research and extension services, technology development and plant and livestock gene banks in order to enhance agricultural productive capacity in developing countries, in particular least developed countries. Correct and prevent trade restrictions and distortions in world agricultural markets, including through the parallel elimination of all forms of agricultural export subsidies and all export measures with equivalent effect, in accordance with the mandate of the Doha Development Round. • Adopt measures to ensure the proper functioning of food commodity markets and their derivatives and facilitate timely access to market information, including on food reserves, in order to help limit extreme food price volatility."

Table 1.2 (Continued)

Goal	Focus	Targets
3	Good Health and Well-Being	• "By 2030, *reduce the global maternal mortality ratio* to less than 70 per 100,000 live births. • By 2030, *end preventable deaths of newborns and children under 5 years of age*, with all countries aiming to reduce neonatal mortality to at least as low as 12 per 1,000 live births and under-5 mortality to at least as low as 25 per 1,000 live births. • By 2030, *end the epidemics* of AIDS, tuberculosis, malaria and neglected tropical diseases and combat hepatitis, water-borne diseases and other communicable diseases. • By 2030, *reduce by one third premature mortality* from non-communicable diseases through prevention and treatment and promote mental health and well-being. • Strengthen the prevention and treatment of substance abuse, including narcotic drug abuse and harmful use of alcohol. • By 2020, *halve the number of global deaths* and injuries from road traffic accidents. • By 2030, ensure *universal access to sexual and reproductive health-care services*, including for family planning, information and education, and the integration of reproductive health into national strategies and programmes. • Achieve *universal health coverage*, including financial risk protection, *access to quality essential health-care services and access to safe, effective, quality and affordable essential medicines and vaccines for all.* • By 2030, substantially reduce the number of deaths and illnesses from hazardous chemicals and air, water and soil pollution and contamination. • Strengthen the implementation of the World Health Organization Framework Convention on Tobacco Control in all countries, as appropriate. • Support the research and development of vaccines and medicines for the communicable and noncommunicable diseases that primarily affect developing countries, provide access to affordable essential medicines and vaccines, in accordance with the Doha Declaration on the TRIPS (Trade-Related Aspects of Intellectual Property Rights) Agreement and Public Health, which affirms the right of developing countries to use to the full the provisions in the Agreement on Trade Related Aspects of Intellectual Property Rights regarding flexibilities to protect public health, and, in particular, provide access to medicines for all. • Substantially increase health financing and the recruitment, development, training and retention of the health workforce in developing countries, especially in least developed countries and small island developing States. • Strengthen the *capacity of all countries*, in particular developing countries, for early warning, risk reduction and management of national and global health risks."
4	Quality Education	• "By 2030, ensure that *all girls and boys complete free, equitable and quality primary and secondary education* leading to relevant and Goal-4 effective learning outcomes. • By 2030, *ensure that all girls and boys have access to quality early childhood development*, care and preprimary education so that they are ready for primary education. • By 2030, ensure equal access for all women and men to affordable and quality technical, vocational and tertiary education, including university. • By 2030, substantially increase the number of youth and adults who have relevant skills, including technical and vocational skills, for employment, decent jobs and entrepreneurship. • By 2030, *eliminate gender disparities in education and ensure equal access to all levels of education and vocational training for the vulnerable, including persons with disabilities, indigenous peoples and children in vulnerable situations.* • By 2030, ensure that all youth and a substantial proportion of adults, both men and women, achieve literacy and numeracy. • By 2030, ensure that *all learners acquire the knowledge and skills needed to promote sustainable development*, including, among others, through education for sustainable development and sustainable lifestyles, human rights, gender equality, promotion of a culture of peace and non-violence, global citizenship and *appreciation of cultural diversity* and of culture's contribution to sustainable development.

(Continued)

Table 1.2 (Continued)

Goal	Focus	Targets
		• *Build and upgrade education facilities that are child, disability and gender sensitive and provide safe, nonviolent, inclusive and effective learning environments for all* • By 2020, *substantially expand globally the number of scholarships available to developing countries*, in particular least developed countries, small island developing States and African countries, for enrolment in higher education, including vocational training and information and communications technology, technical, engineering and scientific programmes, in developed countries and other developing countries. • By 2030, substantially *increase the supply of qualified teachers*, including through international cooperation for teacher training in developing countries, especially least developed countries and small island developing states."
5	Gender Equality	• "*End all forms of discrimination* against all women and girls everywhere. • *Eliminate all forms of violence against all women and girls* in the public and private spheres, including trafficking and sexual and other types of exploitation • *Eliminate all harmful practices*, such as child, early and forced marriage and female genital mutilation. • *Recognize and value unpaid care and domestic work* through the provision of public services, infrastructure and social protection policies and the promotion of shared responsibility within the household and the family as nationally appropriate. • Ensure women's full and effective participation and equal opportunities for leadership at all levels of decision making in political, economic and public life. • *Ensure universal access to sexual and reproductive health and reproductive rights* as agreed in accordance with the Programme of Action of the International Conference on Population and Development and the Beijing Platform for Action and the outcome documents of their review conferences. • Undertake reforms to give women equal rights to economic resources, as well as access to ownership and control over land and other forms of property, financial services, inheritance and natural resources, in accordance with national laws. • Enhance the use of enabling technology, in particular information and communications technology, to promote the empowerment of women. • Adopt and strengthen sound policies and enforceable legislation for the promotion of gender equality and the empowerment of all women and girls at all levels."
6	Clean Water and Sanitation	• "By 2030, achieve universal and equitable access to safe and affordable drinking water for all. • By 2030, *achieve access to adequate and equitable sanitation* and hygiene for all and end open defecation, paying special attention to the needs of women and girls and those in vulnerable situations. • By 2030, *improve water quality* by reducing pollution, eliminating dumping and minimizing release of hazardous chemicals and materials, halving the proportion of untreated wastewater and substantially increasing recycling and safe reuse globally. • By 2030, *substantially increase water-use efficiency* across all sectors and ensure sustainable withdrawals and supply of freshwater to address water scarcity and substantially reduce the number of people suffering from water scarcity. • By 2030, *implement integrated water resources management* at all levels, including through transboundary cooperation as appropriate. • By 2020, protect and restore water-related ecosystems, including mountains, forests, wetlands, rivers, aquifers and lakes. • By 2030, *expand international cooperation* and capacity-building support to developing countries in water- and sanitation-related activities and programmes, including water harvesting, desalination, water efficiency, wastewater treatment, recycling and reuse technologies. • Support and strengthen the participation of local communities in improving water and sanitation management."

Table 1.2 (Continued)

Goal	Focus	Targets
7	Affordable and Clean Energy	• "By 2030, ensure universal access to affordable, reliable and modern energy services. • By 2030, increase substantially the share of renewable energy in the global energy mix. • By 2030, double the global rate of improvement in energy efficiency. • By 2030, *enhance international cooperation* to facilitate access to clean energy research and technology, including renewable energy, energy efficiency and advanced and cleaner fossil-fuel technology, and promote investment in energy infrastructure and clean energy technology. • By 2030, expand infrastructure and upgrade technology for supplying modern and sustainable energy services for all in developing countries, in particular least developed countries, small island developing States, and land-locked developing countries, in accordance with their respective programs of support."
8	Decent Work and Economic Growth	• "*Sustain per capita economic growth* in accordance with national circumstances and, in particular, at least 7 per cent gross domestic product growth per annum in the least developed countries. • Achieve higher levels of economic productivity through diversification, technological upgrading and innovation, including through a focus on high value added and labor-intensive sectors. • *Promote development-oriented policies* that support productive activities, decent job creation, entrepreneurship, creativity and innovation, and encourage the formalization and growth of micro-, small- and medium-sized enterprises, including through access to financial services. • Improve progressively, through 2030, global resource efficiency in consumption and production and endeavor to decouple economic growth from environmental degradation, in accordance with the 10-year framework of programs on sustainable consumption and production, with developed countries taking the lead. • *By 2030, achieve full and productive employment and decent work for all women and men, including for young people and persons with disabilities, and equal pay for work of equal value.* • By 2020, substantially reduce the proportion of youth not in employment, education or training. • Take immediate and effective measures to eradicate forced labor, end modern slavery and human trafficking and secure the prohibition and elimination of the worst forms of child labor, including recruitment and use of child soldiers, and by 2025 end child labor in all its forms. • Protect labor rights and promote safe and secure working environments for all workers, including migrant workers, in particular women migrants, and those in precarious employment. • By 2030, devise and implement policies to promote sustainable tourism that creates jobs and promotes local culture and products. • Strengthen the capacity of domestic financial institutions to encourage and expand access to banking, insurance and financial services for all. • Increase Aid for Trade support for developing countries, in particular least developed countries, including through the Enhanced Integrated Framework for Trade-Related Technical Assistance to Least Developed Countries. • By 2020, *develop and operationalize a global strategy for youth employment* and implement the Global Jobs Pact of the International Labor Organization."
9	Industry, Innovation and Infrastructure	• "*Develop quality, reliable, sustainable and resilient infrastructure*, including regional and transborder infrastructure, to support economic development and human well-being, with a focus on affordable and equitable access for all. • *Promote inclusive and sustainable industrialization* and, by 2030, significantly raise industry's share of employment and gross domestic product, in line with national circumstances, and double its share in least developed countries. • Increase the access of small-scale industrial and other enterprises, in particular in developing countries, to financial services, including affordable credit, and their integration into value chains and markets. • By 2030, *upgrade infrastructure and retrofit industries to make them sustainable*, with increased resource-use efficiency and greater adoption of clean and environmentally sound technologies and industrial processes, with all countries taking action in accordance with their respective capabilities.

(*Continued*)

Table 1.2 (Continued)

Goal	Focus	Targets
		• Enhance scientific research, upgrade the technological capabilities of industrial sectors in all countries, in particular developing countries, including, by 2030, encouraging innovation and substantially increasing the number of research and development workers per 1 million people and public and private research and development spending. • *Facilitate sustainable and resilient infrastructure development* in developing countries through enhanced financial, technological and technical support to African countries, least developed countries, landlocked developing countries and small island developing States. • Support domestic technology development, research and innovation in developing countries, including by ensuring a conducive policy environment for, inter alia, industrial diversification and value addition to commodities. • Significantly *increase access to information and communications technology* and strive to provide universal and affordable access to the Internet in least developed countries by 2020."
10	Reduced Inequality	• "By 2030, progressively achieve and sustain income growth of the bottom 40 per cent of the population at a rate higher than the national average. • By 2030, empower and promote the social, economic and political inclusion of all, irrespective of age, sex, disability, race, ethnicity, origin, religion or economic or other status. • Ensure equal opportunity and reduce inequalities of outcome, including by eliminating discriminatory laws, policies and practices and promoting appropriate legislation, policies and action in this regard. • Adopt policies, especially fiscal, wage and social protection policies, and progressively achieve greater equality. • Improve the regulation and monitoring of global financial markets and institutions and strengthen the implementation of such regulations. • Ensure enhanced representation and voice for developing countries in decision-making in global international economic and financial institutions in order to deliver more effective, credible, accountable and legitimate institutions. • Facilitate orderly, safe, regular and responsible migration and mobility of people, including through the implementation of planned and well-managed migration policies. • Implement the principle of special and differential treatment for developing countries, in particular least developed countries, in accordance with World Trade Organization agreements. • Encourage official development assistance and financial flows, including foreign direct investment, to States where the need is greatest, in particular least developed countries, African countries, small island developing States and landlocked developing countries, in accordance with their national plans and programs. • By 2030, reduce to less than 3 per cent the transaction costs of migrant remittances and eliminate remittance corridors with costs higher than 5 per cent."
11	Sustainable Cities and Communities	• "By 2030, ensure access for all to adequate, safe and affordable housing and basic services and upgrade slums. • By 2030, provide access to safe, affordable, accessible and sustainable transport systems for all, improving road safety, notably by expanding public transport, with special attention to the needs of those in vulnerable situations, women, children, persons with disabilities and older persons. • By 2030, enhance inclusive and sustainable urbanization and capacity for participatory, integrated and sustainable human settlement planning and management in all countries. • Strengthen efforts to protect and safeguard the world's cultural and natural heritage. • By 2030, significantly reduce the number of deaths and the number of people affected and substantially decrease the direct economic losses relative to global gross domestic product caused by disasters, including water-related disasters, with a focus on protecting the poor and people in vulnerable situations. • By 2030, reduce the adverse per capita environmental impact of cities, including by paying special attention to air quality and municipal and other waste management

Table 1.2 (Continued)

Goal	Focus	Targets
		• By 2030, provide universal access to safe, inclusive and accessible, green and public spaces, in particular for women and children, older persons and persons with disabilities. • Support positive economic, social and environmental links between urban, peri-urban and rural areas by strengthening national and regional development planning • By 2020, substantially increase the number of cities and human settlements adopting and implementing integrated policies and plans towards inclusion, resource efficiency, mitigation and adaptation to climate change, resilience to disasters, and develop and implement, in line with the Sendai Framework for Disaster Risk Reduction 2015-2030, holistic disaster risk management at all levels. • Support least developed countries, including through financial and technical assistance, in building sustainable and resilient buildings utilizing local materials."
12	Responsible Consumption and Production	• "Implement the *10-year framework of programs on sustainable consumption and production*, all countries taking action, with developed countries taking the lead, taking into account the development and capabilities of developing countries. • By 2030, achieve the *sustainable management and efficient use of natural resources*. • By 2030, *halve per capita global food waste* at the retail and consumer levels and reduce food losses along production and supply chains, including post-harvest losses. • By 2020, *achieve the environmentally sound management of chemicals and all wastes throughout their life cycle*, in accordance with agreed international frameworks, and significantly reduce their release to air, water and soil in order to minimize their adverse impacts on human health and the environment. • By 2030, *substantially reduce waste generation through prevention, reduction, recycling and reuse.* • *Encourage companies*, especially large and transnational companies, *to adopt sustainable practices* and to integrate sustainability information into their reporting cycle • *Promote public procurement practices* that are sustainable, in accordance with national policies and priorities. • By 2030, ensure that people everywhere have the relevant information and awareness for sustainable development and lifestyles in harmony with nature. • *Support developing countries* to strengthen their scientific and technological capacity to move towards more sustainable patterns of consumption and production. • *Develop and implement tools to monitor sustainable development impacts* for sustainable tourism that creates jobs and promotes local culture and products. • *Rationalize inefficient fossil-fuel subsidies that encourage wasteful consumption by removing market distortions*, in accordance with national circumstances, including by restructuring taxation and phasing out those harmful subsidies, where they exist, to reflect their environmental impacts, taking fully into account the specific needs and conditions of developing countries and minimizing the possible adverse impacts on their development in a manner that protects the poor and the affected communities."
13	Climate Action	• *"Strengthen resilience and adaptive capacity to climate-related hazards* and natural disasters in all countries. • *Integrate climate change measures into national policies, strategies and planning.* • *Improve education, awareness-raising and human and institutional capacity on climate change mitigation, adaptation, impact reduction and early warning.* • *Implement the commitment undertaken by developed-country parties to the United Nations Framework Convention on Climate Change* to a goal of mobilizing jointly $100 billion annually by 2020 from all sources to address the needs of developing countries in the context of meaningful mitigation actions and transparency on implementation and fully operationalize the Green Climate Fund through its capitalization as soon as possible. • *Promote mechanisms for raising capacity for effective climate change-related* planning and management in least developed countries and small island developing States, including focusing on women, youth and local and marginalized communities."

(*Continued*)

Table 1.2 (Continued)

Goal	Focus	Targets
14	Life Below Water	• "By 2025, *prevent and significantly reduce marine pollution of all kinds*, in particular from land-based activities, including marine debris and nutrient pollution. • By 2020, *sustainably manage and protect marine and coastal ecosystems* to avoid significant adverse impacts, including by strengthening their resilience, and take action for their restoration in order to achieve healthy and productive oceans. • Minimize and address the impacts of ocean acidification, including through enhanced scientific cooperation at all levels. • By 2020, *effectively regulate harvesting and end overfishing*, illegal, unreported and unregulated fishing and destructive fishing practices and implement science-based management plans, in order to restore fish stocks in the shortest time feasible, at least to levels that can produce maximum sustainable yield as determined by their biological characteristics. • By 2020, conserve at least 10 per cent of coastal and marine areas, consistent with national and international law and based on the best available scientific information. • By 2020, *prohibit certain forms of fisheries subsidies which contribute to overcapacity and overfishing*, eliminate subsidies that contribute to illegal, unreported and unregulated fishing and refrain from introducing new such subsidies, recognizing that appropriate and effective special and differential treatment for developing and least developed countries should be an integral part of the World Trade Organization fisheries subsidies negotiation. • By 2030, *increase the economic benefits to Small Island developing States* and least developed countries from the sustainable use of marine resources, including through sustainable management of fisheries, aquaculture and tourism. • Increase scientific knowledge, develop research capacity and transfer marine technology, taking into account the Intergovernmental Oceanographic Commission Criteria and Guidelines on the Transfer of Marine Technology, in order to improve ocean health and to enhance the contribution of marine biodiversity to the development of developing countries, in particular small island developing States and least developed countries. • Provide access for small-scale artisanal fishers to marine resources and markets. • Enhance the conservation and sustainable use of oceans and their resources by implementing international law as reflected in UNCLOS (United Nations Convention on the Law of the Sea), which provides the legal framework for the conservation and sustainable use of oceans and their resources, as recalled in paragraph 158 of The Future We Want."
15	Life on Land	• "By 2020, *ensure the conservation, restoration and sustainable use of terrestrial and inland freshwater* ecosystems and their services, in particular forests, wetlands, mountains and drylands, in line with obligations under international agreements. • By 2020, promote the implementation of sustainable management of all types of forests, halt deforestation, restore degraded forests and substantially increase afforestation and reforestation globally. • By 2030, *combat desertification*, restore degraded land and soil, including land affected by desertification, drought and floods, and strive to achieve a land degradation-neutral world. • By 2030, *ensure the conservation of mountain ecosystems*, including their biodiversity, in order to enhance their capacity to provide benefits that are essential for sustainable development. • *Take urgent and significant action to reduce the degradation of natural habitats*, halt the loss of biodiversity and, by 2020, protect and prevent the extinction of threatened species. • *Promote fair and equitable sharing of the benefits arising from the utilization of genetic resources* and promote appropriate access to such resources, as internationally agreed. • *Take urgent action to end poaching and trafficking of protected species of flora and fauna* and address both demand and supply of illegal wildlife products.

Table 1.2 (Continued)

Goal	Focus	Targets
		• By 2020, introduce measures to prevent the introduction and significantly *reduce the impact of invasive alien species* on land and water ecosystems and control or eradicate the priority species. • By 2020, *integrate ecosystem and biodiversity values into national and local planning*, development processes, poverty reduction strategies and accounts. • Mobilize and significantly *increase financial resources from all sources* to conserve and sustainably use biodiversity and ecosystems. • Mobilize significant resources from all sources and at all levels to *finance sustainable forest management* and provide adequate incentives to developing countries to advance such management, including for conservation and reforestation. • Enhance global support for efforts to combat poaching and trafficking of protected species, including by increasing the capacity of local communities to pursue sustainable livelihood opportunities."
16	Peace and Justice Strong Institutions	• "Significantly *reduce all forms of violence and related death rates* everywhere. • *End abuse, exploitation, trafficking and all forms of violence against and torture of children.* • *Promote the rule of law at the national and international levels* and ensure equal access to justice for all. • By 2030, significantly *reduce illicit financial and arms flows*, strengthen the recovery and return of stolen assets and combat all forms of organised crime. • Substantially *reduce corruption and bribery* in all their forms. • Develop effective, accountable and transparent institutions at all levels. • *Ensure responsive, inclusive, participatory and representative decision-making at all levels.* • *Broaden and strengthen the participation of developing countries* in the institutions of global governance. • By 2030, provide legal identity for all, including birth registration. • *Ensure public access to information and protect fundamental freedoms*, in accordance with national legislation and international agreements. • Strengthen relevant national institutions, including through international cooperation, for building capacity at all levels, in particular in developing countries, to prevent violence and combat terrorism and crime. • Promote and enforce non-discriminatory laws and policies for sustainable development."
17	Partnerships to achieve the Goal	a) Finance • "Strengthen domestic resource mobilization, including through international support to developing countries, to improve domestic capacity for tax and other revenue collection. • Developed countries to implement fully their official development assistance commitments, including the commitment by many developed countries to achieve the target of 0.7 per cent of ODA/GNI to developing countries and 0.15 to 0.20 per cent of ODA/GNI to least developed countries ODA providers are encouraged to consider setting a target to provide at least 0.20 per cent of ODA/GNI to least developed countries • Mobilize additional financial resources for developing countries from multiple sources. • Assist developing countries in attaining long-term debt sustainability through coordinated policies aimed at fostering debt financing, debt relief and debt restructuring, as appropriate, and address the external debt of highly indebted poor countries to reduce debt distress. • Adopt and implement investment promotion regimes for least developed countries.

(Continued)

Table 1.2 (Continued)

Goal	Focus	Targets
		b) Technology • Enhance North-South, South-South and triangular regional and international cooperation on and access to science, technology and innovation and enhance knowledge sharing on mutually agreed terms, including through improved coordination among existing mechanisms, in particular at the United Nations level, and through a global technology facilitation mechanism. • Promote the development, transfer, dissemination and diffusion of environmentally sound technologies to developing countries on favorable terms, including on concessional and preferential terms, as mutually agreed. • Fully operationalize the technology bank and science, technology and innovation capacity-building mechanism for least developed countries by 2017 and enhance the use of enabling technology, in particular information and communications technology. c) Capacity building • Enhance international support for implementing effective and targeted capacity-building in developing countries to support national plans to implement all the sustainable development goals, including through North-South, South-South and triangular cooperation. d) Trade • Promote a universal, rules-based, open, non-discriminatory and equitable multilateral trading system under the World Trade Organization, including through the conclusion of negotiations under its Doha Development Agenda. • Significantly increase the exports of developing countries, in particular with a view to doubling the least developed countries' share of global exports by 2020. • Realize timely implementation of duty-free and quota-free market access on a lasting basis for all least developed countries, consistent with World Trade Organization decisions, including by ensuring that preferential rules of origin applicable to imports from least developed countries are transparent and simple, and contribute to facilitating market access. e) Systemic issues *Policy and institutional coherence* • Enhance global macroeconomic stability, including through policy coordination and policy coherence. • Enhance policy coherence for sustainable development. • Respect each country's policy space and leadership to establish and implement policies for poverty eradication and sustainable development. *Multi-stakeholder partnerships* • Enhance the global partnership for sustainable development, complemented by multi-stakeholder partnerships that mobilize and share knowledge, expertise, technology and financial resources, to support the achievement of the sustainable development goals in all countries, in particular developing countries. • Encourage and promote effective public, public-private and civil society partnerships, building on the experience and resourcing strategies of partnerships. *Data, monitoring and accountability* • By 2020, enhance capacity-building support to developing countries, including for least developed countries and small island developing States, to increase significantly the availability of high-quality, timely and reliable data disaggregated by income, gender, age, race, ethnicity, migratory status, disability, geographic location and other characteristics relevant in national contexts. • By 2030, build on existing initiatives to develop measurements of progress on sustainable development that complement gross domestic product, and support statistical capacity-building in developing countries."

Figure 1.7 United Nations Sustainable Development Goals (UN-SDGs); reproduced from [28] license under CC BY 4.0.

energy, climate, oceans, urbanization, transport, science and technology, the Global Sustainable Development Report (GSDR), partnerships, and Small Island Developing States. The main targets for each SDG can be found in the United Nations SDG reports.

Packaging systems can be instrumental in helping to achieve all the SDGs. For example, the packaging program at the United States Agency for International Development (USAID) works to support and provide nutritional food to needy populations, promoting a path to recipient self-reliance and resilience; this effort has a direct connection between packaging and SDGs 1, 2, 3, and 10. A direct link can also be attributed to packaging and SDG 6 – clean water by providing reusable or single-use bottles. More than 250,000 people work in the packaging industry in the US alone, which relates to SDG 8. For SDG 9, the world average compound annual growth rate between 2019 and 2024 for the packaging industry is expected to be 3.0%. Packaging is a growing industry helping people to earn a decent living. For SDG 12, new packaging systems are aligned with responsible consumption and production. Packaging industries are increasingly focusing on reducing waste generation through prevention, reduction, reuse, and recycling.

1.7 Sustainability Indicators (SIs)

To determine if a system, product, or package is sustainable, indicators should be assigned to each social, economic, and environmental evaluation area. Table 1.3 shows an example of popular SIs used by governments and NGOs to evaluate sustainable systems. It has been expressed that "one of the major criticisms of SIs is that they attempt to agglomerate complex and diverse problems in a few simple measures" [8]. Mostly, this is referred to as a reductionist approach, which has received criticism from many who consider that scientists and policymakers are obsessed with quantification. Biologists have used indicators to measure ecosystem health for many years, such as the Shannon-Wiener Index [29]. However, sustainability incorporates many more dimensions than the environment, such as quality of life and economic returns, and the evolution of sustainable development unquestionably leads to the quantification of sustainability.

One of the many problems of sustainability measurement is to know how many indicators for each category are enough. Bell and Morse [8] provide an excellent discussion of SIs and their

Table 1.3 Selected examples of indicators of ecosystem health and sustainability.

Index	Initial Year	Acronyms	Source	Ref.
City Development Index	2015	CDI	United Nations Human Settlements Programme (UN–Habitat)	[30]
Ecological Footprint		EF	Weidman and Barrett	[31]
Environmental Adjusted Net Domestic Product		EDP	UN Statistics Division	[32]
Environmental Performance Index	2018	EPI	Yale and Columbia Universities	[8]
The Environmental Vulnerability Index	2004	EVI	South Pacific Applied Geoscience Commission (SOPAC), the United Nations Environment Programme (UNEP) and their partners	[33]
Genuine Progress Indicator	After 1989	GPI	Based on the Daly and Cobb (1989) and Philip Lawn work	[34]
Genuine Saving Index	2004	GSI	Economic indicator – World Bank	[35]
Harger and Meyer	1996	ESSD	Harger and Meyer, 1996	[36]
Human Development Index		HDI	Amartya, Sen and Haq, Mahbub ul	
Index of Sustainable Economic Welfare	1989	ISEW	Daly and Cobb (1990)	[37]
Inequality-adjusted Human Development Index	2020	IHDI	Human Development Report, 2020	[38]
Living Planet Index	1970	LPI	Living Planet Index 2020 Report	[39, 40]
Maximum sustainable index*	Early 1930	MSY	Russell (1931)	[41]
OECD Better Life Index	2011	WI	OECD (2021)	[42]
Shannon-Wiener Index or Diversity Index	1948	H	Shannon (1948)	[43]
Total Factor Productivity Index	1956	TFPI	Solow (1956)	[44]

* Used mostly for fisheries.

creation. Primarily, they are based on a system of indicators created to measure driving force (D) and state response (SR) (i.e., DSR), and may include some additional areas such as impact (I) -DSIR, or pressure (P) in the system to be evaluated (DSPIR). These methods are mostly based on identifying the leading indicators for each area of interest to be appraised. With the help of stakeholders, find a mechanism of determining the main indicators to be mapped in the social, economic, and environmental areas of interest. One well-known example of SI in the US is the Environmental

Performance Index (EPI), created by Yale and Columbia Universities researchers. The EPI provides a ranking on a national basis with values between 0 (most unsustainable) and 100 (most sustainable). They used a general DSIR system to gauge the sustainability of countries based on 76 variables aggregated into 22 indicators and normalized for quantification; see Figure 1.5 [8]. Although all these variables and indicators are used, as Bell and Morse [8] expressed, "SIs are still a classic reductionist set of tools based on quantification." In general, they are a top-down approach and not stakeholder-driven methodologies.

In general, the construction of SIs is subject to many pressures, agendas, and biases. So, transparency in the process and detailed method description are the recommended procedures for SI construction and implementation.

1.8 Life Cycle Thinking

The criteria of life cycle thinking (LCT) are based on moving away from a linear production system to a circular one, including environmental, social, and economic impacts of a product life cycle so that the number of resources used by a product and the amount of emissions released during its production and use can be optimized. LCT also accounts for improving socioeconomic performance throughout its life cycle. So, resources can be optimized and impact minimized. Figure 1.8 shows a typical product life cycle diagram when considering LCT. A product life cycle can begin with the extraction of raw materials from natural resources in the ground and energy generation. Then,

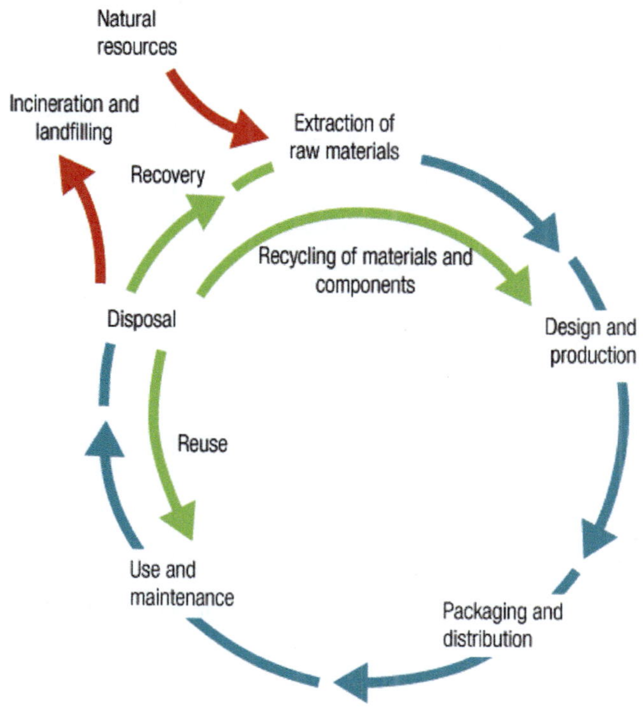

Figure 1.8 Life cycle thinking (LCT) is used for a product cycle.

materials and energy are part of production, packaging, distribution, use, maintaining and reuse, recycling, recovering, and finally disposal to landfill. LCT and life cycle approaches have been part of the world agenda for the last two decades to produce clean energy and goods with a lower environmental footprint. All these activities are part of the green economy.

LCT is mostly made operational through life cycle management (LCM). Some of the primary regulations to make LCM operational have been created through the International Organization for Standardization (ISO) 9000 and 14000, which favor a cyclical approach of plan-do-check-act (PDCA) and provide a basis for continual improvement (Figure 1.9).

ISO has developed standards that help organizations take a proactive approach to manage environmental issues and LCT initiatives; the ISO 14000 family of environmental management standards can be implemented in any organization in the public or private sector. ISO states that although the ISO 14000 standards are designed to be mutually supportive, they can also be used independently to achieve environmental goals. For example, life cycle assessment (LCA), later discussed in Chapter 9, is implemented in the "Do" step of the PDCA cycle as a tool to evaluate the environmental impacts of products or systems.

LCA is a methodology to compile and evaluate the inputs, outputs, and potential environmental impacts of a product system throughout its life cycle; ISO 14040:2006 describes the principles and framework for LCA, and ISO 14044:2006 provides the requirements and guidelines to conduct an LCA. The LCA framework comprises four phases: goal and scope definition, life cycle inventory (LCI) analysis, life cycle impact assessment (LCIA), and interpretation. All four phases

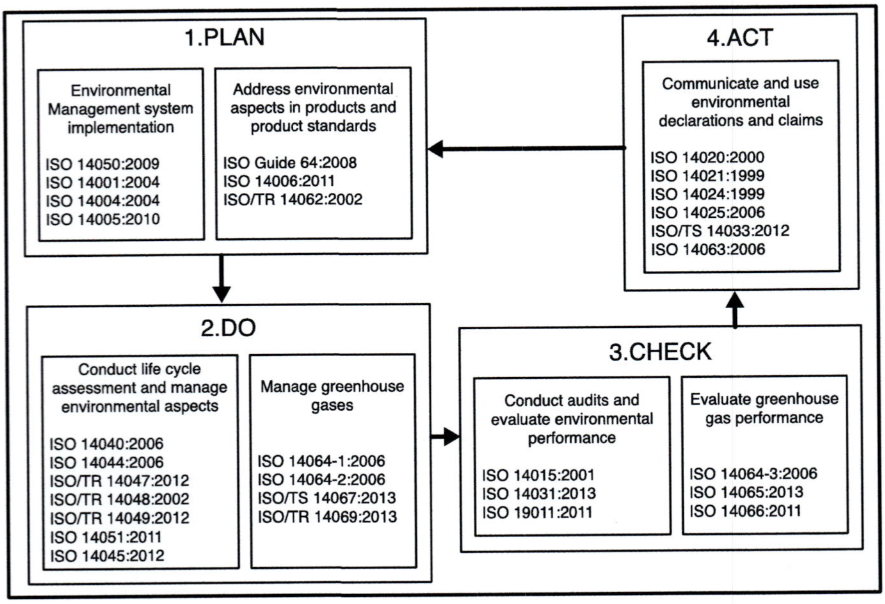

Figure 1.9 The ISO 14000 family of standards within the Plan-Do-Check-Act (PDCA) cycle; adapted from [45].

are interconnected and performed iteratively. Details of how to conduct an LCA, LCI, and LCIA are covered in Chapter 9.

1.9 Circular Economy

A circular economy (CE) is defined as "a regenerative system in which resource input and waste, emission, and energy leakage are minimized by slowing, closing, and narrowing material and energy loops. This can be achieved through long-lasting design, maintenance, repair, reuse, remanufacturing, refurbishing, and recycling" [46]. The CE contrasts with the linear economy, which is a "take, make, dispose of" production model. The CE is like the LCT process. Pearce and Turner [47] initially defined the term CE as a regenerative system where resources are used to produce consumer and capital goods. Then, they are used for consumption and the creation of utility.

Finally, waste is produced, which can be used again for producing capital goods resources. Until then, the standard line of thought was looking at the economy as a linear system, ignoring the environment-economy interactions, and considering the environment as a source of resources and as a sink for waste. They said that the main differences between economic and natural systems are that natural systems tend to recycle their waste (e.g., leaves are decomposed in fall and create organic fertilizer). However, economic systems mostly do not have built-in systems to recycle waste. So, Pearce and Turner [47], looking at the First Law of Thermodynamics – energy and matter cannot be created or destroyed, stated that whatever is used up by way of resources must end up somewhere in the environment in a specific time frame. For example, coal consumption will be equaled to the embedded used energy plus the gases and solid waste produced. This was not the first time that the First Law of Thermodynamics was used to express this concept; in 1966, Kenneth Boulding made the same analogy in his book *The Economics of the Coming Spaceship Earth*. However, he also acknowledged that entropy plays a crucial role in dissipating energy, so the Second Law of Thermodynamics is also involved. As resources are introduced in the economic system, not everything can be recycled. For example, the aluminum of a car can be recycled, but the lead in the gasoline is not. The environment's assimilation capacity is a major economic function in handling this waste. So, if this function of the environment is not damaged, the economic system should work. However, as we harm the natural environment by disposing of more waste than it can absorb, the natural ability of the environment to sink waste gets impaired, conditions that it seems we have reached. Finally, Pearce and Turner [47] concluded that the environment has three main economic functions – "as resource supplier, as waste assimilator, and as a direct source of utility – commodity." They are considered "economic functions" since they all create a positive value when bought and sold in the marketplace. Economists spend much time determining if equilibrium between the economy and the environment is possible, evaluating these economic functions.

Based on this and other initial work, the current industries of the material economy, such as wood, metal, paper, and plastics, are striving to create CE models. For example, the plastics industry, with the current waste problems of white and marine pollution an increasing societal concern [48], is proposing the creation of new models to reduce plastic reaching the environment, as shown in Figure 1.10.

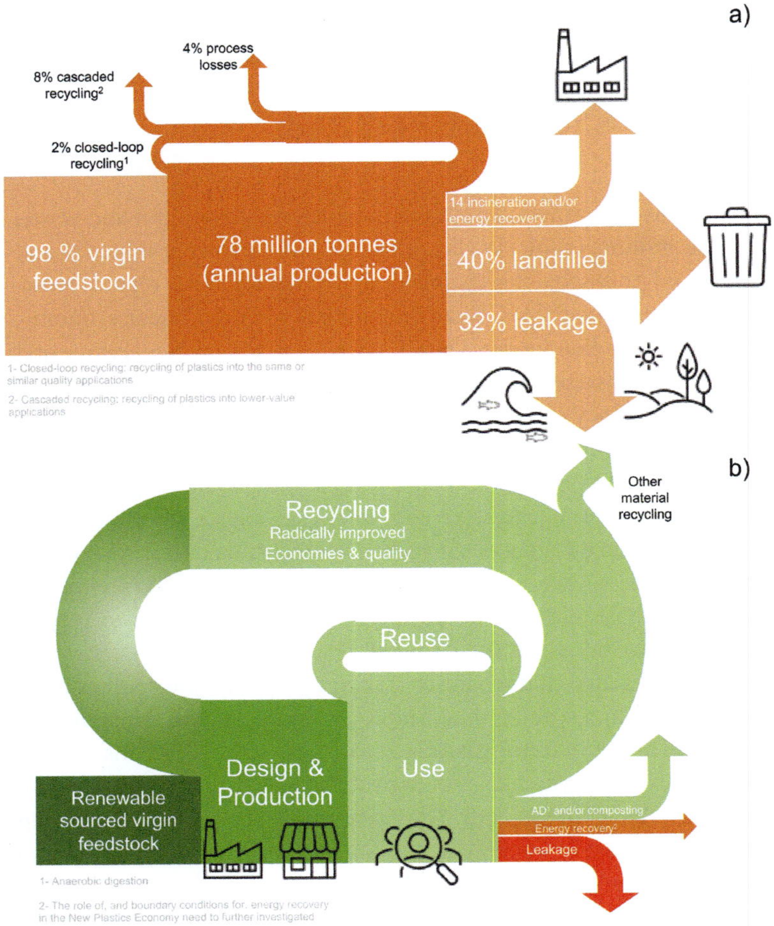

Figure 1.10 (A) Current flow of plastic and (b) new proposed system based on the circular economy approach; adapted from [49].

1.10 Packaging for Sustainable Development

There has been a public perception and widespread belief that packaging is responsible for pollution and a sizeable environmental footprint in the supply chain [50]. So, pressure has been rising from the public and governments to combat packaging as an evil necessity without fully understanding its benefits. Around the year 2000, to counter this perception and to react to mandates that may hinder the packaging industry and its development, some countries and associations such as the Sustainable Packaging Coalition (SPC), the European Organization for Packaging and the Environment (EUROPEN), and the Sustainable Packaging Alliance (SPA) (Table 1.4) came together to create an agenda for packaging for sustainable development. Table 1.4 shows a tentative timeline of the main events driving the creation of these organizations.

By 1970, it was clear that significant environmental pressures and changes were pushing the environment beyond its human carrying capacity and planetary boundaries [52]. Therefore, in the US and Europe, several initiatives and regulations were established to counteract the uncontrolled development generated after the Second World War. Significant pieces of legislation in the US concerning the environment, such as the Clean Water Act (1972) and the Clean Air Act (1963),

Table 1.4 Development of the sustainability concept over time.

Century	Year	Authority	Major Ideas and/or Events	Era
20th	1948		The Federal Water Pollution Control Act – Clean Water Act (1972)	Sustainability revolution
	1955/67/70	Eisenhower/ Johnson/Nixon	Air Pollution Control Act/Air Quality Act/ Clean Air Act	
	1980	Carter Administration	The Comprehensive Environmental Response, Compensation, and Liability Act (CERCLA)	
	1980	Robert Allend	How to save the world	
	1981	Lester Brown	Building a Sustainable Society	
	1994	European Parliament	Directive 94/62/EC – Packaging and packaging waste directive (PPWD)	
	1995	McDonough and Braungart	The next industrial revolution	
	2000		The Hannover principle "cradle-to-cradle"	
	2002	GreenBlue	Nexus project at McDonough Braungart Design Chemistry (MBDC)	
	2002-2017	Australia	Sustainable Packaging Alliance (SPA)	
	2003	SPC	Sustainable Packaging Coalition – 8 founding members, now more than 500 members	
	2004	CEN	Umbrella standard – EN 13427:2004 Prevention – EN13428:2004 Reusable packaging – EN13429 Material recycling – EN 13430 Energy recovery – EN 13431	
	2005	Europen	The European Organization for Packaging and the Environment	
	2007	Walmart	Scorecard	
	2012	Ellen MacArthur Foundation	Circular Economy	
	2018	Alliance to end plastic waste	End white pollution	
	2020	Ellen MacArthur Foundation	Global Commitment 2020. Progress Report [51]	

were already established, and the Comprehensive Environmental Response, Compensation, and Liability Act (CERCLA) were enacted by the Carter administration. About packaging, in 1994, the European Parliament established the Packaging and Packaging Waste Directive (94/62/EC) (PPWD), generating one of the leading modern legislative approaches to the regulation of packaging at end-of-life. By 2000, enough momentum was built among brand owners, converters, and packaging suppliers in the US, Europe, and Australia that several organizations such as the SPC, the EUROPEN, and the SPA were established to promote and to create sustainable packaging systems. In 2004, the 1994 94/62/EC standards were revised and published as a new set of CEN standards under the umbrella standard EN 13427:2004 to deal with packaging waste.

1.11 Sustainable Packaging Organizations around the World and Their Criteria

By the year 2000, it was clear that brand owners and suppliers must coordinate their efforts around developing sustainable packaging systems if they did not want to be mandated by governments to comply with regulations that might not be optimal for the implementation of sound packaging systems. Therefore, much momentum was created around creating economically, socially, and environmentally responsible packaging systems.

The main organization driving the sustainable packaging agenda in the US was the SPC. Although there were only nine founding members of the SPC, it quickly grew to become an association of more than 500 members. From the beginning, it took an active approach, offering two stakeholder meetings per year, coordinating a number of projects, including development of briefs on different materials, and examining recyclable and compostable packaging. With concepts from sustainability, industrial ecology, and business in mind, the SPC defined sustainable packaging as meeting the following criteria:

A) "Is beneficial, safe & healthy for individuals and communities throughout its life cycle
B) Meets market criteria for performance and cost
C) Is sourced, manufactured, transported, and recycled using renewable energy
D) Optimizes the use of renewable or recycled source materials
E) Is manufactured using clean production technologies and best practices
F) Is made from materials healthy throughout the life cycle
G) Is physically designed to optimize materials and energy
H) Is effectively recovered and utilized in biological and/or industrial closed loop cycles." [53] (No ranking is implied in the order of the criteria.)

In addition to the SPC, several retailers such as Walmart®, Target®, Costco®, and Amazon® have developed sustainability initiatives, among them the Walmart® scorecard, which was implemented in 2007.

In 2004 in Europe, EUROPEN was created. EUROPEN's main concept of sustainable packaging is:

- "be designed holistically with the product in order to optimize overall environmental performance be made from responsibly sourced materials
- be designed to be effective and safe throughout its life cycle, to protect the product
- meet market criteria for performance and cost
- meet consumer choice and expectations
- be recycled or recovered efficiently after use." [54]

One of the latest initiatives of EUROPEN was to lead an effort involving 128 national and EU industries to examine the crucial role of implementing the packaging and packaging waste directive (PPWD) and in guaranteeing the fair trade of packaged goods in the EU [54]. Additionally, they are working on creating common requirements for extended producer responsibility in the EU.

One of the earliest organizations related to sustainable packaging was the SPA in Australia. Although some initiatives were developed through this organization, it ran out of momentum and is no longer operational as this chapter was written. The SPA criteria for sustainable packaging were also based on similar ideas that packaging must support sustainable development by being *effective* – adds real value to society, *efficient* use of materials and energy, *cyclic* – materials are recovered through biological or technical cycles, and *safe* – components do not pose risks to human health or ecosystems; when in doubt the precautionary principle applies [55]. Currently, the Australian Packaging Covenant, an agreement between the Australian states, the federal environment ministers, and the Australian Packaging Covenant Organization Ltd. (APCO), an industry-led initiative, and the government is the principle national instrument to reduce the environmental impacts of consumer packaging in Australia since 1999. Mostly, this is done through collaboration among the industries, and the Covenant allows businesses a more straightforward and cost-effective way to demonstrate that they are responsibly dealing with packaging.

On the other hand, it allows the government to require auditable compliance in all the territories where goods are sold. Their latest initiative was the APCO Strategic Plan for 2017 to 2022 [56]. The vision of the plan is "An Australia where packaging resources are used and disposed of wisely to minimize our environmental impact," and the mission is based on a packaging supply chain that: "can collaborate, lead and learn to: efficiently use resources; divert packaging waste from landfill; and share knowledge to improve the packaging sustainability performance of Australian industry" [56].

Other initiatives on packaging have been implemented around the world. However, a general strategy is often developed for large multinational corporations, with adjustments made as required to follow guidelines in specific areas. For example, Nestle®, the largest food company in the world, has its global headquarters in Lausanne, Switzerland, where the company's global sustainability initiatives are devised and then implemented in different regions such as the Middle East, Africa, and South America.

1.12 Tools to Evaluate Sustainable Packaging

Although tools to evaluate the overall sustainability of packaging systems have not yet been created, some tools have been developed to assess the environmental footprint of systems based on LCA methodology. The major full LCA software systems that allow running LCA that is fully compliant with ISO 14040/14044 are:

- SimaPro® developed by PRé Consultants,
- GaBi ts® developed by PE International,
- Umberto® NXT developed by ifu™ Hamburg.
- openLCA an open-source LCA platform managed by GreenDelta in Berlin, with a core development team.

These software systems provide complete LCI and LCIA capabilities. Additionally, some streamlined LCA software has been developed for many industries based on full LCA software.

For packaging, the leading software systems to evaluate the environmental footprint of packaging systems are:

- COMPASS® (Comparative Packaging Assessment) is an online streamlined LCA tool for packaging design evaluations. The SPC initially developed COMPASS® with support from the US EPA and input from the packaging industry and LCA experts. Trayak™ currently administers COMPASS®.
- PackageSmart is streamlined LCA software that can be used as a simplified, web-based tool for evaluating the environmental impacts of packaging and is commercialized by EarthShift Global.
- PIQET®. In association with the SPA, the Centre for Design at RMIT University (Melbourne, Australia) developed the Packaging Impact Quick Evaluation Tool (PIQET®) to provide rapid assessments of the environmental impacts of packaging.

In addition, several companies have developed their in-house systems for analysis of their products such as AMCOR's tool ASSET™.

1.13 Case Study 1.1. The Living Planet Index (LPI)

As discussed in Section 1.7, several indicators have been defined to quantify the development of cities, improvement of human life quality, and deterioration of environments and biodiversity. Among them, the Living Planet Index (LPI), reported since 1970, is a biodiversity index reported by the World Wildlife Foundation (WWF) together with the Institute of Zoology (ZSL) from the Zoological Society of London, both foundations working on wildlife preservation. The main goal of the LPI is to measure the state of the world's biological diversity accounting for population trends of animal species living on land, freshwater, and marine environments. The LPI has been widely accepted and adopted by several institutions, such as the Convention of Biological Diversity, as the main indicator for measuring improvement and defining restoration targets for wildlife habitats. Figure 1.11 shows the LPI from 1970 to 2016. The 2020 report indicated an average 68% worldwide decrease in population sizes of mammals, amphibians, birds, reptiles, and fish between 1970 and 2016. A 94% decline was reported for the tropical subregions of the Americas [40].

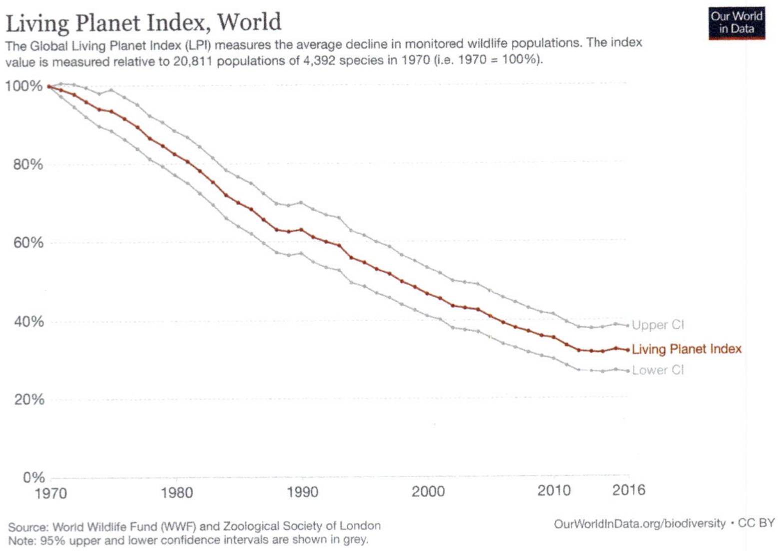

Figure 1.11 Living Planet Index has decreased since 1970 when it was first reported; reproduced from [40], licensed under CC BY 4.0.

Measuring animal population trends is essential since they are a good indicator of overall ecosystem health. Unfortunately, in the last 50 years, human population growth trends, overconsumption, global trade, and massive urbanization have diminished Earth's regeneration rate. We overuse Earth's resources and do not allow biological environments to regenerate, driving relentless destruction. As a result, well-known species such as polar bears, pandas, elephants, and tigers have declined. However, the declines of the larger species are just an indicator of much more significant problems for millions of smaller animals inland and marine environments. The loss of these species threatens biodiversity, ecosystems, and food security. In addition, many of these species, such as fish, feed the world and are regular food staples for billions of people. Therefore, the mismatch between modern economic systems and their incentives and Earth's living carrying capacity must be reevaluated.

1.14 Case Study 1.2. Doughnut Economics

Some of the new economic models emerging are considering the life of every person and their impact on Earth's carrying capacity. A new economic model known as the Doughnut contributes to the main idea of allowing every human being to have a life of dignity and opportunity, meanwhile safeguarding Earth's living carrying capacity. The main model is generated through the constraints of social and planetary boundaries, better known as the Doughnut, as shown in Figure 1.12 [57].

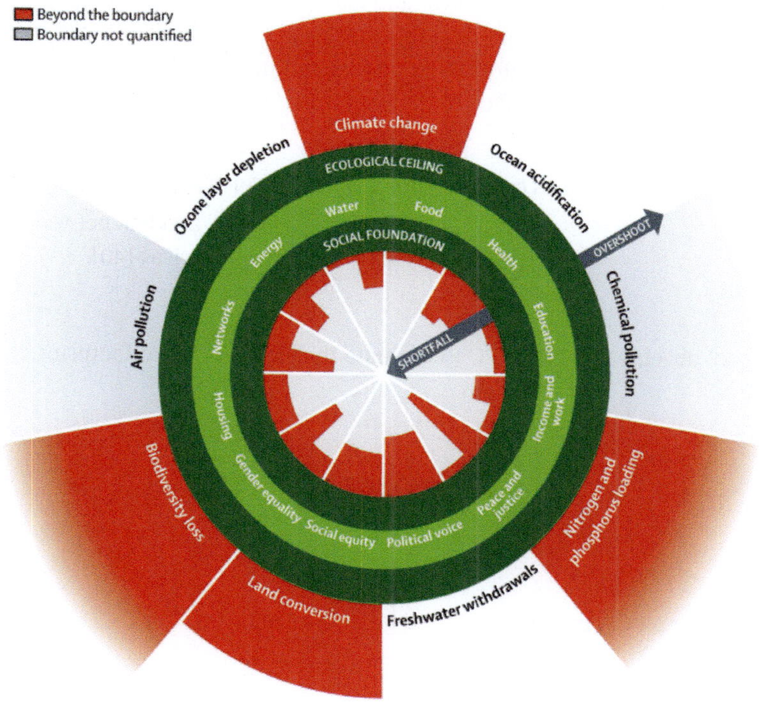

Figure 1.12 Concept of the Doughnut economics, representation of social foundation shortfall and ecological ceiling overshoot of global society in the twenty-first century. The dark green circles represent the social foundation and environmental boundary, encircling humankind's safe and just domain. The red wedges represent shortcomings in the social foundation or overextension of the ecological boundary. The center of the circle indicates critical human deprivation, and the outer ring outside the ecological boundary indicates critical planetary degradation; reproduced from [57] license under CC BY 4.0.

The main idea of the concept is to provide a framework for economics to balance gain between critical human and planetary needs before reaching deprivation and degradation, respectively. These two boundary areas are known as the Doughnut's social and ecological boundaries. The inner periphery provides the main criteria for a social foundation. Below this boundary lie humankind's shortfalls, such as illiteracy, ill health, hunger, that detract from a just and equitable society. These dimensions are well established and are currently being measured by the UN-SDGs. The outer boundary represents the environment ceiling; beyond this edge lies an overpressure on Earth's life-carrying capacity. The planetary boundary framework defines all the exterior boundaries with nine indicators: climate change, novel entities, stratospheric ozone depletion, atmospheric aerosol loading, ocean acidification, biogeochemical flows, freshwater use, land system change, and biosphere integrity [58].

The main idea of the Doughnut is to quantify and visualize the global scale of the shortcomings and overshoots so that a current state of human wellbeing can be assessed. The four main points raised by the Doughnut are as follows:

1) It shows the dependence of human wellbeing and Earth health or living carrying capacity – a symbiosis between the social foundation and the ecological ceiling.
2) The many social shortfall cases and ecological overshoot identify significant inequalities in areas such as income and wealth and gender and race.
3) The Doughnut concept requires a change of current economic utilitarian theories and policymaking. Petrifying concepts such as GDP growth should be replaced by new financial ideas looking at transforming the market from global to local and from linear to circular, regenerative, and distributive economic.
4) It could be used as guidelines for the new millennium to understand the complex correlations among social, economic, and environmental pillars to deliver human wellbeing and planetary health.

More information about Doughnut economics can be found in Kate Raworth's book [59].

1.15 Study Questions

1 What is the name of the tool developed by the SPC to evaluate the environmental footprint of packaging systems?
A PackageSmart
B Compass
C Scorecard
D PIQET
E None of the above

2 The main ISO standards detailing LCA requirements and guidelines are:
A ISO 14049/14043
B ISO 14040/14052
C ISO 14044/14458
D ISO 14044/14049
E None of the above

3 What are the four steps of LCA?
4 What is sustainable development as defined in the Brundtland Commission Report?
5 What are the components of sustainable development?
   A Environment
   B Social
   C Economic
   D a and b
   E a, b, and c
6 Are the following statements correct?

LCI stands for Life Cycle Input	T	F
LCA stands for Life Cycle Cost Assessment	T	F

7 Any basic concept of sustainable packaging systems should at least consider:
   A People, Poverty, and Profit
   B Social Equity, Economic Feasibility, and Ecological Viability
   C Paper, Plastic, and Metal
   D Logistics and the Walmart scorecard system
   E None of the above
8 How do you think that sustainability principles can be implemented at the federal, state, and local government level? Open question.
9 What are the four components of sustainable packaging framework defined by SPA?
10 Describe the main events of the sustainability timeline.
11 What is the main difference between LCT and CE? Open question.
12 What does it mean to use the PDCA approach? Open question.

1.16 Additional Resources

1 UN Sustainability goals
https://www.un.org/development/desa/disabilities/envision2030.html
2 Life Cycle Thinking
https://www.lifecycleinitiative.org/starting-life-cycle-thinking/what-is-life-cycle-thinking/
3 Circular Economy – Ellen Macarthur Foundation
https://www.ellenmacarthurfoundation.org/circular-economy/concept
4 A healthy economy should be designed to thrive, not grow – Kate Raworth
https://www.ted.com/talks/kate_raworth_a_healthy_economy_should_be_designed_to_thrive_not_grow
5 Sustainable Packaging Coalition
https://sustainablepackaging.org
6 EUROPEN
https://europen-packaging.eu
7 Sustainability
https://www.epa.gov/sustainability; https://www.environmentalscience.org/sustainability
8 ISO Standards
https://www.iso.org/standards.html

References

1 Platt, D. (2020). *The Future of Global Packaging to 2020*. Vol. 44. London, UK.

2 Mohan, A.M. Five trends impacting the North American packaging market. *Packag World 2013*. https://www.packworld.com/home/article/13363847/five-trends-impacting-the-north-american-packaging-market (accessed November 12, 2021).

3 Kaza, S., Yao, L., Bhada-Tata, P., and Van Woerden, F. (2018). *What a waste 2.0: A global snapshot of solid waste management to 2050*. The World Bank. DOI: 10.1596/978-1-4648-1329-0.

4 US Environmental Protection Agency. (2020). *Advancing sustainable materials management: 2018 tables and figures assessing trends in materials generation and management in the United States*. U.S. EPA.

5 Jambeck, J., Geyer, R., Wilcox, C., Siegler, T.R., Perryman, M., Andrady, A. et al. (2015). Plastic waste inputs from land into the ocean. *Science (80-)* 347: 768–771.

6 Gold, B.M., Mika, K., Horowitz, C., Herzog, M., and Leitner, L. (2013). *Stemming the Tide of plastic Marine Litter: A global action Agenda*.

7 *U.S. Bureau of labor statistics. Packaging labour information 2020*. https://www.bls.gov/oes/current/oes537064.htm (accessed November 12, 2021).

8 Wendling, Z.A., Emerson, J.W., de Sherbinin, A., Etsy, D.C. (2020). et al. *Environmental Performance Index*. New Haven, CT. DOI: 10.4324/9781315226675-5.

9 Lynam, J.K. and Herdt, R.W. (1989). Sense and Sustainability: Sustainability as an Objective in International Agricultural Research. *Agricultural Economics* 3: 381–398. https://doi.org/10.1016/0169-5150(89)90010-8

10 Fresco, L.O. and Kroonenberg, S.B. (1992). Time and spatial scales in ecological sustainability. *Land Use Policy* 9: 155–168. DOI: 10.1016/0264-8377(92)90040-4.

11 Munro, D.A. (1991). *Caring for the Earth: A Strategy for Sustainable Living*. Gland, Switzerland: IUCN, UNEP, WWF.

12 Bruntland, G.H., Khalid, M., Agnelli, S., Al-Athel, S.A., Chidzero, B., Fadika, L.M. et al. (1987). *Report of the World Commission on Environment and Development. "Our Common Future"*. New York, NY: United Nations. https://sustainabledevelopment.un.org/content/documents/5987our-common-future.pdf

13 Kidd, C. (1992). The evolution of sustainability T. *Journal of Agriculture Environment Ethics* 5: 26. 10.1007/BF01965413.

14 Schaller, N. (1993). Sustainable agriculture and the environment the concept of agricultural sustainability. *Agriculture, Ecosystems & Environment* 46: 89–97. https://doi.org/10.1016/0167-8809(93)90016-I

15 National Research Council. (2011). *Sustainability and the U.S. EPA*. Washington DC: Teh National Academies Press. DOI: 10.17226/13152.

16 Piketty, T. and Goldhammer, A. (2014). *Capital in the Twenty-First Century*. Cambridge, MA: Harvard University Press. https://www.hup.harvard.edu/catalog.php?isbn=9780674979857

17 Edwards, A.R. (2005). *The Sustainability Revolution: Portrait of a Paradigm Shift*. 1e. Gabriola Island: New Society Publishers.

18 Dresner, S. (2008). *The Principles of Sustainability*. 2e. Sterling, VA: Earthscan. https://www.amazon.com/Principles-Sustainability-Simon-Dresner/dp/184407496X

19 Muir, J. and Rowell, G.A. (1989). *The Yosemite : The Original John Muir Text*. San Francisco: Sierra Club Books.

20 Bacon, F. (1627). *The New Atlantis*. Salt Lake City, UT: Proj Gutenb Eb. https://www.gutenberg.org/files/2434/2434-h/2434-h.htm

21 von Carlowitz, H.C. (1713). *Silvicultura oeconomica.* Franklin Classics. https://www.bookdepository.com/Sylvicultura-Oeconomica-Hans-Carl-Von-Carlowitz/9780343516000?ref=grid-view&qid=1648237059347&sr=1-1 https://www.amazon.com/Sylvicultura-Oeconomica-Hans-Carl-Carlowitz/dp/0343516012

22 Malthus, T.R. (2011). *Thomas Robert Malthus, An essay on the principle of population [1798, 1st ed.] [1798].* Lib Fund, Inc. 1–108.

23 Muir, J. (1901). *Our national parks.*

24 Leopold, A. (2000). *A Sand County Almanac.* New York, NY: Oxford University PressOxfO. https://www.amazon.com/Sand-County-Almanac-Sketches-There/dp/0197500269/ref=pd_lpo_1?pd_rd_i=0197500269&psc=1&asin=0197500269&revisionId=&format=4&depth=1

25 Carson, R., Darling, L., and Darling, L. (n.d.). *Silent Spring.* Cambridge, Mass.: Houghton Mifflin Company.

26 Meadows DH, Rome. C of. (1982). *The Limits to Growth : A Report for the Club of Rome's Project on the Predicament of Mankind.* New York: Universe Books.

27 International Union for Conservation of Nature and Natural Resources (IUCN). (1980). *World Conservation Strategy. Living Resource Conservation for Sustainable Development.* Gland, Switzerland. (IUCN), I. U. for C. of N. and N. R. (1980). World conservation strategy. Living resource conservation for sustainable development. In Journal of Planning & Environment Law (Issue September). 77.

28 United Nations. *The 17 Goals | Sustainable development 2015.* https://sdgs.un.org/goals (accessed January 13, 2021).

29 Spellerberg, I.F. and Fedor, P.J. (2003). A tribute to Claude Shannon (1916–2001) and a plea for more rigorous use of species richness, species diversity and the 'Shannon–Wiener' Index. *Glob Ecol Biogeogr* 12: 177–179. DOI: 10.1046/j.1466-822X.2003.00015.x.

30 Orva, R. The city prosperity initiative. *2015 Global City Report 2015.* https://smartnet.niua.org/sites/default/files/resources/cpi_2015_global_city_report.pdf (accessed October 5, 2019).

31 Wiedmann, T. and Barrett, J. (2010). A review of the ecological footprint indicator—Perceptions and methods. *Sustainability* 2: 1645–1693. DOI: 10.3390/su2061645.

32 Connor, M.O. (2000). A portfolio of "Environmentally – Adjusted" National Income Measures: Key concepts and policy application. *ABCDE Eur. Conf.*, Paris, France.

33 Kaly, U.L., Pratt, C.R., and Mitchell, J. (2004). *The Environmental Vulnerability Index (EVI).* 384: 2004.

34 Talberth, J., Cobb, C., Slattery, Noah. (2007). *The genuine progress indicator 2006.* A Tool for Sustainable Development. https://sustainable-economy.org/wp-content/uploads/GPI-2006-Final.pdf (accessed October 5, 2019).

35 Lin, G.T.R. and Hope, C. (2004). Genuine saving measurement and its application to the United Kingdom and Taiwan. *Dev Econ* 42: 3–41. https://doi.org/10.1111/j.1746-1049.2004.tb01014.x.

36 Harger, J.R.E. and Meyer, F.-M. (1996). Definition of indicators for environmentally sustainable development. *Chemosphere* 33: 1749–1775. https://doi.org/10.1016/0045-6535(96)00194-4

37 Daly, H.E. and Cobb, J.B. (1990). *For the Common Good : Redirecting the Economy toward Community, the Environment, and a Sustainable Future.* London: Green print.

38 United Nations Development Programme (UNDP) (2020). Human development report 2020. *The next frontier Human development and the Anthropocene Human Development Report 2020.*

39 World Wide Fund for Nature. (2016). Living planet report 2016 risk and resilience in a new era. https://awsassets.panda.org/downloads/lpr_2016_full_report_low_res.pdf (accessed October 5, 2019).

40 Almond, R.E.A., Grooten, M., and Petersen, T. (2020). *The Living Planet Report 2020 – Bending the Curve of Biodiversity Loss.* Gland, Switzerland.

41 Russell, E. (1931). Some theoretical considerations on the "Overfishing"_ Problem. *Ices Journal Marine Science* 6: 3–20.

42 OECD. (n.d.). *OECD better life index.* http://www.oecdbetterlifeindex.org (accessed October 10, 2019).

43 Shannon, C.E. (1948). A mathematical theory of communication. *Bell System Technical Journal* 27: 379–423. https://doi.org/10.1002/j.1538-7305.1948.tb01338.x.

44 Cardarelli, R. and Lusinyan, L. (2015). *U.S. Total factor productivity slowdown: Evidence from the U.S. States.*

45 ISO. (n.d.). ISO 14000 family. *Environmental Management.* https://www.iso.org/iso-14001-environmental-management.html.

46 Geissdoerfer, M., Savaget, P., Bocken, N.M.P., and Hultink, E.J. (2017). The circular economy – A new sustainability paradigm? *J Clean Prod* 143: 757–768. https://doi.org/10.1016/j.jclepro.2016.12.048.

47 Pearce, D.W. and Turner, R.K. (1991). *Economics of natural resources and the environment.* Baltimore, Md.: Johns Hopkins University Press.

48 Ellen MacArthur Foundation. (2020). *The new plastics economy global commitment.*

49 Ellen MacArthur Foundation. (2016). *The new plastics economy – Rethinking the future of plastics.*

50 Licciardello, F. (2017). Packaging, blessing in disguise. Review on its diverse contribution to food sustainability. *Trends Food Sci Technol* 65: 32–39. 10.1016/J.TIFS.2017.05.003.

51 Ellen MacArthur Foundation. *The global commitment 2020 2020:75.* https://www.ellenmacarthurfoundation.org/assets/downloads/Global-Commitment-2020-Progress-Report.pdf (accessed November 15, 2021).

52 Rockström, J., Steffen, W., Noone, K., Perrson, A., Chapin, F.S., Lambin, E.F. et al. (2009). A safe operating space for humanity. *Nature* 461: 472–475. DOI: 10.1038/461472a.

53 Sustainable Packaging Coalition. (2011). Definition of sustainable packaging. *Renew Energy* 1–10.

54 Europen. *Packaging & the environment 2021.* https://www.europen-packaging.eu/sustainability.

55 Lewis, H., Fitzpatrick, L., Verghese, K., Sonneveld, K., Jordon, R., and Alliance, SP. (2007). *Sustainable packaging redefined.* http://www.helenlewisresearch.com.au/wp-content/uploads/2012/03/Sustainable-Packaging-Redefined-Nov-2007.pdf (accessed October 10, 2019).

56 APCO. (n.d.). *Australian packaging covenant organization.* https://apco.org.au (accessed November 12, 2021).

57 Raworth, K. (2017). A Doughnut for the Anthropocene: Humanity's compass in the 21st century. *Lancet Planet Heal* 1: e48–9. DOI: 10.1016/S2542-5196(17)30028-1.

58 Steffen, W., Richardson, K., Rockström, J., Cornell, S.E., Fetzer, I., Bennett, E.M. et al. (2015). Planetary boundaries: Guiding human development on a changing planet. *Science (80-)* 347: 10.1126/science.1259855.

59 Raworth, K. (2017). *Doughnut Economics.* White River Junction, Vermont, U.S.: Chelsea Green.

2

Design Thinking

The Packaging Design Process

Euihark Lee

School of Packaging, Michigan State University, East Lansing, MI

2.1 Learning Objectives

After learning from this chapter, you are expected to:

- Understand the concept of innovation.
- Apply new packaging design criteria, including sustainability considerations, to design packaging systems.
- Implement the design thinking process into packaging system design.
- Understand various elements of the packaging development process.
- Implement the packaging development process to develop a new packaging system.
- Understand packaging design considerations for recyclability.

2.2 Introduction

Packaging influences customers' buying behavior, and packaging design is an essential decision feature for protection and communication [1–3]. Packaging design must be engaged early in the product development process since several design components require iterative tasks. Packaging design can be defined into two main categories, structure and graphic design. The structure design relates to the packaging shape and features, and the graphic design relates to qualities such as color, typography, and different illustrations on the packaging. The structure and graphic designs should be considered together early in the development process to increase the final packaging's benefit and impact. This chapter describes design features for the packaging development process. First, the design thinking process is explained with detailed examples; then, the packaging design process is introduced as well as various design elements for both structural and graphical design.

2.2.1 Creativity vs. Innovation

Before discussing the design thinking process, it is essential to understand the difference between creativity and innovation. Creativity and innovation both involve ideas, but there is a significant difference

Life Cycle of Sustainable Packaging: From Design to End of Life, First Edition. Rafael A. Auras and Susan E. M. Selke.

Companion Website: www.wiley.com/go/Auras/lifecycleofsustainablepackaging

between them. Creativity can be defined as "unleashing the potential of the mind to conceive new ideas" [4], and it is subjective, making it difficult to measure. On the other hand, innovation is implementing a new idea with "measurable values" [4]. The value can represent economic, environmental, and social benefits, and it can be measured by looking at the improvements of these benefits when the innovative idea is implemented. In other words, creativity can be a necessary condition but not sufficient to provide innovation. Figure 2.1 provides an example of the difference between creativity and innovation. The left figure is a very creative car design; however, it is difficult to envision how this creative design could contribute value to our society if it is not just for an art auction. On the other hand, the car shape of the right figure is very similar to typical car shapes. This car is innovative because it uses electricity to run the vehicle instead of fossil fuel. Using electricity as power provides environmental benefits for society, and it can be measured by comparing CO_2-eq emissions with regular fuel-based cars. So, in short, innovation is putting ideas to work that provide value to society.

When discussing the concept of innovation in packaging, we should think about a new packaging design to provide a service and how the generation of a new design can increase the value of, for example, a current packaging system. Figure 2.2 shows an example of the Planters peanut packaging with reduced weight of 84% by converting the material from glass to plastic [5]. So, the new design provides economic benefit due to reducing costs while decreasing the environmental impact. When considering packaging systems, packaging professionals should concentrate on how to implement innovative ideas into the packaging design process.

Figure 2.1 Example of a difference between creativity (left) and innovation (right).

PILLARS	Measure Items		Descriptions
Economic	Reduce Materials	●	Reduce 84% less weight
	Enhance Transport	○	Reduce 84% less weight
	Energy Usage		
Environment	Reduce Pollution	○	Require 25% fewer trucks
	Recyclability		
	Water Use		
	Product Mile		
	Reduce Waste	○	
Social	Land Usage		
	Social Condition		
	Waste Management		

● Directly Effected ○ Indirectly Effected

Figure 2.2 Sustainable packaging example: Planters peanuts packaging; adapted from [5].

2.2.2 Design of Packaging for Sustainability

In the last few decades, attitudes about sustainability have dramatically changed, and companies have needed to react to these new requirements. Figure 2.3 shows an increase in global sustainability reporting from the 100 and 250 largest companies in the world by revenue [6]. By 2020, at least 80% of these companies generated a sustainability report.

As a result, conventional packaging design objectives are being impacted and expanded to consider sustainability in packaging. In traditional packaging design, there are four main objectives: cost, technical function, appearance, and regulatory compliance. *Cost* is one of the most important criteria in deciding if a new package design is acceptable or not, especially for commodity products such as milk and flour packaging. Therefore, cost-effectiveness must be considered to meet the requirements of the targeted market. Material reduction, finding alternative materials, reducing volume, and many other cost-saving factors can be studied to find a way to enhance cost-effectiveness. The *technical function* represents the performance of the packaging, such as protection, containment, and convenience. *Appearance* is also an important design objective for packaging, so the designer should consider how the design communicates effectively to the customer. Last but not the least, the design should comply with all regulatory requirements.

Consideration of sustainability brings additional design objectives. For example, the SPC, as presented in Chapter 1, defined new sustainable design objectives as shown in Table 2.1 [7]. The basic concept of these additional objectives is expanding design considerations into the packaging life cycle and overall sustainability.

When optimizing resources in a new packaging design, overall source reduction should be one of the first objectives. Source reduction reduces the amount of material entering the waste stream by redesigning the packaging or the product [8]. Source reduction can reduce the environmental impact of the packaging and reduce the cost. However, source reduction can result in under-packaging problems or packaging system failure. Therefore, the packaging designer should balance source reduction to avoid under- or over-packaging issues (more in Chapter 11). Using recycled content helps optimize resources and reduce overall environmental impact. Recycled materials have been recovered or diverted from MSW through recycling and are used as feedstock for a new product. Using recycled content offsets the need for virgin resources and usually consumes significantly less energy [7]. However, adding more recycled content could affect the packaging functions. Therefore, packaging professionals should find the right amount of recycled content for new packaging designs (Chapter 13).

Moreover, designing a more efficient package for transportation is another simple approach to optimizing resources since it can result in energy and material reduction. Sustainable management

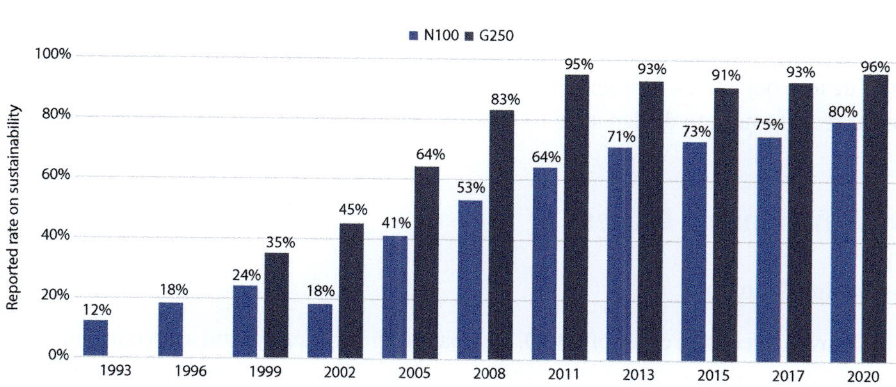

Figure 2.3 Growth in global sustainability reporting rates since 1993: N100 and G250; adapted from [6].

Table 2.1 Sustainable packaging design objectives; adapted from [7].

Design Objectives	Strategies
Optimize resources	• Practice source reduction • Use recycled content • Design for transport
Responsible sourcing	• Design with environmental best practices • Design with fair labor and trade practices • Design with renewable virgin materials from sustainably managed sources
Resource recovery	• Design for reuse • Design for recycling • Design for composting • Design for energy recovery
Material health	• Know the chemistry of your package

practices should be implemented to manage packaging materials and ensure a new packaging design meets best environmental practices. So, responsible sourcing emphasizes how fair regulations and policies follow the used resources and materials. In this category, fair labor and trade practices should be considered to ensure that a new package's production meets social equity. For example, if the supplier is in the US, the practitioner should check if the supplier complies with all Occupational Safety and Health laws and regulations.

Material health is an essential issue for sustainable design consideration. The packaging material and the material used in the manufacturing process should be healthy for humans and the environment. Therefore, knowing the chemistry of the packaging material is important, so potential environmental impacts can be analyzed during the packaging design process (see Chapter 9 discussion about mid- and endpoint indicators).

Finally, yet importantly, packaging professionals should design for the end of life of the package, including resource recovery as a new design objective. The packaging professional should focus on how to extend the package life by enhancing reusability. Providing refilling or multiple usage features can be implemented into a new package design process when economically and environmentally feasible. Designing for recycling is a critical feature, which is gaining importance to increase the recovery rate of packaging materials. Recycling generates a multitude of benefits, including supplying the industry with valuable materials and the conservation of natural resources and energy [7].

Sustainability in packaging design has been a crucial issue in the last few decades as explained in Chapter 1. Still, the biggest challenge for sustainable packaging design is how to balance different aspects of the design objectives. Therefore, understanding the basic concept of design objectives is an essential requirement to design sustainable packaging system.

2.3 The Design Thinking Process

2.3.1 What Is Design Thinking?

According to Tim Brown, executive chair of IDEO, a global design company, and a design thinking pioneer, "*Design thinking is a human-centered approach to innovation that draws from the designer's toolkit to integrate the needs of people, the possibilities of technology, and the requirements for business success*" [9]. The main idea of the design thinking process is trying to solve human needs through the design process, and it can be implemented, so that the innovation process provides social values. This principle can also be applied to the packaging development process. The goal of the packaging professional is to solve many of the issues that people are facing when dealing with packaging such as communicate, carry, open, dose, dispose, and others. Design thinking can be used to initiate the innovation process so that professionals can deliver selected values to our society. To understand the application of design thinking in the packaging development process, we first need to expand on design thinking.

2.3.2 The Five Stages of Design Thinking

There are many different steps during the design thinking process, and the five stages of the design thinking process introduced by the Hasso Plattner Institute of Design at Stanford University is one of the best-known methods. The major description of the stages of the design process presented in this section was adapted from the work titled "An Introduction to Design Thinking Process Guide" [10] as illustrated in Figure 2.4.

The first stage of design thinking process is "*empathize*," which can be described as "to create meaningful innovations, you need to know your users and care about their lives" [10]. In other words, the main goal of the empathize stage is understanding the current issues that the consumers are experiencing. Most products have target customers and understanding current packaging issues for these consumers is the starting point of the packaging development process. During the empathize stage, the packaging professional should collect feedback information from various sources such as obtaining information from suppliers and vendors, investigating the current packaging system, conducting consumer survey and focus interviews, and collecting field data. More recently as the evolution of technology is making it possible, data mining or web scraping tools are being widely used. With these data, the packaging professional should be able to identify the main issues of the current packaging systems.

The second stage is "*define*," stated as "*the define mode of the design process is all about bringing clarity and focus to the design space*" [10]. This is a crucial stage since it determines the direction of the innovation. During this stage, the problems should be narrowed down to issues that the packaging professional wants to and can solve from the collected feedback data in previous stages. The professional should also understand the feasibility of the design. Packaging professionals face many

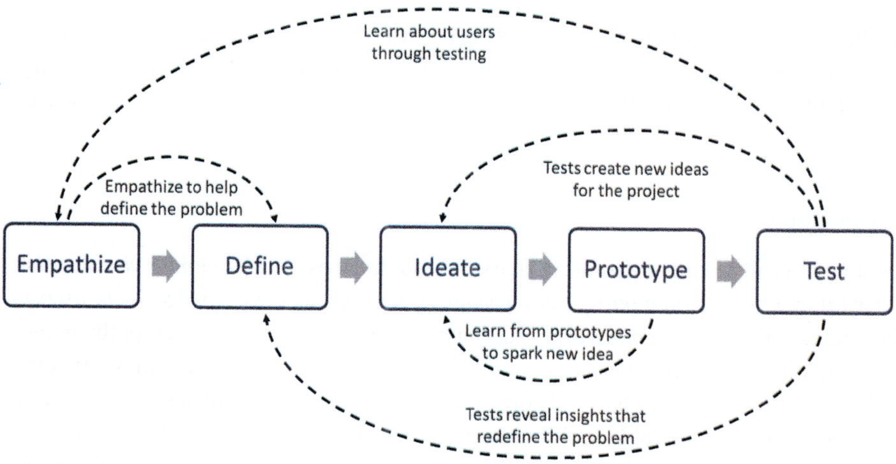

Figure 2.4 The design thinking process stages; adapted from [11].

packaging problems, but not all the issues can be solved using current technologies. Also, in several situations problems are coupled with other issues. Therefore, it is vital to define the critical problem that can solve many packaging issues. For example, excessive material usage for packaging results in increased cost, resource use, and other environmental impacts. Therefore, reducing packaging materials can reduce not only the packaging cost but also adverse environmental impacts.

The third stage of the design thinking process is *"ideate,"* stated as *"Ideate is the mode of the design process in which you concentrate on idea generation"* [10]. Once the problem statement is defined in the define stage, in the ideate stage the packaging professional investigates solutions to translate the packaging problems to solutions. During this stage, research helps to generate ideas about several solutions and to narrow these down to the most optimal solutions, including the modification of the current design, adapting designs from a different market segment, and think outside of the box solutions.

The fourth stage of the design thinking process is *"prototype,"* which can be defined as *"the prototype mode is the iterative generation of artifacts intended to answer questions that get you closer to your final solutions"* [10]. The main goal of prototyping is to iterate the designs and to find out early on about any issues with the new design. Therefore, the quality of the prototyping model is not required to be at the manufacturing level, and it can be made from anything that can interact with the user and the user experience. For example, a common paper-folding carton prototype can be created using a digital cutting table; however, it can also be made by hand if a table is not available. With the prototyped folding carton model, users can try to use all the features implemented into the folding carton and get feedback from these experiences.

The last stage of the design thinking process is *"test,"* which can be defined as: *"The test mode is when you solicit feedback, about the prototypes you have created, from your user and have another opportunity to gain empathy for the people you are designing for"* [10]. The testing stage helps to find out the possible problems of the new idea and redefine the solutions. It is an essential process to make sure your idea is a good fit for your targeted users. The prototyped model can be tested by users to get feedback from users' experiences. Moreover, in the packaging industry, various testing methods are applied to ensure the performance of the packaging meets specifications. Depending on the testing results, the design can be iterated and go back to the early stages – the ideate, define or even empathize stage and redo the entire process until the new design meets all requirements. Figure 2.4 shows the five design stages and the relation of these stages.

2.4 Tools for Thinking about Innovation

One of the most challenging parts of the design thinking process is having enough innovative ideas to implement into the design thinking process, so many idea generation strategies are necessary. Many tools are available to help generating innovative ideas. A few of the main tools are presented in this section.

2.4.1 Empathy Mapping

An empathy mapping is a collaborative visualization tool. It is defined as *"a collaborative visualization used to articulate what we know about a particular type of user. It externalizes knowledge about users to 1) create a shared understanding of user needs and 2) aid in decision making"* [12]. The empathy map is composed of four main questions: what the user says, thinks, does, and feels. Figure 2.5 shows that the targeted user is placed at the center and lists out corresponding keywords as feedback, research, and thoughts. Empathy mapping is a valuable tool to enumerate ideas from the user's end perspective; however, it is possible that the feedback data obtained from the users may not reflect the actual needs.

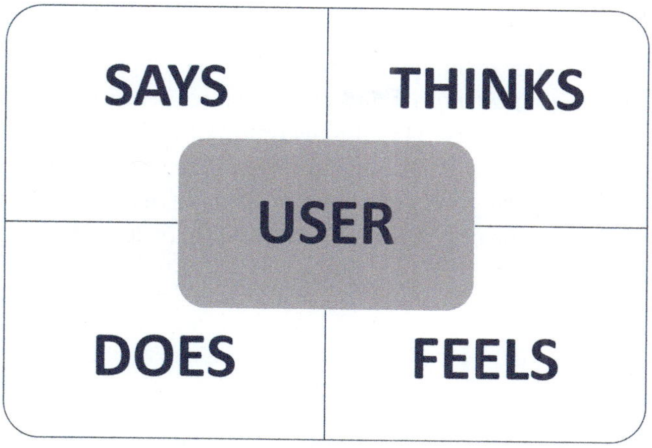

Figure 2.5 Empathy map example; adapted from [12].

2.4.2 Mind Map

A mind map is "*a graphical way to associate a new concept at learning stage with the former ones, and to integrate and detail it by using lines, colours, pictures, symbols and key words*" [13]. There are many different ways to create mind mappings – however, one of the most common starts from a central idea. The central idea can be the problem, a title, or the objective of the design. Then, this central idea triggers a connection to the other keywords called associations. Each association can produce other associations in a further iteration, so the mind map keeps spreading out until all ideas are connected. As shown in Figure 2.6, the word packaging is chosen for the central idea. It is spread out to different packaging aspects such as materials, features, applications, and research extending further from these aspects. Mind mapping is a handy tool to visualize an idea, but it can get complicated very rapidly.

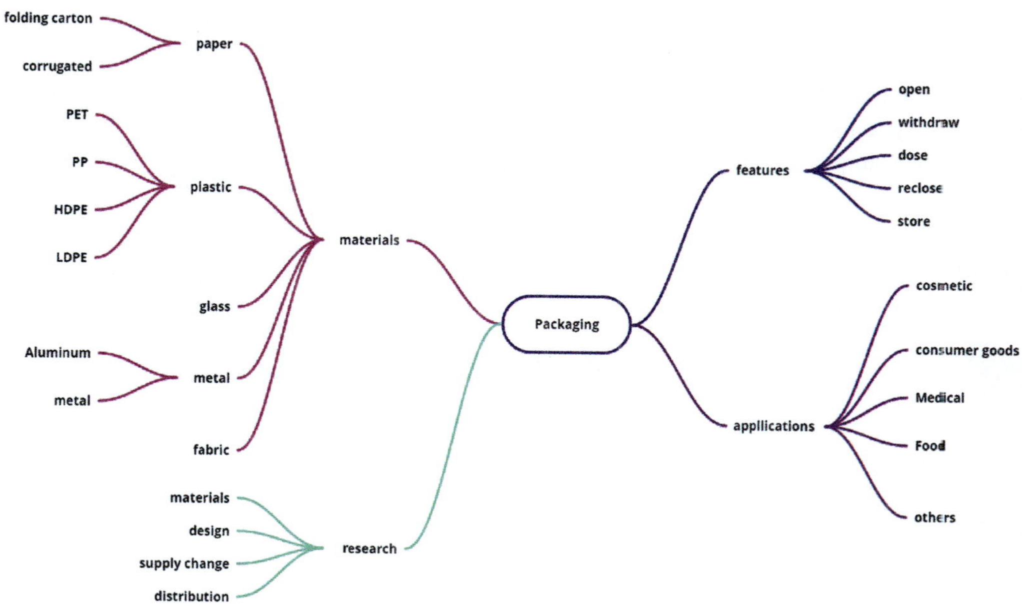

Figure 2.6 Mind mapping example for a central idea such as packaging.

2.4.3 Brainstorming

In 1942, Alex Osborn, the author of this creativity technique, introduced the concept of brainstorming as "*a conference technique by which a group attempts to find a solution for a specific problem by amassing all the ideas spontaneously by its members*" [14]. Brainstorming has been widely applied in several industries since its introduction. A significant advantage of the brainstorming technique is finding new perspectives while considering every idea equally. However, it is time-consuming, and the suggested ideas may not be feasible or practical. There are seven rules of brainstorming, as shown in Figure 2.7 [15].

2.5 Packaging Design Process

Even though the five stages of the design thinking process were previously explained, this section concentrates on the define and ideate stages for the packaging design development process. Each subsection describes different design elements such as materials, features, shape, color, and graphics for each packaging component.

2.5.1 Applying the Design Process to the Packaging System

Once the packaging issues are identified, for example, using the empathy stage of the design thinking process, the problem should be narrowed down to the most significant issues, considered at the defined stage. To narrow down the problem correctly, the packaging professional should fully understand the packaging system's life cycle and technical background, as shown in Figure 2.8.

First, the packaging professional should understand all the included manufacturing technologies used to produce the package, including raw material extraction, packaging conversion, filling, and assembly, depending on the targeted packaging form. The available machinery systems are from the company and suppliers and knowing them is essential information. New technology trends should be understood and researched as well.

Next, the distribution system is an essential factor for the new packaging development. Current distribution systems are more complicated than traditional systems due to the expansion of e-commerce. So, the packaging system should be designed for both brick-and-mortar and e-commerce distribution systems. However, both distribution systems have different levels of environmental hazards. For example, e-commerce packaging systems have at least three times more handling

Figure 2.7 Rules of brainstorming; adapted from [15].

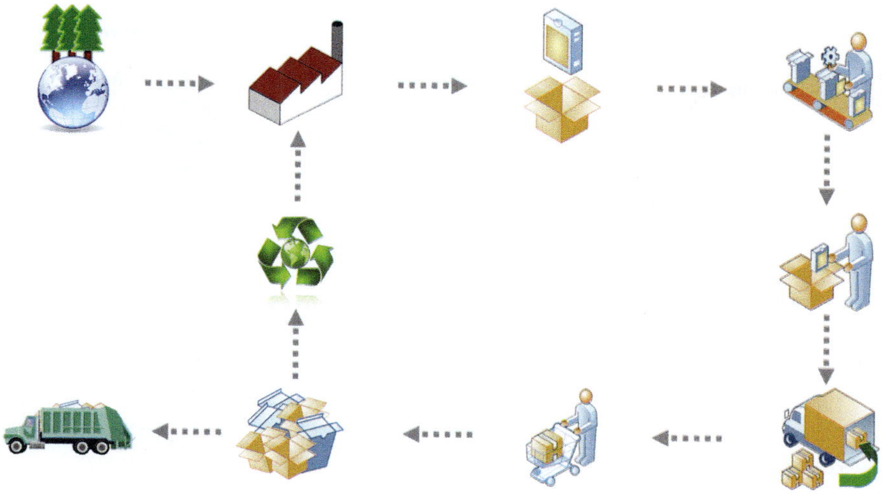

Figure 2.8 Representation of the life cycle of packaging.

points than regular supermarket distribution systems [16], which can cause more damage to the product during the distribution. The packaging professional should conduct an in-depth study of the distribution environment of the targeted market.

Lastly, the end of life of the packaging should be considered. The possible MSW management end-of-life scenarios of the packaging are recycling, incineration with energy recovery, and landfill. However, some can leak and end up as litter, which should not be considered an MSW management technique. So, a new packaging design should be designed to be recovered through the geographically available MSW management end-of-life scenarios. Understanding the accessible MSW management end-of-life scenarios is crucial to designing packaging systems for the circular economy.

2.5.2 Material Selection

Material selection is one of the routes to narrow down the design and selection of packaging since each material character limits the packaging design and the end-of-life. Figure 2.9 shows the typical packaging materials and examples of corresponding packaging formats. Chapter 3 will provide a detailed discussion of each material.

There are many directions to implement an innovative idea into the material selection process, such as material reduction, which is a simple approach that effectively reduces environmental footprint (EFP). For example, the weight of a 500 mL plastic water bottle has been reduced by about half from the early 2000s while meeting its required performance [17]. The weight of the aluminum beer can has also been reduced through advances in can manufacturing technology. In 1972, 21.75 cans could be made from one pound of aluminum. Today, one pound of aluminum can produce about 33 cans [18]. The material reduction is beneficial to reduce not only cost but also the adverse environmental impacts of using nonrenewable resources [18].

Material substitution can be another option for innovation. For example, many rigid plastic packages are converted into flexible packaging forms because of the many benefits such as lightweight, cost-reduction, easy distribution, and others (see for example Case Study 11.1). Material substitution is not only limited to form such as flexible packaging, but it can also substitute materials from glass

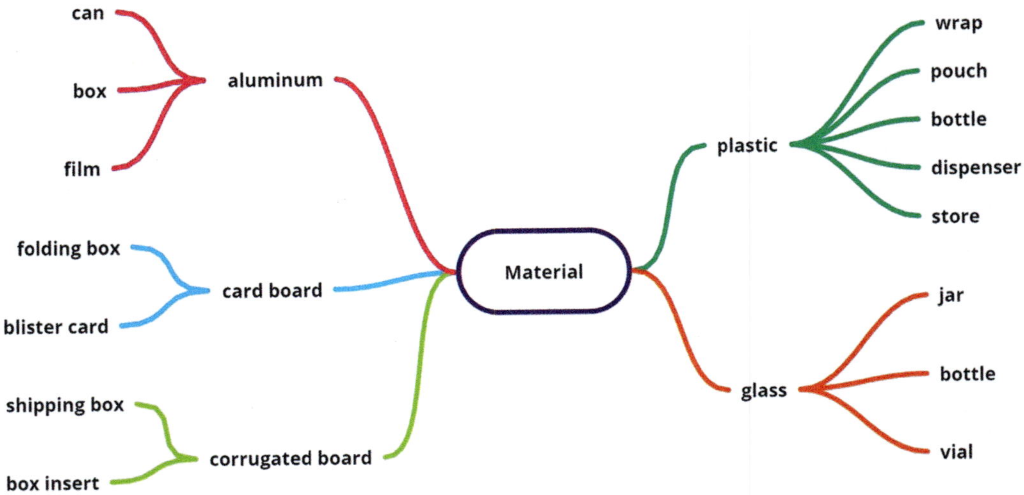

Figure 2.9 Packaging materials and common formats of selected applications.

to rigid plastic, rigid plastic to folding carton or folding carton to flexible plastic packaging as long as the reduction in weight and environmental footprint is obtained. Figure 2.10 shows a schematic diagram of how packaging materials can be substituted from one material to other materials

Introducing and implementing new packaging materials is challenging for innovation. It requires more time and additional cost than just material reduction or substitution, but it is worth keeping as a consideration.

2.5.3 Determining Packaging Features

The main features of packaging are key factors to enhance the user experience of the products. There are five major features, including opening, withdrawing, dosing, reclosing, and storing, as shown in Figure 2.11. These features are places where innovation can be implemented by professional designers.

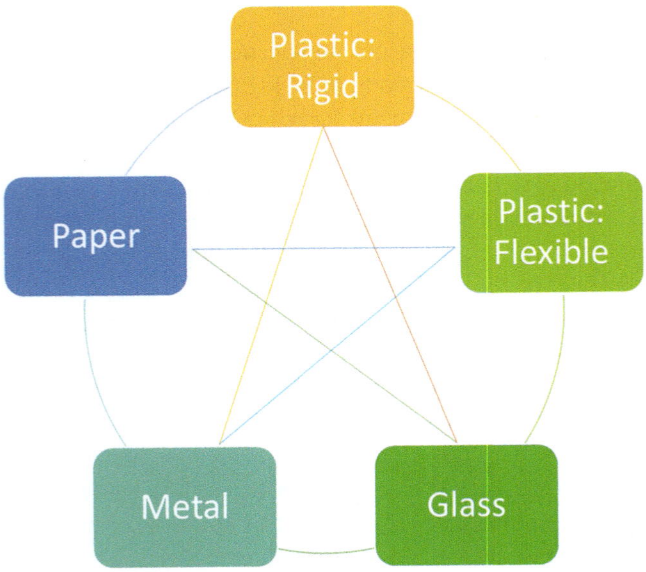

Figure 2.10 Schematic diagram of the different packaging materials substitution.

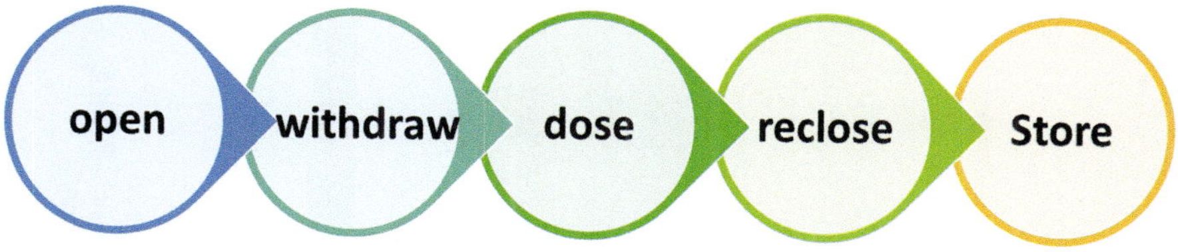

Figure 2.11 The main features of packaging.

The design objective of a feature is to improve the convenience of packaging. By implementing these features into the packaging, the product can be used more effectively. For example, a tear notch is a simple but effective design to make it easy to open a pouch. A simple push mechanism can be used in a container for delivering candies. Resealable film provides a reclose function for cookies and chocolates. Several innovative ideas have been implemented into packaging using these features (Figure 2.12).

However, the key needed features should be identified during the empathy and define stages in the design thinking process. Once the key features are identified, then the process moves on to the ideate stage. Using the tools explained in the previous section, packaging professionals can create many different ideas about the key features. As shown in Figure 2.13, most lip glosses require the use of two hands to apply, one hand to hold the cap and the other hand to apply the lip gloss. Figure 2.13 shows that the packaging professional can identify one hand lip gloss application as a key feature to be improved and shows the development process including brainstorming and the final design. Once the key features are identified, the development process can advance to the design shape stage.

2.5.4 Design Shape

The basic functions provided by the packaging shape are containment and protection. Depending on the shape of the packaging, the amount of the contained product varies. Moreover, the packaging shape can provide room for brand identity. Many customers identify the brand by the typical form of the package. For example, the Coca-Cola® bottle is one of the most representative examples of how the shape of a package can inform the brand image. Therefore, the packaging design not only provides containment but also can provide brand identification.

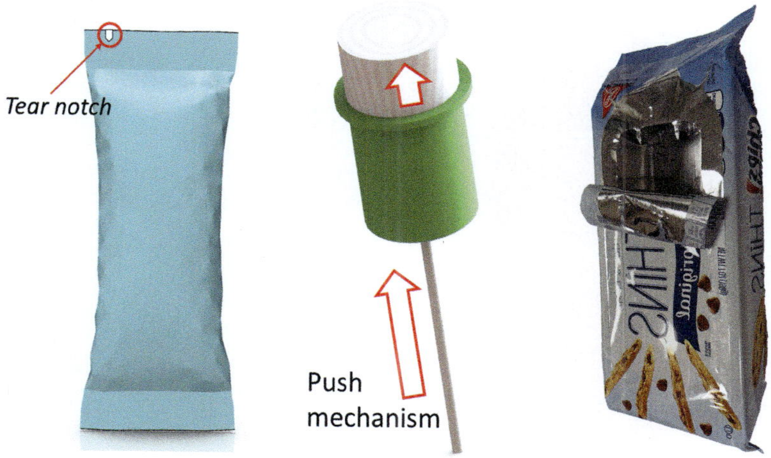

Figure 2.12 Examples of packaging features: tear notch (left), push mechanism (middle), reclose feature (right).

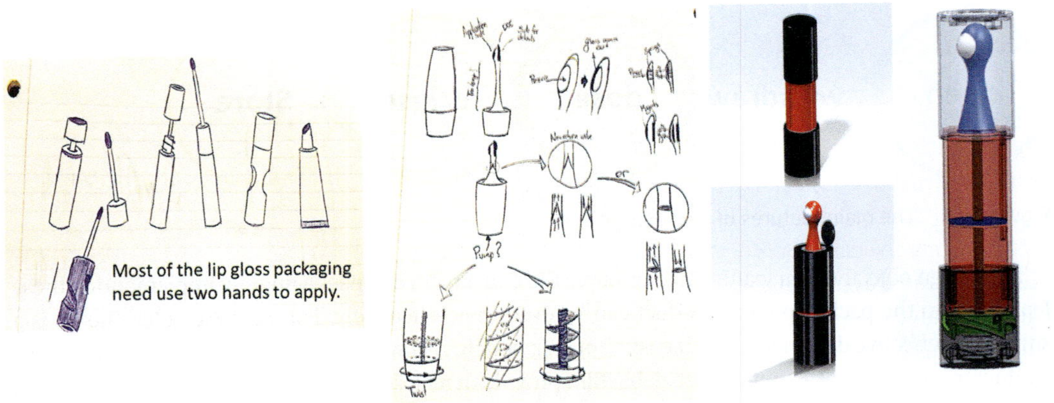

Figure 2.13 One-hand handling lip gloss application: Define key feature (left), brainstorming (middle), and final design (right).

Before creating a new packaging shape, the main material should be selected as explained. The selection of each material limits the packaging shape, so defining the material will narrow down the possibilities for the packaging shape. Moreover, packaging professionals should understand the current technologies to convert the material to the packaging shape. The converting process and machinery can constrain the form. For example, PET – poly(ethylene terephthalate) – bottles are made by injection stretch blow molding (ISBM), and the mold cavity shape determines the shape of the bottle. Therefore, the bottle shape is limited by the constraints on the design of the mold cavity. As a result, some bottle designs could not be made if they could not fit a mold cavity design.

Even though most packaging shapes are 3D, it is worth seeing the shape's silhouette in a 2D configuration. A packaging silhouette can be decomposed into several components, and each component can be simplified as a typical 2D shape, as shown in Figure 2.14.

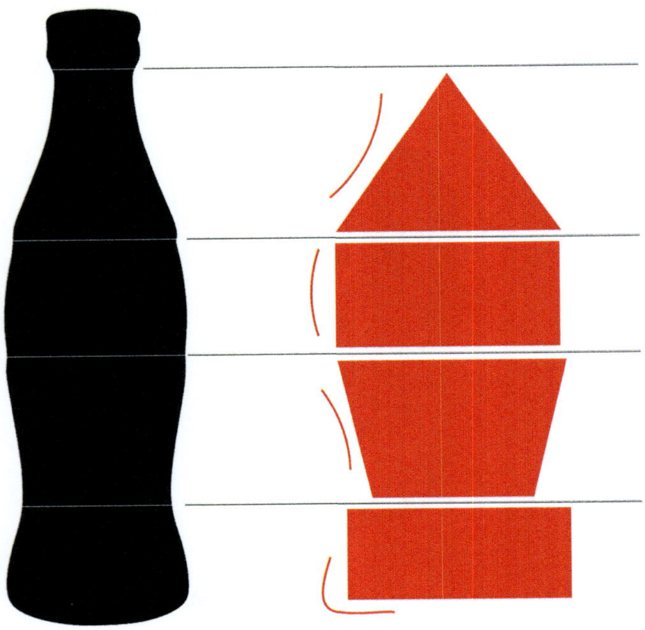

Figure 2.14 PET bottle silhouette decomposition.

The rectangular and triangular shapes are the most common silhouettes for packaging. A rectangular shape gives reliability and security. A triangle shape provides stability and balance as long as it is not inverted [19]. As you can see in Figure 2.14, 2D silhouettes can be decomposed into several shapes, and each shape creates a perception and maybe an emotional feeling. Combining these shapes into the final packaging shape can provide an overall consumer perception and expressive feeling of the packaging.

Another approach of shape designing is implementing universal design or inclusive design concepts. Both universal design and inclusive design provide the most product accessibility to customers. However, the ways to approach the customers are different. The universal design focuses on a solution that can serve the most extensive potential users. The inclusive design provides various solutions for different needs based on culture, age, gender, language, and other factors [20]. In the packaging world, most packaging design is based on the universal design concept, basically, one design for all. However, the inclusive design concept should be considered when considering specific target groups. As shown in Figure 2.15, Grace Beauty designed makeup products that have special features for easy grip, hold and use, providing additional help for a person with a disability [21].

At the same time, ergonomic designs have been implemented into packaging design considerations. Ergonomic design features include a design for human environments, making it easy to handle the packaging and the product, such as making a package easier to grip. De la Fuente et al. (2015) introduced an affordance-based package design theory, an extended concept from the ergonomic design principles. Figure 2.16 shows how aerosol design can be improved using the affordance-based methodology [21].

2.5.5 Color and Packaging

Color is one of the first attributes perceived when we look at an object, and it is a crucial factor that helps distinguish a package on a shelf. In general, we associate each color with some feeling or perception, and some colors have different perceived attributes in different cultures. Table 2.2 shows perceived attributes for colors in western cultures, as reported by [23].

Figure 2.15 Grace Beauty: Disability-friendly makeup tool; adapted from [21].

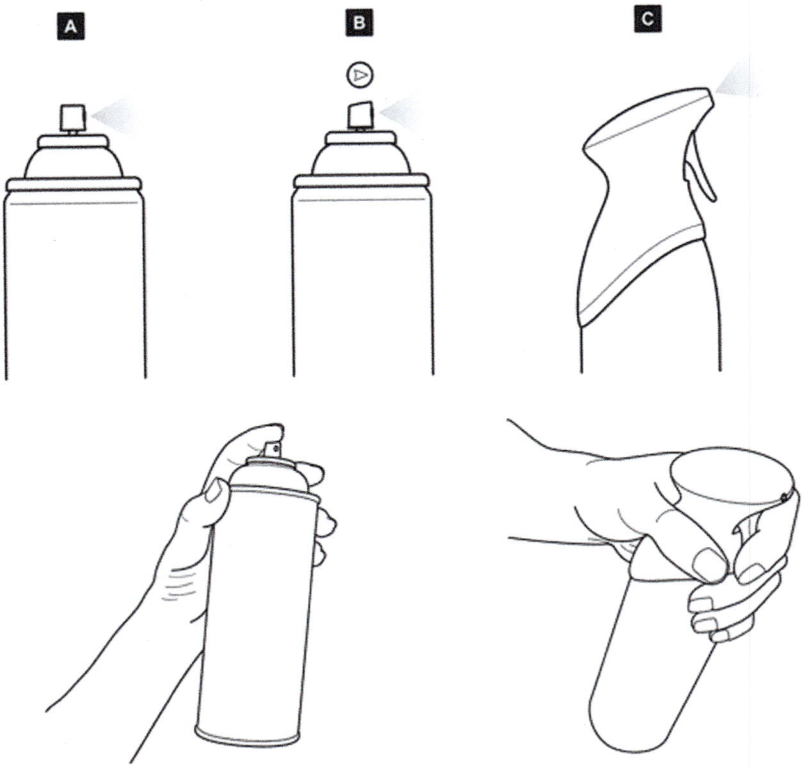

Figure 2.16 Aerosol design development through the affordance design concept; reproduced from [22] license under CC BY-NC-ND 3.0.

Moreover, we should keep in mind that the perception of colors can differ depending on personal background, geographical region, religion, and age. Therefore, packaging professionals should consider selecting color depending on the targeted market. For example, red color represents danger, love, or passion in western culture, but in Asian cultures, it represents happiness, joy, or celebration. Figure 2.17 summarizes the different perceptions or meaning of colors by different cultures.

Colors are also related to capturing the attention from the customer, so it is important to choose the right color for the right products. In general, yellow is the most visible color, and yellow, orange, and red tend to make the object perceived larger and closer. On the other side, green, blue, and violet make the object perceived smaller and further away [25].

So, color characteristics can be used to define brand images. For example, once the product's market target is defined, it is recommended to select three primary colors: base, accent, and neutral. The base color is the dominant color of your package. It shows the brand image and should attract the targeted customers. The accent color should be paired with the base color, keeping the brand image identity. The neutral color is not for gaining attention but for the background, helping with base and accent color balance. Figure 2.18 shows the process of color selection for a brand.

2.5.6 Graphics in Packaging

Graphics in packaging include two main elements, typography and illustration. Squire defines "typography is first and foremost about the communication of information" [26]. There are many

Table 2.2 Perceived attributes for colors in western cultures.

Color	Perception
Green	Balance, harmony, love, communication, social, nature, acceptance
Blue – indigo	Calmness, peace, love, honesty, peace, kindness, truth, inner peace, emotional depth, devotion
Violet	Intuition, imagination, universal flow, meditation, artistic qualities
Red	Physical energy, vitality, stamina, grounding, spontaneity, stability, passion
Orange	Creativity, productive, pleasure, optimism, enthusiasm, emotional expression
Yellow	Fun, humor, lightness, personal power, intellect, logic, creativity
Green	Lower environmental impact, sustainable packaging

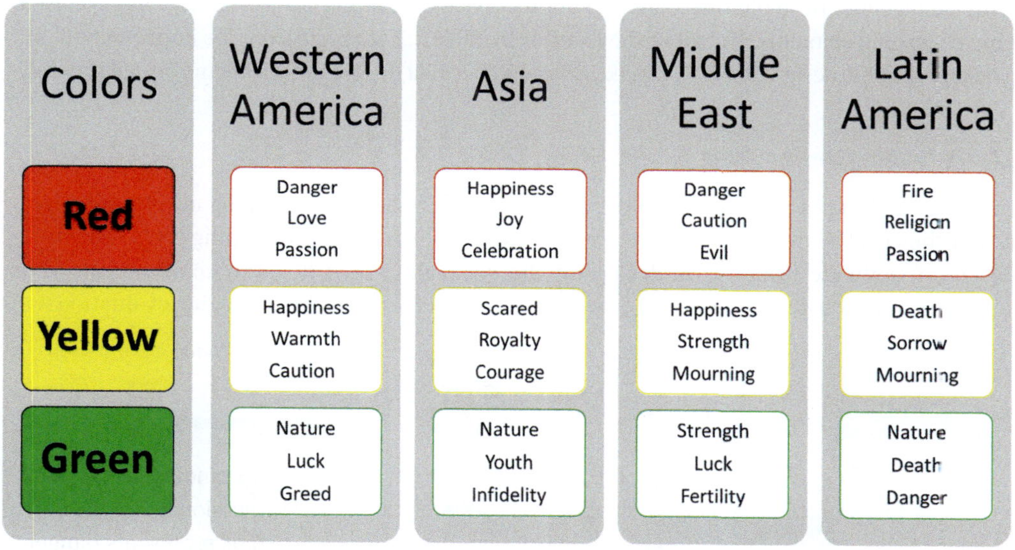

Figure 2.17 Suggested perceived meaning of colors across cultures [24].

Figure 2.18 Brand color example.

objectives for typography in packaging, such as getting customer attention, delivering necessary information, and identifying the product. The typography on packaging can be determined by the following factors: targeted market, packaging size, product type, label type, and illustration on the packaging. In general, packaging typography can be decomposed into three major sections: brand/

product name, the type of product, and other additional information (ingredients, instructions, contact information) [27].

The illustration is another element defining the packaging graphic, and it guides consumers where to focus on the packaging. A widely used strategy is selecting the most important thing that customers should remember about your product and placing it as an illustration or typography at the center of the packaging face. Then, you can place a few more graphic elements around the centerpiece to deliver the message you want to reach the customers.

The purpose of the graphic is not only to grab attention but also to deliver critical information. Nutrition labels are representative examples, and many studies have been done to understand the influence of graphics on the packaging [28]. Figure 2.19 shows graphic elements of the nutrition information for both Front of Pack (FoP) label and the Nutrition Facts Panel (NFP).

There are many different graphic elements, and the packaging professionals should implement proper graphic elements for their packaging applications. For example, the graphic requirements for food packaging differ from pharmaceutical or another packaging. Table 2.3 provides a complete list of graphic packaging elements. Not all of the features in Table 2.3 are required to be implemented into the package graphic design, but they can be used as a checklist during the packaging design process.

2.5.7 Packaging Design Tools

The packaging design tools are one of the essential elements of the packaging development process, and they can be divided into two groups: design software and prototyping tools. The packaging design software visualizes a design for the packaging, so it helps to communicate your design. Moreover, the design files can be used in many ways, such as the packaging evaluation and

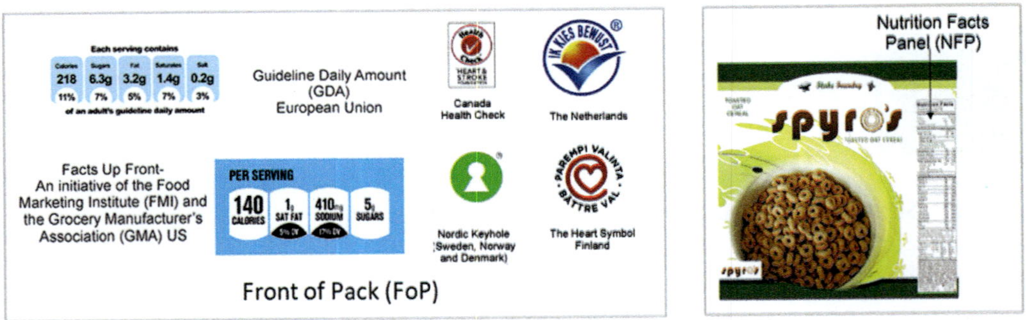

Figure 2.19 Packaging graphic elements for nutrition; reproduced from [28] with permission from Elsevier.

Table 2.3 Elements provided by the packaging graphics.

List of the Graphic Elements		
Brand name	Product name	Company name
Logos or icons	Package "persona"	Visibility and impact
Color selection	Ingredient list	Price panel
UPC code	Opening instruction	Use instruction
Disposal instructions	Emergency instruction	Hazard warning
Legal text requirement	Environmental information	Number of facings
Language requirement		

prototyping processes. In this section, packaging tools for both design and prototyping are described.

The packaging design can be composed into three different groups: paper packaging, rigid packaging, and flexible packaging, and a list of packaging design software is shown in Table 2.4. The paper packaging includes a folding carton, corrugated box, and a stand-up display. Both ArtiosCAD and Impact software include standard packaging libraries such as ECMA (folding cartoon) and FEFCO (corrugated). The rigid packaging includes bottles, closures, containers, and any rigid types of packaging. Most of the Computer Aided Modeling (CAD) software can be used to design 3D rigid packaging. Three CAD software programs are listed in the table, but a lot of other CAD software are available these days such as Inventor and Fusion 360 from Autodesk®, and CATIA. The CAD software is specialized for 3D modeling, and it includes various features that can convert 2D sketches into 3D objects. One of the benefits of the CAD modeling is that the CAD files can be implemented into the Finite Element Analysis (FEA) software to analyze the mechanical behavior of the design under given boundary conditions. The results of FEA can be applied to identify packaging design problems without physical testing. Therefore, it enhances the efficiency of the packaging development process. In the case of flexible packaging, the studio toolkit by Esko has been used widely. This studio toolkit provides a standard shape for flexible packaging that helps create the design more efficiently. Also, the studio toolkit includes an Adobe Illustrator plugin, so packaging professionals can work on the graphics more effectively.

Prototyping can be done in many ways – from simply using scissors and glue for a quick model to using high-tech equipment such as a digital cutting table and 3D printer. In this section, industrial levels of prototyping are explained. To create a paper prototype model, a digital cutting table is widely used as shown in Figure 2.20. Die-cut information is transferred from the design file to the machine; then the cutting head starts to move around on the table to cut or scoreline the

Table 2.4 A list of packaging design software.

Packaging Type	Software	Company	Features
Paper packaging	ArtiosCAD	Esko	• Offer FEFCO and ECMA library • Plugin with Adobe illustrator • Integrate with other Esko software
	Impact	Arden Software	• Offer FEFCO and ECMA library • Drag and drop design component • Able to import a wide range of CAD and non-CAD drawing
Rigid container	SolidWorks	Dassault Systèmes	• 3D modeling features • Various simulation tools • Plug-in rendering and visualization tools
	AutoCAD	Autodesk	• Most popular CAD software • Industry-specific toolset
	Onshape®	Onshape Inc	• Cloud computing system • Easy to collaborate
Flexible packaging	Studio toolkit	Esko	• Provide a custom 3D shape • Suitable for flexible and shrink sleeve design • Plug-in with Adobe illustrator

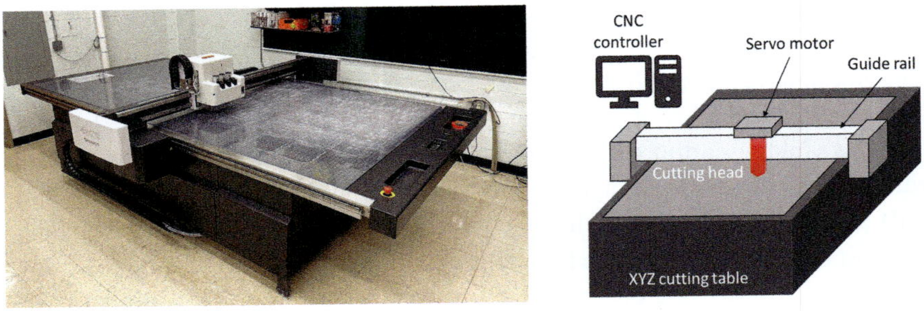

Figure 2.20 Digital cutting table: (Left) picture of digital cutting table; (right) representation of the main parts of the digital cutting table.

paper sheet. Once the process is done, the paperboard or paper sheet can be folded or glued manually to convert the design from a 2D flat sheet to a 3D carton.

Additive manufacturing or 3D printing are commonly used to create rigid packaging prototype models. 3D printing "*is the construction of a three-dimensional object from a CAD model or a digital 3D model*" [29]. It can produce the prototype model precisely and quickly at a relatively low cost. There are many different types of 3D printers, and three types are the most established underlying methods: Fused Deposition Modeling (FDM), Stereolithography (SLA), and selective Laser Sintering (SLS). For FDM, the material is fed into the extruder, which melts and pushes out the material layer by layer, as shown in Figure 2.21. FDM is the most common 3D printing method used these days. However, the quality is not the best since it has the lowest resolution compared to the other methods. As a result, FDM 3D printing is appropriate for basic concept proofing or simple prototyping. Unlike FDM, SLA cures liquid resin using a laser. It has high resolution and can create a watertight model. Therefore, SLA is ideal for precise prototype modeling and smooth surfaces. Lastly, SLS solidifies polymer particles into a solid structure through a high-power laser. One of the benefits of the SLS method is that the unfused powder is used as a support structure during the 3D printing process. As a result, the SLS is ideal for complex designs, including interior, undercut, thin wall, and negative features. Moreover, SLS has low cost, high productivity and has become a favored 3D printing method [30].

2.6 Case Study 2.1. Heinz Single-serve Ketchup Dip and Squeeze

In this section, a development story of the Heinz dip and squeeze ketchup is explained using the five different stages of the design thinking process: *Empathize*, *Define*, *Ideate*, *Prototype*, and *Test*.

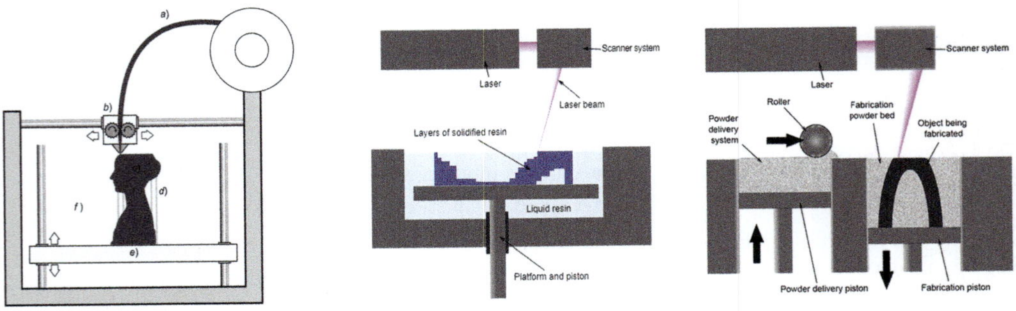

Figure 2.21 3D printing types: FDM (left), SLA (middle), and SLS (right); adapted from [31–33]; reproduced from [31] license under CC BY 4.0 and 3.0 from [32, 33], respectively.

The first step was understanding the current problem of the single-serve ketchup, the *Empathize* process. As shown in Figure 2.22, empathy mapping is implemented to understand how the users think about the single-serve ketchup package. From the empathy map, the packaging professionals found out that the traditional single-serve ketchup is not convenient, especially in the car. It spills, leaks, and the consumer needs to find someplace to squeeze as well as several other issues.

In addition to the empathy mapping, other information can be collected, such as the drive-through journey in a fast-food restaurant, feedback from suppliers, distribution conditions, and insights into the user's experiences.

Once all the necessary information is collected, it can be analyzed through the *define* stage to find out the core problem and design objective. Similar issues can be grouped and merged into different core principles or design objectives. Figure 2.23 shows how the design objectives can be redefined.

The next step is the *ideate* stage, where ideas related to the design objectives are generated. The new ideas can be generated by brainstorming with team members and other subjects. The ideas should focus on the design objectives defined in the previous stage (Figure 2.23). Figure 2.24 shows how ideas can be generated through brainstorming, visualization, and prototyping for a new single-serve ketchup design and experience. Once all possible ideas are collected, the design options should be narrowed down to more feasible designs by considering the material, manufacturing, price, usability, and other aspects.

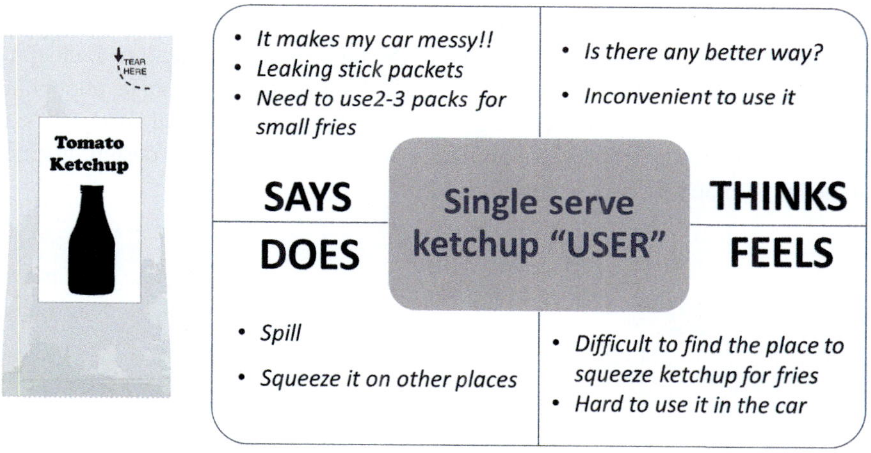

Figure 2.22 Empathy mapping from the single-serve ketchup sachet user feedback.

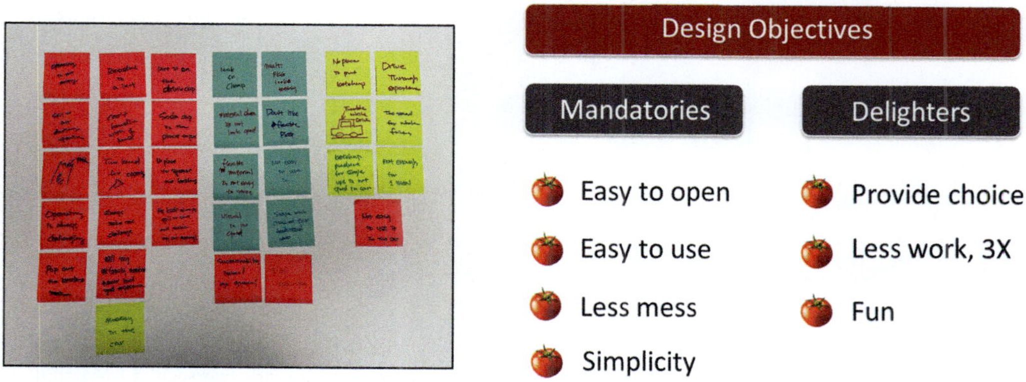

Figure 2.23 Design objectives for single-serve ketchup, images courtesy of J. Marshall.

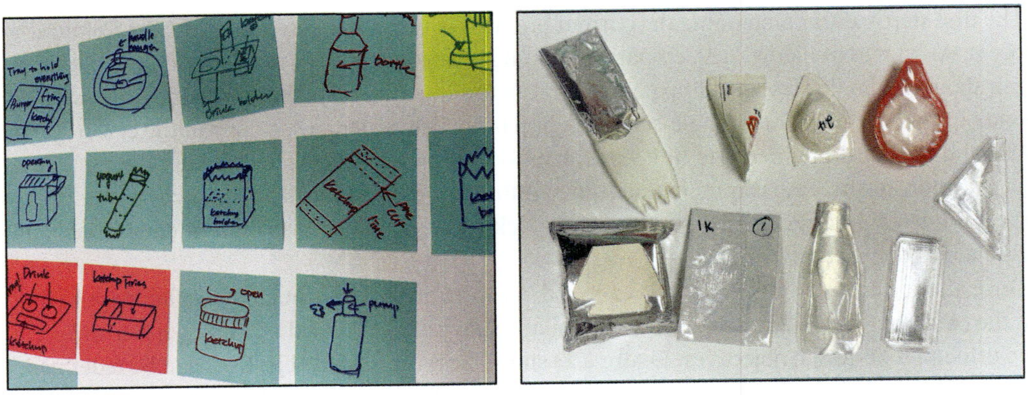

Figure 2.24 Ideate examples of single-serve ketchup.

The feasible designs are then *prototyped* using scaled-down and inexpensive techniques. The prototype models bring the design from the *ideate* stage to the real-life stage. They help to provide feedback from consumer experiences and early *testing*. The input can be used to improve the proposed designs, so prototyping and testing can be done iteratively. Figure 2.25 shows the prototype models of a single-serve ketchup packaging.

As a result, the design can be narrowed down to one design, as shown in Figure 2.26. This design has unique features that allow the users to dip and squeeze ketchup as they wish.

The next step is moving from the lab-scale prototype to the larger-scale production. Packaging professionals should identify the needed technology to enable the process. In this stage, pilot plant machines can be used to evaluate the manufacturing process, and use of Horizontal Form Fill Seal (HFFS) machines was considered to assess the manufacturing process of the final design. A schematic diagram of the HFFS machine is shown in Figure 2.27.

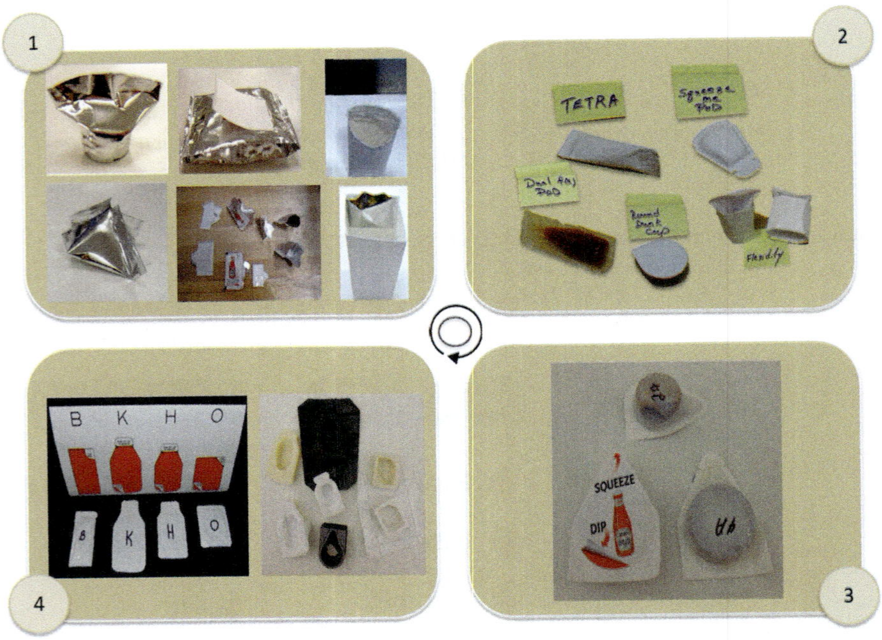

Figure 2.25 Feasibility checks with a prototype for the single-use ketchup, images courtesy of J. Marshall.

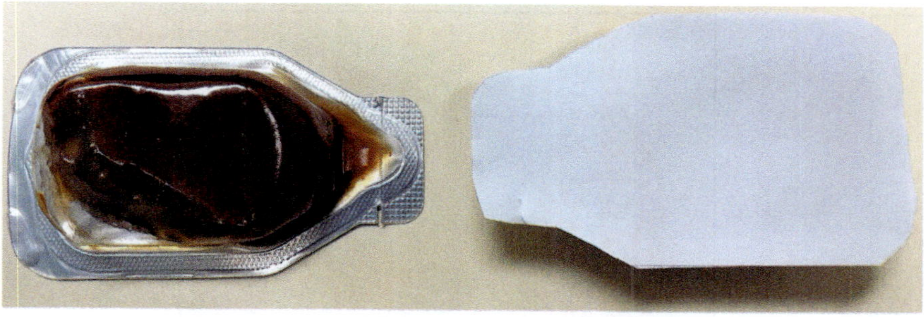

Figure 2.26 Small-scale prototype models for a final design.

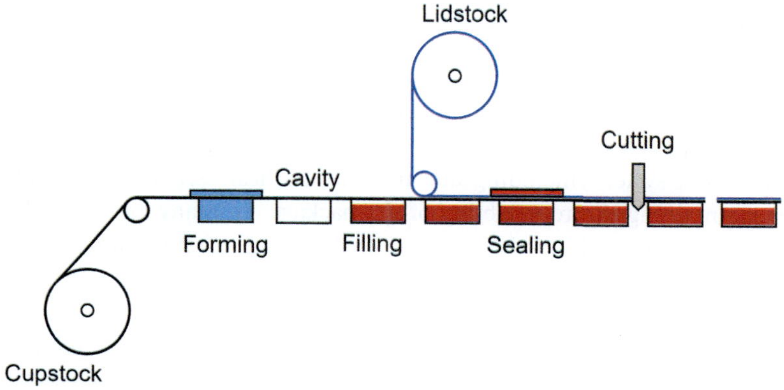

Figure 2.27 A schematic diagram of the Horizontal Form Fill Seal (HFFS) machine; adapted from [34].

Finally, engineering drawings should be completed to finalize the shape and graphic elements to convert the small-scale prototype to mass production scale products. Figure 2.28 shows a single-serve ketchup final design, which is the current Heinz dip and squeeze ketchup packaging in the market.

This new design provides convenience and use versatility. This user-friendly design increased customer satisfaction. This Heinz dip and squeeze ketchup can currently be found in grocery shops and fast-food restaurants.

2.7 Case Study 2.2. Design for Recyclability

This case study focuses on the design considerations for increasing recyclability in polymer packaging. The main content is from the Association of Packaging Recycling (APR) design guide [35]. The first consideration for recyclability is polymer selection. Depending on the packaging applications, different polymers are used, and they have generally been identified by the ASTM international resin identification coding system [36] – see Chapter 13 for further information. In general, homogenous polymer materials used in packaging are favorable for the recycling process. However, many polymer packages require the addition of more layers, coating, additives, labels, and attachments making the recycling process more challenging. If the layers and coatings are different from the base polymer material, additional testing is recommended to make sure that they are not hindering recycling of the entire structure. Incorporation of designs facilitating the easy separation of the layers and coating can help increase the recycling process efficiency. The main color of the polymer used for the packaging is another factor affecting recyclability. Table 2.5 lists the preferred colors for the different polymers to increase recycling.

Figure 2.28 The final design for the Heinz dip and squeeze ketchup packaging.

For the packaging closures, polypropylene and polyethylene closures or same polymers as the base polymer materials are the most preferred materials for enhancing recycling.

The label and the different labeling systems including label materials, adhesives, and inks should be considered carefully for maintaining the recyclability of the package. In many cases, labels interfere with the sorting process, especially in the near infrared sorting machine. Moreover, removing adhesive is costly, so labels should be designed to be separated easily with a minimal amount of adhesive.

Size and shape of the packaging must be considered for the recycling process since material recovery facilities (MRFs) sort items based on size and shape first, then by material. The standard screen size for the recycling process is 2 by 2 inches (5 by 5 cm) 2 dimensions. Therefore, small packages are often captured and accumulated as different materials for recycling, which are mostly considered contamination or waste. Furthermore, any package greater than 2 gallons (7.6 L) in

Table 2.5 The preferred color for polymer recycling.

Resin Code	Material		Preferred Color
1	PET	Poly(ethylene terephthalate)	Clear unpigmented PET Transparent light blue packaging Transparent green packaging
2	HDPE	High Density Polyethylene	Unpigmented HDPE translucent and opaque colors
3	PVC	Poly(vinyl chloride)	Clear unpigmented PVC
4	LDPE, LLDPE	Low Density Polyethylene, Linear Low Density Polyethylene	Unpigmented PE (poly(ethylene)) White, buff, or lightly colored colors Dark colors, particularly blues and greens are detrimental to recycling
5	PP	Poly(propylene)	Unpigmented PP translucent and opaque colors
6	EPS	Expanded Poly(styrene)	Unpigmented or white polymer Light pink or light blue colors
	PS	Poly(styrene)	Clear unpigmented polymer
7	PLA	Poly(lactic acid)	All nondark colors

volume should be further evaluated to determine the overall recyclability. Typical automatic sorting equipment used in MRFs cannot work well if the package volume is too large. As a result, the large items are sorted manually in the MRF, which can decrease the efficiency of the recycling process. Further information about increasing recycling through design can be found in the workbook "APR design guide for plastics recyclability" [35].

2.8 Study Questions

1 What is innovation, and how is innovation different from creativity?
2 What are the four conventional packaging design objectives?
3 What are the sustainable packaging design objectives?
4 List the five stages of design thinking.
5 What stage in the design thinking process focuses on current issues and consumer experience?
6 What are the seven rules of brainstorming?
7 What are the three options for material innovation?
8 List five main packaging features.
9 What is the difference between universal design and inclusive design?
10 What are the two major elements in packaging graphics?
11 What design concept focuses on human environments in the design process?
12 Please list paper packaging (folding carton and corrugated board) design software.
13 What are the three typical 3D printing types?

2.9 Additional Resources

1 Empathy map template
https://servicedesigntools.org/tools/empathy-map
2 Free online mind map tool
https://www.mindmup.com
https://www.mindmaps.app
3 Free online brainstorming tool
https://ideaboardz.com
4 3D model library
https://grabcad.com/library
5 Cloud-based CAD software
https://www.onshape.com/en
Free access for student and educators
https://www.onshape.com/en/education

References

1 Zekiri, J. and Hasani, V.V. (2015). The role and impact of the packaging effect on consumer buying behaviour. *Ecoforum Journal* 2015 Aug 12;4. http://www.ecoforumjournal.ro/index.php/eco/article/view/189. (accessed August 18, 2021).

2 Deliya, M.M. and Parmar, B.J. (2012). Role of packaging on consumer buying behavior–patan district. *Global Journal of Management and Business Research* Jun 18 12 (10):49–67. https://journalofbusiness.org/index.php/GJMBR/article/view/732. (accessed August 18, 2021).

3 Clement, J. (2007). Visual influence on in-store buying decisions: An eye-track experiment on the visual influence of packaging design. *Journal of Marketing Management* Nov 26 23 (9–10): 917–928. DOI: 10.1362/026725707X250395.

4 Primed Associated. (2013). *There's a critical difference between creativity and innovation.* https://www.businessinsider.com/difference-between-creativity-and-innovation-2013-4. (accessed March 25, 2022).

5 Packaging Digest. (January 2014). *Planters launches reduced-weight plastic jars.* https://www.packagingdigest.com/smart-packaging/planters-launches-reduced-weight-plastic-jars.

6 KPMG. (2020). The KPMG survey of sustainability Reporting 2020. *KPMG IMPACT* December. https://assets.kpmg/content/dam/kpmg/xx/pdf/2020/11/the-time-has-come.pdf.

7 Sustainable Packaging Coalition. (2006). *Design guidelines for sustainable packaging.* version 1.0, https://recycledcontent.org.

8 Walmart. *Sustainable Packaging Playbook – A Guidebook for Suppliers to improve packaging sustainability.* https://cdn.corporate.walmart.com/c0/02/c8b62e684567bb16e0877d6614de/new-wm-packagingplaybook.pdf (accessed March 25, 2022).

9 IDEO. (n.d.) *Design thinking.* https://designthinking.ideo.com (accessed March 25, 2022).

10 Plattner, H. An introduction to design thinking process guide. *Institute of Design at Stanford.* https://web.stanford.edu/~mshanks/MichaelShanks/files/509554.pdf (accessed March 25, 2022)

11 Dam, R. and Siang, T. (n.d.) *5 stages in the design thinking process.* https://www.interaction-design.org/literature/article/5-stages-in-the-design-thinking-process (accessed March 25, 2022).

12 Gibbons, S. (2018). *Empathy mapping: The first step in design thinking.* https://www.nngroup.com/articles/empathy-mapping.

13 Buzan, T. and Buzan, B. (1996). *The Mind Map Book: How to Use Radiant Thinking to Maximize Your Brain's Untapped Potential.* New York: Plume.

14 Osborn, A.F. (1942). *How to think up.* New York: McGraw-Hill.

15 Boyle, B. (n.d.) *7 Simple Rules of Brainstorming.* https://www.ideou.com/blogs/inspiration/7-simple-rules-of-brainstorming (accessed March 25, 2022).

16 Cooper, J. (2020). E-comerce packaging has differfent needs. *Netrush.* https://netrush.com/blog/e-commerce-packaging-has-different-needs.

17 Mohan, A. (2017). *Does PET lightweighting really reduce the carbon footprint?* https://www.packworld.com/design/materials-containers/news/13372469/does-pet-lightweighting-really-reduce-the-carbon-footprint.

18 *Renewable Can.* https://www.cancentral.com/can-stats/history-of-the-can/renewable-can-0 (accessed March 25, 2022).

19 Arhipova, A. (n.d.) Knock design into shape. *Psychology of Shapes.* https://blog.tubikstudio.com/knock-design-into-shape-psychology-of-shapes (accessed March 25, 2022).

20 Mourichon, A. (2020). *What are the differences between universal design, accessibility, and inclusive design?* https://sayyeah.com/digital-insights/universal-design-accessibility-inclusive-design.

21 Elle by Kristina Rodulfo. (March 2019). *This new beauty brand is making disability accessible makeup a reality.* https://www.elle.com/beauty/makeup-skin-care/a26959611/grace-beauty-disability-friendly-beauty-products.

22 de la Fuente, J., Gustafson, S., Twomey, C., and Bix, L. (2015). An affordance-based methodology for package design. *Packaging Technology and Science* Feb 28 (2): 157–171. DOI: 10.1002/pts.2087.

23 Resmi. (2013). *Importance of colors in explainer video Marketing.* http://www.bodeanimation.com/blog/importance-of-colours-in-explainer-video-marketing.

24 Summalinguae. *Colours Across Cultures: A Colour Psychology Guide for Brands*. (2021). https://summalinguae.com/language-culture/colours-across-cultures.

25 Branding Strategy Insider. (n.d.) Brand Identity: The Importance of Color. https://www.brandingstrategyinsider.com/brand_identity_/#.YQAZzo5KhPY. (accessed March 25, 2022).

26 Squire, V., Forssman, F., and Willberg, H.P. (2006). *Getting it right with type: The dos and don'ts of typography*. Laurence King Publishing.

27 Suleman, A. (2016). *The science of typography in packaging to Grab customer attention*. https://www.zilliondesigns.com/blog/science-of-typography-in-packaging. (accessed March 25, 2022).

28 Becker, M.W., Bello, N.M., Sundar, R.P., Peltier, C., and Bix, L. (2015). Front of pack labels enhance attention to nutrition information in novel and commercial brands. *Food Policy* Oct 1 56: 76–86. DOI: 10.1016/j.foodpol.2015.08.001.

29 3D printing. (n.d.) https://en.wikipedia.org/wiki/3D_printing (accessed March 25, 2022).

30 formlabs. (n.d.) *Powerful, Affordable 3D Printers for Professionals*. https://formlabs.com/3d-printers.

31 Scopigno, R., Cignoni, P., Pietroni, N., Callieri, M., and Dellepiane, M. (2017). Digital fabrication techniques for cultural heritage: A survey. *Computer Graphics Forum* 36 (1): 6–21. DOI: 10.1111/cgf.12781.

32 Materialgeeza. (2013). *Stereolithograthy apparatus schematic (vector version made with Inkscape)*. https://commons.wikimedia.org/wiki/File:Stereolithography_apparatus.svg.

33 Materialgeeza. (n.d.) *SLS system schematic*. https://commons.wikimedia.org/wiki/File:Selective_laser_melting_system_schematic.jpg (accessed March 25, 2022).

34 Hughes, H.A. (2019). Chapter 27 – Food Packaging Machinery. In: *Handbook of Farm, Dairy and Food Machinery Engineering*, 3e (ed. M. Kutz), 717–739. Academic Press. ISBN 9780128148037.

35 APR events. (n.d.) *The APR design guide for plastics recyclability crash course*. https://plasticsrecycling.org/web-seminars/the-apr-design-guide-for-plastics-recyclability-crash-course. (accessed March 25, 2022).

36 ASTM. (2022). ASTM D7611/D7611-21. International. *Standard practice for coding plastic manufactured articles for resin identification*. https://www.astm.org/d7611_d7611m-21.html. (accessed January 21, 2016).

3

Packaging in the Upstream and Downstream Supply Chains

3.1 Learning Objectives

After learning from this chapter, you are expected to:

- Understand the differences among packaging used in the upstream and downstream supply chains.
- Describe the basic environmental characteristics of glass, metal, paper, and plastic materials used in packaging.
- Understand the main sources and materials used to produce energy.
- Explain the differences between primary, secondary, and tertiary packaging systems.
- Calculate the basic geometrical parameters for quantifying the packaging environmental footprint (EFP).

3.2 Introduction

Packaging is an intrinsic part of our current economy; it is a synchronized system of preparing products for transportation, distribution, storage, retailing, use and final disposal. Packaging facilitates the protection, movement, and storage of goods. Over centuries, the science and technology of packaging materials and systems have evolved and adapted to the needs of the global supply chain from traditional materials such as the ancient Roman clay amphora for olive oil and wooden barrels for wine through refrigeration and on to novel technologies for extending the shelf life of products such as dairy, meat, and fresh produce [1, 2]. An evaluation of packaging through time is also a study of civilization, since packaging has evolved and changed as society desired and need more products and goods. Since packaging is so connected to everything that we do daily, the interdisciplinary and multidisciplinary science of packaging is in a continuous state of change. Lockhart initially described packaging as a socio-scientific discipline, which operates in society to ensure the delivery of goods to the consumer in the best conditions intended for their use. He explained that a package operates in a system of product, packaging and distribution, within three environments: physical, atmospheric, and human. Among these three environments packaging performs three main functions: protection, utility, and communication, as shown in Figure 3.1 [3]. The initial functions of protecting, carrying, and dispensing a product have been in place since the early amphorae and wooden barrels. The other requirements, such as extending shelf life, enhancing utility and communicating to the consumer, are more modern concepts.

Life Cycle of Sustainable Packaging: From Design to End of Life, First Edition. Rafael A. Auras and Susan E. M. Selke.

Companion Website: www.wiley.com/go/Auras/lifecycleofsustainablepackaging

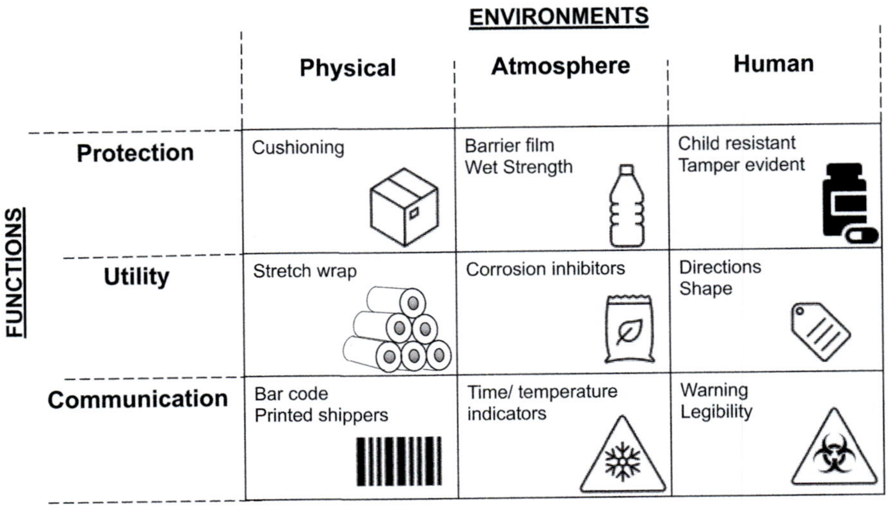

Figure 3.1 A typical generic environment and function matrix for packaging; adapted from [3].

3.3 Resource Use

As for any human activity, resources are needed to create goods. The level of use of nonrenewable resources and their scarcity is always a concern. To evaluate the level of consumption, it is useful to divide it into consumption of materials and consumption of energy. As discussed in Chapter 1, life cycle thinking is always recommended when analyzing the use of materials and energy. Specifically, for packaging, we must, at a minimum, consider the following stages:

- Extraction of resources
- Production of packaging materials
- Manufacture of materials and packages
- Production operations such as filling
- Distribution of the packages and their contents
- Use and storage of the packaged product
- Disposal of the package when it has served its purpose

Materials and energy are used in all these stages. In this chapter, we will provide a short overview of the materials used in packaging, and the main types of energy used to produce them. Books dealing extensively with each of the packaging materials are available, and they are an excellent source for full coverage and additional insights on packaging materials such as paper and paperboard [4], plastic [5], and glass and metal [6]. A great source for energy information around the world can be found in the World Energy Outlook [7].

3.4 Packaging Materials

The main materials used in packaging are metal – aluminum and steel, glass, paper and paperboard, wood, and plastics. Figure 3.2 shows the different packaging materials present in the MSW in the United States, which in 2018 consisted of 41,900 thousand tons of paper and paperboard,

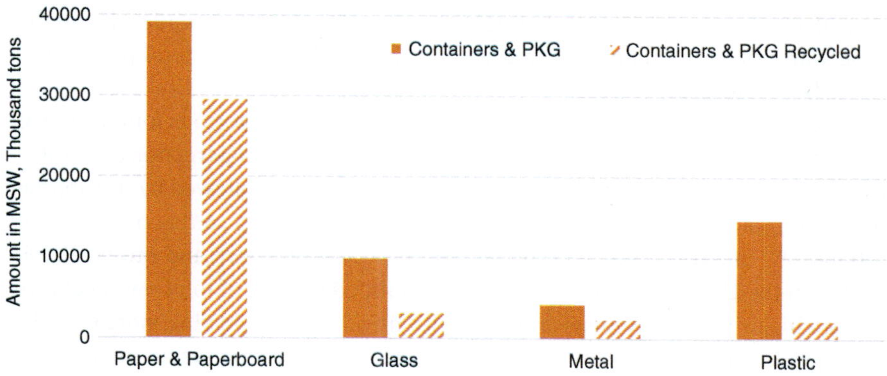

Figure 3.2 Main packaging materials and their recycling in the United States in 2018 [8].

mostly corrugated board 79.4% wt., 9,790 thousand tons of glass, 4,130 thousand tons of metal, and 14,530 thousand tons of plastic [8]. Paper and paperboard were the most preeminent material in the MSW in the United States, and corrugated board was the most recycled one. If we look at the percentages, paper and paperboard, glass, metal and plastic represented 58%, 14%, 6%, and 22%, respectively. Similar percentage ratios by weight of packaging materials were reported in the EU MSW, with 48% paper and paperboard, 22% glass, 7% metal, and 22% plastic [9].

3.4.1 Metal

Metal has been used in packaging since metals were first created, to contain different products such as spices, olive oil, and grains. However, they were not formally developed as one-way metal containers and packaging until the extraction and processing cost was reduced. Initially, metal was used to make returnable or long-term storage containers. Metals account for around 24% of the mass of our planet, and can be classified as ferrous metals (iron, carbon, and stainless steels), nonferrous metals, and their metallic alloys. Many of the unique metal properties such as strength, ductility, high melting point, thermal and electrical conductivity, and toughness made them useful for durable and consumer goods. The main characteristic that distinguishes metals from nonmetals is their type of bonding. Metals share metallic bonds that allow the free movement of electrons from atom to atom, often referred to as a "sea of electrons." This special type of bond gives metals unique properties, such as their high thermal and electrical conductivities. For packaging, the main metals used are iron/steel and aluminum, with small amounts of tin. Other metals such as copper, nickel, and chromium are used in packaging, but for small and specific applications.

Steel is created from iron ore, the most concentrated iron source in our planet's crust, with between 0.05 and 2 wt.% of carbon, and other elements such as nickel, manganese, chromium, and vanadium. Steel with different alloying elements has different properties such as malleability and anticorrosion and usually melts at temperatures lower than cast iron, which melts at about 1,375°C (2,507°F). Addition of alloying elements not only results in different properties but also allows melting at lower temperatures, typically around 1,100°C (2,012°F). Production of steel requires heating the iron ore and imbedded carbon in a large furnace in a low-oxygen environment to produce an iron alloy commonly called pig iron. This initial alloy contains too much carbon to be useful for most applications, so subsequent processes are used to convert it to steel by reducing the amount of carbon and adding selected alloying elements. After the initial ingots are produced, they are heated and hot rolled to produce slabs to roll into sheet metal. Billets – solid lengths of material extruded into shape – are rolled into bars, rod, and wire for production of structural pieces. The density of steel varies between 7.5 and 8 g cm^{-3}.

The US domestic production in 2019 from the iron and steel industry and ferrous foundries was valued at about \$98 billion, with an 11% decrease from \$103 billion in 2018. Construction accounted for an estimated 44% of total domestic shipments by market classification, followed by transportation (predominantly automotive), 28%; machinery and equipment, 9%; energy, 6%; appliances, 5%; and other applications, 8%. In 2019, China was the largest producer of pig iron and raw steel, followed by Japan and then the US. Within the US, Indiana accounted for 26% of total raw steel production, Ohio 12%, Michigan 5%, and Pennsylvania 5%, with no other states having more than 5% of the total raw steel production. Sources of scrap iron and steel for recycling are abundant [10]. This will be discussed in detail in Chapter 13.

Stainless steel may not need protection from the environment, but any other type of steel used in packaging needs adequate protection from rust. Figure 3.3 shows an array of steel containers used in packaging. Spurred by the offer by Napoleon of 12,000 francs for a method of preserving food to sustain his army, the canning process developed by Nicolas Appert in 1809 became the main method for preserving food. A year later, Peter Durand invented the "tin canister," which was brought to the US by William Underwood in 1817. It was not until 1819 that Thomas Kensett and his father-in-law Ezra Daggett started the first partnership for can production in New York using a slow and laborious handmade process. Mechanization of the can-making process did not happen until 1896, introduced by Max Ams [6].

Figure 3.3 Assorted selection of metal containers made of steel; adapted from [6].

All the cans produced until the early nineteenth century had a side seam. The creation of the "draw and iron" D&I process (also called draw and wall-ironed, DWI) allowed the production of two-piece aluminum cans without a side seam, beginning in the mid-1960s. For steel cans, the draw-and-redraw (DRD) process helped to increase the production speed of steel cans. The two-piece can had a large impact in the beverage industry, displacing almost completely the three-piece can (see Figure 3.4).

Some of the main advantages of steel cans are that they can be produced and filled at very high speeds, they are impermeable to moisture and gases, and they are produced from material that is highly recyclable. Some of the disadvantages are the higher weight compared to plastic and paper, large initial installation cost for assembly, and the need for rust protection. Tin coating was for a long time the main way to protect regular steel to avoid corrosion. However, as tin became more expensive, "tin-free steel" with a chromium/chrome oxide layer coupled with a polymer coating increasingly displaced tinplate. Even for tinplate, the use of coatings such as epoxy is common to protect the surface of the steel.

Aluminum as an element is 8% wt. of the earth's crust. Al metal uses bauxite ore as a feedstock, found mostly in topsoils in tropical and subtropical regions in Africa, Oceania, and South America. It is refined to pure aluminum oxide, or alumina (Al_2O_3). Alumina is then smelted using an electrolytic process to produce pure Al, which is converted to ingots, films, or structural materials. The United States domestic consumption of Al in 2016 was 4,840 thousand metric tons, of which 20% was used in packaging. Production of Al in the US is declining, with the majority of imported Al coming from Canada [11]. The density of aluminum is around 2.7 $g cm^{-3}$. In packaging most of the aluminum is used as coils to produce containers such as cans in the draw and iron process, or for aluminum foil.

Environmental footprint of steel and aluminum. Table 3.1 shows the EFP of 1 kg of cast iron, iron pellet, pig iron, and aluminum. The production of 1 kg of cast iron, iron pellet, pig iron, and aluminum results in emissions of 1.75, 0.11, 1.69, and 18.95 kg CO_2-eq, respectively. The large emission for aluminum is mostly due to the energy required to reduce aluminum oxide to metallic aluminum. Table 3.1 also shows that other EFP indicators of Al are higher than cast iron, iron pellet, and pig iron. We should recognize that the density of Al is much lower than steel, so a comparison of 1 kg of Al is just relative without describing an application. Chapter 9 provides additional EFP information for materials and their application considering specific functional units.

3.4.2 Glass

Glass is an inorganic material produced by fusing a combination of silica sand (silicon oxide, SiO_2, between 68 and 73% wt.), limestone (calcium carbonate, $CaCO_3$, between 10 and 13% wt.), soda ash

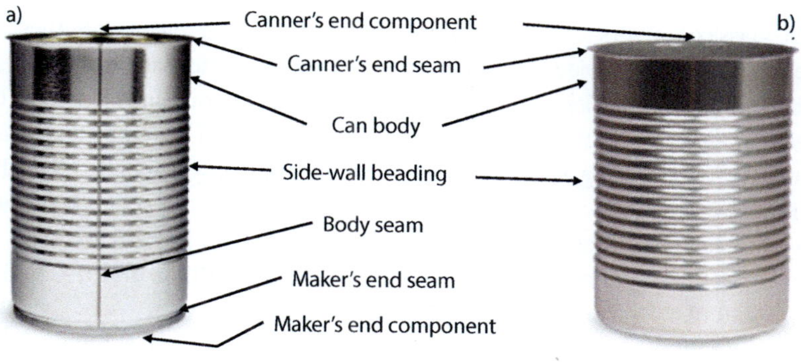

Figure 3.4 Three- and two-piece cans; adapted from [6].

Table 3.1 Environmental footprint of 1 kg of cast iron, 1 kg of iron pellet, 1 kg of pig iron, and 1 kg of aluminum, global average data for market, Impact assessment method: ReCiPe 2016 Midpoint (H) V1.04/ World (2010)/ Characterization/Excluding long-term emissions – data obtained from Ecoinvent 3 – allocation, cut-off by classification – system.

Impact category	Unit	Cast iron {GLO}| market for | Cut-off, S	Iron pellet {GLO}| market for | Cut-off, S	Pig iron {GLO}| market for | Cut-off, S	Aluminium, primary, cast alloy slab from continuous casting {GLO}| market for | Cut-off, S
Global warming	kg CO2 eq	1.75	0.11	1.69	18.95
Stratospheric ozone depletion	kg CFC11 eq	4.27E-07	9.56E-08	3.18E-07	4.62E-06
Ionizing radiation	kBq Co-60 eq	6.24E-03	7.96E-04	2.39E-03	3.79E-02
Ozone formation, Human health	kg NOx eq	4.69E-03	7.60E-04	4.31E-03	4.83E-02
Fine particulate matter formation	kg PM2.5 eq	3.60E-03	6.03E-04	3.34E-03	3.83E-02
Ozone formation, Terrestrial ecosyst	kg NOx eq	5.01E-03	7.71E-04	4.72E-03	4.86E-02
Terrestrial acidification	kg SO2 eq	5.58E-03	7.50E-04	5.14E-03	8.34E-02
Freshwater eutrophication	kg P eq	8.45E-05	1.56E-05	8.47E-05	6.24E-04
Marine eutrophication	kg N eq	2.79E-05	4.96E-05	3.67E-05	3.04E-05
Terrestrial ecotoxicity	kg 1,4-DCB	9.81	0.28	2.10	15.32
Freshwater ecotoxicity	kg 1,4-DCB	7.95E-03	4.61E-04	9.45E-04	1.74E-02
Marine ecotoxicity	kg 1,4-DCB	1.66E-02	7.96E-04	2.54E-03	3.44E-02
Human carcinogenic toxicity	kg 1,4-DCB	4.65E-01	1.70E-03	5.91E-03	9.41E-01
Human non-carcinogenic toxicity	kg 1,4-DCB	0.95	0.06	0.22	8.63
Land use	m2a crop eq	2.85E-02	2.63E-03	2.30E-02	1.94E-01
Mineral resource scarcity	kg Cu eq	5.89E-02	5.11E-02	8.05E-02	2.29E-01
Fossil resource scarcity	kg oil eq	0.44	0.03	0.39	3.99
Water consumption	m3	6.19E-03	-4.46E-03	3.94E-04	1.04E-01

(sodium carbonate, Na_2CO_3, between 12 and 15% wt.), and aluminum silicates (aluminum oxide, Al_2O_3, between 1.5 and 2% wt.) at temperatures around 1,500°C (2,732°F). The initial melting process is a complicated procedure that can be divided into three main stages of melting, fining, and conditioning, which are all required to properly melt each material, remove residual gases, and to obtain the specified glass for production. After melting, glass is cooled down to around 1,300°C to be distributed to the different forming processes (i.e., pressing, blowing, casting, rolling, drawing, fiber-forming, and dry gaging) that are done at around 800 to 1,000°C. After those processes, the parts are annealed at about 500°C to remove residual stresses; then, they are inspected, tested, and packed. The produced gob – specific amount of molten glass – can be used for several production techniques such as blow-and-blow, press-and-blow, and narrow-neck press-and-blow. These are the main techniques used to produce glass containers ranging from wide-mouth jars to narrow neck bottles.

Several coloring agents are used to provide color to glass. For example, cupric oxide is used for a red color and manganese is used for a violet color. Colors provide different light filtering and protection. Glass density is around $2.5 gcm^{-3}$, and it has high tensile strength, a very smooth surface, high gloss, and low wettability. Glass advantages for packaging applications include inertness, clarity, strength, recyclability, excellent barrier, and microwavability. However, it does have some disadvantages, such as being heavy, fragile, energy-intensive, and high cost. Most of the glass in packaging is used for jars and bottles.

The global glass manufacturing market size was valued at $ 127.1 billion in 2019, and it is expected to grow at a CAGR of 4.1% from 2020 to 2027 [12].

In 2007, world glass production was 115 million metric tons. Glass production is highly concentrated, with four companies producing over 50% of the global glass in 2010 [13]. Glass production in the US has declined since 1990. It is energy-intensive, with energy costs a large percentage

(~22%) of the production cost per ton of glass. Since glass production is done at a very high temperature, glass manufacturing is a significant emitter of GHG.

Of the glass produced, flat glass is used for buildings and automobiles; container glass is used in glass packaging for food and beverages and pharmaceutical applications; filament glass is used to produce composite materials. Other minor US domestic glass production mostly involves manufacturing glass tableware, cookware, and decorative items, and special types of glass are used for items such as lighting, laboratory glassware, and as parts of liquid crystal displays [13].

Specifically, container glass, which refers to packaging glass, has the highest demand in the Asia-Pacific region including China with 37% of world demand (Asia-Pacific: 37%; Europe: 33%; North America: 9%; South America: 16%; others: 5%). The total demand for container glass in 2012 was around 46 million tonnes, and it was expected to grow at 3.1% per year from 2013 to 2019, driven mainly by the Asia-Pacific demand. In the US, more than 120,000 people were employed by the glass industry in 2011 [13]. Container glass is divided by sector: 52% beer, 9% beverages, 3% flavored alcoholic beverages, 22% food, 4% liquor, 6% wine, and 4% other.

A special category of glass is used for pharmaceutical packaging. This is divided into several types: type I – borosilicate glass, which has the most stringent extractable standards with a very high melting point; type II – soda-lime glass treated in an annealing oven with sulfur to reduce alkali solubility; and type III – conventional soda-lime glass having specified extractive levels. Soda-lime glass not meeting type III qualifications is classed as USP (US Pharmacopeia) type NP (nonparenteral) – general purpose.

The environmental footprint of glass: Table 3.2 shows the EFPs of 1 kg of packaging glass, white, brown (amber), and green. Production and bringing to the market 1 kg of glass emit around 1 kg CO_2-eq and about 16 $MJkg^{-1}$ of primary energy demand (PED). This number closely matches the global warming impact reported for 75% of North American glass containers as 1.25 kg CO_2-eqkg^{-1} glass and PED of 16.6 $MJkg^{-1}$ [14]. Although this report is from 2010, significant changes in the glass industry have not happened in the last decade. So, we can expect that these EFP values represent the current glass packaging market.

3.4.3 Wood

Wood and packages made from woody fibers are among the oldest packaging materials. Initially, wood was used as a container, but soon baskets made of different fibers became widely used. Wood as lumber is mainly used for wooden containers such as barrels, boxes, crates, and pallets. Strong containers made from wood are used for shipping machinery, military packaging, and high-value gift containers. Wood is a renewable resource, and it can, if properly managed, be harvested indefinitely. However, proper management practices are needed to avoid the depletion of soil nutrients (see Chapter 5). The wood used in packaging is low cost, and it should be specially treated by heat or chemical treatment to avoid pest infestation during national and international commerce. Densities of wood used in packaging range from 0.3 gcm^{-3} for bamboo to 0.74 gcm^{-3} for the American red oak [4].

Most seasoned dry wood is used for container applications, with specific moisture content – between 12% and 19%. Hardwoods and softwoods are used for different applications. The wood used for container production is classified into groups 1 to 4 according to its nail-holding power and tendency to split. Different types of wooden boxes can be produced for packaging applications, such as standard nailed box styles, wirebound boxes, and pallet bins. Additionally, crates can be made to hold and transport large equipment or structures. Another packaging structure produced from wood is barrels, one of the strongest wood containers, which is generally used for storing dry and liquid products.

Table 3.2 Environmental footprint of 1 kg of packaging glass, white, brown, and green, global average data for market, Impact assessment method: ReCiPe 2016 Midpoint (H) V1.04/World (2010)/ Characterization/Excluding long-term emissions – data obtained from Ecoinvent 3 –allocation, cut-off by classification – system.

Impact category	Unit	**Packaging glass, white {GLO}| market for | Cut-off, S**	**Packaging glass, brown {GLO}| market for | Cut-off, S**	**Packaging glass, green {GLO}| market for | Cut-off, S**
Global warming	kg CO2 eq	1.04	1.04	1.02
Stratospheric ozone depletion	kg CFC11 eq	6.19E-07	6.21E-07	6.16E-07
Ionizing radiation	kBq Co-60 eq	5.73E-03	5.73E-03	5.68E-03
Ozone formation, Human health	kg NOx eq	3.65E-03	3.66E-03	3.62E-03
Fine particulate matter formation	kg PM2.5 eq	2.39E-03	2.40E-03	2.37E-03
Ozone formation, Terrestrial ecosystems	kg NOx eq	3.69E-03	3.70E-03	3.66E-03
Terrestrial acidification	kg SO2 eq	6.27E-03	6.28E-03	6.19E-03
Freshwater eutrophication	kg P eq	2.17E-05	2.18E-05	2.10E-05
Marine eutrophication	kg N eq	9.76E-06	9.76E-06	9.88E-06
Terrestrial ecotoxicity	kg 1,4-DCB	3.60	3.61	3.53
Freshwater ecotoxicity	kg 1,4-DCB	1.05E-03	1.05E-03	1.04E-03
Marine ecotoxicity	kg 1,4-DCB	3.56E-03	3.57E-03	3.51E-03
Human carcinogenic toxicity	kg 1,4-DCB	8.94E-03	8.96E-03	8.85E-03
Human non-carcinogenic toxicity	kg 1,4-DCB	0.25	0.25	0.25
Land use	m2a crop eq	8.63E-02	8.64E-02	8.53E-02
Mineral resource scarcity	kg Cu eq	5.21E-03	5.29E-03	5.10E-03
Fossil resource scarcity	kg oil eq	0.30	0.30	0.30
Water consumption	m3	6.09E-03	6.15E-03	5.82E-03
Total primary energy demand*	MJ	16.27	16.31	16.12

* Total primary energy demand is reported according to midpoint indicator Cumulative Energy Demand V1.11.

Wooden pallets, horizontal rigid platforms that can hold products, are the largest use of wood in packaging. Pallets facilitate assembling, storing, handling, and transporting goods as a unit load. The majority of the material handling in the world today is done using pallets since one person with a forklift or a pallet jack can manage an entire warehouse. Two-thirds of the wood pallets are made from hardwood and the remainder from softwood. Oak is the largest species used for wood pallets, with around 27% of the market [4].

3.4.4 Paper and Paperboard

To make paper and paperboard, wood must be pulped to obtain the fibers. Other types of materials can also be used as a source of fibers, such as bagasse from sugar cane and straws from wheat, rice, oats, and barley. China is the largest producer of alternative papers. The North American pulping industry has a long innovation tradition in converting wood fibers to paper and paperboard. Other countries such as Brazil, Chile, Argentina, and Russia are important paper producers due to the large growth of wood such as pine specially planted for paper making. To produce paper, first, the wood must be debarked and chipped. Then, the chips are broken down, releasing the cellulosic compounds and removing the lignin, the glue holding the fibers together. Once the fibers are separated, they are suspended in water, sometimes bleached, and refined for paper production. Packaging paper is mainly made from softwood since long and strong fibers can be obtained from it.

The strength of paper relies on its interlocking network of fibers. Paper is mainly produced using mechanical and chemical pulping processes. One of the main differences between mechanical and chemical pulping is that most of the lignin is not removed in the mechanical pulping process, so that paper produced by mechanical pulping is less strong and is prone to yellowing. However, through mechanical pulping higher yields can be obtained, lowering the cost of virgin fiber. Mechanical pulping is energy-intensive. So, mechanical pulp mills are located near inexpensive energy sources such as hydroelectric power facilities. Seventy percent of all produced paper is made through chemical pulping, the majority through the Kraft process. During the chemical pulping process, most of the lignin is removed. Therefore, chemically produced paper is less susceptible to light-induced UV degradation.

A large amount of wastewater is produced by chemical pulping, so a larger amount of chemical contamination needs to be controlled and monitored in this process. In addition to paper production from these two processes, a large amount of fiber is recovered through recycling. Recycling is only a mechanical process. In the US, five categories of recycled papers are recognized: (a) *Pulp substitutes*, a high grade of unprinted uncoated white sheets; (b) *Mixed Paper* which has mostly office waste and boxboard cuttings; (c) *Old Newspaper* (ONP), declining but still important; (d) *Old Corrugated Containers* (OCC), recycled in highest volume; (e) *High-Grade Deinked*, recycled in smaller quantities. Other countries divide these materials into slightly different categories [4].

The environmental footprint of wood, paper, paperboard, and corrugated board: Table 3.3 shows the EFP of 1 kg of corrugated board, kraft paper bleached and unbleached, and tissue paper. One kilogram of corrugated board emits 0.67 kg CO_2-eq, which is a little higher than the average value reported in a 2014 US LCA study about the US average corrugated global warming emissions as 0.533 kg CO_2-eq per kg of corrugated board, considering an average corrugated consisting of 66.8% linerboard and 33.2% corrugated medium with an average basis weight of 131.6 lb/thousand square feet (msf, 0.643 kgm^{-2}) [15]. The industry-average containerboard utilizes about 52% recovered fiber, primarily OCC, with the balance supplied mostly by kraft and semi-chemical pulp. Values of global warming for kraft paper bleached and unbleached are almost twice that of the corrugated board since very little recycled fiber is included. Tissue paper has a large EFP in each midpoint impact category.

3.4.5 Plastics

Modern synthetic plastics are macromolecules formed by long chains of monomers joined by covalent bonds. They are mostly made of carbon, with minor amounts of hydrogen, oxygen, and nitrogen. Some special plastics contain some metallic compounds [5]. Polymers are lighter tailored materials with a density below 2.2 gcm^{-3}, corrosion-resistant, low electrical and thermal conductivity, and low melting points. Several advantages of plastic for packaging applications are cost, lightweight, malleability, and tailored properties. However, it does have some disadvantages, such as being difficult to recycle when combined with other polymers.

Plastics are widespread and have become indispensable to our everyday life. Global production of plastics reached around 9,500 million tons from 1950 through 2019 [16, 17]; production was estimated at 400 million tons in 2020 and is expected to reach 600 million tons annually by 2030. The global value of the plastics industry was $570 billion (USD) in 2019 and is expected to increase to $850 billion (USD) by 2030 if business continues as usual [18]. Figure 3.5 shows that nearly all

Table 3.3 Environmental footprint of 1 kg of corrugated board box, kraft paper bleached and unbleached, and tissue paper, global average data (GLO) and rest of the world (Row) for market, Impact assessment method: ReCiPe 2016 Midpoint (H) V1.04/World (2010)/ Characterization/Excluding long-term emissions – data obtained from Ecoinvent 3 – allocation, cut-off by classification – system.

Impact category	Unit	**Corrugated board box {RoW}| market for corrugated board box | Cut-off, S**	**Kraft paper, bleached {GLO}| market for | Cut-off, S**	**Kraft paper, unbleached {GLO}| market for | Cut-off, S**	**Tissue paper {GLO}| market for | APOS, S**
Global warming	kg CO2 eq	0.67	1.48	1.12	2.55
Stratospheric ozone depletion	kg CFC11 eq	6.25E-07	8.19E-07	1.25E-06	1.43E-06
Ionizing radiation	kBq Co-60 eq	2.69E-03	1.21E-02	1.12E-02	5.81E-01
Ozone formation, Human health	kg NOx eq	2.09E-03	5.04E-03	4.55E-03	8.10E-03
Fine particulate matter formation	kg PM2.5 eq	1.09E-03	4.06E-03	2.76E-03	6.16E-03
Ozone formation, Terrestrial ecosystems	kg NOx eq	2.14E-03	5.14E-03	4.62E-03	8.22E-03
Terrestrial acidification	kg SO2 eq	2.35E-03	5.34E-03	5.20E-03	1.49E-02
Freshwater eutrophication	kg P eq	3.98E-05	9.22E-05	1.25E-04	4.12E-04
Marine eutrophication	kg N eq	1.30E-04	4.54E-05	5.39E-05	1.88E-04
Terrestrial ecotoxicity	kg 1,4-DCB	2.63	4.81	5.91	19.81
Freshwater ecotoxicity	kg 1,4-DCB	3.01E-03	1.69E-03	1.41E-03	1.64E-02
Marine ecotoxicity	kg 1,4-DCB	4.35E-03	4.70E-03	4.94E-03	3.18E-02
Human carcinogenic toxicity	kg 1,4-DCB	4.58E-03	1.61E-02	9.30E-03	1.19E-01
Human non-carcinogenic toxicity	kg 1,4-DCB	0.21	0.34	1.02	2.45
Land use	m2a crop eq	3.01E-01	1.39E+00	2.13E+00	2.94E+00
Mineral resource scarcity	kg Cu eq	1.96E-03	4.10E-03	3.57E-03	6.17E-01
Fossil resource scarcity	kg oil eq	0.19	0.43	0.30	0.59
Water consumption	m3	8.00E-03	4.58E-02	5.29E-02	2.15E-01

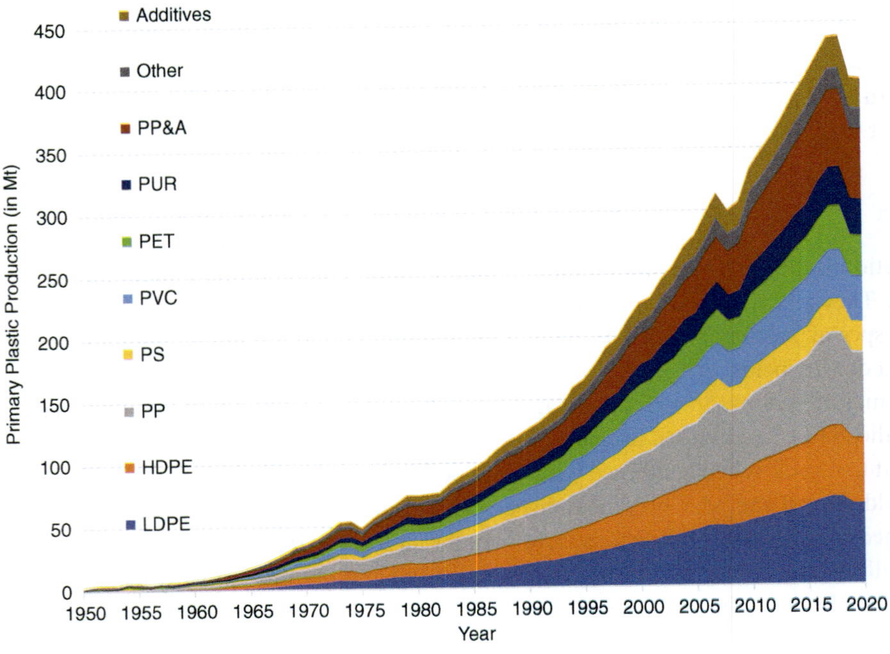

Figure 3.5 Global production of plastics since 1950; adapted from [16].

plastic production until now has been primarily derived from fossil resources. The leading producer of plastic is Asia, with China currently the top producing country at 31% of the current world total. The region of NAFTA and Europe produce 19% and 16%, respectively [19].

Of the currently produced plastics, polyolefins (i.e., LDPE, LLDPE, HDPE, and PP) represent more than 50% of the total; PVC is the second most-produced type of resin, followed by PS and PET (Figure 3.6).

Of the plastic consumed in 2017 worldwide, 30% was used for packaging, and around 17% for building and construction (Figure 3.7). Due to plastic's flexibility and low cost, single-use packaging has increasingly been made with plastic. In the US, more than 940,000 people work directly in the plastics industry, and more than 780,000 people work in downstream industries supplied by the plastics industry [20].

The environmental footprint of the main plastics used in packaging. Table 3.4 shows the EFP of 1 kg of polyethylene terephthalate amorphous and bottle grade-, high- and low- density polyethylene,

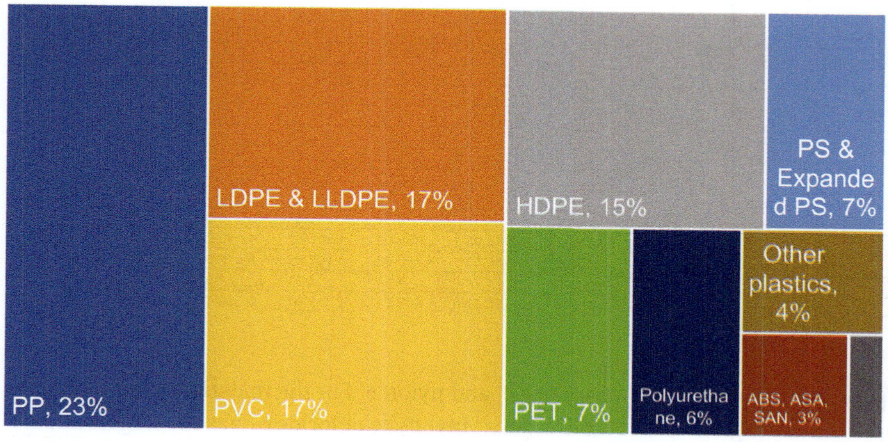

Figure 3.6 World plastics demand 2019 by polymer types; adapted from [19].

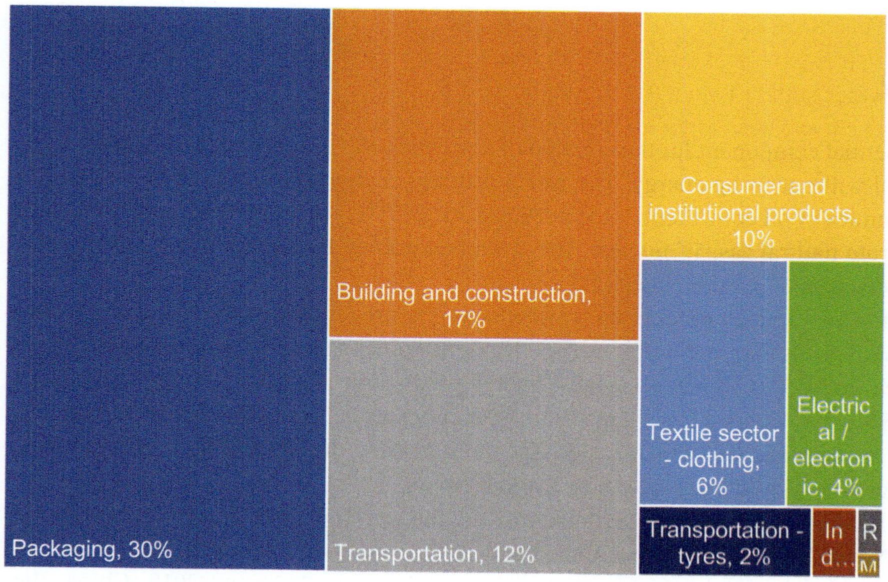

Figure 3.7 Share of total world polymer production by industrial use sector; adapted from [19].

Table 3.4 Environmental footprint of 1 kg of polyethylene terephthalate amorphous and bottle grade-, high- and low- density polyethylene, polypropylene, polystyrene, polyvinylidene chloride, and nylon 6, global average data (GLO) and rest of the world (Row) for market, Impact assessment method: ReCiPe 2016 Midpoint (H) V1.04/World (2010)/ Characterization/Excluding long-term emissions – data obtained from Ecoinvent 3 – allocation, cut-off by classification – system.

Impact category	Unit	Polyethylene terephthalate, granulate, amorphous {GLO}| market for | Cut-off, S	Polyethylene terephthalate, granulate, bottle grade {GLO}| market for | Cut-off, S	Polyethylene, high density, granulate {GLO}| market for | Cut-off, S	Polyethylene, linear low density, granulate {GLO}| market for | Cut-off, S	Polypropylene, granulate {GLO}| market for | Cut-off, S	Polystyrene, general purpose {GLO}| market for | Cut-off, S	Polyvinylidene chloride, granulate {RoW}| market for polyvinylidene chloride, granulate | Cut-off, S	Nylon 6 {RoW}| market for nylon 6 | Cut-off, S
Global warming	kg CO2 eq	3.21	2.97	2.39	2.33	2.37	3.93	5.39	9.88
Stratospheric ozone depletion	kg CFC11 eq	6.93E-07	6.43E-07	5.13E-07	5.40E-07	4.21E-07	5.62E-08	5.67E-05	9.48E-05
Ionizing radiation	kBq Co-60 eq	1.22E-02	1.11E-02	5.84E-03	5.41E-03	4.62E-03	6.57E-04	1.35E-03	7.41E-04
Ozone formation, Human health	kg NOx eq	6.81E-03	6.41E-03	5.17E-03	4.99E-03	4.87E-03	6.48E-03	1.12E-02	2.06E-02
Fine particulate matter formation	kg PM2.5 eq	4.10E-03	3.77E-03	2.62E-03	2.52E-03	2.47E-03	3.12E-03	9.03E-03	7.85E-03
Ozone formation, Terrestrial ecosys	kg NOx eq	7.13E-03	6.72E-03	5.53E-03	5.33E-03	5.18E-03	6.83E-03	1.17E-02	2.14E-02
Terrestrial acidification	kg SO2 eq	9.53E-03	8.50E-03	6.54E-03	6.32E-03	6.40E-03	9.66E-03	2.29E-02	2.42E-02
Freshwater eutrophication	kg P eq	8.25E-05	7.92E-05	4.95E-05	4.67E-05	4.73E-05	4.93E-06	2.08E-05	1.10E-04
Marine eutrophication	kg N eq	4.53E-05	4.88E-05	9.40E-06	6.24E-06	9.49E-06	1.21E-05	7.64E-06	3.09E-04
Terrestrial ecotoxicity	kg 1,4-DCB	13.09	12.28	4.85	4.94	4.52	1.98	2.76	2.06
Freshwater ecotoxicity	kg 1,4-DCB	2.14E-03	2.28E-03	1.30E-03	1.23E-03	1.16E-03	1.71E-03	1.46E-02	4.08E-03
Marine ecotoxicity	kg 1,4-DCB	9.99E-03	9.71E-03	4.39E-03	4.33E-03	4.01E-03	3.29E-03	2.17E-02	6.51E-03
Human carcinogenic toxicity	kg 1,4-DCB	2.51E-02	2.43E-02	1.78E-02	1.35E-02	1.31E-02	1.57E-02	2.70E-01	3.91E-02
Human non-carcinogenic toxicity	kg 1,4-DCB	0.64	0.64	0.33	0.33	0.33	0.07	0.84	0.12
Land use	m2a crop eq	4.69E-02	4.00E-02	2.03E-02	2.06E-02	1.90E-02	4.41E-03	1.07E-02	5.26E-03
Mineral resource scarcity	kg Cu eq	1.20E-02	1.21E-02	4.86E-03	5.75E-03	5.78E-03	8.81E-04	1.96E-03	9.97E-04
Fossil resource scarcity	kg oil eq	1.60	1.55	1.68	1.64	1.71	1.87	1.44	2.51
Water consumption	m3	3.67E-02	3.40E-02	2.38E-02	4.62E-02	2.06E-02	5.25E-02	5.68E-02	6.91E-02

polypropylene, polystyrene, polyvinylidene chloride, and nylon 6. For the majority of the midpoint indicators, nylon 6 has the highest EFP. PVDC – poly(vinylidene chloride) – has the largest impact in human carcinogenic and noncarcinogenic, terrestrial, and freshwater ecotoxicity, mostly due to the emissions of dioxins at the end-of-life.

3.5 Energy

Energy is an essential component for the creation of any goods. We cannot produce, transform, or transfer material without using energy. The world primary energy demand in 2018 was 14,314 Mtoe (million tonnes of oil equivalent; 1 toe = 39,683,207.2 BTU) with 27% from coal, 31% from crude oil, 23% from natural gas, 5% nuclear, 10% from renewables, including hydro, modern bioenergy, and others, and 4% solid biomass [21].

The main sources for energy production worldwide are nonrenewables: oil, coal, gas, nuclear, and renewables: hydro, solar, and wind sources. Nonrenewables represented around 81% of the energy produced in 2018; however, renewable energy production is growing, mostly hydro and solar (Figure 3.8). Latin America has a large share of energy production derived from renewables, mainly coming from hydro. Still, Europe and Asia are increasing the renewable use of resources for energy production derived from solar, wind, and biomass.

According to the 2019 World Energy Outlook, in 2018, the countries with the largest energy use were the US in North America and China in the Asia Pacific, as shown in Figure 3.9 [23]. The Asia Pacific region doubled its total primary energy demand between 2000 and 2018. China, the European Union, and Japan had the largest increase in renewables during this period, mostly from solar energy followed by wind (data not shown).

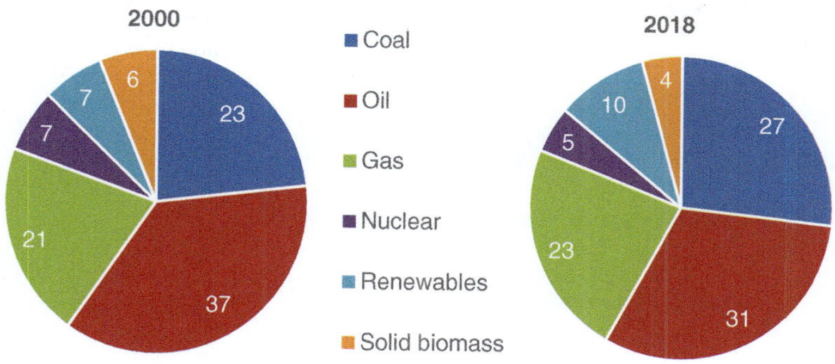

Figure 3.8 Comparative primary energy consumption over the past 18 years worldwide in %; adapted from [22].

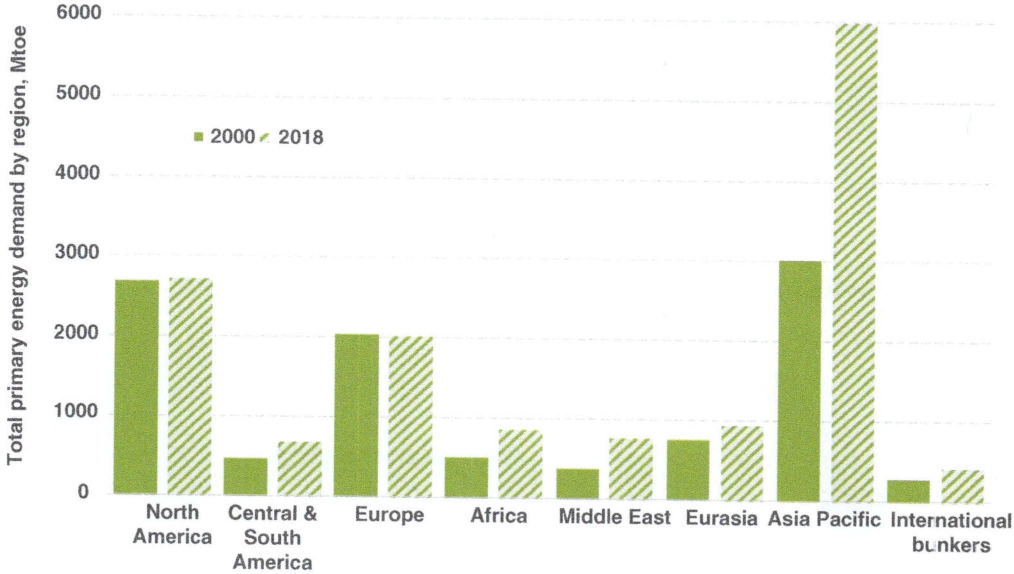

Figure 3.9 Total primary energy demand by region; adapted from [23]. Note: International bunkers refer to energy consumed by ships and aircraft on international routes.

In 2020, the US consumed 92.9 Quads of energy (1 Quad is one quadrillion BTUs). The significant sources of energy were petroleum, natural gas, and coal. Many resources are used for electricity generation to power the residential, commercial, and industrial sectors. To power transportation, mainly petroleum and biomass (mostly in the form of ethanol added to gasoline) are used. Electricity is a minor component. Of the total energy produced, 92.9 Quads, 62.3% of the energy, is lost as waste heat (Figure 3.10).

In the US in 2018, around 5,268 million metric tons of carbon dioxide emissions were emitted, with the following breakdown: 36.2% transportation, 19.0% industrial, 6.3% residential, and 4.8% commercial [25]. The Lawrence Livermore National Laboratory (LLNL) published a similar breakdown for carbon dioxide emissions (Figure 3.11) [26]. In 2018, the US electric power sector emitted around 1,764 million metric tons of carbon, of which 65% corresponded to coal, 33% to natural gas, and the remainder to petroleum, biomass, and geothermal.

3.5.1 Nonrenewables

Nonrenewable sources of energy can be defined as energy extracted from the earth that is not replaceable in reasonable time frames. It includes carbon-based resources (fossil fuels) such as petroleum, coal, and (most) natural gas or minerals such as uranium.

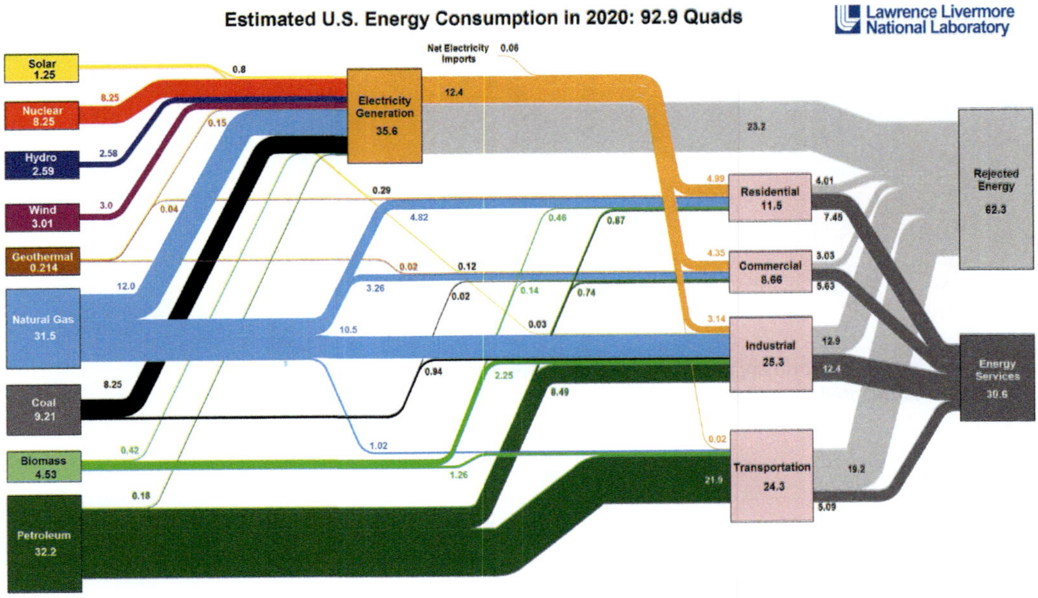

Figure 3.10 Estimated US energy consumption in 2020:92.9 Quads; reproduced from [24] license under CC BY 4.0.

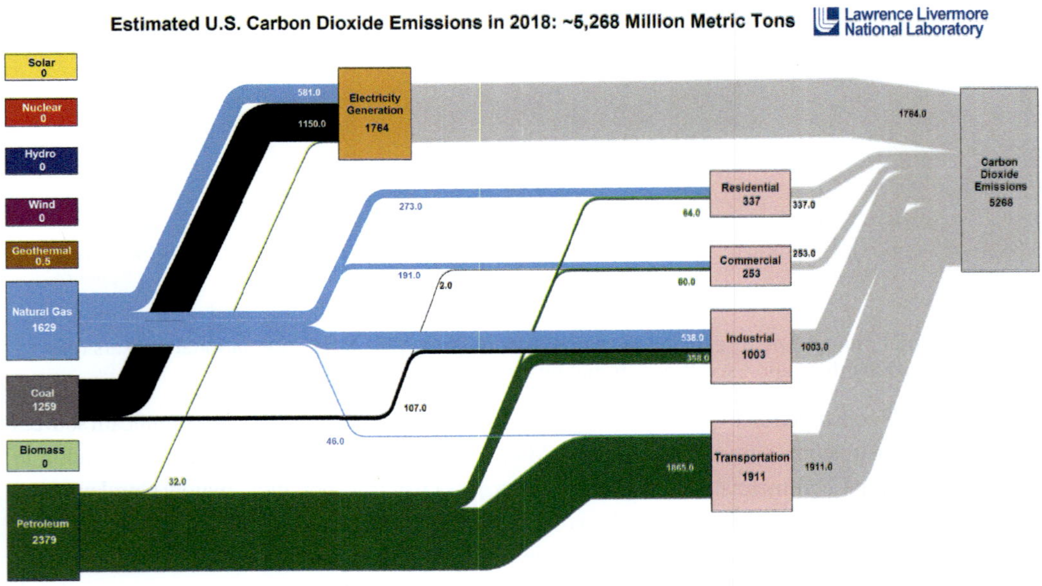

Figure 3.11 US carbon emissions in 2018; reproduced from [26] license under CC BY 4.0.

3.5.1.1 Petroleum

Petroleum originates as crude oil, a mixture of hydrocarbons mainly produced from biomass degradation from millions of years ago. Crude oil is defined as fossil fuel, and it exists underground in reservoirs in small spaces within sedimentary rocks and tar sands. Several products are made from petroleum such as gasoline, distillates such as diesel fuel and heating oil, jet fuel, waxes, asphalt, lubricating oils, and petrochemical feedstocks. In 2020, the US refinery produced from one barrel of crude oil (42 gallons) on average 45 gallons of petroleum products due to refinery processing gain and additional inputs such as fuel ethanol, blending components, and hydrocarbon gas liquids (i.e., 19–20 gal of motor gasoline, 4 gal jet fuel, 11–12 gal ultra-low sulfur distillate – diesel fuel, 9 gal of miscellaneous products) [25].

Petroleum, the main source of energy in the US, in the past largely originated as crude oil imports; however, in 2020, the majority of crude oil used in the US was produced domestically [25]. The largest sources of US petroleum imports in 2020 were Canada (52%), Mexico (11%), Saudi Arabia (7%), Russia (7%), and Colombia (4%) [27]. Although the US imports crude oil, it also exports crude oil in the form of refined products since it makes more sense, for example, for refiners in the US Gulf Coast region to export some of their gasoline to Mexico rather than shipping it to the East Coast of the US. Petroleum is still the major material consumed for energy production, 35% of the total in 2020. The primary use is transportation, 66% of the total. Industrial uses are next, at 28%. Natural gas is close behind, at 34% of total consumption in 2020 [27].

3.5.1.2 Coal

Coal is a combustible black or brownish-black sedimentary rock that contains a large amount of carbon and hydrocarbons. It is a nonrenewable source of energy since it was formed by the pressure and time of plants living in swampy forests and covered by dirt and rock for millions of years. Coal can be classified as anthracite, bituminous, subbituminous, and lignite. Anthracite is mostly used in the metal industry and has 86% to 97% carbon. Anthracite was less than 1% of US coal production in 2019, and the mines are all in northeastern Pennsylvania. Bituminous coal containing between 45% and 86% carbon is the most abundant coal in the US; it is between 100 and 300 million years old and represented around 48% of US coal production in 2019. Subbituminous coal contains between 35% and 45% carbon; it is around 100 million years old and represented about 44% of coal production in 2019. Lignite contains between 25% and 35% carbon; it is a relatively young material (though still millions of years old) and accounted for 7.5% of US coal production in 2019 [25]. Bituminous and subbituminous coal are used for electricity generation in the US. Just 3% is used for industrial heating and other energy-related use. Most of the coal in the US is obtained from surface mines (65% of the total in 2019), which are primarily found in the Appalachian coal region (including Alabama, Eastern Kentucky, Maryland, Tennessee, Ohio, Pennsylvania, Virginia, and West Virginia), and the interior coal region (including Arkansas, Kansas, Illinois, Louisiana, Missouri, Mississippi, Texas, Oklahoma, Western Kentucky, and Indiana). The Western coal region includes Arizona, Alaska, New Mexico, Wyoming, Colorado, North Dakota, Utah, and Washington. Wyoming (41%), West Virginia (11%), Pennsylvania (6%), Illinois (6%), and Kentucky (6%) derive coal almost exclusively from surface mining [28]. By far, the major use for coal is in electricity production, over 90% of the total in 2019 [29].

3.5.1.3 Natural Gas

Natural gas is mostly composed of methane (~94%) and a mix of low molecular weight hydrocarbons such as ethane, propane, butane, pentane, and hexane, along with small amounts of carbon dioxide, hydrogen sulfide, nitrogen, and a minimal amount of water. The decomposition of plants and animals produced natural gas; with time and pressure, gas moved into large cracks, spaces between rocks, and tiny pores within shale and other sedimentary rock formations. This gas is referred to as shale gas or tight gas. Natural gas can also occur in coal deposits and is called coalbed methane. After the natural gas is extracted, it is processed, transported, and stored for consumption. Methane and other valuable gases are separated from other gases to produce dry or consumer-grade natural gas. In the US in 2019, natural gas came from shale gas wells (39%), natural gas wells (25%), crude oil wells (16%), coalbed wells (3%), and the rest from storage and imports. Natural gas production in the US in 2019 was mostly from Texas (23.9%), Pennsylvania (20%), Louisiana (9.3%), Oklahoma (8.5%), and Ohio (7.7%) [30]. After the successful exploitation of the

Barnett Shale located in north-central Texas, natural gas extraction from shales became a reality in the US and has been a game-changer with respect to energy production. Additional shale gas sources include the Fayetteville Shale in northern Arkansas, the Haynesville in eastern Texas and north Louisiana, the Woodford in Oklahoma, the Eagle Ford in southern Texas, and the Marcellus and Utica shales in northern Appalachia. The US has abundant shale gas, with domestic production in 2020 exceeding domestic use by about 10% [30]. Production is expected to continue to grow. Natural gas uses in 2020 were 38% for electric power, 33% for industrial use, 15% residential, 10% commercial, and 3% for transportation [31]. Shale gas is also being used to produce polymers [32].

3.5.1.4 Nuclear

Nuclear energy is produced from uranium, the most widely used fuel for nuclear plants, especially U-235, since it can be easily split apart. Uranium is obtained through mining and then processed and enriched to increase the normal 0.7% of U-235 to 3–5% before being used for the energy generation [33]. While there is some uranium production from deposits in the western US, about 90% of the uranium used for nuclear power is imported [34].

Nuclear energy generation is based on using the heat released by splitting U-235 (fission) to boil water into steam, which then is used to turn turbines and generate electricity. Nuclear energy in the US is almost exclusively used for electricity generation. Nearly 20% of US electricity (805 billion kilowatt-hours in 2017) comes from nuclear energy from 98 nuclear plants, which are located in 30 states [35]. France is the second-largest nuclear energy generator (381.8 billion kilowatt-hours in 2017) representing 71.5% of the electricity generation for France [35].

After the uranium is mined, milled, and enriched, it is formed into nuclear fuel pellets and packed into fuel rods which are further combined into fuel assemblies. At the nuclear power plant, the fuel assemblies are loaded into the reactor, with water where the nuclear reactions occur. After they have been used, the fuel assemblies become highly radioactive, and when they are removed from the reactor replaced by a new fuel assembly, they must be stored in a special place for several years before they can be moved to underground storage in a permanent underground repository. The US has no permanent repository, so all spent fuel in the US remains in temporary storage. It is possible to reprocess spent fuel to generate plutonium as a new fuel source, but this is not legal in the US.

After the nuclear power plant disasters in Chernobyl, April 25 and 26, 1986, and Fukushima Daiichi, March 11, 2011, global perception and acceptance of nuclear energy for electricity generation have declined. Germany has shut down eight reactors and pledged to close all the remaining ones by 2022. Switzerland and Spain have banned the construction of any new nuclear energy power plants. Many countries such as Japan, Taiwan, Belgium, and Mexico have decided to be less dependent on nuclear energy. Countries like Italy, Australia, Denmark, Greece, Austria, Ireland, Latvia, Liechtenstein, Malta, Luxemburg, Israel, Serbia, Portugal, Norway, and Malta do not have nuclear reactors to generate energy, and they are opposed to them.

3.5.2 Renewables

Renewable energy comes from naturally replenished sources such as biomass, wind, and solar. Most of these can be continually generated, but the amount of energy available per unit of time varies with conditions, e.g., wind speed or solar intensity.

3.5.2.1 Biomass

Biomass is derived from organic materials such as wood, crops, landfill gas, alcohol fuels, and garbage that comes from plants and animals. Biomass contains stored energy from the sun absorbed through the photosynthesis process. When it is burned, the energy is released as heat, which can be used directly for energy production; or the carbohydrates present in biomass can be converted to liquid biofuels. Overall, wood, wood processing wastes, agricultural crops and wastes, food, yard, and wood waste in the garbage all can be burned to generate electricity in power plants or converted to biogas in landfills. Animal manure and human sewage can also be converted to biogas and burned for fuel. Besides burning, biomass can be converted to other forms of energy, such as methane or transportation fuels – ethanol and biodiesel. Biomass in the US provided around 5% of the primary energy in 2019. About 30% was used as fuel for transportation and 10% for electricity generation. Since biogas production from landfills is counted as renewable energy, some landfill operators have pushed for a larger amount of organics disposed of in a landfill, which has been controversial [25]; further discussion is provided in Chapter 16 – Case Study 16.1.

3.5.2.2 Hydropower

Hydropower is generated by exploiting the kinetic energy provided by flowing water. The amount of available energy in moving water is a function of the elevation and the volume of the water flow. In the US, in 2016, hydropower was the largest renewable energy source for electricity generation. However, most dams in the US were constructed for irrigation and flood control. The top five states in the US for hydropower capacity for electricity generation are Washington (27%), California (13%), Oregon (11%), New York (6%), and Alabama (4%). Although hydropower is considered a renewable energy source, dams and reservoirs can affect the natural water temperature, water chemistry, river flow characteristics, and silt loads. All these issues can affect the physical characteristics of the river and its natural ecology (i.e., fish reproduction and river vegetation) and result in soil, water, and air pollution. Droughts in the western US are becoming a significant concern for hydropower generation.

3.5.2.3 Wind

Wind is the second-largest renewable energy source in the US and is nearly all used for electricity generation. Wind is created by the changes in the temperature of the earth's surface during daily cycles. Wind energy is harvested by wind turbines, which use the wind's kinetic energy to move a transmission shaft that rotates an electric generator, producing electricity. Since 2000 the amount of electricity generation from wind has grown more than 3,000%. Wind harvest varies across the US, with five states, Texas, Oklahoma, Kansas, Iowa, and California, producing around 55% of the total US wind electricity generation in 2016. Although wind has almost no emissions and an individual wind turbine has a small environmental and physical footprint, groups of wind turbines, called wind farms, can have visual and noise impacts and affect birds and bats [25].

3.5.2.4 Solar Energy

Solar energy is increasingly being used. For thousands of years, we have used solar energy to provide heat and dry products such as fruit, grains, and meat. Devices have been developed to use solar thermal energy to heat water, homes, greenhouses, and buildings and to generate solar power. A large portion of solar electricity generation is done through solar photovoltaic systems, which convert sunlight into electricity. Sunlight is composed of photons of different amounts of energy depending on the wavelength, so photovoltaic cells absorb the photons and convert the solar incident energy into electricity. Solar energy has many advantages regarding pollution; however, the amount of sunlight that arrives at the earth's surface in 24 h is not constant [25].

3.5.2.5 Geothermal Energy

Geothermal energy is heat from the earth that can be used for electricity production or heating. It is clean and has low emissions. Geothermal electricity production is done using steam produced from the introduction of water and heated by geothermal energy. Geothermal energy is used for electricity production, space, or water heating. Several types of geothermal power plants are available running on dry steam (i.e., obtained from underground resources of steam), flash steam (i.e., using reservoirs of very hot water underground that are converted into steam as it flows to the surface), and binary cycle (i.e., underground water at lower temperatures around 225–360 °F (107–182 °C) is used to warm and vaporize a secondary organic fluid). All these types of steam are used to turn turbines and produce electricity.

3.6 Components of the Packaging System

Packaging structures vary depending on their functions (i.e., protection, utility, and communication – Figure 3.1) and the stage of the implementation of the supply chain. Packaging systems are used to deliver products such as food, pharmaceuticals, or automotive parts from the harvest or production stage of a product to the next stage, such as the consumer or another production site. Packaging systems take many configurations, but they have three main components: (a) primary, (b) secondary, and (c) tertiary or distribution packaging. Figure 3.12 shows an example of the different forms that the packaging structures can take to ship bottles. The entire collection of packages from the bottle to the pallet is called the packaging system.

3.6.1 Primary Packaging

Primary packaging is identified as the stage and form of packaging in contact with the product, such as bottles, cans, tubes, and bags. It can take many forms and can be made of several materials. It can also include more than one component, for example, bottles, labels, and caps.

3.6.2 Secondary Packaging

The secondary packaging considered as bundlings, such as the folding carton containing the bottle of medicine or the folding carton containing a bunch of candy bars, is the stage of packaging used to deliver the primary product and packaging to the distribution center (DC) and the stores, such as a corrugated box, tape, and labels.

3.6.3 Tertiary or Distribution Packaging

The tertiary or distribution packaging is the structure or materials used to deliver the primary and secondary packages to the DC, assembly plant, and/or the supermarket. It is generally cataloged as the pallet and the restraining material such as shrinkwrap or straps. The North American or GMA – Grocery Manufacturers Association – pallets are 48″ × 40″, with 48″ length stringers and 40″ width deckboards. The height of the pallet is usually 3.75″. This is the standard pallet in the US. Still, the automotive industry uses a 48″ × 45″ standard. The most common international standard

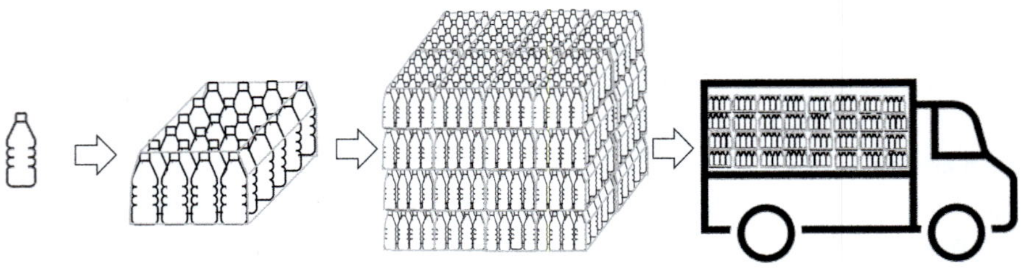

Figure 3.12 Packaging system structures and stages.

pallet has dimensions of 1,200 mm × 1,000 mm (47.2″ × 39.5″). The Euro-pallets are 1,200 mm × 800 mm × 144 mm. Euro pallets must comply with the UIC 435-2 specification approved by the International Union of Railways, with a pallet weight between 20 and 25 kg. There are several other pallet dimensions, as described elsewhere [4]. Pallet dimensions are relevant for the truck's design, railroad containers, and the entire distribution system.

3.7 Parameters for Quantifying the Environmental Footprint (EFP) of Packaging Systems

Several parameters must be measured or estimated to assess the EFP of packaging systems, such as the number of components, mass, volume, density, cube efficiency, secondary arrangement, and pallet configuration. Several tools and software are used to collect this data and estimate this configuration, or they can be calculated by hand.

Mass: the mass of the primary, secondary, and tertiary packages and the products is its weight, and it can be measured with a simple balance. For example, the approximate weight of a 355 mL PET bottle is 7 g (0.25 oz) plus the weight of the label/s (around 0.3 g) and the cap (around 1 g).

Volume: the volume of a package can be estimated using several methods. In the case of water-resistant containers, water can be added, and the volume can be obtained. In the case of rectangular containers, the volume can be determined by measuring the length (L), width (W), and height (H) of the package containing the product.

Density: the density is the ratio of the mass to the volume. Density is a fundamental material property. Figure 3.13 shows the density of regular packaging materials. The density of water is

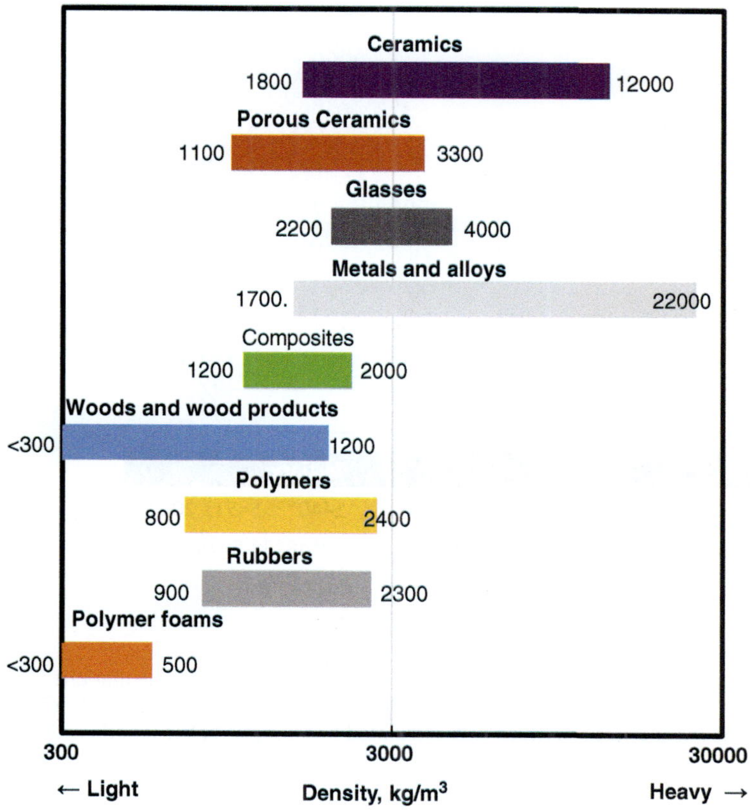

Figure 3.13 Density of regular packaging materials.

1,000 kgm^{-3}, and the density of plastics, for example, crosses this range. Metals and glass have a much higher density than water.

Cube efficiency: the cube efficiency of a package is defined as how efficiently a package fills the space. It is calculated as the proportion of the volume of a cuboid containing the package occupied by the package and its contents. Figure 3.14 shows the estimation of the cube efficiency of a PET water bottle for the primary, secondary, and tertiary packaging system. For the primary, we estimate the bottle's volume and divide by the cuboid of the height (7 in.) and width and length equal to the diameter (2.5 in.). This gives a cube efficiency for the primary of 49.5%. In the case of the secondary, the volume is equal to the 24 bottles times 355 cc, and the cuboid volume is $15 \times 10 \times 7$ in.^{3} = 1,050 in.^{3} or 17,206 cc. This also provides a cube efficiency of 49.5%. In the case of the tertiary, we need to consider all the bottles in a pallet and the dimensions of the pallet for the volume of the cuboid (i.e., $40 \times 48 \times 38.5 = 73{,}920$ in.^{3} or 1,211,332 cc). In this case, the cube efficiency for the tertiary is 42.2%.

3.8 Case Study 3.1. Cube Efficiency Estimation Using CAPE®

Several tools are used to estimate the best configuration to arrange the primary packaging in the secondary and the tertiary packaging systems. Among these, CAPE® and TOPS® are the most used. Figure 3.15 shows the estimation of the cube efficiency of a polypropylene jar used for marmalade estimated by CAPE®, software used to recommend packaging configurations. The same estimation

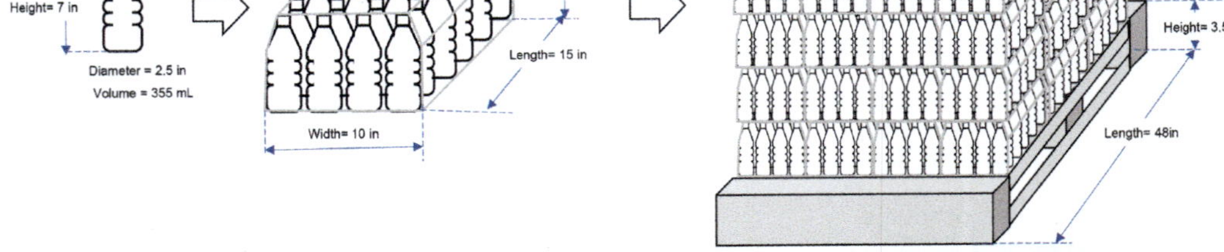

	Primary	Secondary	Tertiary
Volume, cc	355	24*355= 8,520	1440*355=511,200
Volume cube, in^3 (cc)	43.75 (716.9)	15*10*7=1050 (17,206)	40*48*38.5=73920 (1211331.8)
Cube efficiency, %	49.5	49.5	42.2

Figure 3.14 Estimation of the cube efficiency of primary, secondary, and tertiary packages. Calculations were done by estimating the volume occupied by the primary, secondary, and tertiary packages, and dividing the cube size of the primary, secondary, and tertiary, which was calculated by multiplying the length, width, and height of a cube.

Friday, January 31, 2020

```
Product Name         MO_PPJar
Product Code
Datafile Name        mo_ppjar (1/31/2020)
Load Ref.             1 C                      32       Gable End / Case
Cube Used             65.5 %                 1152       Gable End / Load
Area Used             80.1 %                   12            Case / Layer
Pallet type          48X40                      3           Layer / Load
                                               36            Case / Load
Truck Ref.            1  C                    864            Case / 53footer
Truck Area Used      75.5 %                    24            Load / 53footer
Truck Cube Used      31.0 %                 27648       Gable End / 53footer

               Length    Width   Height         Net     Gross
Gable En(OD)    2.750    2.750    5.750 in    1.466     1.538 lb
Case     (OD)  11.320   11.320   12.140 in   49.216    50.042 lb
Product        45.280   33.960   36.420 in 1771.776  1801.512 lb
Load           48.000   40.000   41.920 in 1801.512  1851.512 lb
Product       576.000   80.000   41.920 in  43236.5   44436.5 lb
53footer      636.000   96.000  102.000 in  44436.5   54436.5 lb
```

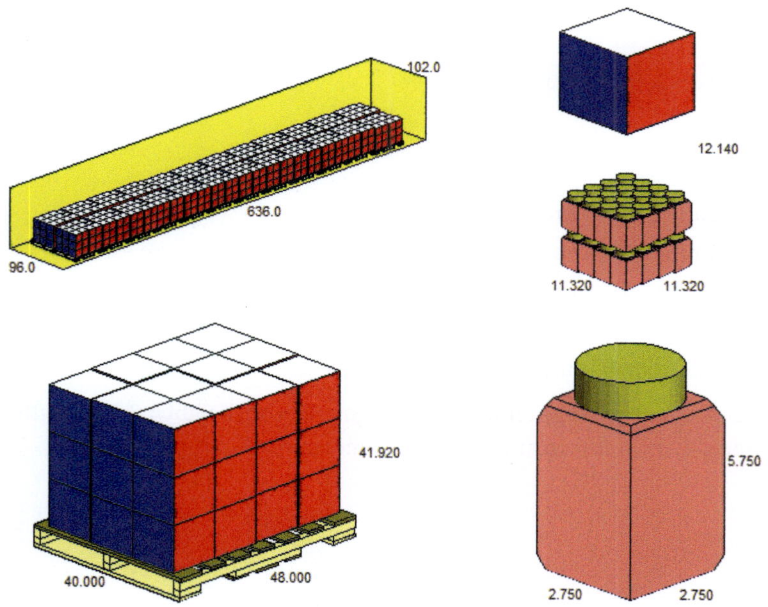

Figure 3.15 Estimation of the cube efficiency of the polypropylene jar used to contain marmalade.

can be conducted by hand calculation as shown in Figure 3.14. The cube efficiency of the jar is 65.5% and the truck cube efficiency as represented in the figure is 31.0% – the truck weight according to US transportation rules reaches its maximum with this amount of marmalade.

3.9 Study Questions

1 What are the main life cycle stages of package from cradle to cradle?

2 What are the main packaging materials?

3 Three-piece cans are primarily made of:

   A Aluminum

   B Steel

- **C** Paperboard
- **D** Plastic
- **E** None of the above

4 The main component of glass is:

- **A** Limestone
- **B** Soda ash
- **C** Alumina
- **D** Silicon oxide
- **E** None of the above

5 What is the main use of wood as lumber in packaging?

6 Which are the main plastics used in packaging?

7 An example of a renewable energy source is

- **A** Coal
- **B** Petroleum
- **C** Nuclear
- **D** Wind
- **E** None of the above

8 Which of the following statements are correct?

Hydropower is generated through the combustion of petroleum	T	F
Wind is the main source of renewable energy	T	F
Solar energy is becoming increasingly more used	T	F
Geothermal energy is derived from heath extracted from earth	T	F

9 Estimate the cube efficiency of the package shown in the figure below (height= 3.5"; length: 1.55"; width: 0.65"; volume=40 mL; closure: high=0.75"; diameter=0.65).

10 What are the main sources of energy production?

11 Which are the main parts of a packaging system?

3.10 Additional Resources

1 Introduction to Materials – NDT Resource Center
https://www.nde-ed.org/EducationResources/CommunityCollege/Materials/Introduction/introduction.htm

2 International Energy Agency and US Energy Information Administration
https://www.iea.org

3 Lawrence Livermore National Laboratory – Energy Flow Charts
https://flowcharts.llnl.gov

4 Fundamentals of Packaging Technology – Walter Soroka – 5th edition, IoPP, ISBN-13: 978-0615709345

5 The four things you need to know about the energy you use? By Jordan Wirfs-Brock
https://www.ted.com/talks/jordan_wirfs_brock_the_four_things_you_need_to_know_about_the_energy_you_use?language=en

6 How much energy electricity does take to power the world?
https://www.ted.com/talks/ted_ed_how_much_electricity_does_it_take_to_power_the_world/transcript?language=en

References

1 Twede, D. (2005). The cask age: The technology and history of wooden barrels. *Packaging Technology and Science* 18: 253–264. DOI: 10.1002/pts.696.

2 Twede, D. (2002). The packaging technology and science of ancient transport amphoras. *Packaging Technology and Science* 15: 181–195. DOI: 10.1002/pts.597.

3 Lockhart, H.E. (1997). A Paradigm for Packaging. *Packaging Technology and Science* 10: 237–252. DOI: 10.1002/(SICI)1099-1522(199709/10)10:5<237::AID-PTS395>3.0.CO;2-#.

4 Twede, D., Selke, S.E.M., Kamdem, D.P., and Shires, D.B. (2014). *Cartons, crates and corrugated board: Handbook of paper and wood packaging technology*. DEStech Publications, Inc., Lancaster, PA, US.

5 Selke, S., Culter, J., Auras, R.A., and Rabnawaz, M. (2021). *Plastics Packaging: Properties, Processing, Applications, and Regulations*, 4e. Carl Hanswer Verlag. Cincinnati, OH, US.

6 Hanlon, J.F., Kelsey, R.J., and Forcinio, H.E. (1998). *Handbook of Package Engineering*. Technomic Pub. Co.

7 International Energy Agency. (2017). *World Energy Outlook*.

8 US Environmental Protection Agency. (2020). *Advancing Sustainable Materials Management: 2018 Tables and Figures Assessing Trends in Materials Generation and Management in the United States*. U.S. EPA.

9 Eurostat. *Packaging waste statistics 2019*. https://ec.europa.eu/eurostat/statistics-explained/index.php?title=Packaging_waste_statistics. (accessed January 27, 2021).

10 U.S. Geological Survey. (2020). *Mineral commodity summaries 2020*. https://pubs.usgs.gov/periodicals/mcs2020/mcs2020.pdf. (accessed January 27, 2021).

11 Ore, I., Pigments, I.O., Rock, P., Crystal, Q., Earths, R., and Ash, S. (2019). *Mineral Commodity Summaries*. Washington, DC: U.S. Geological Survey. https://pubs.usgs.gov/periodicals/mcs2020/mcs2020.pdf. (accessed January 27, 2021).

12 Grand View Research. (2020). *Glass Manufacturing Market Size, Share & Trends Analysis Report By Application (Construction, Transportation), By Product (Container, Flat, Fiber), By Region, And Segment Forecasts, 2020 – 2027*. San Francisco, CA.

13 Wintour, N. (n.d.) *The glass industry: Recent trends and changes in working conditions and employment relations. Sectoral Policies Department*. https://www.ilo.org/wcmsp5/groups/public/---ed_dialogue/---sector/documents/publication/wcms_442086.pdf. (accessed January 27, 2021).

14 Glass Packaging Institute. (2010). *Environmental overview complete Life Cycle Assessment of North American Container Glass*. https://assets.noviams.com/novi-file-uploads/gpi/pdfs-and-documents/Learn_About_Glass/LCA_-_GPI2010_-_compressed.pdf. (accessed January 27, 2021).

15 Cary, N. (2017). *2014 Life Cycle Assessment of U.S. Average Corrugated Product*. http://www.corrugated.org/wp-content/uploads/PDFs/LCA/NCASI_2014_LCA_Final.pdf. (accessed January 27, 2021).

16 Geyer, R., Jambeck, J.R., and Law, K.L. (2017). Production, use, and fate of all plastics ever made. *Science Advances* 3: 1–5. DOI: 10.1126/sciadv.1700782.

17 Skoczinski, P., Chinthapalli, R., Carus, M., Baltus, W., de Guzman, D., Käb, H., Raschka, A., Ravenstijin, J. (nova-Institut GmbH). (2020). *Bio-based Building Blocks and Polymers – Global Capacities, Production and Trends 2019 – 2024*. www.bio-based.edu/reports. (accessed date January 28, 2021).

18 Mano, K. (2019). *KPMG FAS Newsletter*. 04.

19 Statista. (2020). *Plastic industry worldwide*. https://www.statista.com/statistics/282732/global-production-of-plastics-since-1950/. (accessed date January 27, 2021).

20 Plastics Industry Association. (2018). *Size & Impact*. Washington, DC: Plastic Industry Association. 2005.

21 International Energy Agency. (2018). *World Energy Outlook 2018*.

22 International Energy Agency. (2020). *World Energy Outlook 2020*. 1–461.

23 International Energy Agency. (2019). *World Energy Outlook 2019*.

24 Lawrence Livermore National Laboratory. *Estimated U.S. Energy Consumption in 2019: 100.2 Quads 2020*.

25 U.S. Energy Information Administration. (2019). *Annual energy outlook 2019 with projections to 2050*.

26 Lawrence Livermore National Laboratory. (2014). *Carbon flow charts*. vol. 11.

27 Energy Information Administration (EIA). (n.d.) *Oil imports and exports - U.S.* https://www.eia.gov/energyexplained/oil-and-petroleum-products/imports-and-exports.php (accessed January 27, 2021).

28 Energy Information Administration (EIA). (n.d.) *Where our coal comes from – U.S.* https://www.eia.gov/energyexplained/coal/where-our-coal-comes-from.php (accessed January 27, 2021).

29 Energy Information Administration (EIA). (n.d.) *Use of coal – U.S.* https://www.eia.gov/energyexplained/coal/use-of-coal.php (accessed January 27, 2021).

30 Energy Information Administration (EIA). (n.d.) *Where our natural gas comes from – U.S.* https://www.eia.gov/energyexplained/natural-gas/where-our-natural-gas-comes-from.php (accessed January 27, 2021).

31 Energy Information Administration (EIA). (n.d.) *Use of natural gas – U.S.* https://www.eia.gov/energyexplained/natural-gas/use-of-natural-gas.php (accessed January 27, 2021).

32 Shell Global. (n.d.) *Pennsylvania Petrochemicals Complex*. https://www.shell.com/about-us/major-projects/pennsylvania-petrochemicals-complex.html (accessed June 11, 2021).

33 Energy Information Administration (EIA). (n.d.) *The nuclear fuel cycle – U.S.* https://www.eia.gov/energyexplained/nuclear/the-nuclear-fuel-cycle.php (accessed January 27, 2021).

34 Energy Information Administration (EIA). (n.d.) *Where our uranium comes from – U.S.* https://www.eia.gov/energyexplained/nuclear/where-our-uranium-comes-from.php (accessed January 27, 2021).

35 Energy Information Administration (EIA). (n.d.) *Nuclear power plants – U.S.* https://www.eia.gov/energyexplained/nuclear/nuclear-power-plants.php (accessed January 27, 2021).

4

Pollution and Risk Management

4.1 Learning Objectives

After learning from this chapter, you are expected to:

- Understand the basic concepts about pollution and the factors affecting pollution.
- Understand factors affecting pollution and pathways to humans.
- Explain the main disciplines involved in pollution science.
- Understand risk management and assessment.
- Explain the main steps of risk assessment.
- Differentiate between carcinogen and noncarcinogen compounds.
- Explain acute, subchronic, and chronic exposure.
- Understand dose–response curves.
- Calculate lifetime risk for carcinogenic and noncarcinogenic compounds.
- Describe ecological and microbial risk assessment.

4.2 Introduction

Pollution can be stated as the accumulation of pollutants or contaminants that can have adverse effects on human health and welfare, and/or the environment [1]. Pollution is ubiquitous, and it can be found in many compounds, forms, and shapes such as metals, inorganic and organic compounds, radioactive materials such as radon, pathogens such as *salmonella* and *Escherichia coli O157:H7*, genetically modified microbes, and solar radiation. Pollutants can be airborne, waterborne, soilborne, foodborne, and human-borne. Multiexposure pathways can be routes of pollution for humans (Figure 4.1). Contaminants can result from many different natural and human sources, such as water, fires, and microorganisms. The environment (i.e., land, water, atmosphere) and the atmospheric conditions play an important role in the fate of pollutants since pollutants can be initially introduced in one or several of these environments. After introduction and interaction with the environment, pollutants experience chemical and physical changes. However, they may not be destroyed, rather just converted from one form to another. So, assessing the rate that these contaminants are introduced to the environment, their transformation changes, and where they end up is very important for animal and human health. In pollution, the environment is considered as a continuum since many of the changes and rates of

Life Cycle of Sustainable Packaging: From Design to End of Life, First Edition. Rafael A. Auras and Susan E. M. Selke.

Companion Website: www.wiley.com/go/Auras/lifecycleofsustainablepackaging

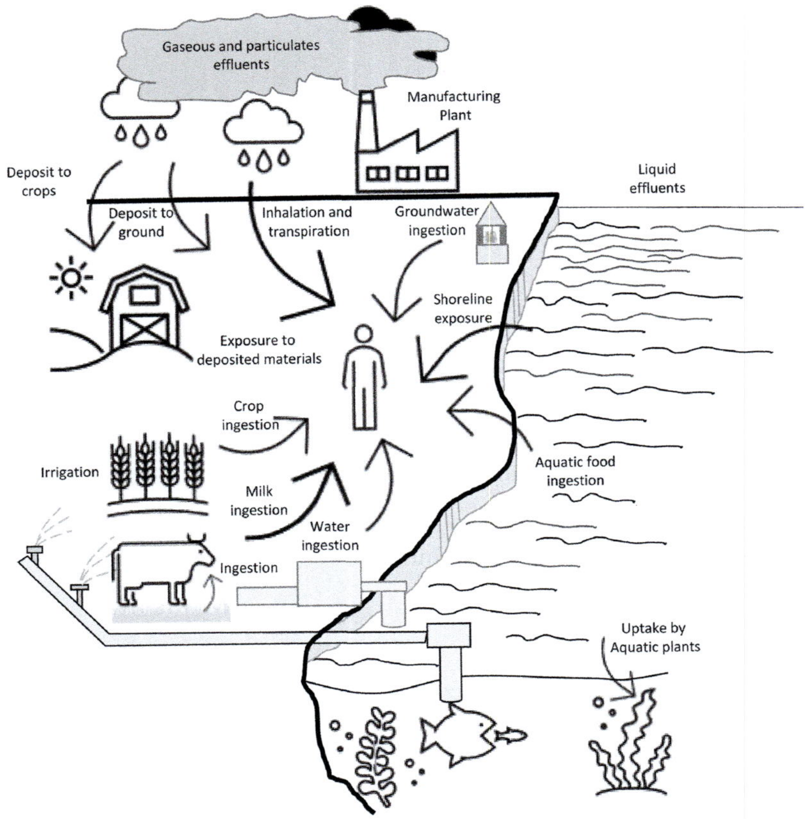

Figure 4.1 Multiexposure pathways to humans.

pollutant modifications occur at the interface among many of these environments. Historically, natural waste was generated and then transformed into other useful compounds such as compost. The industrial period caused the production of compounds that are extraneous to natural ecosystems, so they cannot easily be degraded or in some cases their toxicity and accumulation are enhanced. Some pollutants, although they are not toxic at low concentrations, are toxic when added to the ecosystem at high concentrations, such as nitrate fertilizers added to land at high levels. If these nitrates reach water systems, they can cause diseases, such as methemoglobinemia, better known as "blue baby disease" in infants.

4.3 Pollution Science

Pollution science is defined as the "study of the physical, chemical, and biological processes fundamental to the transport, fate, and mitigation of contaminants that arise from human activities as well as natural processes" [1]. Due to the complexity of pollution, several scientific disciplines are amalgamated in the study of pollution science (e.g., hydrology, toxicology, soil science, municipal waste management, atmospheric science, occupational health, hazardous waste management, biotechnology, packaging, environmental chemistry, and microbiology; see Figure 4.2.) Several factors have been combined to increase current pollution, such as population growth. Increased population

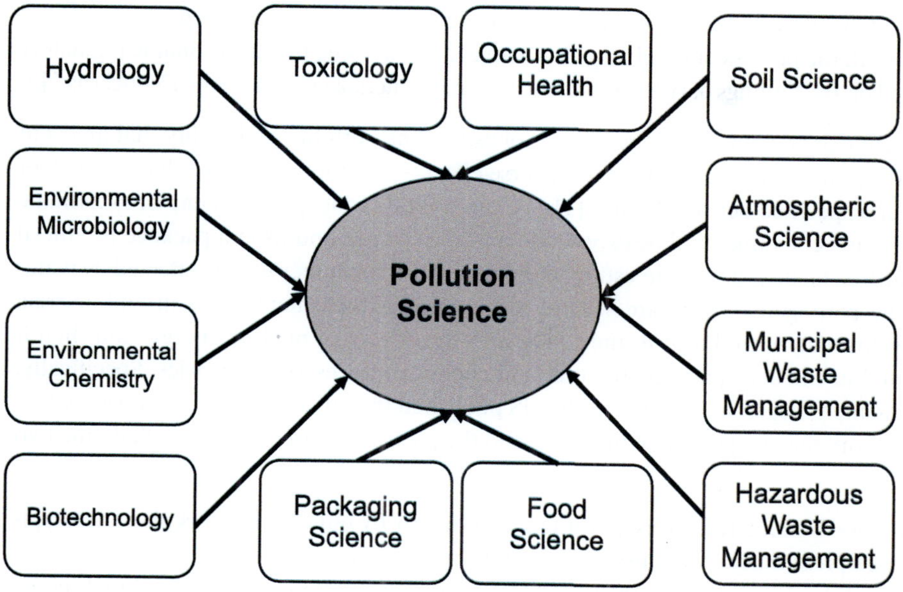

Figure 4.2 Example of disciplines involved in understanding pollution science; adapted from [1].

causes the production of waste that at low concentrations may not be hazardous but becomes hazardous as amounts increase and population grows. For example, the high demand for protein to feed the increasing population gave rise to concentrated animal feedlot operations (CAFOs) where sizable numbers of animals are raised in close proximity, producing concentrated wastes (see Chapter 5). Furthermore, people nowadays tend to live in "mega-cities" such as Tokyo, Mexico City, São Paolo, Buenos Aires, and New York. These urbanized areas consume more resources and produce more concentrated waste, so new or innovative systems to control pollution and human waste must become available.

As pollution can enter the human body through multiexposure pathways as shown in Figure 4.1, in this chapter, we will concentrate on understanding pollutants and the risk of these pollutants reaching living organisms from air, water, and soil, as well as the basic concepts of a risk-based approach to pollution science. These concepts are relevant to understanding environmental footprint indicators for assessment of packaging systems as later discussed in Chapter 9.

4.4 Risk Assessment and Management

Risk is an innate concern of human life, and it is an important factor in decision making. Risk is a notion used to define situations posing hazards to people or belongings that they value. Risk is characterized mathematically as the probability that an adverse incident will happen, as well as the magnitude of the consequences of that adverse event. It is measured in terms of probability (e.g., 1:1,000,000) of a specific harmful endpoint (e.g., accident, fatality, cancer.)

The National Academic of Science defines the following key terms as related to risk assessment:

"*Hazard*: capability of a substance to cause an adverse effect.
Risk: probability that the hazard will occur under specific exposure conditions.

Risk assessment: the process by which hazard, exposure, and risk are determined.

Risk management: the process of weighting alternatives and choosing the most suitable regulatory actions grounded on the findings of risk assessment and economic, social, and political concerns." [2]

Risk assessment can be defined as "the process of estimating both the probability that an event will occur, and the probable magnitude of its adverse effects – economic, health/safety-related, or ecological – over a specified period of time" [1]. We can use this concept, for example, to estimate the probability that a consumer will be exposed to a peanut compound from a package produced in a non-peanut-free facility, or we can estimate the risk that a consumer will be affected by water contaminated by pathogens. There are several forms of risk assessment. As a discipline, risk assessment was formally introduced in the 1940s with the advent of nuclear energy, and then it was implemented in the aerospace, petroleum, and chemical processing industries. Specifically, health-risk assessment began in 1976 when the US published the guidelines for carcinogenic risk assessment [3]. From the perspective of this book and the packaging discipline, we will define two types of risks:

Health-based risks (HRs): HR focuses on human health, mostly outside the workplace. HR typically involves high probability, low consequence, and chronic exposure.

Ecological risks (ERs): ER focuses on the multiple interactions among communities, inhabitants, and micro and macro ecosystems (including food chains). They are concerned with events such as oil spills, and long-term exposure to chemicals.

To conduct risk assessment and management, the *Silver Book* for Advancing Risk Assessment published by the National Academy of Sciences describes three main phases as shown in Figure 4.3.

The first phase requires identification of the problem and its scope. The second phase is the planning and conducting of the risk assessment, and the last phase is the risk management portion.

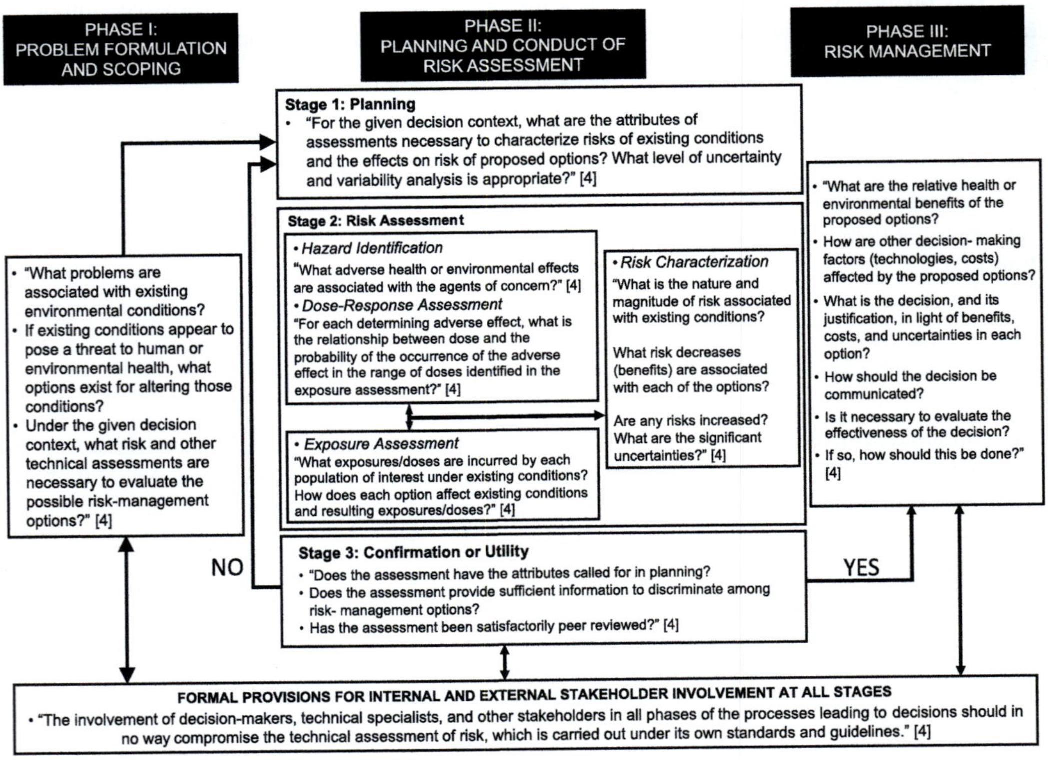

Figure 4.3 Framework for risk-based decision making; adapted from [4].

Conduction of the risk assessment consists of four main steps. Table 4.1 shows these four main steps: (a) *Exposure assessment* focusing on identifying the type and concentration of the contaminant and the frequency of exposure; (b) *hazard identification* consisting of defining the nature of the harm such as the source of harm; (c) *dose–response assessment* quantifying the adverse effect resulting from the exposure; and (d) *risk characterization* estimating the possible effect of the hazards grounded in the severity of the effects and the quantity of the exposure.

The final step after risk characterization is risk management, where regulatory and other options to reduce risk are considered. One of the basic components of risk management is risk communication, which requires aggregation of all the risk assessment process information in an effective and collaborative manner, to evaluate the tradeoffs between risk and cost. The US FDA, EPA, and the Occupational Safety and Health Administration (OSHA) regularly use risk analysis and risk assessment. Figure 4.4 shows the basic interactions between risk assessment and risk management.

To evaluate risk, we are inevitably constrained to compare a particular risk with other risks due to current life conditions and several other circumstances. However, comparison of different risks cannot be translated as the acceptability of a risk. Table 4.2 shows an example of different common mortality risks in the US in 2014. For example, there is lifetime risk mortality that for every four people that smoke one will develop cancer during the person's lifetime. When considering risk, we must understand that voluntary risk is generally more accepted than involuntary risk. For instance, some workers such as mine workers are accepting higher risk by working in a mine where they will be exposed to air pollutants in exchange for remuneration. It is generally accepted that a lifetime risk of 10^{-6} or 1:1,000,000 is a low acceptable risk for the general public. Although this number is not scientifically based, it is generally ingrained in our day-to-day policy and behavior. For example, the US EPA cited that "A 10^{-4} risk level corresponds to the upper end of EPA's generally acceptable risk range of 10^{-6} to 10^{-4} as discussed in the National Contingency Plan (NCP), 40 CFR 300.430" [5].

4.4.1 Exposure Assessment

This method consists of measurement of the intensity, frequency, and length of human contact with an environmental agent, such as chemical compounds and radiation. Exposure assessment is a main step in the risk assessment method. A well-known expression in toxicology is no exposure is equal to no risk. Exposure to contaminants can happen through many routes such as inhalation, drinking of water, consumption of food, and sorption through the skin. As stated, transport and fate of a contaminant are crucial information for understanding exposure.

Table 4.1 Four basic steps of the risk assessment process.

Step	Description
1)	*Exposure assessment*: determine the amount of contaminant in the environment and the rate of intake.
2)	*Hazard identification*: define the threat and the type of harm.
3)	*Dose-response assessment*: quantify the adverse effects coming from exposure to a hazardous compound based on the exposure level.
4)	*Risk characterization*: estimate the probable impact of a contaminant based on the gravity of its effects and the amount of exposure.

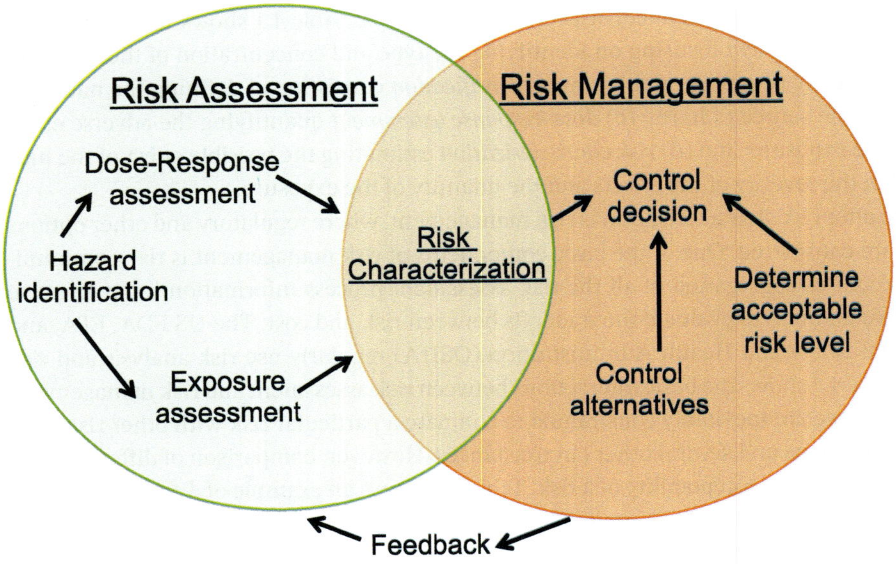

Figure 4.4 Interaction between processes of risk assessment and risk management; adapted from [4].

Table 4.2 Examples of some commonplace risks in the US in 2014.

Risk	Lifetime Risk of Mortality	Reference
Cancer from smoking one package of cigarettes per day	1:4	[1]
Death in a car accident	2:100	[1]
Heart disease	193:100,000	[6]
Cancer	186:100,000	[6]
Stroke	42:100,000	[6]
Firearms assault	3.4:100,000	[6]
Pedestrian death due to a car collision	2:100,000	[6]
Death from a firework accident	1:340,733	[1]
Attacked and killed by a shark	1:3,748,067	[1]

Exposure pathways as shown in Figure 4.1 are the routes that hazardous agents take from a source to a receptor, humans or animals, via environmental transporters such as air or water. The only exception is radiation, which does not need a medium. The exposure route or the intake pathway is the means by which the transfer occurs, such as inhalation, skin contact, and/or ingestion. The exposure pathway has five parts: (1) a source of exposure; (2) a transport medium or mechanism such as groundwater; (3) a point of exposure such as a person's house; (4) a route of exposure such as breathing, eating, and/or drinking; and (5) a receptor population such as a group of people. Several factors are considered when estimating exposure, intake, or potential dose involving variables such as concentrations of chemicals, exposure rates, and understanding special characteristics of receptors (e.g., body weight, and sorption capacity of chemicals). Ideally, exposure concentrations should be estimated at the point of contact. However, this is not always possible so to calculate exposure concentrations; a combination of monitoring and modeling data plus judgment are used. To estimate exposure rates, we need to consider all the routes or pathways that can result in exposure to the substance, for example water consumption. Estimating water consumption is not always easy since we consume water in several different ways such as tap water, soft drinks, and other types of water. Moreover, the consumption of water changes drastically with age and demographics, so these parameters are not simple. Additional routes of exposure should also be considered, such as consumer products (many during a day), clothing, water, beverages, indoor and outdoor air, food, packaging, furniture, toys, carpeting, tobacco or other smoke, outdoor soil and soil tracked indoors, as shown in Table 4.3.

Exposure assessment is considered a three-step process: (1) Characterization of the point of exposure location and scenario such as daily running through a contaminated path, for example a person eating a particular product such as cereal every morning for a period of his/her life; (2) identification of exposure pathways such as indoor air and groundwater; (3) quantification of the exposure such as miligram of compound inhaled per kilogram of body weight per day.

The food supply is an important pathway of exposure. Since estimating the concentration of contaminants in water tends to be straightforward, we can use this amount to estimate the amount reaching the food supply by using a bioconcentration factor (BCF), which allows translation of the amount of contaminant in water to the expected amount in a food such as fish. BCF are expressed in Lkg^{-1}. For example, the BCF of trichloroethylene (TCE) is 10.6 Lkg^{-1} and if the concentration in the sample of water is 40 mgL^{-1}, the concentration of TCE in fish is $10.6\ \text{Lkg}^{-1} \times 40\ \text{mgL}^{-1} = 424\ \text{mgkg}^{-1}$.

4.4.2 Hazard Identification

During this first stage of risk assessment, the main goal is to identify the hazard, which could be a specific chemical compound, a physical agent (e.g., radiation and sound), or a microorganism

Table 4.3 EPA typical default exposure factors; modified and adapted from [1].

Land Use	Exposure pathway	Daily intake	Exposure frequency (day/year)	Exposure duration (years)
Residential	Drinking of drinkable water	2 Ld^{-1}	350	30
	Consumption of soil and dust	200 mg (child) 100 mg (adult)	350	6 24
	Inhalation of contaminants	20 m^3 (total) 15 m^3 (indoor)	350	30
Industrial and commercial	Drinking of drinkable water	1 L	250	25
	Consumption of soil and dust	50 mg	250	25
	Inhalation of contaminants	20 m^3 (workday)	250	25
Agricultural	Ingesting of homegrown produce	42 g (fruit) 80 g (vegetable)	350	30
Recreational	Consumption of locally caught fish	54 g	350	30

identified as posing a potential hazard to humans and that can cause an illness or a disease. During the hazard identification stage, all relevant information such as the Guidelines for Carcinogen Risk Assessment is reviewed. Information on animal and human exposure is gathered and correlated to weight-to-evidence classification (i.e., relation between concentration and exposure and human weight). Additional emerging methods are also currently used for hazard identification such as computer models and chip testing (e.g., "organ on a chip" or "lung-on-a-chip" [7]). Different materials can be used for identifying different risks. Clinical studies help to identify large risks (1:10 or 1:100). Epidemiological studies help to identify risk between 1:1,000 and 1:10,000. Lower risks on the order of 1:100,000 are generally obtained by extrapolating effects of high doses given to animals. Another approach for hazard identification is read-across, which can be quantitative or qualitative and is defined as:

> Qualitative read-across, the presence (or absence) of a property/activity such as a particular type of toxic effect for the chemical of interest is inferred from the presence (or absence) of the same property/activity for one or more other chemicals. This qualitative approach provides a "yes/no" answer.
>
> Quantitative read-across uses information for one or more chemicals to estimate what the chemical of interest will be like. Thus, quantitative read-across can be used to obtain a quantitative value for an endpoint, such as a dose-response relationship. [4]

However, the emergent method for hazard identification is using the adverse outcome pathways (AOPs), which translate from high-dose studies in laboratory animals to *in vitro* methods to assess changes in normal cellular signaling pathways commonly encountered in human cells or tissues. The National Toxicology Program recognized in 2004 that with the advances in molecular biology and computer science, *in vitro* biochemical, nonrodent animal models, and cell-based assays for toxicological testing can be used for higher output of data at lower cost. Using these methods, numerous chemicals can be tested in a day [8].

4.4.3 Dose–Response Assessment

Contaminants have different capacities to cause adverse effects, so we need to understand their capacity to cause harm. Health benchmark (HB) values quantify a chemical's likely toxicity and its capability of provoking a health effect. HB values are obtained from toxicity data. They signify a "dose-response" approximation that communicates the probability and severity of adverse health effects as determined by exposure and dose. HB is used together with individual exposure levels to estimate if a risk can happen. Health effects are associated with the exposure path and the timing of the exposure, and HBs are specific to the route and the length of the exposure (i.e., acute, subchronic, or chronic).

- *Acute* is defined as exposure of less than one day.
- *Subchronic* or *intermediate* is defined as an exposure of a few weeks to a few months.
- *Chronic* exposure is defined as an exposure that can span several months to years.

To determine the health benchmarks, some toxicity data are directly obtained from clinical, occupational, and epidemiological reports. However, the majority of the toxicological data are derived from animal studies, mostly mice and rats, so higher concentrations can be tested to obtain the dose–response relationship, which is a quantitative relationship between an agent and its toxicity. The dose is standardized as the weight of the substance or pathogen ingested, inhaled, or ad/

absorbed through the skin per kilogram of body weight per day and it is expressed as $mgkg^{-1}d^{-1}$. The effects can vary largely depending on the amount of substance from no observable effects, to temporary and reversible effects, to permanent organ injury, to chronic impairment, to death. Figure 4.5 shows an example of a dose–response assessment relationship between the quantity of a toxicant and the risk of an adverse outcome effect. Dose–responses are derived from animal studies. So, caution must be taken since animal studies are conducted at high doses with few animals due to the study cost and then extrapolated to low doses and to humans. There are many approaches to extrapolate data from high doses to low doses. The principal controversy is selection of the method to extrapolate to low doses. The choice of model is a policy decision.

For noncarcinogenic chemicals, the supposition is that there is a threshold below which there is no observable toxicity. The lowest dose obtained from studies on animals or humans is divided by safety factors to offer a margin of safety.

For carcinogenic chemical responses, the assumption is that there is no threshold and that exposure to any amount will have some likelihood of cancer. So, for carcinogenic compounds, the only safe amount is zero, and the dose–response curve crosses the origin (0,0).

There are many mathematical models, and all of them produce different lifetime risks of cancer derived from the same compound (Table 4.4).

The US EPA uses a modified version of the multistage model called the *Linearized Multistage Model (LMS)*. The model assumes that there are multiple stages for cancer involving many carcinogens, cocarcinogens, and promoters, which are modeled by a sequence of mathematical functions.

At low dose–response, the curve produced is called the potency factor (PF) or slope factor (SF). This is a toxicity value that quantifies the correlation between the dose and the response. The SF is the reciprocal of the concentration of the chemical measured in $mgkg^{-1}d^{-1}$ or the risk produced by a lifetime average dose (AD) of 1 $mgkg^{-1}d^{-1}$. "The cancer SF is a plausible upper-bound estimate of the probability that an individual will develop cancer if exposed to a chemical for a lifetime of 70 years. The cancer slope factor is expressed as $mgkg^{-1}d^{-1}$" [4]. Table 4.5 shows SF values for select carcinogens.

The dose–response equation for a carcinogen, as shown in Figure 4.5 is:

$$\text{Lifetime Risk} = AD \times SF \tag{4.1}$$

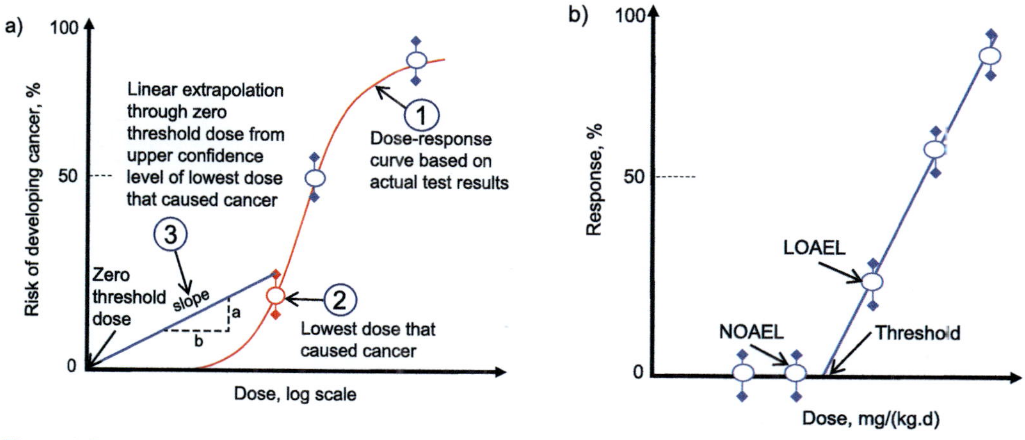

Figure 4.5 Dose–response relationship for (a) zero threshold from the multistage model (Linearized Multistage Model - LMS) to infer cancer risk from a dose–response curve using the cancer slope factor (Potency factor (PF)): Slope of dose–response of the curve produced by the multistage model), and (b) dose–response for noncarcinogenic effects (NOAEL: No observed adverse effect level; LOAEL: Lowest observed adverse effect level); adapted from [7].

Table 4.4 Primary models used for assessment of nonthreshold effects.

Model	Description	Lifetime Risk ($1.0\ mgkg^{-1}d^{-1}$)	
One-hit – Very conservative	A single compound is able to induce a malign stage and this change occurs in one single stage	6.0×10^{-5}	1 in 17,000
Multistage	Tumors are the results of a sequence of events	6.0×10^{-6}	1 in 167,000
Multihit	A number of dose-related events are required before a cell turn out to be malignant	4.4×10^{-7}	1 in 2.3 million
Probit	Assumes that the probability of the cancer is a direct function of the log of the dose (i.e., log normal distribution)	1.9×10^{-10}	1 in 5.3 million

The probability of getting cancer (not of dying of cancer) is estimated for an assumed 70-year human lifetime. It is called the lifetime average daily dose or chronic daily intake. The US EPA uses the LMS model in cancer risk assessments, and the model yields a cancer SF, better known as q1* (Q1-star), used to predict cancer risk at a specific dose.

The dose–response for noncarcinogens needs to consider the existence of thresholds as shown in Figure 4.5, which indicates that below the no observed adverse effect level (NOAEL), exposure does not pose a risk. When a NOAEL is not available, the contaminant's LOAEL (lowest observed adverse effect level) is used, but additional uncertainty factors need to be incorporated. The reference dose (RfD) can be estimated with this information, defined as the intake or dose of a substance per unit of body weight per day ($mgkg^{-1}d^{-1}$). A RfD for oral exposure and reference concentration (RfC) for inhalation exposure are used to evaluate noncancer effects. The RfD and RfC are determined from the daily exposure level to a person that does not have an appreciable risk of harmful effects during a lifetime and is reported as $mgkg^{-1}d^{-1}$ for RfD or mgm^{-3} for RfC. Substances with high slopes and lower RfD are associated with high toxicity. Table 4.5 shows selected RfD for some chemical compounds. Equation 4.2 shows the general formula to estimate RfD.

Table 4.5 Toxicity data as chemical RfDs for chronic noncarcinogenic effects and PF for selected potential carcinogens.

Chemical	RfD ($mgkg^{-1}d^{-1}$)	Chemical	PF – Oral Route ($mgkg^{-1}d^{-1}$)
Acetone	0.1	Arsenic	1.75
Cadmium	0.0005	Benzene	2.9×10^{-2}
Chloroform	0.01	Chloroform	6.1×10^{-3}
Phenol	0.06	DDT	0.34
Polychlorinated biphenyl	0.0001	PCBs	7.7
Toluene	0.3	Trichloroethylene (TCE)	1.1×10^{-2}
Xylene	2.0	Vinyl chloride	2.3

$$\mathrm{RfD} = \frac{\mathrm{NOAEL\ or\ LOAEL}}{VF_1 \times VF_2 \cdots \times VF_n} \tag{4.2}$$

where VF_i are the safety factors. For example, a 10-fold safety or uncertainty factor is used to account for differences in sensitivities between different human populations such as children, pregnant women, and the elderly. Another factor of 10 is used if the data is obtained from questionable or limited human and animal exposure data. The more uncertainty about the information obtained, the more significant number of uncertainty factors are used (i.e., a factor for each type of uncertainty). Equation 4.3 can be used to estimate risk assessment of noncarcinogens as:

$$\text{Lifetime Risk} = PF(\mathrm{ADI} - \mathrm{RfD}) \tag{4.3}$$

where ADI – acceptable daily intake – is the quantity of a chemical to which a person could be exposed each day for a long time, such as a lifetime without suffering harmful effects, and PF is the chronic daily intake.

$$\mathrm{ADI}\left(\mathrm{mgkg^{-1}d^{-1}}\right) = \frac{\text{Average daily dose}\left(\mathrm{mgd^{-1}}\right)}{\text{Body weight}\left(\mathrm{kg}\right)} \tag{4.4}$$

In general, the lifetime risk for noncarcinogens is not estimated, and RfD is used as the main indicator.

A more preferred method since 1980 is the benchmark dose method (BMD), which addresses some of the limitations of the NOAEL method, and is less reliant on dose selection and time. This method considers the shape of the dose–response curve, as shown in Figure 4.6. Estimating a BMD 95% lower bound confidence limit (BMDL) results in a point of departure from animal toxicology data that properly accounts for the study sample size. With the availability of BMD software such as the EPA's Benchmark Dose Software (BMDS), the BMD is becoming the method of choice adopted for several health organizations worldwide [9].

4.4.4 Risk Characterization

The final stage of risk assessment is to estimate the risk characterization, which consists of predicting the frequency and severity of effects in exposed populations. The main goal of this stage is to obtain probabilities for occurrence of a specific effect. Quantitative risk is estimated for appropriate media and pathways. The information is then used to create standards and guidelines for specific compounds.

4.4.4.1 Carcinogenic Risks

Carcinogenic risks are associated with exposure to chemicals that might initiate cancer (e.g., 1 person in 10^6). Potential human carcinogenic risks are communicated as an increased probability of developing cancer during a human lifetime. For instance, a 10^{-6} increased cancer risk signifies an increased lifetime risk of one in one million for developing cancer. For carcinogenicity, the likelihood of a person developing cancer over a lifetime is estimated by multiplying the cancer slope factor ($\mathrm{mgkg^{-1}d^{-1}}$) for the substance by the chronic (70-year average) daily intake ($\mathrm{mgkg^{-1}d^{-1}}$).

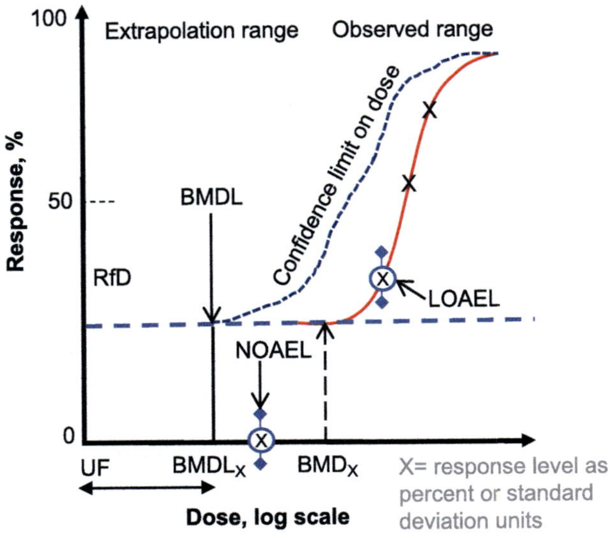

Figure 4.6 Extrapolated values using the benchmark dose method reflecting the shape of the dose–response curve; adapted from [4].

Dose–response curves for carcinogen compounds are linear, so the incremental lifetime risk of cancer is defined as:

$$\text{Incremental lifetime risk of cancer} = \text{LADD} \times SF \tag{4.5}$$

where LADD can be estimated as:

$$\text{LADD} = \frac{C \times \text{CR} \times \text{EF} \times \text{ED}}{\text{BW} \times \text{AT}} \tag{4.6}$$

where

- LADD is defined as the lifetime average daily dose for carcinogens in $\text{mgkg}^{-1}\text{d}^{-1}$
- C is the chemical concentration in the media such as air or water
- CR is the contact rate per unit of time in Ld^{-1} or m^3d^{-1}
- EF is the exposure frequency in dyear^{-1}
- ED is the exposure duration in years
- BW is the average body mass during the exposure in kg
- AT is the average time in days for the exposure

SF values can be obtained from the US EPA Integrated Risk Information Systems (IRIS) [10], which is an electronic database that provides a list of carcinogen and noncarcinogen safety concentrations such as NOAEL, LOAEL, and RfD.

Example of Health Benchmarks for Acrylonitrile (CH_2CHCN)

Chronic:

- Inhalation LADD: 0.24 $\text{mgkg}^{-1}\text{d}^{-1}$
- Oral LADD: 0.54 $\text{mgkg}^{-1}\text{d}^{-1}$

- RfC: 0.002 mgm^{-3}
- RfD: 0.001 $mgkg^{-1}d^{-1}$

Subchronic:
- RfC: 0.02 mgm^{-3}

Acute:
- ATSDR MRL: 0.22 mgm^{-3}

4.4.4.2 Noncarcinogenic Risks

Noncancer risks encompasses all the categories of remaining health effects from chemical exposure (e.g., enzyme changes, weight loss, reproductive and developmental abnormalities, and respiratory reactions). The exposure duration as previously defined as acute, intermediate, and chronic effects are used for exposure time. In the case of noncancer risks, they are expressed as hazard quotient (HQ) for a single substance or hazard index (HI) for many substances and/or exposure pathways.

$$HQ = \frac{AD\left(mgkg^{-1}d^{-1}\right)}{RfD\left(mgkg^{-1}d^{-1}\right)} \tag{4.7}$$

In the case of noncarcinogens, the toxicity is only important during the exposure. HQ less than 1 is defined as not significant, and above 1 could represent a risk. For more than one chemical or multiexposure, HI can be estimated as:

$$HI = \sum_{i=1}^{i=n} HQ_i \tag{4.8}$$

Example 4.1: The average concentration of 1,2-dichlorobenze in a water supply is 2.0 μgL^{-1}. Estimate the chronic daily intake for a 70 kg adult. Two liters of water are consumed per day by the person. Example adapted from [1].

<u>**Answer:**</u>

$$LADD = \frac{C \times CR \times EF \times ED}{BW \times AT} = \frac{0.002\ mg.L^{-1} \times 2\ L.d^{-1} \times 365\ d.y^{-1} \times 30\ y}{70\ kg \times 365\ d.y^{-1} \times 30y} = 5.71 \times 10^{-5}\ mg.kg^{-1}.d^{-1}$$

Example 4.2: A water supply contains 0.2 mgL^{-1} of acetone and 0.1 mgL^{-1} of choroform. An adult person of 70 kg drinks 2 L of this water per day for 5 years. Estimate the hazard index and the carcinogenic risk for the person drinking this water? Example adapted from [1].

For Acetone:

$$AD = \frac{0.2\ mgL^{-1} \times 2\ L.d^{-1}}{70\ kg} = 0.005\ \ mgkg^{-1}d^{-1}$$

The RfD of acetone is $0.1 \text{ mgkg}^{-1}\text{d}^{-1}$

$$\text{HQ} = \frac{0.005 \text{ mgkg}^{-1}\text{d}^{-1}}{0.1 \text{ mgkg}^{-1}\text{d}^{-1}} = 0.05$$

For Chloroform:

$$AD = \frac{0.1 \text{ mgL}^{-1} \times 2 \text{ L.d}^{-1}}{70 \text{ kg}} = 0.0029 \text{ mgkg}^{-1}\text{d}^{-1}$$

The RfD of chloroform is $0.01 \text{ mgkg}^{-1}\text{d}^{-1}$

$$\text{HQ} = \frac{0.0029 \text{ mgkg}^{-1}\text{d}^{-1}}{0.01 \text{ mgkg}^{-1}\text{d}^{-1}} = 0.29$$

Hazard index: $0.05 + 0.29 = 0.34$
Because the HI is below 1.0, the water is safe. We do not need to consider the 5 years.

Risk: LADD × SF

$$\text{LADD} = \frac{C \times \text{CR} \times \text{EF} \times \text{ED}}{\text{BW} \times \text{AT}} = \frac{0.1 \text{ mgL}^{-1} \times 2 \text{ L.d}^{-1} \times 365 \text{ d.y}^{-1} \times 5 \text{ y}}{70 \text{ kg} \times 365 \text{ d.y}^{-1} \times 70\ y} = 2.04 \times 10^{-4} \text{ mgkg}^{-1}\text{d}^{-1}$$

Now, the SF for chloroform is 6.1×10^{-3}.

Risk: $2.04 \times 10^{-4} \text{ mgkg}^{-1}\text{d}^{-1} \times 6.1 \times 10^{-3} \text{ mgkg}^{-1}\text{d}^{-1} = 1.24 \times 10^{-6}$
So, from a cancer standpoint, the risk for this period is around the 10^{-6} goal.

After estimating the initial lifetime risk, it is important to understand the uncertainty of the estimated lifetime health risk values. Sensitivity analysis can provide important information about the uncertainty in the magnitude of the risk.

The final stage of risk assessment is the risk characterization. In this stage, exposure and dose–response information are aggregated to yield probabilities of effects happening in humans under specific exposure conditions. For example, the risk of being exposed to lead in water can be estimated over a set period of time or for a lifetime. Assuming that a person drinks 2 L of water per day over 70 years, and with the different concentrations of lead in water, we can estimate the exposure and the risk.

4.5 Ecological Risk Assessment

Ecological risk assessment estimates the probability that adverse ecological effects will occur due to exposure to one or more stressors (i.e., a substance, circumstance, or an energy field having an inherent ability to produce adverse health effects), for example, contamination with a chemical coming from groundwater flow. In the US, the CERCLA (better identified as the SUPERFUND), the Resource Conservation and Recovery Acts (RCRA), and many other laws and regulations require an ecological risk assessment as part of remediation investigations and feasibility studies. One of the main objectives of this legislation is to define and to document the actual or potential effects of contaminants in humans and environments.

There are four stages of an ecological risk assessment: (a) problem formulation and hazard identification – identification of the stressors and characterization endpoints; (b) exposure assessment – determination of the environmental concentration; (c) ecological effects/toxicity assessment – identification and quantification of the adverse health effects produced by the stressors; and (d) risk characterization – compiling the exposure and stressor-response profiles to determine the probability of effects and their distribution.

4.6 Microbial Risk Assessment

Microbial risk assessment is a process that permits estimation of dose–response in relation to risk of microbial infection. The steps used to assess the health-based risk are very similar to the other methods: (a) hazard identification; (b) exposure assessment; (c) dose–response; and (d) risk characterization. The US EPA establishes that human populations should not be subjected to microbial risks higher than 1:10^4 for a yearly exposure. In the case of water consumption, it requires that less than 1 microorganism per 1,000 liters is present.

Summarizing, risk assessment is a major tool for governments and communities to make decisions about toxic chemical, ecological or microorganism exposure. The process allows risk managers to use the information in complex decision processes to keep a population safe.

4.7 Case Study 4.1. Estimation of the Health Risk of Dichloro diphenyl trichloroethane (DDT) and Polybrominated Diphenyl Ether (PBDE)

A water supply has 0.3 mgL^{-1} of DDT and 0.1 mgL^{-1} of tetraBDE. A 70 kg person drinks 2 L of water per day for 5 years. Calculate the hazard index and the carcinogenic risk for this person from drinking this water?

DDT and tetraBDE are both carginocenics. So,

$Risk_{DDT}$: LADD × SF

$$\text{LADD} = \frac{C \times \text{CR} \times \text{EF} \times \text{ED}}{\text{BW} \times \text{AT}} = \frac{0.3\ \text{mgL}^{-1} \times 2\ \text{L.d}^{-1} \times 365\ \text{d.y}^{-1} \times 5\ \text{y}}{70\ \text{kg} \times 365\ \text{d.y}^{-1} \times 70\ \text{y}} = 6.1 \times 10^{-4}\ \text{mgkg}^{-1}\text{d}^{-1}$$

Now, the SF for DDT is 5E-4 $mgkg^{-1}d^{-1}$.

$$\text{Risk: } 6.1 \times 10^{-4}\ \text{mgkg}^{-1}\text{d}^{-1} \times 5.0 \times 10^{-4}\ \text{mgkg}^{-1}\text{d}^{-1} = 3.06 \times 10^{-7}$$

So, for a cancer standpoint the risk of consuming DDT for 5 years is below the 10^{-6} goal. If the water is consumed for 30 y, the risk is about the 10^{-6} goal.

$Risk_{tetraBDE}$: LADD × SF

$$\text{LADD} = \frac{C \times \text{CR} \times \text{EF} \times \text{ED}}{\text{BW} \times \text{AT}} = \frac{0.1\ \text{mgL}^{-1} \times 2\ \text{L.d}^{-1} \times 365\ \text{d.}y^{-1} \times 5\ \text{y}}{70\ \text{kg} \times 365\ \text{d.y}^{-1} \times 70\ \text{y}} = 2 \times 10^{-4}\ \text{mgkg}^{-1}\text{d}^{-1}$$

Now, the SF for tetraBDE is 0.0001 $mgkg^{-1}d^{-1}$.

$$\text{Risk: } 2 \times 10^{-4}\ \text{mgkg}^{-1}\text{d}^{-1} \times 1.0 \times 10^{-4}\ \text{mgkg}^{-1}\text{d}^{-1} = 2.1 \times 10^{-8}$$

So, for a cancer standpoint the risk of consuming tetraBDE for 5 years is below the 10^{-6} goal. However, if the water is consumed for 30 y, the risk is highly increased, but still below the 10^{-6} goal.

4.8 Study Questions

1 Define pollution. What are the main sources of pollutants?
2 What are the main disciplines that are part of pollution science?
3 Define risk. What do we understand as risk assessment?
4 What are the main steps of risk assessment? Explain each step.
5 What is BCF? Provide an example.
6 Define acute, subchronic, and chronic exposures.
7 What is the main difference for risk assessment between carcinogen and noncarcinogen compounds? What are RfD, RfC, NOAEL, and LOAEL?
8 The mean phenol concentration in a water supply is 1 μgL^{-1}. Calculate the lifetime average daily dose intake for a 70 kg adult. Assume that 2 L of water are consumed per day.
9 A drinking water supply is found to contain 0.1 mgL^{-1} of xylene, 0.1 mgL^{-1} of DDT, and 0.05 mgL^{-1} of chloroform. An adult of 70 kg drinks 2 L per day of this water for a period of 10 years. Calculate the hazard index and the carcinogenic risk from drinking this water.
10 Define ecological and microbial risk assessment.

4.9 Additional Resources

1 Integrated Risk Information System
https://www.epa.gov/iris
2 Possible Exposure Pathways During Emergencies – US EPA
https://www.epa.gov/emergency-response/possible-exposure-pathways-during-emergencies
3 Rachel Carson. (1962). Silent Spring. Houghton Mifflin Company
4 United Nations – Division of Pollution
http://web.unep.org/environmentassembly/dimensions-pollution
5 Public Health Assessment Guidance Manual (2005 Update) – Agency for Toxic Substances and Disease Registry (ATSDR)
https://www.atsdr.cdc.gov/hac/phamanual/toc.html
6 ToxTutor – US National Library Medicine
https://toxtutor.nlm.nih.gov/06-003.html
7 Exposure Science in the 21st Century; A Vision and a Strategy (2012)
https://www.nap.edu/download/13507

References

1 Brusseau, M.L., Pepper, I.L., and Gerba, C.P. (2019). *Environmental and Pollution Science*, 2e. London, UK: Academic Press Publications.
2 National Academic of Sciences. (1983). *Risk Assessment in the Federal Government: Managing the Process*. Vol. 25. Washington DC. DOI: 10.1080/00139157.1983.9931232.
3 US Environmental Protection Agency *Guidelines for Carcinogen risk assessment 2005*. https://www.epa.gov/sites/production/files/2013-09/documents/cancer_guidelines_final_3-25-05.pdf (accessed July 26, 2019).

4 National Research Council. (2009). *Science and decisions: Advancing risk assessment.* Washington, DC: The National Academies Press. https://doi.org/10.17226/12209.

5 U.S. Government (2011). *40 CFR 300.430 – Remedial investigation/feasibility study and selection of remedy.*

6 Geggel, L. (n.d.). *The odds of dying.* https://injuryfacts.nsc.org/all-injuries/preventable-death-overview/odds-of-dying/data-details.

7 U.S. National Library of Medicine *ToxTutor – Dose-response assessment 2018.* https://toxtutor.nlm.nih.gov/06-003.html (accessed September 2, 2019).

8 National Toxicology Program U.S. Department of Health and Human Services. (2004). *A national toxicology program for the 21 st century roadmap to achieve the NTP vision 2004.* https://ntp.niehs.nih.gov/ntp/about_ntp/ntpvision/ntproadmap_508.pdf (accessed September 3, 2019).

9 US Environmental Protection Agency. (n.d.). *Benchmark dose tools* https://www.epa.gov/bmds (accessed September 3, 2019).

10 US Environmental Protection Agency. (n.d.). *IRIS assessments.* https://cfpub.epa.gov/ncea/iris_drafts/atoz.cfm?list_type=alpha (accessed July 31, 2019).

5

Soil Pollution

5.1 Learning Objectives

After learning from this chapter, you are expected to:

- Understand soil pollution.
- Identify the types of soil pollutants.
- Describe the main natural and anthropogenic sources of soil pollutants.
- Identify the consequences of human activities on the generation of soil pollutants.
- Describe the main body of regulations about soil pollutants in the US.

5.2 Introduction

Human activities such as mining, agriculture, and deforestation use an increasing amount of energy and, at the same time, directly and indirectly impact the soil and pollute land. Mining produces large quantities of structureless geologic materials, such as crushed rock containing significant amounts of toxic metals such as lead, cadmium, and some salts. Wastes from mines are generally stored near the mining operations. Modern agricultural production requires large amounts of fertilizers, pesticides, and chemicals, polluting land, and runoff into water streams. Land deforestation affects land and water quality due to soil and sediment transport erosion, accelerating the loss of nutrient-rich soil. All these activities can increase soil salinity and acidity of surrounding lands due to the release of unwanted metals. Additionally, these compounds can be released in water and air. The majority of this soil contamination has occurred and has been increasing since the industrial revolution [1].

5.3 Surface Mining

Mining coal and ore was one of the main activities catalyzing the industrial revolution. Converted coal and ore became energy and metals, the building blocks for development. Although chemicals and plastics are a strong component of today's industrial production, metals and coal are still extremely important for constructing and producing modern goods such as cars and appliances.

Life Cycle of Sustainable Packaging: From Design to End of Life, First Edition. Rafael A. Auras and Susan E. M. Selke.

Companion Website: www.wiley.com/go/Auras/lifecycleofsustainablepackaging

Although the main metals for industrial production are iron, aluminum, and copper, several other metals, such as lead, cadmium, nickel, mercury, arsenic, and selenium, are also important. Metallic elements are commonly found in industrial wastes, and their effects on the environment are not fully understood. Mining and process ores produce run-off streams abundant in Na^+, Ca^{++}, Cl^-, SO_4, and $CO_3^{=}$. These elements are common in the environment, but the large quantities produced from mining are a concern. For example, large quantities of residues are produced by mining in the form of mine spoils and mine tailings. Mine spoils are surface materials that are not of interest and are often piled in large areas. Mine tailings are crushed materials that have been processed to extract the target metal/s. Generally, they are deposited in valleys or depressions. These mine tailings can alter the environment and can contain high concentrations of minerals affecting the environment, such as by reducing the pH.

Furthermore, mining operations that use chemicals to extract metals, such as copper and gold, can generate large volumes of acidic wastewater containing metals, which are toxic when released into water streams. Figure 5.1 shows the contamination of the Tinto river in the southwestern part of Spain in the mountains of Andalusia. The Rio Tinto has a unique red and orange color due to the acid levels produced by a high concentration of iron and heavy metal waste.

Smelting and refining of metals are done at high temperatures. Iron melts at 1,536°C, copper at 1,083°C, and aluminum at 660°C. When these metals melt, other metals and metal compounds volatilize since, for example, the boiling point of mercury is 357°C, arsenic 613°C, cadmium 765°C, and zinc 906°C. If smelter and metal refining companies do not have systems to scrub the released gases, significant volatile toxic compounds can be emitted to the atmosphere and then deposited onto the land.

5.4 Deforestation

Deforestation is defined as the cutting of trees and removing wood. Forests are cleared to make space for agricultural production or to harvest wood for papermaking or timber and lumber production. Deforestation decreases soil stability, increases erosion, transports sediments, reduces

Figure 5.1 Mining for copper, silver, gold, and other minerals. (Left) Chuquicamata copper mine, Antofagasta Region, Chile; reproduced from Astro_Alex licensed under CC BY-SA 2.0 (Right) Rio Tinto, Huelva, Spain, reproduced from Lorenmart licensed under CC BY-NC 2.0.

Figure 5.2 Deforestation in North Island, New Zealand; reproduced from Coghlan, M. license under CC BY-SA 2.0.

biodiversity through loss of natural habitat, and changes the local climate due to changes in vegetation, as shown in Figure 5.2. During deforestation, the air quality in surrounding regions may be affected due to burning or wind erosion. Reducing forests can also reduce the amount of CO_2 absorbed and impact the global balance of atmospheric gases since trees draw carbon out of the atmosphere and release oxygen.

Decrease of forest vegetation can hinder nutrient and metal retention by soil. Water run-off can easily strip nutrients and increase the presence of sediments in streams and rivers. Moreover, naturally occurring pockets of metals can be washed out to water streams when vegetation is not present, and these metals can fixate in other places or wash out in water streams.

Typically, initial deforestation is carried out by burning so that space is created for agricultural production or to clear previously harvested forest of the remaining vegetation. The burning of biomass creates air quality issues such as increasing particulate matter, resulting in breathing issues for animals and humans.

5.5 Soil Acidity and Salinity

Acid soils can occur naturally or due to agricultural practices. Naturally, acid soils are found in tropical areas with extensive precipitation since rain can leach basic cations and pH buffering minerals such as carbonates. Soil acidity and salinity transformation produce additional changes in silicon-based minerals, leaving behind acidic iron and aluminum minerals. Agricultural soils with the continuous addition of ammonia and urea from fertilizers can also result in acid soil environments. Coal-burning emissions such as SO_2 and NO_x can cause acid rain and then be deposited in the environment.

Soil salinity or salt affected soils naturally occur worldwide in arid and semiarid regions with precipitation below 27 cm annually. Salt-affected soil regions are created through weathering, involving processes such as hydrolysis, hydration, solution, oxidation, and carbonation. The released salts are transported to groundwater streams and gradually concentrated. Salt moves from more humid to less humid environments during this process, especially to arid regions.

5.6 Soil Erosion

Soil erosion is a natural process that human activities can accelerate. In soil erosion, particles are dislodged by various energy inputs. Then, particles are transported to different environments, such as water streams. An example of human activities causing soil erosion is the decay in agricultural soil quality through reiterative planting of mono-crops.

In the case of soil erosion, when particles move by running water, soil water erosion, the water as rain, for example, provides the force needed to detach soil particles. Raindrops travel at high speed and have enough kinetic energy to release soil particles. As they aggregate, they can travel long distances in run-off water. Smaller particles travel more easily than larger particles. Soil loss from land is a function of the rainfall and erodability potential, the gradient of the land, and the type of cover on the soil. In past decades, US farmers have improved soil erosion by reducing tilling of the land. Cover crops also provide a long-term mechanism to minimize soil erosion in erodable lands.

In the case of particle movement by wind – soil wind erosion – the wind provides the kinetic energy for the detachment and movement of the soil. As the wind blows, particles are dislodged and start rolling and bouncing on the surface. As large particles hit smaller particles, these smaller particles can be suspended in the air. Several measures can be implemented to reduce wind erosion, such as windbreaks (i.e., trees, shrubs, or grasses), increasing surface roughness (i.e., providing creases in the land of 5 to 10 cm in height), and adding soil particles that can bind to the soil.

5.7 Agricultural Activities

Fertilizers and pesticides are compounds increasingly used after World War II to increase crop productivity. Plants need more than 16 basic elements to grow, which are regularly obtained from the soil via mineral weathering, atmospheric inputs, nutrient recycling, and deposition of materials. As a result of continuous and intense cropping, most of these basic elements can be depleted, but N, P, and K are by far the most used nutrients during intensive plant growth, so they are called macronutrients. NPK are often growth-limiting in intense growing regions, so natural and synthetic fertilizers are often added to maintain and increase productivity. In the US, the amounts of N and P emitted into water dramatically increased between 1960 and 1980 due to the increased use of fertilizers. Since then, the amount has plateaued due to lower agriculture profitability, as adding more NPK adds cost without providing additional returns. Excessive amounts of NPK fertilizers create water run-off and pollution due to the solubilization of nitrates and phosphates in water. Figure 5.3 shows the amount of fertilizer (nitrogen, phosphate, and potash) applied in the US since 1960. Figure 5.4 shows the fertilizer products, including nitrogenous, potash, and phosphate fertilizers (including ground rock phosphate) applied per hectare of cropland in 2015 around the world (the most recent year for which the data was available).

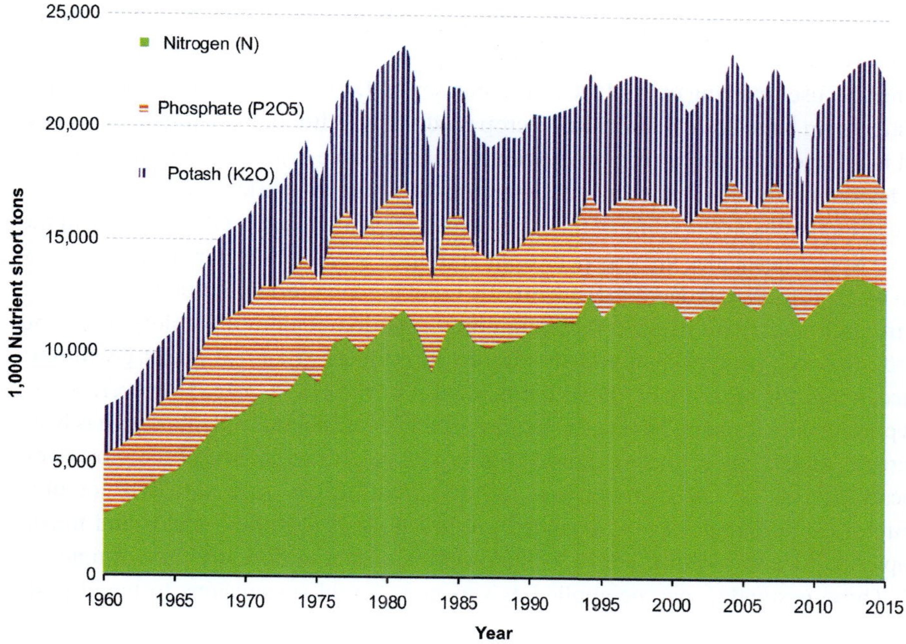

Figure 5.3 Commercial fertilizer use in the US between 1960 and 2015; adapted from [2]. Original data source: USDA, Economic Research Service using data from Tennessee Valley Authority (TVA), Association of American Plant Food Control Officials (AAPFCO), and The Fertilizer Institute (TFI).

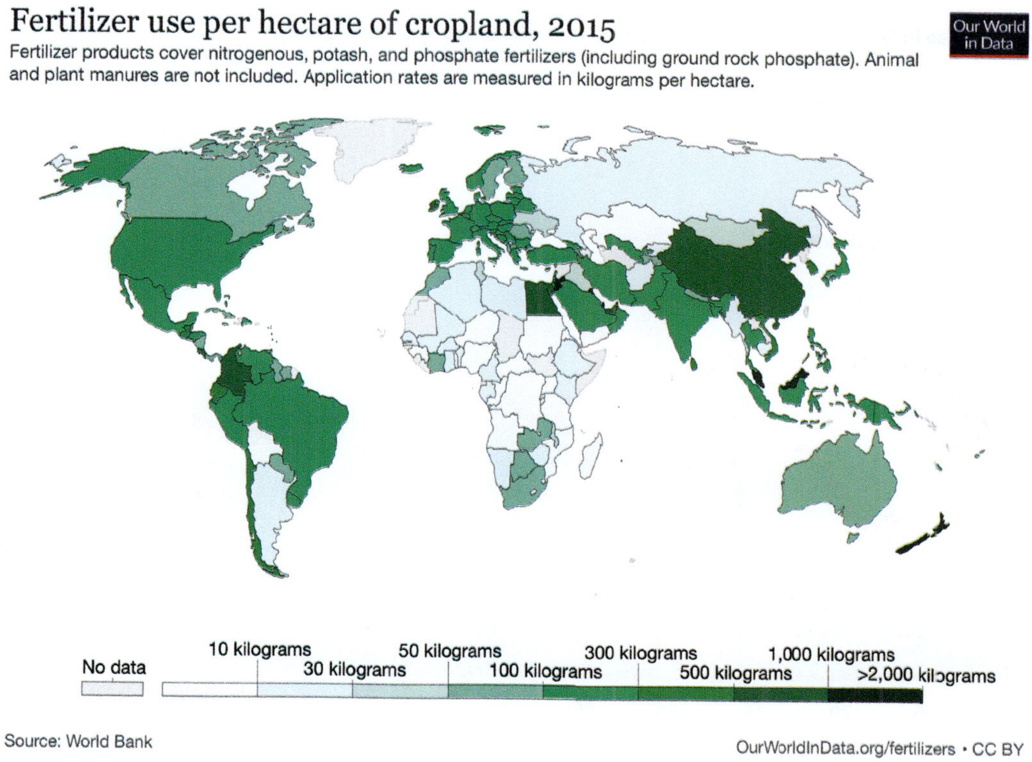

Figure 5.4 Fertilizer use per hectare of cropland, 2015. Fertilizer products include nitrogenous, potash, and phosphate fertilizers (including ground rock phosphate). Animal and plant manures are not included. Application rates are measured in kilograms per hectare; reproduced from Our World in Data license under CC BY 4.0.

Pesticides are defined as any substance used to destroy, prevent, or mitigate insects, rodents, fungi, weeds, or any type of undesirable pests, including plant or insect growth regulators. They have been extensively used starting in the 1940s with the use of DDT to control mosquitoes and the adoption of pesticides in large-scale monocultural crop production. Although the initial use of pesticides resulted in a large increase in crop production, it soon became evident that indiscriminate use of insecticides and herbicides created a new generation of pesticide-resistant insects and plants as well as the persistent presence of some of these pesticides, such as chlorinated chemicals, in the environment [3]. Newer pesticides are less persistent and often soluble in water. They are easily carried away in water run-off and degraded, potentially reducing long-term pollution.

Insecticides are used to control insects. Two commonly used insecticides are chlorpyrifos and malathion. *Herbicides* such as glyphosate and atrazine are formulated to control weeds; glyphosate and atrazine accounted for more than 70% of herbicide use in the US in 2015. However, recent news about glyphosate and atrazine has cast a shadow on these compounds. Glyphosate has been reported to increase cancer risk in those exposed to it by 41% [4]. Atrazine, one of the most commonly used pesticides in the US, can cause sexual abnormalities in frogs and is a potential endocrine disruptor [5, 6]. *Fungicides* are designed to control fungi, including molds and mushrooms. Chloropicrin, metam-sodium, and 1,3-dichloropropene are commonly used fungicides applied to soil. These pesticides may be applied as vapors or liquid suspensions to target pests through different media such as leaves or roots.

All pesticides must be registered with the US EPA, and the Federal Insecticide, Fungicide, and Rodenticide Act (FIFRA) regulates their use [7]. The US EPA is also authorized by the Federal Food, Drug, and Cosmetic Act (FFDCA) to establish tolerance limits for pesticide residues in raw

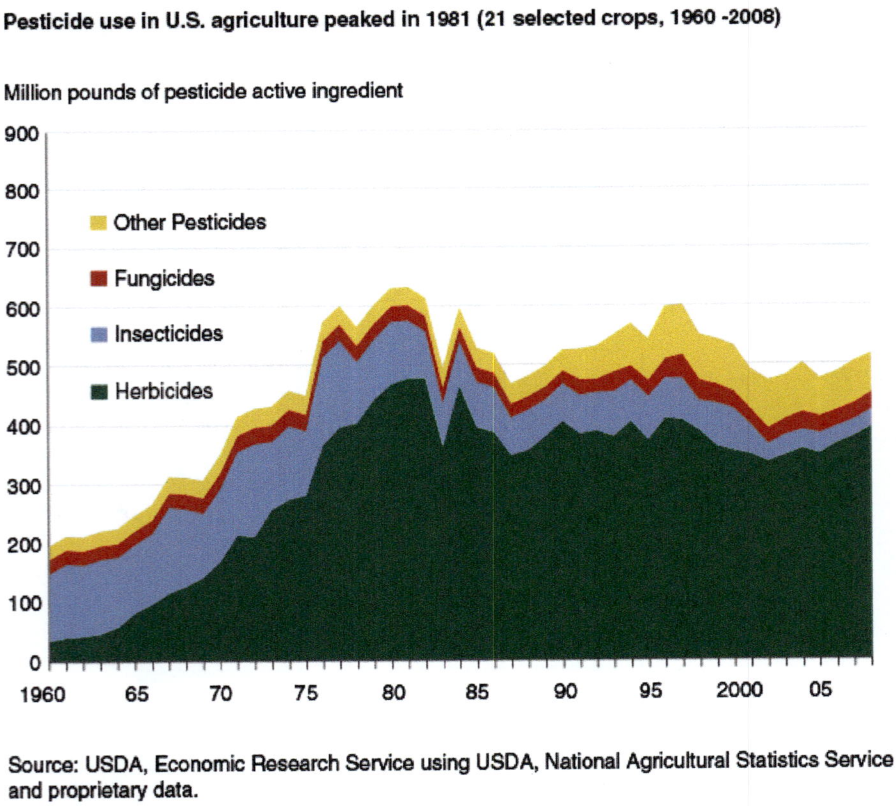

Figure 5.5 Historical annual amount of pesticides in the US since 1960 in selected 21 crops; reproduced from [8].

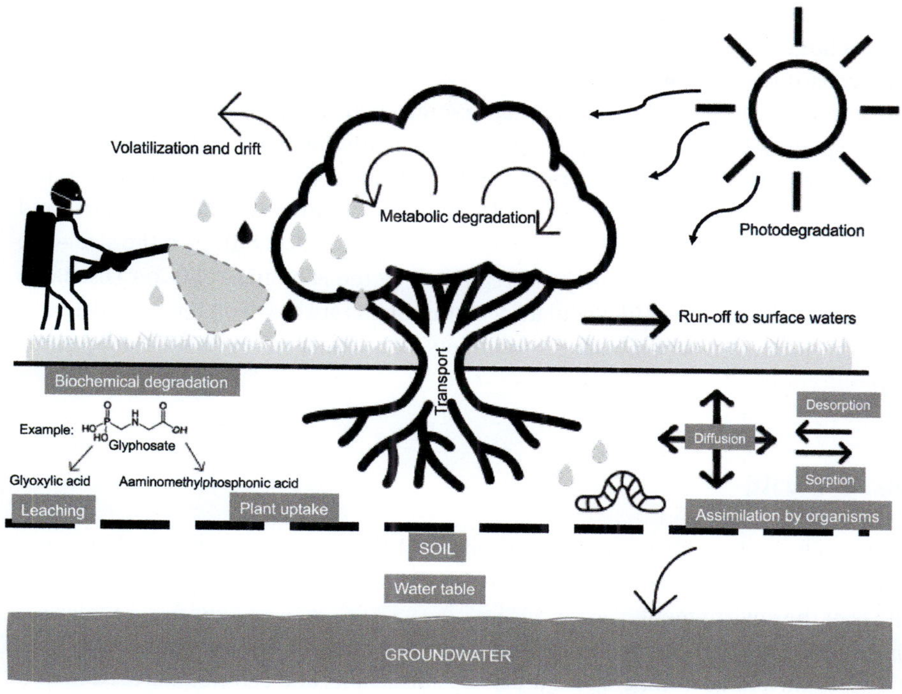

Figure 5.6 Fate of pesticides applied to a field; adapted from [7].

and processed foods. The US FDA enforces the specified tolerances. States like California and Arizona have also compiled lists of different pesticides and sometimes have stricter criteria for their states. Most of the pesticides in the US are used in agricultural production. Figure 5.5 shows the amount of pesticide uses increasing after 1960, but it leveled off in the 1980s. Due to better industrial practices, the amount of pesticide used in agricultural production per kilogram of the product has been reduced; however, the amount of pesticide used at home has increased.

Several fates can result after a pesticide is applied to a field and the soil depending on its persistence and mobility, as illustrated in Figure 5.6. A pesticide may be lost to the atmosphere, enter the soil, may be taken up by the plant, be degraded into other forms, or be leached into the groundwater. So, to understand the contamination potential of a pesticide, it is crucial to understand its persistence and mobility. For example, a persistent pesticide with large mobility produces a large concern for groundwater contamination. Pesticides with lower persistence and medium mobility may be safer and more efficient since they can be active against the targeted pest and not accumulate in the environment. For example, glyphosate and atrazine have different relative mobility and persistence in soil.

5.8 Animal Waste

As increasing amounts of animal protein are demanded by society, animal wastes are becoming a more prevalent land pollutant, increasing the amount of antibiotics, growth hormones, and pesticides applied to land. Animal agricultural wastes can be derived from range and pasture production and CAFO. Range and pasture production, in general, produces much more diffuse and dispersed releases of animal wastes. However, CAFOs where larger number of animals are being raised, such as feedlots, diaries, swine operations, poultry houses, and intensive aquaculture, can discharge large amounts of animal waste in a small land area, requiring systems such as large waste-storage ponds to discharge and treat the wastes.

The production of large numbers of animals on a small land base can result in the stockpiling of wastes due to the construction of large waste-storage reservoirs, as well as waste applications to land in excess of agronomic crop needs. CAFOs and, to some degree, pasture production are considered point source pollution since concentrated pollutants are discharged in a small land region. The most common point source pollution originates in milksheds and barns, dairy and other processing plants, CAFOs, anaerobic and aerobic lagoons, evaporative storage reservoirs, and some intensive aquaculture systems. For the US EPA, some CAFOs – depending on the size – are designated as point source pollutants, so they are regulated by the National Pollution Discharge Elimination System (NPDES) permits [9]. Regulatory requirements are based on the Best Available Control Technology (BACT). Nonpoint source polluters are regulated by Best Management Practices (BMPs).

5.9 Industrial Waste

Several industries, such as oil drilling and fracking, and coal-burning electric power plants, produce large amounts of salts and organics. All these activities release pollutants in different amounts and different areas of the environment.

The oil drilling and fracking process require powerful drill rigs, which drilling fluids must lubricate. These fluids contain a large number of high-density surfactants and compounds used in drilling to facilitate extraction, and they tend to stay in the ground after drilling. So, old oil fields may contain large amounts of these compounds. Nowadays, gas fracking uses a large amount of water and chemicals to facilitate oil and gas extraction. More than 62 different chemicals are used for fracking purposes. Up to 2% of the pumped fluids are chemicals such as zirconium oxide, hydrochloric acid, peroxydisulfate, glutaraldehyde, acetic acid, tetramethyl ammonium chloride, methanol, guar gum, polyacrylate, aromatic hydrocarbons, and isopropanol. All these chemicals provide different functions during extraction, and not all of them are fully recovered [10]. The main concern with these chemicals is groundwater contamination leading to human exposure and health issues.

Coal-burning electric power plants produce fly and bottom ashes (see Chapter 15). When properly collected, fly and bottom ashes are not considered pollutants, and they can be stored in landfills or used for road construction or land recovery. Despite stricter regulations and gas scrubbing technology, sulfur dioxide and nitrous oxides generated during the incomplete combustion of coal are still being released into the environment. Together with these gases, metals such as mercury and lead can be released if the gas scrubbing system is not adequately maintained and monitored. Significant decreases in these gases and metals emissions have been achieved during the last decade. However, they have not yet been completely eliminated.

Besides the oil and coal industries, several other industries such as chemical, pharmaceutical, paper mills, and food processing generate significant amounts of waste, and stricter controls have helped reduce their emissions. Although this is taking place in the US, low-income countries with less stringent regulations are still suffering from pollution from many of these industries.

5.10 Invasive Species

As the globe is more interconnected, invasive species such as kudzu (an invasive plant species in the southeast US), beavers (originally from the US and now an invasive species in Tierra del Fuego, Argentina), Asian carp (originally from China and now an invasive species in the US), and emerald ash borer (native from Asia and now found in 35 US states) are all examples of species producing damage in foreign environments where a natural predator is not available. A key element of these invasive species is that humans introduce them. In most cases, invasive species take advantage of ecosystems. Some invasive species such as kudzu were purposely introduced to provide erosion control without enough assessment of the possible consequences.

5.11 Case Study 5.1. Kudzu as Invasive Species in the Southern US

Kudzu, a perennial vine originally from Asia mainly growing in subtropical regions of China, Japan, and Korea, was introduced in the US to prevent erosion. However, its introduction has produced devastating environmental consequences requiring millions of USD of remediation. Kudzu has spread rapidly in the southern US, covering more than 150,000 acres, as shown in Figure 5.7.

Figure 5.7 Kudzu covering trees in the South of the US, reproduced from Katie Ashdown license under CC BY 2.0 [11].

5.12 Study Questions

1. Explain the main sources of soil pollutants generated by human activities.
2. Give examples of land pollution associated with surface mining, soil erosion, and animal wastes.
3. Provide two examples of regularly used pesticides.
4. Explain how deforestation affects water quality.
5. What does FIFRA regulate?
6. Describe the fate of pesticides.
7. What is a CAFO? What type of pollutants are created by CAFOs?
8. What are the main pollutants released by the oil and coal industries?
9. Provide examples of invasive species.
10. Insecticides and fungicides are considered pesticides T F
11. How are airborne pollutants controlled in coal-burning electric power plants?

5.13 Additional Resources

1. Contaminated Land
https://www.epa.gov/report-environment/contaminated-land
2. Federal Insecticide, Fungicide, and Rodenticide Act
https://www.epa.gov/laws-regulations/summary-federal-insecticide-fungicide-and-rodenticide-act
3. Mining
https://www.pollutionissues.com/Li-Na/Mining.html
4. Animal Feeding Operations (AFOs)
https://www.epa.gov/npdes/animal-feeding-operations-afos
5. Deforestation
https://www.un.org/sustainabledevelopment/blog/tag/deforestation
6. Soil erosion
https://www.scientificamerican.com/article/only-60-years-of-farming-left-if-soil-degradation-continues
7. Fracking
https://www.bbc.com/news/uk-14432401
8. Hydraulic Fracturing Study Fact Sheets
https://www.epa.gov/hfstudy/hydraulic-fracturing-study-fact-sheets
9. Invasive Species
https://www.nwf.org/Educational-Resources/Wildlife-Guide/Threats-to-Wildlife/Invasive-Species
10. The silent drama of photography, Sebastião Salgado
https://www.ted.com/talks/sebastiao_salgado_the_silent_drama_of_photography#t-721

References

1 Brusseau, M.L., Pepper, I.L., and Gerba, C.P. (2019). *Environmental and Pollution Science*, 2e. London, UK: Academic Press Publications.
2 US Environmental Protection Agency. (2022). *Fertilizer Applied for Agricultural Purposes*. https://cfpub.epa.gov/roe/indicator.cfm?i=55. (accessed August 16, 2019).

3 Carson, R., Darling, L., and Darling, L. (n.d.). *Silent Spring*. Cambridge, Mass: Houghton Mifflin Company.

4 Zhang, L., Rana, I., Shaffer, R.M., Taioli, E., and Sheppard, L. (2019). Exposure to glyphosate-based herbicides and risk for non-Hodgkin lymphoma: A meta-analysis and supporting evidence. *Mutation Reviews in Mutation Research* 781: 186–206. DOI: 10.1016/J.MRREV.2019.02.001.

5 Omran, N.E. and Salama, W.M. (2016). The endocrine disruptor effect of the herbicides atrazine and glyphosate on *Biomphalaria alexandrina* snails. *Toxicology and Industrial Health* 32: 656–665. DOI: 10.1177/0748233713506959.

6 Kucka, M., Pogrmic-Majkic, K., Fa, S., Stojilkovic, S.S., and Kovacevic, R. (2012). Atrazine acts as an endocrine disrupter by inhibiting cAMP-specific phosphodiesterase-4. *Toxicology and Applied Pharmacology* 265: 19–26. DOI: 10.1016/J.TAAP.2012.09.019.

7 US Environmental Protection Agency. (n.d.). *Summary of the Federal Insecticide, Fungicide, and Rodenticide Act*. https://www.epa.gov/laws-regulations/summary-federal-insecticide-fungicide-and-rodenticide-act (accessed August 16, 2019).

8 Fernandez-Cornejo, J., Osteen, C., Nehring, R., and Wechsler, S.J. (n.d.). *USDA ERS – Pesticide use peaked in 1981, Then trended downward, driven by technological innovations and other factors*. https://www.ers.usda.gov/amber-waves/2014/june/pesticide-use-peaked-in-1981-then-trended-downward-driven-by-technological-innovations-and-other-factors (accessed August 16, 2019).

9 US Environmental Protection Agency. (n.d.). *National Pollutant Discharge Elimination System (NPDES)*. https://www.epa.gov/npdes (accessed August 3, 2019).

10 Kargbo, D.M., Wilhelm, R.G., and Campbell, D.J. (2010). Natural gas plays in the Marcellus Shale: challenges and potential opportunities. *Environmental Science and Technology* 44: 5679–5684. DOI: 10.1021/es903811p.

11 Ashdown, K. (2021). https://search.creativecommons.org/photos/df44e9ea-344b-4243-bc33-258b2c8ac4c8 (accessed August 10, 2021).

6

Water Pollution

6.1 Learning Objectives

After learning from this chapter, you are expected to:

- Define groundwater and surface water.
- Identify groundwater and surface water pollutants and their main sources.
- Discuss factors affecting groundwater vulnerability to contamination.
- Distinguish between point and diffuse source contamination of groundwater.
- Describe the main body of regulations about water pollutants.

6.2 Introduction

Water covers much more of the earth's surface than does land. The movement of water around the world through evaporation, condensation, precipitation, interruption or interception, infiltration, percolation, transpiration, runoff, and storage is called the hydrologic cycle (Figure 6.1). Table 6.1 shows a detailed description of the physical processes of water movement. This water cycle has taken millions of years to form, but it has been altered due to current human influences such as global climate change, desertification, and excessive groundwater pumping. The evaporation of water moderates temperature. Clouds made of water vapor filter solar radiation, and wind moves water throughout the world, allowing precipitation to release the water from the clouds and allowing life to flourish. The latent heat of water (i.e., the energy released or absorbed as water changes stages) serves to moderate temperatures around the globe, allowing humans and other organisms to thrive. Many processes in the hydrologic cycle help to purify water of contaminants, such as condensation, precipitation, and filtration through porous soils. These processes maintained the water quality before human activities. After the industrial revolution, human impact reached such a large scale that it impacted the ability of natural systems to cleanse water.

Furthermore, many of these new pollutants are persistent. Therefore, the distribution and movement of these pollutants in the hydrologic cycle are important. Historically, the main mechanism to control pollutants was through dilution, although this is no longer a feasible long-term approach.

Global freshwater comprises around 2.5% of all water, while 97.5% is saltwater. Freshwater is divided into 68.6% in glaciers and permanent snow cover, 30.1% as groundwater, soil moisture,

Life Cycle of Sustainable Packaging: From Design to End of Life, First Edition. Rafael A. Auras and Susan E. M. Selke.

Companion Website: www.wiley.com/go/Auras/lifecycleofsustainablepackaging

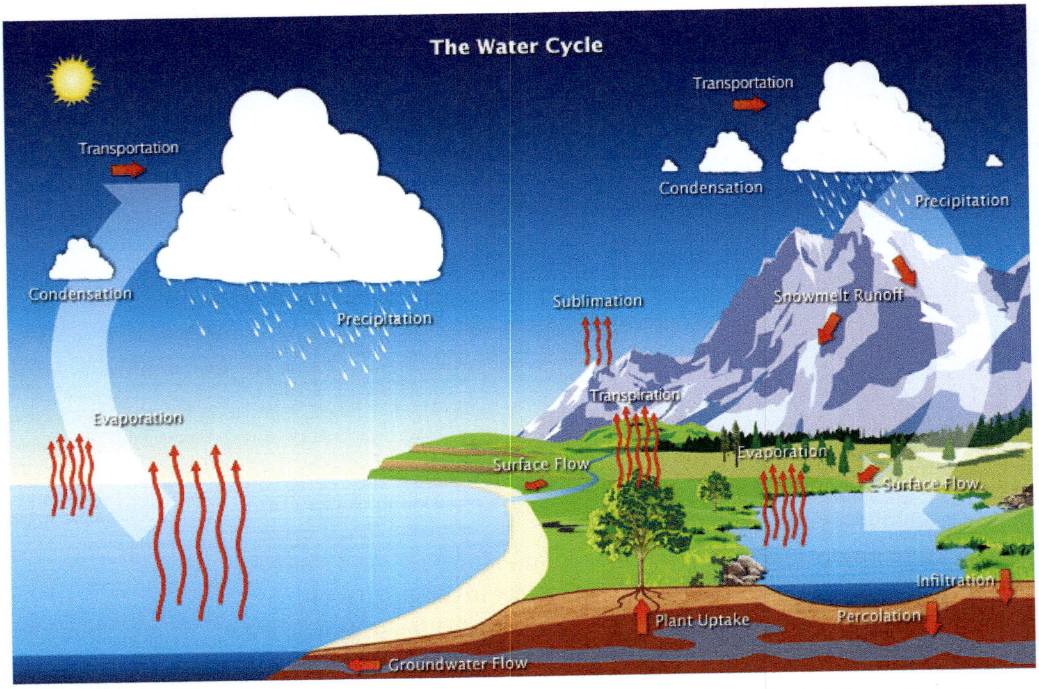

Figure 6.1 The hydrologic or water cycle, image courtesy of AIRS, the Atmospheric Infrared Sounder, a spaceborne instrument on board NASA's Aqua satellite, licensed under CC BY 2.0.

Table 6.1 Description of the major contributing physical processes of water movement.

Physical Process	Definition
Evaporation	Change of the physical state of water from a liquid state to a gaseous state. Around 600 calories of energy are released for each gram of water during this transition.
Condensation	Change of the physical state of water from vapor to most commonly liquid. Examples are the formation of dew, fogs, and clouds.
Precipitation	Fall of water from the atmosphere to the ground such as water, ice, and snow.
Interception	Process of interfering with the movement of water in the transportation flow events leading to streams such as by vegetation cover, reservoir, and land obstructions.
Infiltration	Movement of water between the atmosphere and the soil.
Percolation	Movement of water though the soil, and its layers, by gravity and capillary forces.
Transpiration	Biological process where water inside plants is transferred to the atmosphere.
Runoff	The flow of water from a drainage basin appearing in the surface streams.
Storage	There are three main locations for storing water: in the atmosphere, on the surface of the earth, and in the ground.

swamp water, and permafrost, and 1.3% as surface/other freshwater such as lakes and rivers. Figure 6.2 shows the distribution of Earth's water. Approximately 110,000 km^3 of rain, snow, and ice falls annually on land surfaces, replenishing freshwater resources.

Freshwater can be divided into two main sources:

- *Groundwater* located beneath the ground surface in soil pore spaces or fractures is a critical source of potable water, supporting food and crop production, and used in a multitude of industrial activities.

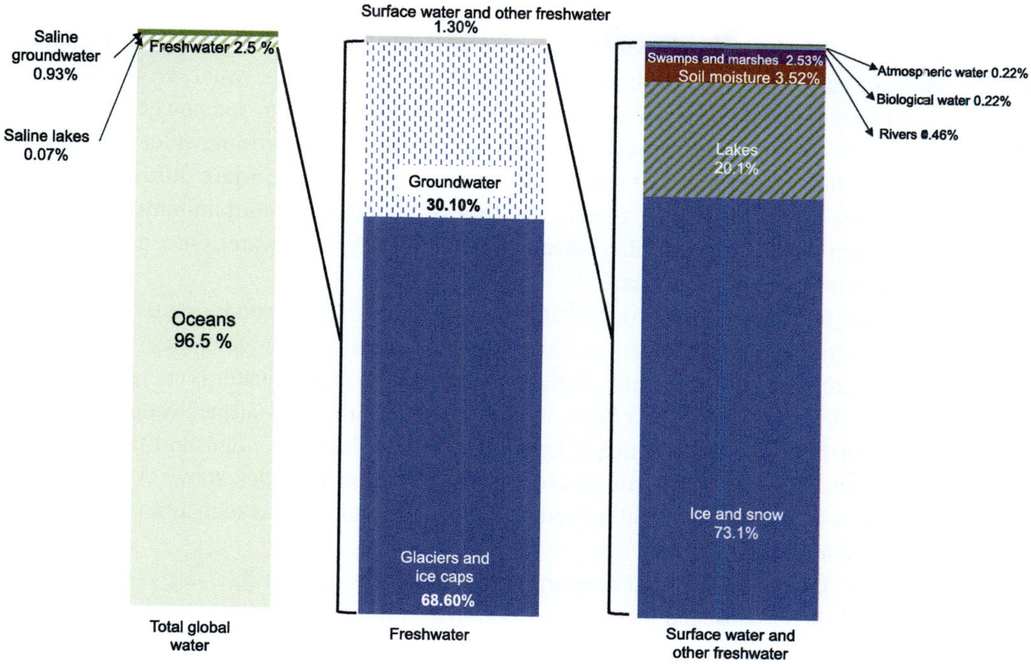

Figure 6.2 Distribution of Earth's water. Initial data source: Igor Shiklomanov's chapter "World fresh water resources" in Peter H Gleick, Ed (1993); figure adapted from [1].

- *Surface water* is collected on the ground or in streams, rivers, lakes, wetlands, or oceans, and it is a crucial resource for transportation and recreation.

The limited supplies of surface water and groundwater receive substantial amounts of pollutants, mostly produced by humans [2]. Water pollution is a term that describes the situation in which contaminants impede the intended use of the water. Pollution of groundwater and surface water is very different since the initial pollutants, their dispersion in the water stream, and the travel of pollutants in each water type are very different. Retention of pollutants in lakes is, for example, around 100 years, but rivers and streams have a much shorter retention time. So, persistent and low movement pollutants could remain trapped and intact for several decades in lakes until they are transported to other areas such as oceans. On the other hand, the retention time of groundwater can be measured in centuries to millennia. So, persistent pollutants can remain intact for long periods since transformation (i.e., the change from one constitutional stage to another) can be constrained. Pollutants in water streams such as lake, rivers, and oceans can then be transferred to the food chain by filter-feeding organisms or sequestered in cold, deep sinks where they are resistant to transformation. Around 40% of the world population lives in coastal areas (i.e., within 100 km of the coastline), so oceans are especially vulnerable to the introduction of pollutants coming directly from surface water and groundwater from the population living around these coastal areas [3, 4].

6.3 Groundwater

Groundwater is a critical resource for potable water, supporting food and crop production, and is extensively used by industry. Groundwater has increasingly been used throughout human history to provide a reliable water source. Wells have been dug since prehistoric times. The Romans and the Arabs created large water distribution systems for groundwater and surface water. Groundwater represents around 15% of the potable water in Australia, 75% in Europe, 29% in central and South America, and 51% in the US [2]. Furthermore, besides supplying potable water, groundwater is used for agricultural applications, such as crop irrigation, industry, mining, and power generation. Groundwater can be contaminated by several routes. Two main routes of groundwater contamination are pollutants introduced by *point sources* and pollutants introduced more diffusely by *non-point sources* (Figure 6.3). Factors such as residence times (i.e., the time a contaminant stays in the water stream) are critical, and dilution effects are less significant in groundwater. Once groundwater is contaminated, it is challenging to clean.

The major groups of pollutants for groundwater are hazardous organic and inorganic chemicals, agrochemical pollutants, and salinization; minor groups include pathogens such as viruses, bacteria, and protozoa; and gasoline and solvent additives. When groundwater is contaminated, it is very difficult to clean since dilution effects such as occur in air and in surface water are much less significant. Furthermore, the absence of light and UV radiation is an important factor, eliminating photochemical reactions in groundwater. Solar radiation provides some of the energy needed for chemical transformation and reduction in aboveground systems such as air and surface water, and it is absent in groundwater.

Since preventing all contaminants from entering groundwater is not possible, risk assessment, as explained in Chapter 4, is a particular focus when it comes to groundwater pollution. The main goal is to assess the risk posed by an activity or event to groundwater resources. Groundwater pollution risk is assessed as two main components: groundwater vulnerability and the possible contamination load. Groundwater vulnerability is defined as the likelihood of an aquifer becoming contaminated, which is

Figure 6.3 Sources of groundwater contamination, reproduced from [5].

Table 6.2 Factors affecting groundwater vulnerability to contamination; adapted from [2].

Factor	Increased Vulnerability	Decreased Vulnerability
Depth of groundwater	Shallow	Deep
Soil type	Well drained (sandy)	Poorly drained (high clay, organic matter content)
Vadose zone physical properties	Preferential flow channels	Horizontal low-permeability values
Recharge	High precipitation, high infiltration	Low precipitation, low infiltration
Subsurface attenuation processes	Minimal attenuation	Significant attenuation

affected by several factors as shown in Table 6.2. Aquifers that are near the ground surface, covered by sandy soil, and located in an area of high precipitation are more vulnerable to contamination than aquifers that are hundreds of meters below the soil surface and located in areas of low precipitation.

The type and concentration of contaminants released, and the timescale and mode of release, affect the contaminant loads in groundwater. The potential for contamination and the magnitude of contamination are controlled by the contaminant transport and fate behavior. Contaminants readily soluble in water and with low sorption in other materials without being modified to a large extent (i.e., persistent pollutants) have large potential to pollute groundwater; examples include DDT and per- and polyfluoroalkyl substances (PFAS). Most of the contamination enters closest to the ground surface, so contaminants must travel through the vadose zone (i.e., the part of Earth between the land surface and the zone where groundwater is located still at atmospheric pressure) to the water table. Processes such as sorption and biodegradation help to decrease and limit the transport of contaminants. For this reason, the vadose zone is known as the "living filter" zone. Table 6.3 shows the main transport and attenuation properties of major subsurface contaminants.

So, the risk of groundwater contamination or pollution is a combination of the intrinsic vulnerability of the aquifer and contaminant load factors (i.e., an aquifer with high vulnerability may not have a high risk of pollution if the contaminant concentration remains low, and *vice versa*). High pollution risks are associated with aquifers with high vulnerability and high contaminant loading. Since not much can be done to alter the vulnerability of an aquifer, controlling the contaminant load by preventing and managing pollution is preferred. Human activities that generate a large concentration of contaminants should not be located near vulnerable aquifers. Aquifer vulnerability maps have been constructed to help assess the chance of pollution. A common tool to create aquifer vulnerability maps is through the DRASTIC model, developed by the US EPA [7]. DRASTIC stands for Depth to water, net Recharge, Aquifer media, Soil media, Topography, Impact of vadose zone, and hydraulic Conductivity. It is a seven-parameter model, which assigns vulnerability scores to each area. Figure 6.4 shows an example of an aquifer vulnerability map.

6.3.1 Point-Source Contamination

Point-source contamination refers to localized sources of contaminants, such as waste-injection wells and leaking storage tanks. The location where the contamination occurs is known as the source zone, which is near the contamination release point. The zone where the contaminant is released in the groundwater is called the dissolved plume (see Figure 6.5), and it is often much

Table 6.3 Transport and attenuation characteristics of major subsurface contaminants; adapted from [2, 6].

		Attenuation Mechanism							
Contaminant	**Source**	**Biochemical Degradation**	**Sorption**	**Filtration**	**Precipitation**	**Permitted Drinking Water Concentration**	**Mobility**	**Persistence**	**Potential to Develop Extensive Groundwater Plume**
Pathogens	Sewage	3	3*	4	1	Very low (<1 per 100 mL)	Low	Low – moderate	Low
Nitrogen (N)	Agriculture, sewage	2	2	1	1	Moderate (10–20 mg NL^{-1})	Very high	Very high	High
Chloride (Cl)	Sewage, industry, road deicer	1	1	1	1	High	Very high	Very high	High
Sulfate (SO_2)	Road-runoff, industry	2**	2	1	2	High	High	High	High
Dissolved organic carbon (DOC)	Sewage, industry (esp. food processing, textiles)	1	4	1	3	Non controlled	Moderate	Low – moderate	Moderate

Heavy metals	Industry	2	2	1***	1	Low (variable)	Generally low unless pH low (except Cr [VI])	High	High
Halogenated solvents	Industry, commercial	2	2	1	1	Low (5–30 μgL^{-1})	High	High	High
Fuels, lubricants, oils, other, hydrocarbons (LNAPLs)	Fuel station, industry	4	3	1	1	Low (10–700 μgL^{-1} BTEX$^{\varnothing}$)	Moderate	Low – moderate	Moderate
Other synthetic organics	Industry, sewage	Variable	Variable	2	2	Low (variable)	Variable	Variable	Variable

Note 1: no attenuation; 2: some attenuation; 3: significant attenuation; 4: highly attenuated

*sorbed ammonia; **can be reduced; ***where occur as organic complexes; BTEX$^{\varnothing}$: benzene, toluene, ethylbenzene and xylene; aromatic compounds with health guidelines limits; LNAPLs: light, non-aqueous phase liquids; Cr[VI]: chromium VI.

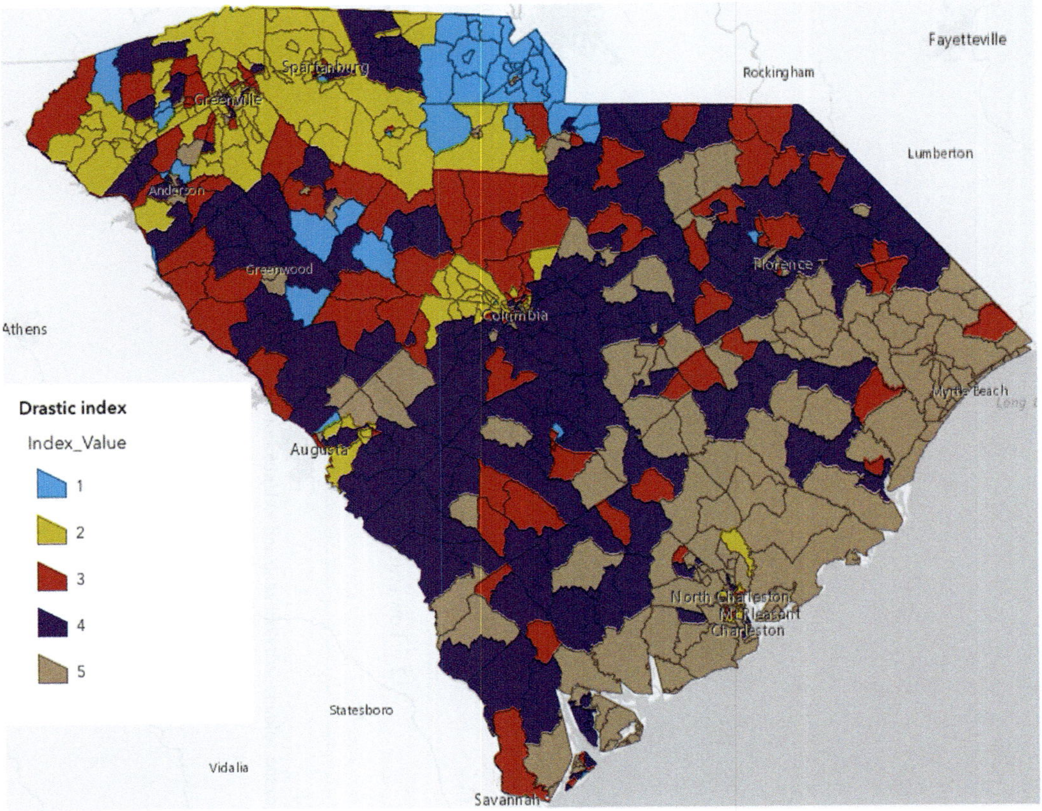

Figure 6.4 Drastic Index of the state of South Carolina. Data from 2010 Census tracts of South Carolina categorized by the DRASTIC index. DRASTIC index values range from 1 to 5, with 1 being the least vulnerable to groundwater pollution, and 5 being the most vulnerable to groundwater pollution; adapted from [8].

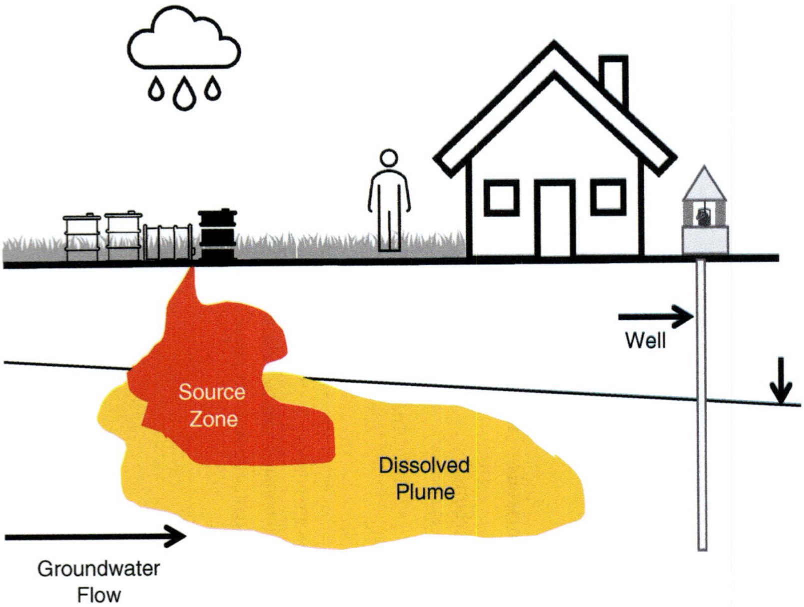

Figure 6.5 Distribution of contaminants in point-source groundwater contamination; adapted from [2].

larger than the source zone. The structure and architecture of the source zone are key to the pollution risk posed by these sites. The groundwater contamination plume is one of the main risks of these zones for human contamination.

6.3.1.1 Hazardous Organic Chemicals

Hazardous organic chemicals are one of the main risks and issues associated with groundwater pollution in the US and many industrialized countries, since when these chemicals are introduced to the groundwater streams, they are difficult to remove. Table 6.4 shows an example list of organic compounds that are prevalent in contaminated groundwater sites. Underground fuel storage tanks are one of the main point sources of contaminants that have been found in the US. One of the main concerns with these contaminated sites is that the contaminants may become trapped in pore spaces. Once chemicals are entrapped, they are difficult to remove, so they are a long-term source of subsurface contamination as they dissolve in the groundwater or soil-pore water structure, greatly increasing the cost of remediation.

The presence of liquids that are not miscible with water is usually the single most significant factor limiting the cleanup of sites contaminated with organic chemicals. The density of the organic chemical is one of the main factors. Organic chemicals denser than water (Dense, Non-Aqueous Phase Liquids – DNAPL) can sink below the water table; on the other hand, organic chemicals less dense than water (light nonaqueous phase liquids – LNAPL) float in the water table. The dissolved plume shape (i.e., distribution and amount of organic chemicals in the groundwater flow) is mostly affected by immiscible contaminants. Table 6.3 shows the general major classes of subsurface contaminants and the potential to form a plume. For example, chlorinated solvents create plumes that are hundreds to thousands of meters long since the compounds are very difficult to biodegrade. However, plumes generated by fuel hydrocarbons are much smaller since they are more easily biodegraded.

6.3.1.2 Landfill

One of the main sources of groundwater contamination is landfills, mostly aging landfills. Figure 6.6 shows the structure of a modern closed landfill and the groundwater monitoring system composed of wells used to evaluate contamination from leachate. Most of the contaminants are released when water enters the landfill and contacts the waste materials. The infiltrated water can carry with it both

Table 6.4 Major classes of hazardous organic chemicals of significance for groundwater pollution; adapted from [2, 6].

Class	Example Compounds	Sources/Uses	Properties
Chlorinated solvents	Trichloroethene Tetrachloroethene	Manufacturing Degreasing Dry cleaning	DNAPL
Coal tar, creosote	Polynuclear Aromatic Hydrocarbons Phenols	Coal gasification Wood treatment	DNAPL
Hydrocarbon fuels	Benzene	Fuel	LNAPL
Polychlorinated biphenyls	Aroclor	Transformer fluid	DNAPL

DNAPL: dense, non-aqueous phase liquids, denser than water; LNAPL: light, non-aqueous phase liquids, less dense than water.

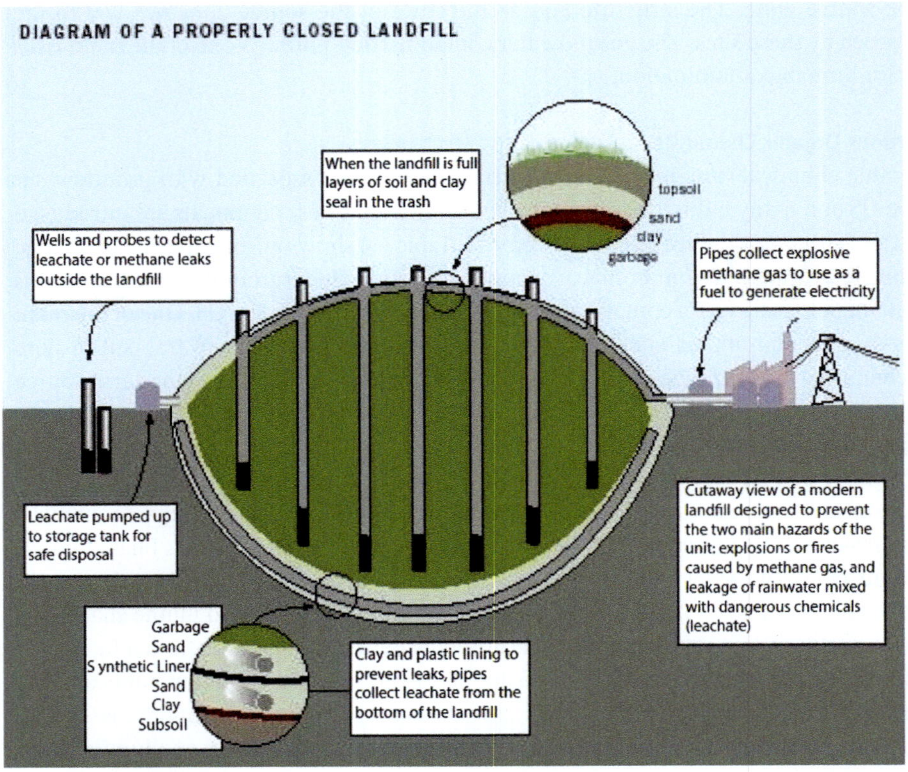

Figure 6.6 Modern landfill site and groundwater monitoring system with a single composite liner system and final cover system; reproduced from the US EPA [9].

hazardous and nonhazardous chemicals. If these chemicals are not captured by the leachate treatment systems, they can percolate through the clay and porous structure between the landfill and the water table. Table 6.5 shows a number of contaminants and their concentrations that can emanate from a landfill. Since leachate can leak to some degree from all landfills, it is crucial to implement a well-designed monitoring system to control for landfill problems. There are a number of factors such as pH, salt concentration, and total dissolved solids that can be used to evaluate for leachate contamination in wells around the landfill.

The US EPA now requires that a well-designed landfill should contain all the disposed materials and contain any potential leachate to the environment as explained in the soil pollution chapter.

6.3.2 Diffuse Source Contamination

Contamination resulting from nonpoint sources (i.e., diffuse sources) refers to contamination occurring over a wide area and is mostly associated with particular types of land use, in contrast to point-source discharges. These diffuse sources can create large, contaminated areas; however, the contaminant concentration is typically much lower than in the case of point-source contamination. Examples of diffuse sources are agrochemical contamination and salt-water intrusion.

6.3.2.1 Agrochemical Contamination

The increasing intensity of agricultural production practices due to the need for increased food production to feed people worldwide has greatly increased the use of fertilizers and pesticides,

Table 6.5 Common contaminants found in landfill leachate in the US; adapted from [10].

Category	Description
Group I – Organic matter	Dissolved organic matter, volatile fatty acids, fulvic-like and humic-like compounds
Group II – Inorganic macro components	Calcium (Ca_2^+), magnesium (Mg_2^+), sodium (Na^+), potassium (K^+), ammonium (NH_4^+), iron (Fe_2^+), manganese (Mn_2^+), chloride (Cl^-), sulfate (SO_4^{2-}), and hydrogen carbonate (HCO_3^-)
Group III – Heavy metals	Cadmium (Cd_2^+), chromium (Cr_3^+), copper (Cu_2^+), lead (Pb_2^+), nickel (Ni_2^+), and zinc (Zn_2^+)
Group IV – Xenobiotic organic compounds (XOCs)	Originating from household or industrial chemicals and present in relatively low concentrations (usually less than 1 mgL^{-1} of individual compounds). These compounds include among others a variety of aromatic hydrocarbons, phenols, chlorinated aliphatic, pesticides, and plasticizers
Other compounds found in lower concentration	Originating from household or industrial chemicals and present in relatively low concentrations (usually less than 1 mgL^{-1} of individual compounds). These compounds include, among others, a variety of aromatic hydrocarbons, phenols, chlorinated aliphatic, pesticides, and plasticizers

collectively known as agrochemicals. Nitrate (commonly present in fertilizers, pesticides, and animal wastes) has become one of the most common and pervasive groundwater contaminants in the US and around the world. It has been reported that many private wells around the US have had levels of nitrate several times as high as the maximum permitted by federal drinking water standards. Rural areas generally have more significant groundwater nitrate contamination than do urban areas. Since groundwater is often being depleted in cultivated areas, it is not uncommon to see a concentration effect of nitrate due to declining groundwater levels and high levels of nitrate contamination. Nitrates are highly mobile in the subsurface, so due to the large areas of fertilizer and pesticide application, large groundwater plumes of nitrate contamination can be found. Some of the main ways to deal with nitrate contamination are to drill deeper wells or blend contaminated water with clean water. However, none of these techniques are sustainable in the long term, contaminating more water. Additional agricultural pollutants include herbicides such as atrazine and diethyl atrazine, which are used around the world to control weeds, and pesticides such as diazinon, carbofuran, and carbaryl to control insects and fungal pests in crops such as cereals, corn, soybean, and sugarcane. Figure 6.7 shows a map of atrazine application in the US in 2016 and the main crops where it was used.

Most pesticides are organic compounds, and their fate and transport behavior are linked to their chemical properties. Some old pesticides, such as DDT, were very persistent in the environment. Newer pesticides such as atrazine are designed to be less persistent. However, the variety of types of pesticides, both old and new, that are available (e.g., atrazine for cereals, chloropropane for soft fruit, carbamates for citrus and horticulture, and organophosphorus for many other types of crops) complicates their detection and quantification.

6.3.2.2 Saltwater Intrusion

Saltwater intrusion or salinization is one of the most widespread types of groundwater contamination (see Figure 6.8). As larger amounts of water are extracted through pumped wells on land, saltwater intrusion into the groundwater can occur. Around 3% of saltwater contamination can

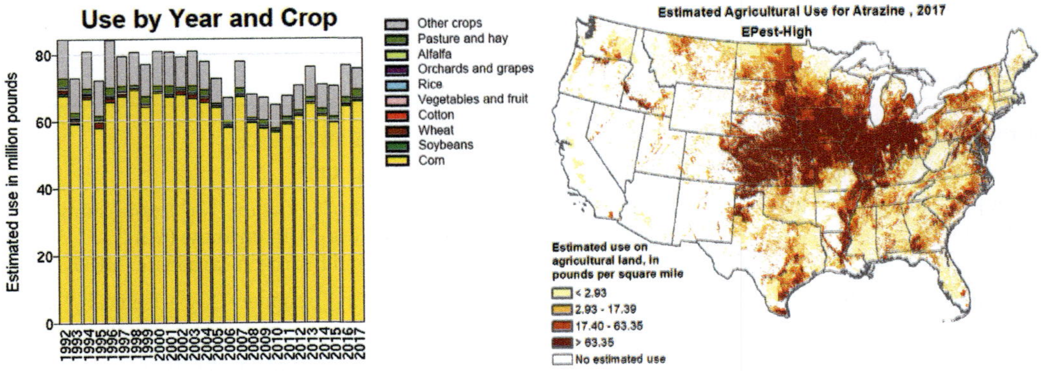

Figure 6.7 (Left) used by year and crop of atrazine; (right) estimated agricultural use of atrazine, 2017, Epest-high, reproduced from the US Department of the Interior [11]. Note: right figure: Epest-high notation estimates include more extensive estimates of pesticide use not reported in surveys, which sometimes include states or areas where use restrictions have been imposed.

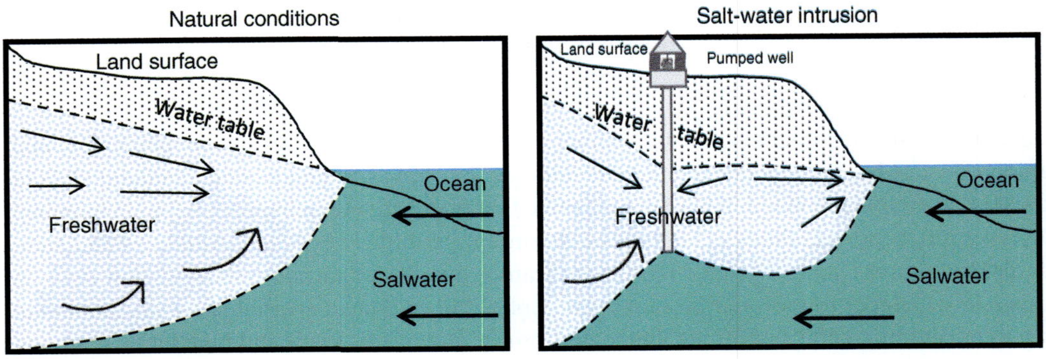

Figure 6.8 Groundwater salinization; adapted from [13].

make groundwater unsuitable for use as freshwater. Once groundwater is contaminated, if positive groundwater recharge conditions (aquifer replenishment) are not established, it will never recover. Coastal regions are most often affected by saltwater intrusion. Still, inland areas can also be affected due to the presence of geological systems where trapped deep-water seas are now part of the continent. These areas are mostly deep, far from the subsurface; however, excessive pumping can create flow from regions with high concentrations of salt or brine water to freshwater aquifers. Although there are no harmful health effects related to low concentrations of chloride coming from salt in drinking water (the US EPA has an unenforceable secondary drinking water limit of 250 mgL^{-1} of chloride [12]), the water supply will be degraded, and people will not find the water acceptable.

In addition to the pollutants discussed above, there are several other groundwater contaminants such as pathogens, gasoline and solvent additives, perchlorate, arsenic, and acid-mine drainage.

6.3.2.3 Microbial Contamination

Microbial pathogens in groundwater, including viruses, bacteria, and protozoa, mostly come from sewage treatment by-products, septic tank systems, and latrines. In general, public water systems are mostly free of pathogen contaminants since the water is treated before release for public consumption. However, water extracted from private wells is highly susceptible to contamination. Proper construction and maintenance of septic and well systems are needed to minimize probable pollution problems.

6.3.2.4 Gasoline Additives

Gasoline additives such as benzene, toluene, and methyl tertiary-butyl ether (MTBE) are common groundwater contaminants deriving from fuel tanks and oil wells. MTBE, an additive used in gasoline for many years that is highly soluble in water, has a dissolved plume that can expand to thousands of meters when it reaches a water stream. MTBE is a volatile, flammable, and colorless liquid at room temperature that dissolves rather easily in water. MTBE was manufactured in very great quantities (over 200,000 barrels per day in the US in 1999) and is mainly used as a fuel additive in motor gasoline. MTBE is mostly used to raise the oxygen content of gasoline. MTBE replaced lead as an octane enhancer (which helps prevent the engine from "knocking"). Between 1992 and 2005, MTBE was used at higher concentrations in some gasoline to meet the oxygenate requirements set by Congress in the 1990 Clean Air Act Amendments [14]. Oxygen helps to complete the burning of the gasoline reducing harmful tailpipe emissions from vehicles. The presence of oxygen dilutes or displaces gasoline components such as aromatics (e.g., benzene) and sulfur. Additionally, oxygen optimizes oxidation during combustion. Most refiners use MTBE over other oxygenates due to its blending features and for cost. Many states such as California and New York have restricted the use of MTBE as a gasoline and fuel additive. Since the use of MTBE is now discouraged as an additive, switching over to the use of ethanol has increased since it provides the needed oxygenation. The US EPA released a Drinking Water Advisory (DWA) regarding MTBE, stating that there was insufficient information to establish firm limits on quantities of MTBE related to health risks [15]. The DWA indicates that there is little likelihood that MTBE will cause adverse health effects at concentrations between 20 and 40 ppb (μgL^{-1}) or below.

One of the main solvent additives found in groundwater has been 1,4-dioxane, commonly referred to simply as dioxane, which is frequently used to stabilize chlorinated solvents such as tetrachloroethene (PCER), trichloroethene (TCE), 1,1,1-trichloroethane (TCA), and paint thinners during the manufacturing process. Dioxane is also used in manufacturing paper, cotton, textiles, automotive coolants, shampoos, and cosmetics. It is fully miscible in water and is resistant to biodegradation in water and soil. A dioxane plume can be larger than the actual initial area of chemical contamination. Dioxane is a likely human carcinogen. Many states have established drinking water and groundwater guidelines with limits below 10 ppb and many below 1 ppb; Alaska allows up to 77 ppb. Methods to detect dioxane are still being developed, and detection at concentrations below 1 ppb is still challenging [16]. The US EPA has not yet declared a national regulatory standard for dioxane.

6.3.2.5 Perchlorate

Perchlorate is another emerging groundwater contaminant. Perchlorate, ClO_4^-, is an inorganic anion often present as a salt. These compounds are mostly used as propellants for their explosive properties, so military purposes are the main application, about 90% of total use. Most perchlorate contamination is around military bases and installations. Perchlorate is highly mobile in groundwater and is very persistent in the environment. The US EPA has found that perchlorate may have an adverse effect on human health, affecting the normal function of the thyroid gland in both children and adults. A new rule for limiting perchlorate to a Maximum Contaminant Level (MCL) and a Maximum Contaminant Level Goal (MCLG) – see Section 6.5 for a detailed explanation of these definitions – between 18 and 90 ppb or μgL^{-1}, respectively, is under consideration [17]. A few states have adopted action levels or advisory levels for perchlorate in the range of 1 to 6 ppb.

6.3.2.6 Arsenic

Arsenic (As) is another highly toxic contaminant in groundwater that can originate in either natural or human activities. Arsenic is a natural element found in soil, rocks, air, plants, and animals, and the inorganic form of arsenic is more predominant than the organic forms. Through erosion, dissolution, and weathering, arsenic can be released into the environment. Human activities such as agriculture, smelting, mining, burning of fuels, paper, glass, and cement manufacturing are potential sources of arsenic. In the US, most of the arsenic originated in use by the wood industry as a preservative for outdoor exposure. Inorganic arsenic in drinking water can create toxic effects after acute or chronic exposure. Chronic exposure to arsenic through water can result in skin cancer, bladder, lung, kidney, nasal passages, liver, or prostate. It can also have noncancerous effects on the cardiovascular, pulmonary, immunological, neurological, and endocrine (e.g., diabetes) systems. The US EPA established MCL of 10 ppb and MCLG of zero for arsenic in drinking water [18].

6.3.2.7 Acid-Mine Drainage

Another source of groundwater contamination to consider is acid-mine drainage. From mine tailings or after a mine is abandoned, rain can discharge highly acidic water. The acidic water can change the oxidation-reduction potential and encourage the release of minerals and metals, as occurred with the Flint water crisis when the water supply was switched to a new water source with a different pH and removed the protective coating in lead pipes in the water supply, exposing the population of Flint, MI, to high concentrations of lead.

Maintaining groundwater resources is part of the United Nations Sustainable Development Goals (SDGs), as discussed in Chapter 1. Goal 6 of the SDGs is defined as access for the world population to clean, accessible water for all [19]. One of the UN targets by 2030 is "to achieve universal and equitable access to safe and affordable drinking water for all." The amount of groundwater extracted should not exceed the amount replenished through recharge to achieve this target. Managing and preventing groundwater contamination is one of the main steps for achieving this target.

6.4 Surface Water

Fresh surface water is mostly located in streams and lakes. Oceans contain most of the saline surface water. So, fresh surface water is a scarce and valuable resource. The control and study of surface water have mostly centered on streams and lakes. Surface water is subjected to heavy stress, and it is easily contaminated since it is used as the water supply for agriculture and industry, as well as for recreational activities.

6.4.1 Marine Water Resources

Oceans contain most of the water on the planet, but they can easily be contaminated. Ocean environments are generally divided into coastal zones and open ocean waters, further subdivided into the upper mixed layer and the abyssal oceanic layer.

The coastal zone extends from the low-tide line to the 200-meter depth contour, which generally coincides with continents' geophysical demarcation (i.e., based on locating and visually correlating floors or anomalies to construct the coastal zone). The coastal zone can vary between less than 1 km and 1,400 km from the shoreline, with an average around the world of 50 km, and comprises around 8% of the water in the ocean. Most of all the water-carried waste from the continent enters the coastal zone through estuaries, which are bodies of water with a free connection to the sea, and whose salinity is diluted through freshwater from rivers or streams. Estuaries provide critical environments and habitats for species to raise and feed their young, so water quality is of utmost importance and concern. Since estuaries must be preserved, treated wastewater effluents are usually discharged offshore through a large pipeline called an outfall. The design of outfalls must consider water movement and flow in and out of the estuaries and the coastal zone.

The open ocean waters include two main components: the upper mixed layer, which varies between 100 and 1,000 m in depth, and the abyssal ocean layer, which reaches the bottom of the ocean. Contamination of the open ocean waters can be caused by atmospheric fallout, oil spills, and dumping of hazardous wastes and sewage sludge, among others. The mixing of water between the upper and deeper layers of the ocean is hampered by density gradients. Particles and contaminants discharged in the upper mixed layer will not move to the deep layers if the density is insufficient. Particles with high density can be transferred and resuspended in the abyssal ocean. Since most of the ocean's life lives in the upper mixed layer, precautions must be taken to avoid contamination of this layer.

6.4.2 Sources of Water Pollution

It can take only a minimal amount of a pollutant to render a water stream unfit for drinking; however, this water may remain satisfactory for other uses such as irrigation. Major sources of surface water contaminations include construction, logging, overgrazing by domestic animals, AFOs or CAFOs, urbanization and road construction, and mining operations. Natural sources can also pollute water, such as plant growth and decay and gases from volcanic activities.

6.4.3 Sediments as Surface Water Contaminants

The presence of suspended solids and turbidity are signs of water contamination. Anthropogenic activities have been the main reason for water pollution since the 1900s. Increasing amounts of sediments have been released to watersheds, often blocking and impounding sediment behind dams. Almost any type of human activity increases the amount of sediment released to a watershed. Logging, deforestation, and wildfires increase sediments by at least twofold for short periods. Overgrazing by domestic animals can reduce vegetation, decreasing its ability to prevent or reduce erosion. Urbanization and road construction can result in a 5- to 20-fold increase in suspended sediments. Impervious materials such as pavement, parking lots, or rooftops can increase the velocity of stormwater runoff and increase erosion. Mining operations such as strip mines can bring extraordinary levels of erosion and subsequent sedimentation in streams. Roughly, rivers discharge 40 million m^3 of water into the world's oceans, and for every 1 m^3 of water reaching the ocean, there is 0.5 kg of sediment removed from the land [2]. The flow of sediments depends on many factors such as particle size, water flow rate, and density. Suspended sediment is a major carrier of pollutants, such as heavy metals, organics, pathogens, and nutrients responsible for eutrophication.

Total suspended solids (TSS) in water are considered as one of the most important water pollutant indicators. TSS levels vary widely around the rivers of the world. TSS is defined as all solids suspended in water that will not pass through a 2.0 um glass fiber filter. Excessive amounts of TSS result in a water reservoir not being able to support life. Furthermore, TSS absorbs heat and increases the temperature of the water while also decreasing the amount of oxygen due to the consumption of organic matter by respiring bacteria. However, the major effect of increasing TSS is indirect due to the pollutants attached to them, such as nutrients, metals, organic compounds such as polycyclic aromatic hydrocarbon (PAH) and PCB, herbicides, and pesticides. Turbidity is related to TSS, but it is not a proxy for TSS measurements.

6.4.4 Metals as Surface Water Contaminants

Metals such as zinc, copper, and chromium are common nutrients for human life; however, they can also be toxic when present in particular oxidation states and concentrations. There are 50 known metals that can be toxic to humans. Some of the common metals known to be toxic are mercury, lead, arsenic, cadmium, copper, and zinc. Heavy metals are environmentally persistent, so their toxicity increases as they accumulate in the environment under certain conditions.

6.4.4.1 Mercury

Mercury (Hg) is one of the well-known pollutants heavily used in the tanning industries in the early 1900s. Currently, the main source of Hg as a pollutant is the burning of coal in power plants. Although many forms of Hg exist, methylmercury ($CH_3Hg^+X^-$; where X stands for an anion) is the form of Hg that is most readily absorbed and bioaccumulated. In fish, Hg as methylmercury can bioaccumulate (i.e., accumulate through time in living organisms). Then, it can be biomagnified (i.e., progressive accumulation through a prey-predator linkage) and enter the food supply chain. Hg can cause adverse developmental effects in the fetus as well as toxic effects in the liver and kidney of adults and children. Hg exposure can affect speaking, learning, feeling, seeing, taste, and moving. Children under 15 are most vulnerable since their brain is still developing. Hg exposure can occur through inhalation, ingestion, or skin contact. Due to biomagnification, the largest exposure to Hg is through fish consumption [20]. The Hg (inorganic) MCLG and MCL are both 2 ppb [21].

6.4.4.2 Arsenic

Arsenic (As) is often introduced into water by the dissolution of minerals and ores. As is used in industry and agriculture and is a by-product of copper smelting, mining, and coal burning. Chromated copper arsenate containing chromium trioxide (CrO_3), arsenic pentoxide (As_2O_5), and copper oxide (CuO) is a common wood preservative containing 22% arsenic. In treated wood, As is believed to be in the form of chromium (III) arsenate ($CrAsO_4$) and copper (II) arsenate ($Cu_3(AsO_4)_2$) or in stable chromium as dimer-arsenic clusters. The most common oxidation states of As in water are As III and V. Arsenic exposure produces vomiting, esophageal and abdominal pain, and bloody diarrhea. Chronic exposure can cause cancer in the skin, lung, urinary bladder, and kidney [18, 22]. The As MCLG is 0 ppb and MCL is 10 ppb [18, 21].

6.4.4.3 Chromium

Chromium (Cr) is found in natural deposits of ore. Cr is used to produce stainless steel, protective coatings, and magnetic tapes. Cr is most commonly found in oxidative stages + 2, + 3 (most stable), and + 6. The toxicity of Cr depends on the oxidative state. Cr + 3 is an essential nutrient, but exposure to Cr + 6 can cause absorption through lungs and intestinal tracts, and it is carcinogenic specifically for the lung and sino-nasal cavity [23]. Different chromium compounds were used in dyes, paints, and leather tanning, so Cr and its compounds are often found in soil and groundwater at active and abandoned industrial sites. These sites need environmental cleanup and remediation. Cr + 6 has been detected in the tap water of several US cities. Primer paint containing Cr + 6 is still broadly used for aerospace and automobile finishing applications. Cr + 6 has been determined genotoxic. Chromium (total) MCLG and MCL are 100 ppb [21].

6.4.4.4 Selenium

Selenium (Se) occurs naturally in the environment as selenide. Se is an essential nutrient needed to produce the amino acid selenocysteine. Se has a small range between being a nutrient 50 μgd^{-1} and becoming toxic at levels higher than 400 μgd^{-1}. Se at high concentration and at its higher toxic and oxidized stage is especially found in alkaline soils. Short-term exposure to Se can produce hair and fingernail changes, nervous system damage, and irritability. Chronic exposure can cause damage to the liver, kidney tissue, nervous system, and circulatory system. Selenium can bioaccumulate; however, Se concentration does not increase through the food chain (does not biomagnify). Se toxicity can affect the survival of young fish. Birds eating fish with Se toxicity succumb or give birth to deformed or dead offspring [24]. Selenium MCLG and MCL are 50 ppb [21].

6.4.5 Nutrients and Eutrophication of Surface Waters

Eutrophication is known as the main cause of impaired water reservoirs. Eutrophication or hypertrophication (from the Greek *eutrophos* "well-nourished") is defined as a body of water becoming overly enriched with minerals and nutrients, which induces excessive growth of algae, which may result in oxygen depletion (Figure 6.9). Eutrophication affects lakes, streams, coastal areas, estuaries, and salt marshes, increasing algal biomass. The nutrients that most limit the production of aquatic life are carbon, nitrogen, and phosphorus, as described in Chapter 4. The specific ratio most often needed for growth is 106C:16N:1P. So, it is uncommon for carbon to be the limiting nutrient due to its abundance on earth. Nitrogen is used in many essential biological steps: ammonification, nitrification, and denitrification. Phosphorus is also a necessary nutrient for life, but it is scarcer in the environment. So, it is often the limiting nutrient in aquatic systems. Soluble reactive phosphorus (SRP) indicates the amount of phosphorus in the water. SRP in unpolluted rivers

Figure 6.9 Eutrophication examples: (left) in a lake, by D. Goehring licensed under CC BY 2.0; (right) harmful algal blooms and dead zones in the Gulf of Mexico, due to the discharge of the Mississippi River, which carries millions of tons of nutrient-rich sediment into the Gulf each year; picture reproduced from NOAA's National Ocean Service and licensed under CC PDM 1.0.

averages about 0.01 mgL^{-1}. Agricultural activities may increase the SRP to 0.05 to 0.1 mgL^{-1}. Municipal effluents may increase this value to more than 1 mgL^{-1}.

Eutrophication, besides increasing algal biomass, results in the depletion of dissolved oxygen, increases in pathogenic bacteria and viruses, production of potentially toxic species of algae, fish kills, and loss of biodiversity.

Harmful algal blooms (HABs) can occur in both marine and freshwater systems. HABs in marine systems mostly occur in coastal zones where inland runoff creates the proliferation of monospecific blooms of toxic algae such as *Dinoflagellates*, which have the potential to produce toxins harmful to humans and wildlife. *Dinoflagellates* are responsible for producing the well-known "Red" tides, which result from particular environmental conditions. Ciguatera is the most commonly reported disease associated with poisoning from fish consumption. The US Centers for Disease Control (CDC) estimates that around 50,000 cases of ciguatera occur globally each year. HABs in freshwater systems can also be produced by similar species like *dinoflagellates*. However, the majority of freshwater toxins are caused by cyanobacteria. Cyanotoxins, produced by cyanobacteria previously referred to as blue-green algae, can greatly affect humans and wildlife. Cyanotoxins can be removed through conventional water treatment methods such as coagulation, sedimentation, filtration, and chlorination [25].

6.4.6 Organic Compounds in Water

Several persistent, bio-accumulative, and toxic (PBT) contaminants can be present in water systems. These compounds can bioaccumulate and biomagnify within aquatic systems, accumulating in surface waters. The most well-known PBT compounds are dichloro-diphenyl-trichloroethane (DDT) and polychlorinated biphenyls (PCBs). DDT was developed as a synthetic insecticide to combat malaria, typhus, and other insect-borne human diseases among both military and civilian populations. It was also used as an insect pest control tool. However, mounting evidence of its environmental and toxicological effects was apparent by the 1960s. Rachel Carson's book *Silent Spring*, published in 1962, finally created enough public awareness, and with the help of the Environmental Defense Fund (EDF), DDT was finally banned in 1972 [26, 27]. DDT is very persistent in the environment, accumulates in fatty tissues, and can travel long distances in the atmosphere. DDT can persist for up to 15 years in the environment, depending on the conditions [27]. PCBs are a group of organic compounds with similar structure and chemistry, ranging from oily liquids to waxy solids. PCBs are formed by adding chlorine (Cl_2) to biphenyl ($C_{12}H_{10}$). There are 209 variations of PCB congeners, which have different toxicities. PCBs have been used in hundreds of industrial applications such as electrical insulation, hydraulic equipment, and plasticizers for paints, plastics, and rubber products. PCBs adhere strongly to particulates in water, and fish are especially susceptible to the accumulation and concentration of PCBs [28]. PCB was banned in the US in 1977. Routes of exposure for PCB and DDT are mostly through aqueous rather than terrestrial environments (see Case Study 6.1). As with DDT, PCB can bioaccumulate and biomagnify. Both DDT and PCB are endocrine disruptors and have chronic effects at low doses. DDT has been regulated at a maximum of 30 ppt in drinking water since 2015 [29]. The PCB MCLG and MCL are zero and 0.5 ppb, respectively [21].

6.4.7 Enteric Pathogens as Surface Water Contaminants

Animals and humans excrete disease-causing intestinal microorganisms. Sources of pathogens in surface waters include urban stormwater, combined sewer and sanitary sewer overflows, animal feeding operations, sewage treatment plants, and septic tanks. Pathogens in surface water can remain infectious for prolonged periods, affecting recreational activities, shellfish harvesting, and drinking water treatment requirements. Generally, the more pathogens present in surface water, the more treatment is needed. Additionally, waterborne disease outbreaks increase as water runoff from storms increases. Most of the failures to meet water standards are due to pathogen loads measured by fecal coliform bacteria.

Stormwater can collect and contain a variety of pathogens originating from feces of wild and domestic animals. These pathogens can be carried out with storm water. Sewage treatment plants are designed to treat these pathogens. However, after heavy rainfall, the capacity of the sewage treatment plant can be exceeded due to increases from combined sewer overflows (CSOs). This happens more often in old cities with inadequate treatment capacity.

AFOs and CAFOs are major sources to produce food protein. In the US, there are more than 450,000 AFOs, and they produce more than 600 million tons of manure annually. Operations larger than 1,000 animals are required, according to the Clean Water Act (CWA), to have a facility permit as a CAFO [2, 30].

Septic tanks are another local source of pathogens. There are more than 4 billion gallons of treated effluent per day emitted from more than 26 million homes, businesses, and recreational sites in the US. Poorly treated effluents can contaminate soil and be transported to water streams. Mainly, these problems occur in areas with a shallow groundwater table.

6.5 Groundwater and Surface Water Legislation

Two main laws, the Safe Drinking Water Act (SDWA) and the CWA, are the basis for the regulation of water use and discharge in the US (Figure 6.10). The SDWA is the US federal law initially passed in 1974 to protect public health by regulating public drinking water supplies throughout the nation. Under the SDWA, EPA sets standards for drinking water quality and, with its partners such as the states, implements various technical and financial programs to ensure drinking water safety. Enforcement of compliance with the SDWA is relegated to the states [12]. The CWA establishes in the US the basic structure for regulating discharges of pollutants into waters and the establishment of quality standards for surface waters. The CWA was initially enacted in 1948 as the Federal Water Pollution Control Act, but the CWA was considerably rewritten, reorganized, and expanded in 1972 [31].

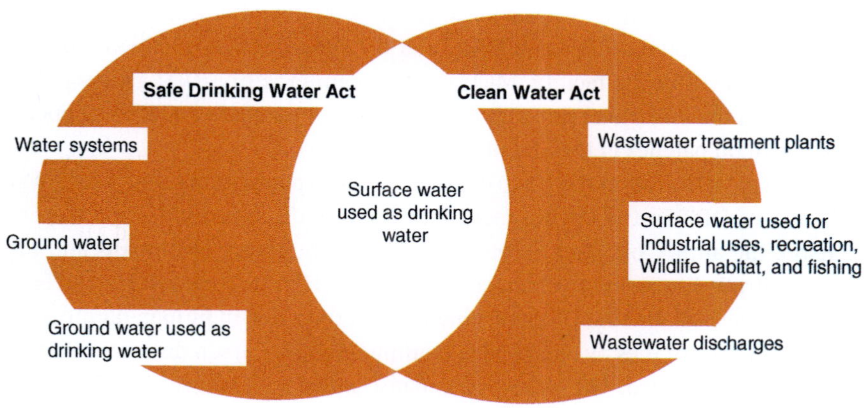

Figure 6.10 The two major water legislation in the US and application areas and overlap.

6.5.1 Total Maximum Daily Load (TMDL)

TMDL is the maximum amount of pollution that a water body can assimilate without violating water quality standards. The TMDL is the sum of the allowable loads of a single pollutant from all contributing points and nonpoint sources so that a water body can meet a designated use, such as swimming or fishing. The TMDL is estimated as follows:

TMDL: point sources of waste allocations (e.g., sewage treatment plant discharge) + nonpoint source load allocations (urban runoff) + natural sources (mineral deposits, wild animals) + growth factor (growth of enteric bacteria) + a margin of safety (to compensate for uncertainties about the link between pollutant loads and impairments).

Fecal coliform bacteria and increase in water temperature (mostly from power plant discharges) are the main contaminants that cause water bodies in the US not to meet water quality standards.

The SDWA establishes regulations for pollution according to maximum contaminant level (MCL) and maximum contaminant level goals (MCLG). Table 6.6 shows examples of microorganisms and chemicals regulated in drinking water. State programs are responsible for enforcing the regulations and developing their own regulations. These could be stricter or at the same level as the federal government requirements, but not more lenient. Local drinking water information can be found in required consumer confidence reports (CCR) as an annual water quality report delivered by community water systems to their customers. "The CCR includes information on source water, the levels of detected contaminants, compliance with drinking water rules, and some educational language. The reports are due to customers by July 1st of each year" [32].

The CWA is aimed to control the discharge of effluents into surface waters, primarily to protect the recreational use of surface waters, regulated by the US EPA according to best available technology for designated point sources and BMPs. The enforcement tool of the CWA is the permit. Anybody involved in construction or operation of a facility that may discharge anything – animal, vegetable, or mineral (or even energy) into navigable waters, tributaries to navigable waters, interstate waters, the oceans out to 200 miles, and intrastate waters must first obtain a permit from the National Pollutant Discharge Elimination System (NPDES) [33]. States are responsible for enforcement of the CWA. The Clean Water Act requires surface water quality to be high enough to support fish and wildlife populations, protect drinking water sources, and allow for human recreation.

6.6 Case Study 6.1. Pine River Contamination Site

The Michigan Chemical Company (MCC) was established in St. Louis, Michigan, in the 1930s, 50 miles north of Lansing, the capital of the state. MCC and the subsequent acquisitions such as Velsicol produced DDT and PCB in the facility. MCC started manufacturing DDT in 1944. The MCC discharged effluents in the Pine River due to poor management practices and factory operation conditions, resulting in oil patches in the St. Louis River.

Due to the extensive use of DDT around the US to control mosquito populations after World War II, concern about the use of DDT in the environment increased, as discussed in Chapter 1. For example, Rachel Carson, the author of *Silent Spring*, wrote about the threat to humans and the ecosystem from the uncontrolled exposure to chemicals, especially DDT. In her book, Carson confronted companies such as Velsicol and the US government about their role in human domination of nature and deciding who lives or dies by chemical poisoning. Mainly, Carson discussed that all systems in nature were interconnected. The Velsicol company gained control over MCC in 1965. The Velsicol company threatened to sue the publisher of *Silent Spring* (Houghton Mifflin) as well as the *New Yorker* over claims about the danger of DDT.

Table 6.6 Safe Drinking Water Act (SDWA) selected examples of primary drinking water regulations; adapted from [21]. Note: For a full list of compounds see the US Environmental Protection Agency - National Primary Drinking Water Regulations [21].

Contaminant	MCLG, mgL^{-1}	MCL or TT, mgL^{-1}	Potential Health Effect from Ingestion of Water	Sources of Contamination in Drinking Water
Microorganisms				
Total coliforms (including fecal coliform and *E.Coli*)	Zero	5.0%[1]	Not a health threat in itself; it is used to indicate whether other potentially harmful bacteria can be present	Coliforms are naturally present in the environment; as well as feces; fecal coliforms and *E.coli* only come from human and animal fecal waste.
Turbidity	n/a	TT	Turbidity is a measure is a measure of the cloudiness of water. It is used to indicate water quality and filtration effectiveness ...	Soil runoff
Legionella	Zero	TT2	Legionnaire's disease, a type of pneumonia	Found naturally in water, multiplies in water systems
Giardia lambia	Zero	TT3	Gastrointestinal illness (e.g., diarrhea, vomiting, cramps)	Human and fecal animal waste
Disinfection By-products				
Bromate	Zero	0.010	Increased risk of cancer	By-product of drinking water disinfection
Chlorite	0.8	1.0	Anemia, infants and young children; nervous system effects	By-product of drinking water disinfection
Haloacetic acids (HAAS)	n/a	0.06	Increased risk of cancer	By-product of drinking water disinfection
....	....	...	...	...
Disinfectants				
Chloramines (as Cl_2)	MRDLG = 4[1]	MRDL = 4.0[1]	Eye/nose irritation; stomach discomfort, anemia	Water additive used to control microbes
Chlorine (as Cl_2)	MRDLG = 4[1]	MRDL = 4.0	Eye/nose irritation; stomach discomfort	Water additive used to control microbes
Chlorine dioxide (ClO_2)	MRDLG = 0.8	MRDL = 0.8	Anemia; infants and young children; nervous system effects	Water additive used to control microbes

(*Continued*)

Table 6.6 (Continued)

Contaminant	MCLG, mgL^{-1}	MCL or TT, mgL^{-1}	Potential Health Effect from Ingestion of Water	Sources of Contamination in Drinking Water
Inorganic chemicals				
Antimony	0.006	0.006	Increase in blood cholesterol; decrease in blood sugar	Discharge from petroleum refineries; fire retardants; ceramics; electronics; solder
Arsenic	0	0.010 as of 01/23/06	Skin damage or problems with circulatory systems, and may have increased risk of getting cancer	Erosion of natural deposits; runoff from orchards, runoff from glass and electronics' production wastes
Asbestos (fiber > 10 micrometers)	7 million fibers per liter (MFL)	7 MFL	Increased risk of developing benign intestinal polyps	Decay of asbestos cement in water mains; erosion of natural deposits
Barium	2	2	Increase in blood pressure	Discharge of drilling wastes; discharge from metal refineries; erosion of natural deposits
Beryllium	0.004	0.004	Intestinal lesions	Discharge from metal refineries and coal-burning factories; discharge from electrical, aerospace, and defense industries
Cadmium	0.005	0.005	Kidney damage	Corrosion of galvanized pipes; erosion of natural deposits; discharge from metal refineries; runoff from waste batteries and paints
Chromium (total)	0.1	0.1	Allergic dermatitis	Discharge from steel and pulp mills; erosion of natural deposits
Copper	1.3	TT[7]; action level = 1.3	Short-term exposure: Gastrointestinal distress Long-term exposure: Liver or kidney damage People with Wilson's Disease should consult their personal doctor if the amount of copper in their water exceeds the action level	Corrosion of household plumbing systems; erosion of natural deposits
Cyanide (as free cyanide)	0.2	0.2	Nerve damage or thyroid problems	Discharge from steel/metal factories; discharge from plastic and fertilizer factories

Fluoride	4.0	4.0	Bone disease (pain and tenderness of the bones); children may get mottled teeth	Water additive which promotes strong teeth; erosion of natural deposits; discharge from fertilizer and aluminum factories
Lead	Zero	TT[7]; action level = 0.015	Infants and children: delays in physical or mental development; children could show slight deficits in attention span and learning abilities Adults: Kidney problems; high blood pressure	Corrosion of household plumbing systems; erosion of natural deposits
Mercury (inorganic)	0.002	0.002	Kidney damage	Erosion of natural deposits; discharge from refineries and factories; runoff from landfills and croplands
Nitrate (measured as Nitrogen)	10	10	Infants below the age of six months who drink water containing nitrate in excess of the MCL could become seriously ill and, if untreated, may die. Symptoms include shortness of breath and blue-baby syndrome.	Runoff from fertilizer use; leaking from septic tanks, sewage; erosion of natural deposits
Selenium	0.05	0.05	Hair or fingernail loss; numbness in fingers or toes; circulatory problems	Discharge from petroleum refineries; erosion of natural deposits; discharge from mines
Thallium	0.0005	0.002	Hair loss; changes in blood; kidney, intestine, or liver problems	Leaching from ore-processing sites; discharge from electronics, glass, and drug factories
<u>Organic chemicals</u>				
Acrylamide	Zero	TT[8]	Nervous system or blood problems; increased risk of cancer	Added to water during sewage/ wastewater treatment
Atrazine	0.003	0.003	Cardiovascular system or reproductive problems	Runoff from herbicide used on row crops

(Continued)

Table 6.6 (Continued)

Contaminant	MCLG, mgL^{-1}	MCL or TT, mgL^{-1}	Potential Health Effect from Ingestion of Water	Sources of Contamination in Drinking Water
1,1-Dichloroethylene	0.007	0.007	Liver problems	Discharge from industrial chemical factories
Dioxin (2,3,7,8-TCDD)	Zero	0.00000003	Reproductive difficulties; increased risk of cancer	Emissions from waste incineration and other combustion; discharge from chemical factories
Polychlorinated biphenyls (PCBs)	Zero	0.0005	Skin changes; thymus gland problems; immune deficiencies; reproductive or nervous system difficulties; increased risk of cancer	Runoff from landfills; discharge of waste chemicals
Styrene	0.1	0.1	Liver, kidney, or circulatory system problems	Discharge from rubber and plastic factories; leaching from landfills
Trichloroethylene	Zero	0.005	Liver problems; increased risk of cancer	Discharge from metal degreasing sites and other factories
Vinyl chloride	Zero	0.002	Increased risk of cancer	Leaching from PVC pipes; discharge from plastic factories
Xylenes (total)	10	10	Nervous system damage	Discharge from petroleum factories; discharge from chemical factories
Radionuclides				
Alpha particles	None ——— - Zero	15 picocuries per liter ($pCiL^{-1}$)	Increased risk of cancer	Erosion of natural deposits of certain minerals that are radioactive and may emit a form of radiation known as alpha radiation
Uranium	Zero	30 ugL^{-1} as of 12/08/03	Increased risk of cancer, kidney toxicity	Erosion of natural deposits

Notes "[1]Definitions:
Maximum Contaminant Level Goal (MCLG) – The level of a contaminant in drinking water below which there is no known or expected risk to health. MCLGs allow for a margin of safety and are nonenforceable public health goals.

Maximum Contaminant Level (MCL) – The highest level of a contaminant that is allowed in drinking water. MCLs are set as close to MCLGs as feasible using the best available treatment technology and taking cost into consideration. MCLs are enforceable standards.

Maximum Residual Disinfectant Level Goal (MRDLG) – The level of a drinking water disinfectant below which there is no known or expected risk to health. MRDLGs do not reflect the benefits of the use of disinfectants to control microbial contaminants.

Treatment Technique (TT) – A required process intended to reduce the level of a contaminant in drinking water.

Maximum Residual Disinfectant Level (MRDL) – The highest level of a disinfectant allowed in drinking water. There is convincing evidence that addition of a disinfectant is necessary for control of microbial contaminants.

[2]Units are in milligrams per liter (mgL^{-1}) unless otherwise noted. Milligrams per liter are equivalent to parts per million (PPM).

[3]EPA's surface water treatment rules require systems using surface water or ground water under the direct influence of surface water to disinfect their water, and filter their water, or meet criteria for avoiding filtration so that the following contaminants are controlled at the following levels: *Cryptosporidium*: Unfiltered systems are required to include *Cryptosporidium* in their existing watershed control provisions. *Giardia lamblia*: 99.9% removal/inactivation. Viruses: 99.99% removal/inactivation. *Legionella*: No limit, but EPA believes that if *Giardia* and viruses are removed/inactivated, according to the treatment techniques in the Surface Water Treatment Rule, *Legionella* will also be controlled. Turbidity: For systems that use conventional or direct filtration, at no time can turbidity (cloudiness of water) go higher than 1 Nephelometric Turbidity Unit (NTU), and samples for turbidity must be less than or equal to 0.3 NTUs in at least 95% of the samples in any month. Systems that use filtration other than the conventional or direct filtration must follow state limits, which must include turbidity at no time exceeding 5 NTUs. Heterotrophic Plate Count (HPC): No more than 500 bacterial colonies per milliliter. Long-Term 1 Enhanced Surface Water Treatment: Surface water systems or groundwater under the direct influence (GWUDI) systems serving fewer than 10,000 people must comply with the applicable Long-Term 1 Enhanced Surface Water Treatment Rule provisions (such as turbidity standards, individual filter monitoring, *Cryptosporidium* removal requirements, updated watershed control requirements for unfiltered systems). Long-Term 2 Enhanced Surface Water Treatment Rule: This rule applies to all surface water systems or ground water systems under the direct influence of surface water. The rule targets additional *Cryptosporidium* treatment requirements for higher risk systems and includes provisions to reduce risks from uncovered finished water storage facilities and to ensure that the systems maintain microbial protection as they take steps to reduce the formation of disinfection by-products. Filter Backwash Recycling: This rule requires systems that recycle to return specific recycle flows through all processes of the system's existing conventional or direct filtration system or at an alternate location approved by the state.

[4]No more than 5.0% samples total coliform-positive (TC-positive) in a month. (For water systems that collect fewer than 40 routine samples per month, no more than one sample can be total coliform-positive per month.) Every sample that has total coliform must be analyzed for either fecal coliforms or *E. coli* if two consecutive TC-positive samples, and one is also positive for *E.coli* fecal coliforms, system has an acute MCL violation.

[5]Fecal coliform and *E. coli* are bacteria whose presence indicates that the water may be contaminated with human or animal wastes. Disease-causing microbes (pathogens) in these wastes can cause diarrhea, cramps, nausea, headaches, or other symptoms. These pathogens may pose a special health risk for infants, young children, and people with severely compromised immune systems.

[6]Although there is no collective MCLG for this contaminant group, there are individual MCLGs for some of the individual contaminants: Trihalomethanes: bromodichloromethane (zero); bromoform (zero); dibromochloromethane (0.06 mgL^{-1}): chloroform (0.07 mgL^{-1}). Haloacetic acids: dichloroacetic acid (zero); trichloroacetic acid (0.02 mgL^{-1}); monochloroacetic acid (0.07 mgL^{-1}). Bromoacetic acid and dibromoacetic acid are regulated with this group but have no MCLGs.

[7]Lead and copper are regulated by a treatment technique that requires systems to control the corrosiveness of their water. If more than 10% of tap water samples exceed the action level, water systems must take additional steps. For copper, the action level is 1.3 mgL^{-1}, and for lead is 0.015 mgL^{-1}.

[8]Each water system must certify, in writing, to the state (using third-party or manufacturer's certification) that when acrylamide and epichlorohydrin are used to treat water, the combination (or product) of dose and monomer level does not exceed the levels specified, as follows: Acrylamide = 0.05% dosed at 1 mgL^{-1} (or equivalent) Epichlorohydrin = 0.01% dosed at 20 mgL^{-1} (or equivalent)"

Due to the significant contamination of the site with DDT and other chemicals, in 1970, a complete fishing ban was imposed in the Pine River. Velsicol started producing polybrominated biphenyls (PBB) in St. Louis, MI facility in 1971. Accidentally, in 1973, PBB was shipped to farms as cattle feed due to mislabeled bags – a packaging error. A cluster of Hodgkin's lymphomas – a cancer of the lymphatic system – was detected in the region with 12.2 times the expected rate in 1974. The regional farmers sued Velsicol. The Velsicol plant closed in 1978 due to the Congressional hearing and the pending lawsuits. In 1982, a settlement between the judge, the Michigan Department of Natural Resources (DNR), US EPA, and Velsicol exempted the company from cleaning up the contamination site. In 1986, the US EPA designated the Velsicol plant site as a Superfund site. By 1989, the amount of DDT in state-collected fish samples was 10.5 ppm (remember, the maximum for DDT in drinking water is 0.00003 ppm). The DDT concentration reached 16.1 ppm by 1995. By 1998, a Community Advisory Group (CAG) was formed, and a task force was created to help in cleaning the Pine River contamination. Since then, several legal battles have taken place between the plants, the US EPA, the MI Department of Environmental Quality (DEQ) – now the new MI department of Environment, Great Lakes, and Energy (EGLE) – and the CAG. In 2005, PCBs, a by-product of DDT, was detected in the St. Louis city drinking water. Table 6.7 shows a timeline of the main related events. Nowadays, the St. Louis community is heavily involved in the river's clean-up through the Pine River Superfund Site, headed by the Pine River Citizen's Advisory Group alongside the EPA and EGLE. The cleanup of the Pine River is expected to cost more than $500 million [34]. Figure 6.11 shows an aerial view in 1975 and in 2021 of contamination and clean-up site, respectively. The aerial view of 1975 shows the location of the chemical storage tanks and the quality of the river water. The aerial view of 2021 shows a cleaned area with the location of the SuperFund operation site.

Table 6.7 Timeline of the main Pine River contamination site events.

Year	Event
1930	Pollution monitoring began.
1934	Construction began of a Mid-West refinery in Alma, Michigan.
1935	Three environmentally hazardous companies opened their doors. 1. The first was the Michigan Chemical Company, which built its factory on the shores of the Pine River in St. Louis, Michigan. 2. The second was Leonard Petroleum Company, which constructed a refinery in Alma. 3. The final development in 1935 was a McClanahan refinery, located in St. Louis. Within the same year, the McClanahan refinery was dumping waste in the Pine River. A research report done by the Institute for Fisheries in August of 1935 states several thousand fish were found dead.
1940	Several times over the summer, the surface of the Pine River was found covered in oil slicks and dead fish. At this time, residents stopped eating fish from the river.
1941	The citizens of St. Louis signed a petition to prompt the city to start investigating the odors emanating from the river.
1944	The Michigan Chemical Company began manufacturing a pesticide known as DDT.
1961	Velsicol threatened Houghton Mifflin and the *New Yorker* in efforts to block publication of Rachel Carson's *Silent Spring*. The book discusses the extensive risks of using DDT and other pesticides.
1963	Velsicol Chemical completely takes over the Michigan Chemical Company.
1965	A new lagoon system is installed at the Alma refinery. This system prevented substantial waste oil from reaching the Pine River.
1967	DNR studies say that Velsicol was having a negative impact on the river. The area of the river near the plant was almost totally devoid of life.

Table 6.7 (Continued)

Year	Event
1968	Northwest Industries buys Reading Railroad and Fruit of the Loom.
1970	A complete fishing ban is imposed on the Pine River by state health officials.
1971	Velsicol starts producing [PBB, polybrominated biphenyls] for their product "Firemaster," a fire retardant used in appliances.
1973	Tons of the poisonous PBB are accidentally shipped in mislabeled bags as a supplement in cattle feed. Over the next few years, cattle are found with deformities and are voluntarily destroyed across Michigan due to the PBB crisis.
1974	A cluster of Hodgkin's disease is detected in Breckenridge, 12.2 times the expected rate.
1976	Congressional hearings and farmer and worker lawsuits lead to the closing of the Velsicol plant in 1978.
1978	A Michigan DNR study shows Velsicol was still impacting the Pine River even after its doors were closed.
1979	Mergers in the oil industry and resource depletion causes Total to lay off most of its workers at the Alma refinery.
1980	The Total headquarters moves from Alma to Denver, Colorado
1982	A consent judgment between the DNR, EPA, and Velsicol results in the company being exempt from river clean up.
1986	The EPA designates the Velsicol plant site as a Superfund Site.
1987	A NorthWest Industries subsidiary is created to own the old chemical plant in St. Louis. Later that year, Fruit of the Loom becomes the owner of NorthWest Industries.
1989	The state-collected area fish samples have an average concentration total DDT of 10.5 parts per million.
1991	Out of court settlement with Velsicol workers.
1994	The state-collected fish samples have an average concentration of 23.3 parts per million, more than double that of 1989.
1995	State-collected fish have average DDT concentration of 16.1 ppm.
1997	EPA comes to St. Louis to inform the community of the pollution levels detected since the departure of Velsicol. At the end of the meeting, the EPA official informs the residents that they can form a Citizen Advisory Group to oversee rehabilitation of the Pine River.
1998	Under EPA regulations, a Community Advisory Group (CAG) is formed – the Pine River Superfund Task Force.
1998	US EPA brings a lawsuit against the Alma Total refinery for violations of the Clean Air and Clean Water Acts. In the same year, the EPA launches an emergency cleanup of the Pine River DDT contamination.
1999	Total refinery shuts down. Fruit of the Loom and NorthWest Industries file for bankruptcy in 1999 as well.
2000	Consent Decree issued for the violations by Total refinery. The Pine River CAG files a bankruptcy claim for Fruit of the Loom assets.
2000	City of Alma and CAG try to prevent signing of Consent Decree.
2001	Consent Decree signed with modifications that the City accepted.
2001	Implementation of Consent Decree, cleaning site.
2001	Oxford automotive announces closure of Alma plant.
2002 to 2007	Many events have occurred with respect to the Pine River. Some of the most important events include the Pine River OU2 Cleanup. * NAPL (Non-Aqueous Phase Liquids) was discovered and characterized to some degree. An interceptor trench was installed to collect the NAPL. The east and west sides of the Pine River cleanup have been completed. The Back-dam (mill pond) was also completed. * Other noteworthy activities include the downstream assessment plan's completion, and the NAPL assessment.
2005	Discovery of PCBa in St. Louis city drinking water. PCBs is an acidic by-product from the manufacturing of DDT.

(*Continued*)

Table 6.7 (Continued)

Year	Event
2006	Six wells are routinely measured around the Superfund area for PCBs and other contaminants. Levels are low but slowly rising. Some residents fear the slurry wall constructed to contain the Velsicol contamination may be leaking new pollutants into the Pine River.
2020	Cleanup is still in progress. On-site groundwater is contaminated with DDT, chorobenzene, carbon tetrachloride, trichloroethylene (TCE), and other chlorinated compounds. The sediments of the Pine River were also contaminated with similar chemicals. Surface water does not seem to have the same impact. However, there is still a potential risk for people eating contaminated fish on the proximity of the site.

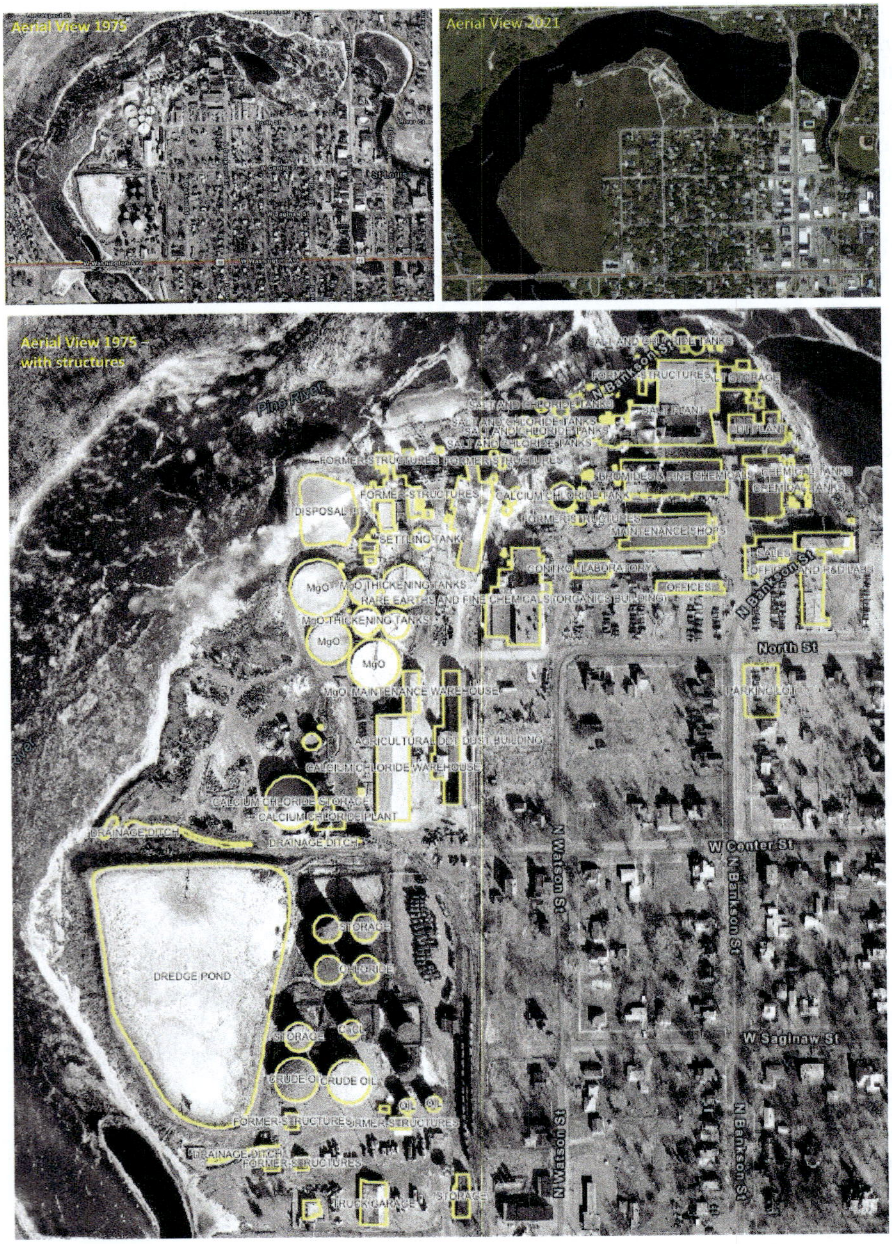

Figure 6.11 Pine river aerial view in 1975 and aerial view in 2021 of contamination and clean-up site; reproduced from the Velsicol Chemical Corp. Superfund Site, St. Louis, MI, US EPA [35].

6.7 Case Study 6.2. The Flint Water Crisis

The city of Flint was receiving water from 1967 to 2013 from the Detroit Water, and Sewerage Department sourced from the Lake Huron and the Detroit River. On April 25, 2014, due to a budgetary crisis and under the emergency manager Michael Brown appointed by Governor Snyder, county officials finalized a delayed pipe to source and obtain water from the Flint River. In June 2014, an outbreak of Legionnaire's disease – atypical pneumonia caused by *Legionella* bacteria – began and continued until November 2015. The outbreak of Legionnaires' disease was responsible for killing 12 people and affecting another 87.

After the switch, residents started complaining about the water's taste, smell, and appearance. The change in the water was due to the lack of corrosion inhibitors needed to avoid the leaching of lead from aging pipes lines. Due to this oversight, 100,000 residents were exposed to elevated lead levels. Among them, between 6,000 and 12,000 children were exposed to a high lead level in the drinking water. Children are at risk from lead poisoning, causing an impact on long-term intellectual functioning and increasing the chance of Alzheimer's disease. Blood test data from children from Flint below five years old showed elevated blood-lead levels up to 10.6% from a baseline of 4% [36].

Until today, considerable disagreements and lack of accountability have plagued the Flint water crisis. Several criminal cases and charges have been filed. On August 20, 2020, a settlement was reached with a combined award amount of $600 million – 80% going to the families with affected children. Currently, an extensive replacement of lead service lines is underway, and by 2021 a large number (>27,000) water services have been inspected with the replacement of more than 10,000 lead pipes [37]. The city of Flint provides a wake-up call to cities around the world with aging infrastructure and a lack of investment in their services. A few additional resources about the Flint water crisis are provided at the end of the chapter – Section 6.9.

6.8 Study Questions

1 Describe the movement of water around the world.
2 How can freshwater be divided?
3 What are the main factors affecting groundwater vulnerability to the contamination?
4 What is DRASTIC?
5 What is the main difference between point-source contamination and diffuse source contamination?
6 Where is surface water mostly located?
7 What are the main sources of surface water pollution?
8 What is TSS, and what does it represent?
9 What metals are the main contaminants of surface water?
   A Fe, Cu, Cr, and Zn
   B Hg, As, Cr, and Se
   C Hg, As, Fe, and Se
   D Zn, As, Fe, and Cr
   E None of the above
10 What is eutrophication?
11 What are the main groundwater and surface water legislation, and what do they regulate?
12 Define MCLG and MCL.
13 Describe the main events taken place in the Pine River contamination site. What lessons have we learned on St. Louis, Mi contamination site that can inform the Flint water crisis?

6.9 Additional Resources

1 The water cycle https://pmm.nasa.gov/education/water-cycle
2 Distribution of Water on the Earth's Surface https://www.e-education.psu.edu/earth103/node/701
3 Groundwater https://www.usgs.gov/faqs/what-groundwater&qt-news_science_products=0#qt-news_science_products
4 Potential Well Water Contaminants and Their Impacts https://www.epa.gov/privatewells/potential-well-water-contaminants-and-their-impacts
5 Ground Water and Drinking Water https://www.epa.gov/ground-water-and-drinking-water
6 Safe Drinking Water Act https://www.epa.gov/ground-water-and-drinking-water/national-primary-drinking-water-regulations%23Disinfectants
7 Clean Water Act https://www.epa.gov/laws-regulations/summary-clean-water-act
https://www.epa.gov/laws-regulations/summary-clean-water-act
8 VELSICOL CHEMICAL CORP. (MICHIGAN) ST. LOUIS, MI https://cumulis.epa.gov/supercpad/SiteProfiles/index.cfm?fuseaction=second.cleanup&id=0502194
https://cumulis.epa.gov/supercpad/SiteProfiles/index.cfm?fuseaction=second.cleanup&id=0502194
9 Michigan Radio podcast – One company's toxic legacy https://www.michiganradio.org/topic/one-companys-toxic-legacy
10 Marc Edwards on Flint Water and Public-Inspired Science https://video.wkar.org/video/msu-video-public-inspired-science-and-modern-land-grant-university
11 Flint's Deadly Water (2019) – PBS's Frontline https://www.pbs.org/wgbh/frontline/film/flints-deadly-water

References

1 US Geological Survey. (2019). *Where is Earth's water?* https://www.usgs.gov/special-topic/water-science-school/science/where-earths-water%3Fqt-science_center_objects%3D0%23qt-science_center_objects (accessed August 1, 2019).
2 Brusseau, M.L., Pepper, I.L., and Gerba, C.P. (2019). *Environmental and Pollution Science*, 2e. London, UK: Academic Press Publications.
3 Socioeconomic Data and Applications Center (SEDAC). *Percentage of total population living in coastal areas 2019.* https://sedac.ciesin.columbia.edu/es/papers/Coastal_Zone_Pop_Method.pdf (accessed August 1, 2019).
4 Neumann, B., Vafeidis, A.T., Zimmermann, J., and Nicholls, R.J. (2015). Future coastal population growth and exposure to sea-level rise and coastal flooding – A global assessment. *PLoS One* 10: e0118571. DOI: 10.1371/journal.pone.0118571.
5 US Environmental Protection Agency. *Safe drinking water act – (Title XIV of public health service act) 2019.* https://www.govinfo.gov/app/collection/comps (accessed May 24, 2019).
6 Morris, B.L., Lawrence, A.R.L., Chilton, P.J.C., Adams, B., Calow, R.C., and Klinck, B.A. (2003). *Groundwater and its Susceptibility to Degradation: A Global Assessment of the Problem and Options for Management.* Noirobi, Keya.
7 US Environmental Protection Agency. (2004). *Drastic: A standardized system for evaluating ground water pollution potential using hydrogeologic settings 2019.* https://cfpub.epa.gov/si/si_public_record_Report.cfm%3FLab=ORD&dirEntryID=35474 (accessed September 6, 2019).

8 drastic index | drastic index | ArcGIS Hub 2019. (2019). *Drastic index*. https://hub.arcgis.com/datasets/SCGIS::drastic-index/explore%3Ffilters=eyJJbmRleF9WYWx1ZSI6WzEsNV19&;location=33.575154%2C-80.910668%2C8.00&style=Index_Value (accessed September 1, 2021).

9 Granger. *Sustainable Waste Collection and Disposal*. (n.d.). https://wwwGrangerwasteservices Com/2019.https://www.grangerwasteservices.com/about/responsible-waste-management (accessed September 2, 2021).

10 Kjeldsen, P., Barlaz, M.A., Rooker, A.P., Baun, A., Ledin, A., and Christensen, T.H. (2002). Present and Long-term composition of MSW Landfill Leachate: A review. *Critical Reviews in Environmental Science and Technology* 32: 297–336. DOI: 10.1080/10643380290813462.

11 US Geological Survey. (2016). Estimated annual agricultural pesticide use. *Pestic Natl Synth Proj 2016*. https://www.usgs.gov/media/images/estimated-annual-agricultural-pesticide-use.

12 US Environmental ProtectionAgency. (2019). *Safe drinking water act (SDWA) 2019*. https://www.epa.gov/sdwa (accessed May 24, 2019).

13 US Geological Survey. (2019). *Groundwater Salinization. Saltwater Intrusion 2018*. https://pubs.usgs.gov/gip/gw/images/Intrusion.gif (accessed August 25, 2019).

14 US Environmental ProtectionAgency. (2016). *Gasoline | Methyl Tertiary Butyl Ether (MTBE) | US EPA 2021*. https://archive.epa.gov/mtbe/web/html/gas.html%23:~:text=WhatisMTBE%3F, fuel additive in motor gasoline. (accessed February 3, 2021).

15 US Environmental ProtectionAgency. (2016). *Drinking water | Methyl Tertiary Butyl Ether (MTBE) | US EPA 2021*. https://archive.epa.gov/mtbe/web/html/water.html (accessed November 15, 2021).

16 US Environmental ProtectionAgency. (2017). *Technical Fact Sheet – 1,4-Dioxane 2017:9*. https://www.epa.gov/sites/default/files/2014-03/documents/ffrro_factsheet_contaminant_14-dioxane_january2014_final.pdf (accessed November 15, 2021).

17 US Environmental ProtectionAgency. (2019). *Perchlorate in drinking water 2019*. https://www.epa.gov/dwstandardsregulations/perchlorate-drinking-water (accessed September 7, 2019).

18 US Environmental ProtectionAgency. (n.d.). *Technical fact sheet: Final rule for Arsenic in water 2001:6*. https://www.wqa.org/Portals/0/Technical/TechnicalFactSheets/2014_Arsenic.pdf (accessed August 21, 2021).

19 United Nations. (n.d.). *Clean water and sanitation 2016:2*. https://www.globalgoals.org/6-clean-water-and-sanitation (accessed August 2, 2019).

20 WHO WHO. (n.d.). *Mercury and health 2019*. https://www.who.int/news-room/fact-sheets/detail/mercury-and-health (accessed August 2, 2019).

21 US Environmental ProtectionAgency. (n.d.). *National primary drinking water regulations*.

22 US National Research Council. (2001). *Subcommittee on Arsenic in Drinking Water. Arsenic in Drinking Water*. Washington, DC: National Academy Press.

23 Niehs. (2019). *Hexavalent chromium factsheet* www.osha.gov/SLTC/hexavalentchromium (accessed August 2, 2019).

24 Agency for Toxic Substances and Disease Registry (ATSDR). (n.d.). *Selenium 2003*. http://www.atsdr.cdc.gov/toxfaq.html. (accessed August 2, 2019).

25 US Environmental ProtectionAgency. (n.d.) *Managing Cyanotoxins in public drinking water systems*. https://www.epa.gov/ground-water-and-drinking-water/managing-cyanotoxins-public-drinking-water-systems (accessed August 2, 2019).

26 Carson, R., Darling, L., and Darling, L. (n.d.). *Silent Spring*. Cambridge, Mass: Houghton Mifflin Company.

27 US Environmental ProtectionAgency. *DDT – A brief history and status 2019*. https://www.epa.gov/ingredients-used-pesticide-products/ddt-brief-history-and-status (accessed August 2, 2019).

28 US Environmental ProtectionAgency.(n.d.). *Learn about Polychlorinated Biphenyls (PCBs) 2019.* https://www.epa.gov/pcbs/learn-about-polychlorinated-biphenyls-pcbs (accessed August 2, 2019).

29 US Environmental ProtectionAgency. (n.d.). *National recommended water quality criteria – Human health criteria table 2019.* https://www.epa.gov/wqc/national-recommended-water-quality-criteria-human-health-criteria-table (accessed September 8, 2019).

30 US Department of Agriculture. (n.d.). *Animal Feeding Operations | NRCS 2019.* https://www.nrcs.usda.gov/wps/portal/nrcs/main/national/plantsanimals/livestock/afo (accessed August 3, 2019).

31 US Environmental ProtectionAgency. (n.d.). *Summary of the clean water act.* https://www.epa.gov/laws-regulations/summary-clean-water-act (accessed August 3, 2019).

32 US Environmental ProtectionAgency. (n.d.). *Consumer Confidence Reports (CCR) 2019.* https://ofmpub.epa.gov/apex/safewater/f?p=136:102:::::: (accessed August 3, 2019).

33 US Environmental ProtectionAgency. (n.d.). *National Pollutant Discharge Elimination System (NPDES) 2019.* https://www.epa.gov/npdes (accessed August 3, 2019).

34 US Environmental ProtectionAgency. (n.d.). VELSICOL CHEMICAL CORP. *(MICHIGAN) Site Profile 2019.* https://cumulis.epa.gov/supercpad/SiteProfiles/index.cfm?fuseaction=second.Cleanup&id=0502194#bkground (accessed September 9, 2019).

35 Velsicol Chemical Corp. *Superfund Site, St. Louis, Michigan.* (2019). https://epa.maps.arcgis.com/apps/MapSeries/index.html?appid=48de7c0961294c4b8699a385e8d916b2 (accessed September 3, 2021).

36 Hanna-Attisha, M., LaChance, J., Sadler, R.C., and Schnepp, A.C. (2016). *Elevated blood lead levels in children associated with the flint drinking water crisis: A spatial analysis of risk and public health response.* https://DoiOrg/102105/AJPH2015303003 106: 283–290. DOI: 10.2105/AJPH.2015.303003.

37 Service Line Replacement Program – City of Flint. (2019). https://www.cityofflint.com/gettheleadout (accessed September 3, 2021).

7

Air Pollution

7.1 Learning Objectives

After learning from this chapter, you are expected to:

- Identify natural and anthropogenic sources of air pollutants.
- Describe the main primary and secondary air pollutants and their sources and formation.
- Understand the goal, framework, and key elements of the Clean Air Act.
- Calculate air quality index (AQI) and describe its use and applications.

7.2 Introduction

Air pollutants are gases or particulates that, at high enough concentrations, could be harmful to life, the environment, and/or property. These pollutants may originate from natural or anthropogenic sources or both (Figure 7.1). The concentration of atmospheric pollutants depends mainly on the total mass of pollution emitted into the atmosphere, as well as the atmospheric conditions affecting their fate and transport. The dispersion of pollutants in the environment is controlled by atmospheric motion and such factors as winds and the vertical temperature variation [1]. Even when emissions of pollutants are relatively constant, the pollutant concentrations at particular locations can change widely due to variations in the atmospheric conditions. Winds determine the horizontal movement of pollutants in the atmosphere. For example, changes in wind direction can significantly reduce or increase pollutant concentration, and changes in temperature between day and night can precipitate pollutants from the air. The UN considers air pollution a foremost global problem. In the US, billions of dollars are spent annually to avert, control, and clean up air pollution. Air pollution poses a health risk, harming the human respiratory system. Breathing polluted air slows lung development in children [1]. Breathing polluted air with dust or particulates is a major cause of respiratory illnesses such as asthma. Polluted air not only impacts humans but also vegetation by causing decreased resistance to diseases and slower plant development.

Air pollution is not a new problem. King Edward I, in thirteenth-century London, prohibited the burning of coal due to the production of soot and sulfur dioxide (SO_2). The industrial revolution significantly increased air pollution in North America and Europe. By the

Life Cycle of Sustainable Packaging: From Design to End of Life, First Edition. Rafael A. Auras and Susan E. M. Selke.

Companion Website: www.wiley.com/go/Auras/lifecycleofsustainablepackaging

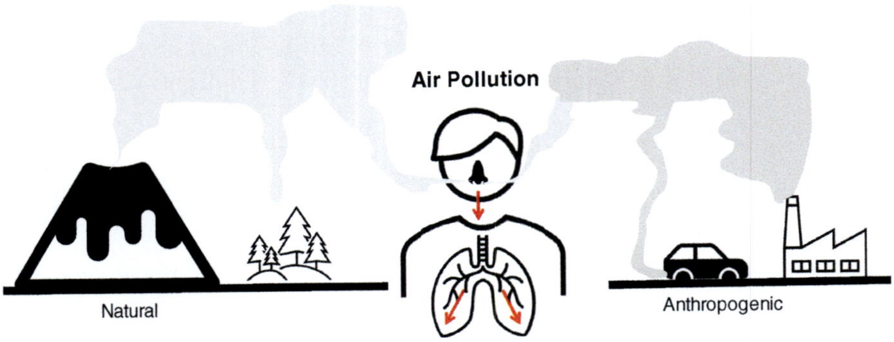

Figure 7.1 Atmospheric pollutants and their sources.

mid-1940s, it was common to have photochemical smog alerts in Los Angeles, CA. Nowadays, almost all the "megacities" such as New York, Delhi, Buenos Aires, and São Paolo are affected by air pollution.

The main components of air are nitrogen and oxygen with small amounts of neon, helium, and krypton, as well as variable concentrations of trace gases (Table 7.1). A small part of the trace gases can be air pollutants, including gases and particulate matter suspended as aerosols.

The main sources of air pollutants are biogenic sources, fires, and anthropogenic stationary and mobile sources. Biogenic sources are mostly vegetation and soil. Fires are due to agricultural field

Table 7.1 Concentration of main and variable components in the atmosphere; adapted from [1].

Gas	Abbreviation	Percentage by Volume of Dry Air, %	Concentration, μLL^{-1}
	Constant atmospheric components		
Nitrogen	N_2	78.1	780,840
Oxygen	O_2	20.9	209,460
Argon	Ar	0.9	9,340
Neon + helium + krypton	Ne + He + Kr	0.002	24
	Variable atmospheric components		<0.01
Water vapor	H_2O		<10,000
Carbon dioxide	CO_2		380
Methane	CH_4		1.5
Hydrogen	H_2		0.50
Nitrous oxide	N_2O		0.31
Ozone	O_3		0.02
Carbon monoxide	CO		<0.05
Ammonia	NH_3		0.004
Nitrogen dioxide	NO_2		0.001
Sulfur dioxide	SO_2		0.001
Nitric oxide	NO		0.0005
Hydrogen sulfide	SH_2		0.00005

burning, prescribed fires, and wildfires. Anthropogenic stationary sources include dust from roads and construction, agriculture, fuel combustion from commercial and institutional entities, industrial boilers, electricity generation, residential and industrial processes, solvents, etc. Mobile sources include different modes of transportation such as aircraft, commercial marine vessels, locomotives, nonroad equipment, and on-road vehicles, and represent around 56% of primary pollutants (Figure 7.2).

There are several types of air pollutants. The US EPA classifies air pollutants in two broad categories, primary and secondary pollutants. Primary air pollutants are pollutants directly emitted from a source such as transportation. Secondary air pollutants are formed when the primary pollutants react in the atmosphere. Additional information about pollutants in the US can be found in the national emission inventory (NEI) website [2, 3].

7.3 Primary Air Pollutants

Primary pollutants are the main pollutants present in the environment. They can be emitted from many sources such as coal-fired and natural gas power plants, biomass burning, cars, natural forest fires, volcanoes, and many others. They can affect humans, plants, and animals, and they contribute to the formation of secondary pollutants. The EPA is required to set National Ambient Air Quality Standards (NAAQS) for six pollutants – known as criteria pollutants – which can be harmful to public health and the environment. The criteria pollutants are carbon monoxide (CO), lead, nitrogen dioxide, ozone, particulate matter (PM 2.5 and 10), and sulfur dioxide.

7.3.1 Carbon Monoxide (CO)

CO is a major "criteria" pollutant in urban air and is the product of incomplete combustion. More than 50% of all anthropogenic CO emissions originate from transportation. CO concentrations are the highest along highways and streets. CO concentrations have been reduced in the US as newer cars have better fuel efficiencies and catalytic convertors. CO is highly toxic to most animals, and the US EPA standard limit for human exposure is a 24 h average of 9 nLL^{-1} (ppb = part per billion) or 1 h average of 35 nLL^{-1} (ppb). When inhaled, CO reduces the capacity of blood

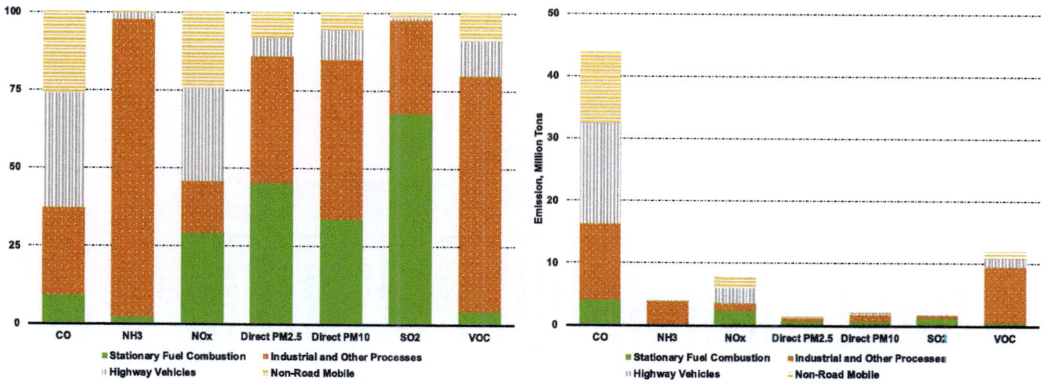

Figure 7.2 Source contributions of primary air pollutants in the US and percentage of total air pollutants by type; reproduced from [4].

hemoglobin to attach oxygen, causing hypoxia. CO is short-lived in the atmosphere since it oxidizes to CO_2 by reacting with hydroxide radicals. Figure 7.3 shows a reduction of 81% in ambient CO between 1980 and 2020.

7.3.2 Hydrocarbons (HCs)

HCs or volatile organic carbons (VOCs) are compounds composed of H and C. CH_4 is the most plentiful hydrocarbon in the atmosphere and is a GHG. Nonmethane hydrocarbons (NMHCs), such as benzene and its by-products such as formaldehyde, phenol, and toluene, can be carcinogenic and contribute to ozone production during the formation of photochemical smog (i.e., reaction of UV light from the sun with NO_2 in the atmosphere). HCs are also produced by the natural decomposition of organic matter and by certain plants, such as pines and creosote bushes. A large portion of HCs and NMHCs are due to human activity. For example, emission of formaldehyde can result from the production of new carpeting. HCs are also produced by combustion in gasoline engines and by evaporation of gasoline during fueling cars, especially in summer with high temperatures. Therefore, several states such as California, Maine, and Florida require the installation of special traps on gasoline pumps to collect HC vapors. As for CO, concentration of HCs is largest near heavily traveled highways and streets.

7.3.3 Particulate Matter (PM)

PM is composed of small particles or liquid droplets small enough that they can be suspended in air and is also a "criteria" pollutant. These small particles are not defined by a specific composition;

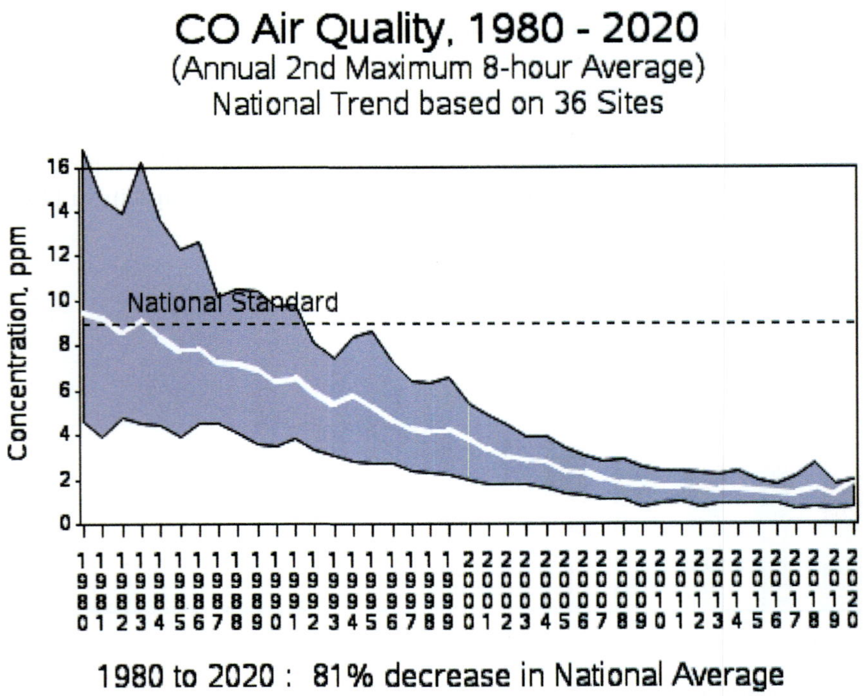

Figure 7.3 CO air quality between 1980 and 2020; reproduced from the US EPA [5].

they can be of many origins and include a complex of compounds such as soot, smoke, dust, asbestos fibers, pesticides, acid, and organic chemicals, as well as metals such as Hg, Fe, Cu, and Pb. Particulate matter is characterized by size. PM_{10} or "inhalable coarse particles" are particles below 10 μm in size that may settle out of the atmosphere in less than 24 h. PM_{10} particles can pass through the throat and nose, entering the lungs. After inhalation, PM_{10} particles can cause health effects in the heart or lungs. They are mostly found near roadways and dusty industries. Smaller particles 2.5 μm or less in size are known as $PM_{2.5}$ or fine particles, and typically require several days to settle out of the atmosphere. $PM_{2.5}$ particles are found in smoke and haze. They can be emitted by forest fires or be formed by gases emitted by power plants, industries, and transportation reacting with air. PM_{10} and $PM_{2.5}$ are regulated by the US EPA and must be reported yearly to the US EPA by every state [6]. According to the US EPA, there was a 26% and 41% decrease of PM_{10} and $PM_{2.5}$ particles between 1990 and 2020 and between 2000 and 2020, respectively (Figure 7.4). Around 60% of PM comes from transportation and around 26% from agriculture, forests, and fires. Construction is also a contributor. Ultrafine particles (UFPs), below 0.1 μm or 100 nm in diameter, are not specifically regulated yet, but they are believed to have more severe health implications than PM_{10} and $PM_{2.5}$ since they are small enough to pass through the lung tissues into the bloodstream and they may be able to translocate through cell membranes [7]. UFPs can be intentionally fabricated or come from volcanic lava, ocean spray, and smoke. Several studies have been conducted that found high concentrations of UFPs near airports and identified them as mostly derived from anthropogenic emission sources [8].

7.3.4 Sulfur Dioxide (SO_2)

SO_2 is mostly derived from burning fossil fuels such as coal that contain sulfur. Coal can contain as much as 6% sulfur, and when burned if not properly filtered can release the sulfur, which can react in the atmosphere producing SO_2 and acid rain (i.e., combination of SO_2 with H_2O to produce sulfuric acid, H_2SO_4). Additional sources of SO_2 emissions are ore smelters and oil refineries. SO_2 can cause respiratory problems and at high concentrations can harm plant tissues. Figure 7.5 shows a 94% reduction in average SO_2 between 1980 and 2020 in the US.

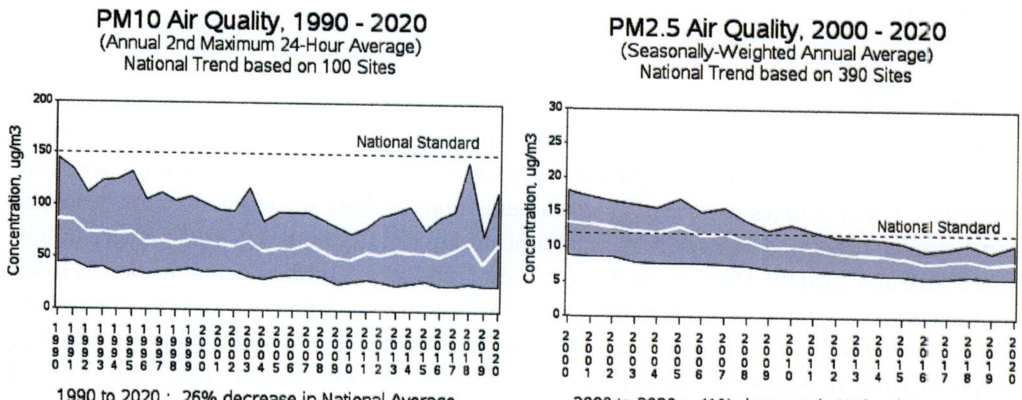

Figure 7.4 PM_{10} and $PM_{2.5}$ air quality in the US between 1990 and 2020 and between 2000 and 2020, respectively, based on 100 and 390 monitored sites, respectively; reproduced from the US EPA [5].

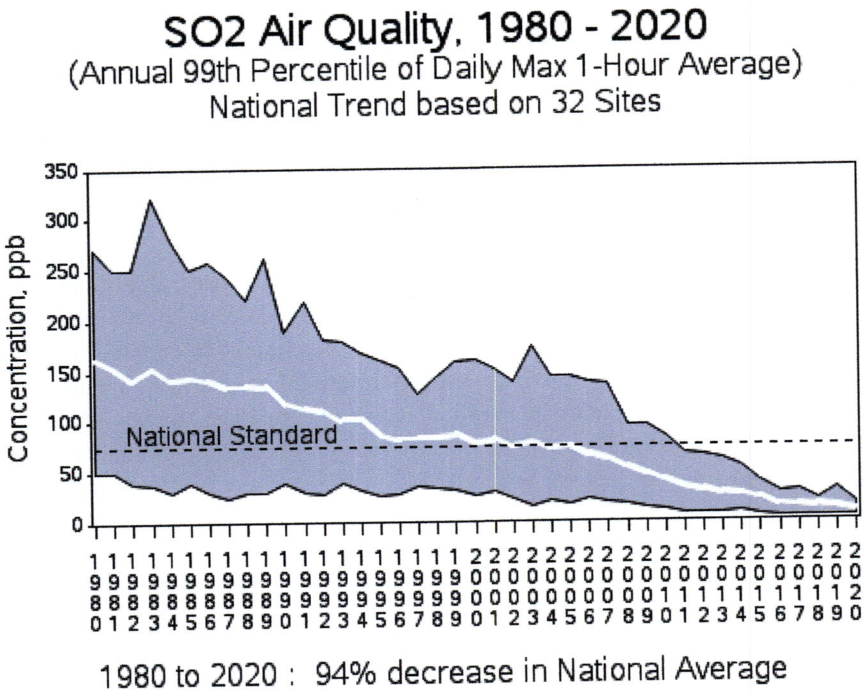

Figure 7.5 SO_2 air quality between 1980 and 2020; reproduced from the US EPA [5].

7.3.5 Nitrogen Oxides (NOx)

NO_x is a mixture of NO and NO_2 that is formed mostly by the combination of N_2 and O_2 during high-temperature combustion, such as burning fuel in cars or power plants. In motor vehicles, catalytic convertors are used to reduce these emissions. NO is responsible for the reddish-brown haze in many cities, and it contributes to heart and lung disease and may be carcinogenic. NO_x is responsible for the formation of nitric acid ($3NO_2 + H_2O = 2HNO_3 + NO$) and other acids contributing to acid rain. Figure 7.6 shows the reduction of NO_2 monitored in 20 sites in the US between 1980 and 2020.

7.3.6 Lead (Pb)

Lead is a highly toxic compound. Lead emissions vary at the national level, but in air they come mostly from ore and metal processing and in the past from piston-engine aircraft using lead-containing fuels. Other sources contributing to lead emissions include waste incinerators, lead-acid battery manufacturers, and lead smelters. Lead concentrations in air in the US were reduced by 86% between 2010 and 2020 [9]. Once taken into the body, lead reaches the blood and the bones. It can affect the nervous system, kidney function, immune system, reproductive and development systems, and the cardiovascular system. It is especially concerning when infants and young children are exposed since it produces neurological and cardiovascular effects and affects learning and lowers IQ. In the US, lead-based paints are still a concern where children are exposed. Houses constructed before 1978 may have lead-based paint, being a health hazard when chipped or removed. Figure 7.7 shows the concentration of lead in the air between 2010 and 2020 monitored at 83 sites.

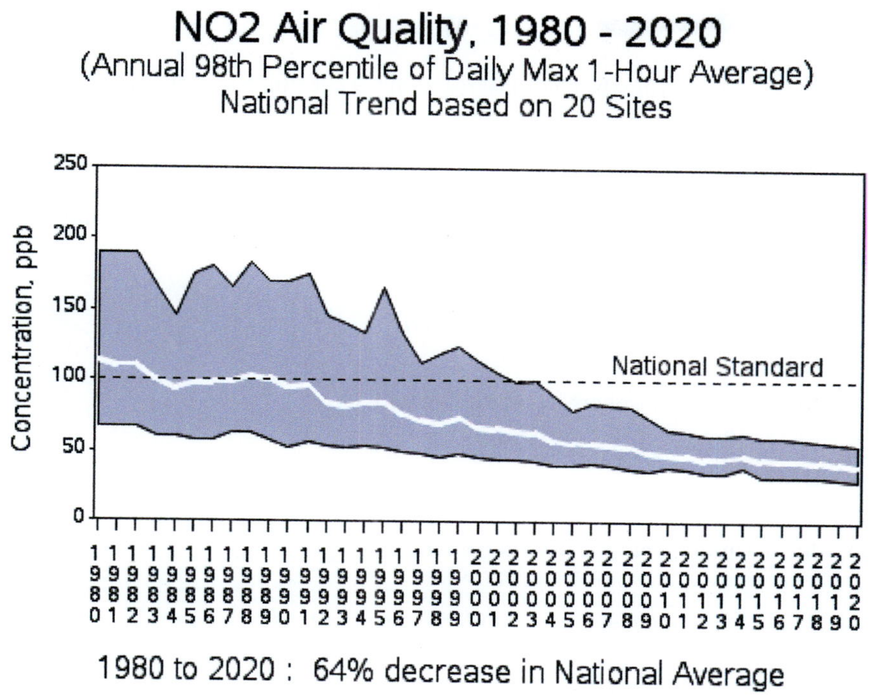

Figure 7.6 NO_2 air quality between 1980 and 2020 from 20 sites; reproduced from the US EPA [5].

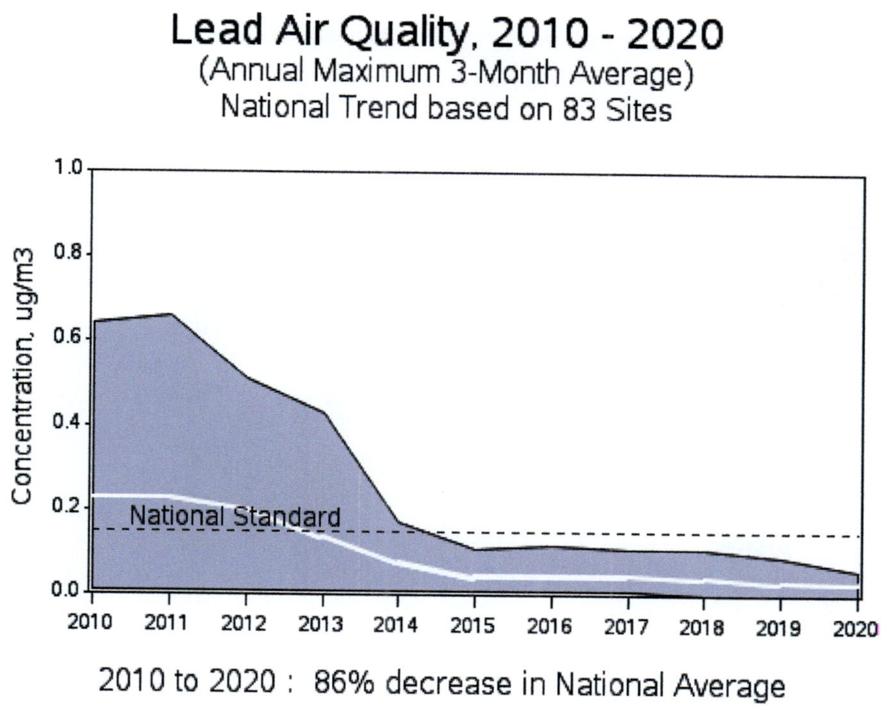

Figure 7.7 Lead air quality between 2010 and 2020 measured in 83 sites; adapted from the US EPA [5].

7.4 Secondary Pollutants

Secondary pollutants are created during the chemical reaction between primary air pollutants and the atmosphere, which contains water vapor and gases (Figure 7.8.) Most of these reactions are facilitated by sunlight, so they ultimately produce photochemical smog, mostly in urban areas and megacities where solar radiation is strong, such as Los Angeles and Mexico City.

Several reactions occur during the formation of photochemical smog, which is mostly composed of O_3, peroxyacetyl nitrate (PAN), and other oxidants as shown in the scheme below.

$$N_2+O_2 \rightarrow 2NO \text{ (inside an engine, nitrogen is oxidized)}$$

$$2NO+O_2 \rightarrow 2NO_2 \text{ (in the atmosphere, combination with } O_2\text{)}$$

$$NO_2+h\nu \rightarrow NO+O$$

$$O+O_2 \rightarrow O_3 \text{ (due to sunlight, atomic oxygen combines with oxygen gas)}$$

$$NO+O_3 \rightarrow NO_2 + O_2 \text{ (reaction depends on sunlight and elevated temperature)}$$

$$HC+ NO+O_2 \rightarrow NO_2 + PAN \text{ (PAN:peroxyacetyl nitrate)}$$

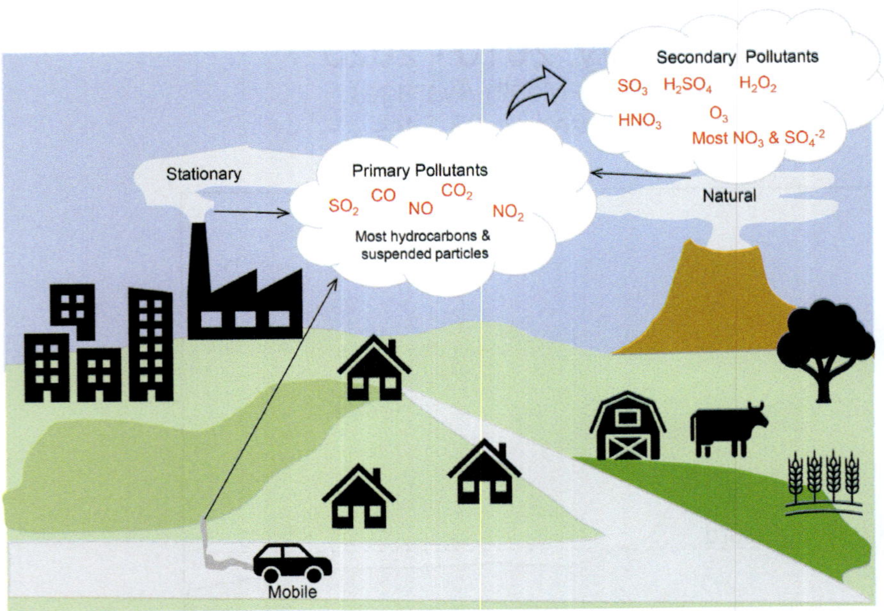

Figure 7.8 Formation of secondary pollutants through the reaction of primary pollutants and compounds in the atmosphere.

If the air temperature exceeds 32°C, with low winds, intense solar radiation, and low precipitation, conditions are favorable for O_3 formation. Unfortunately, in the last decade, many US cities have exceeded the federal air quality standard for O_3 of <80 nLL^{-1} for 1 h 1 d per year averaged over a 3-year period. HCs are necessary for the formation of O_3 in the atmosphere. In the absence of HCs, UV radiation breaks down NO_2 into NO and O, and the O atoms combine with O_2 forming O_3, which combines with NO to form NO_2 and O_2. In the absence of HCs, O_3 will not accumulate in the atmosphere. So, the reduction of emissions from transportation plays a critical role in reducing O_3 formation. For western US cities, photochemical smog is referred to as *brown cloud* (O_3 + PAN + NO_x). Since eastern and midwestern US cities receive less solar radiation, the reaction shown above is not complete and the smog in these areas is referred to as *gray air* due to particles, mostly $PM_{2.5}$ [10]. The presence of O_3 can be beneficial or deleterious depending on its location. Smog O_3 reduces normal functioning of the lungs due to inflammation of cells and increase of asthma. Smog O_3 also damages plant life, slowing growth and reducing yield. O_3 in the ozone layer (high in the stratosphere – formed by the reaction of O_2 and H_2O) – is beneficial since it absorbs damaging UV radiation. Great concerns still exist due to the high concentration of ozone in urban, suburban, and rural areas of the US affecting health despite efforts to reduce the emission of ozone precursors, NOx, and HCs (Figure 7.9).

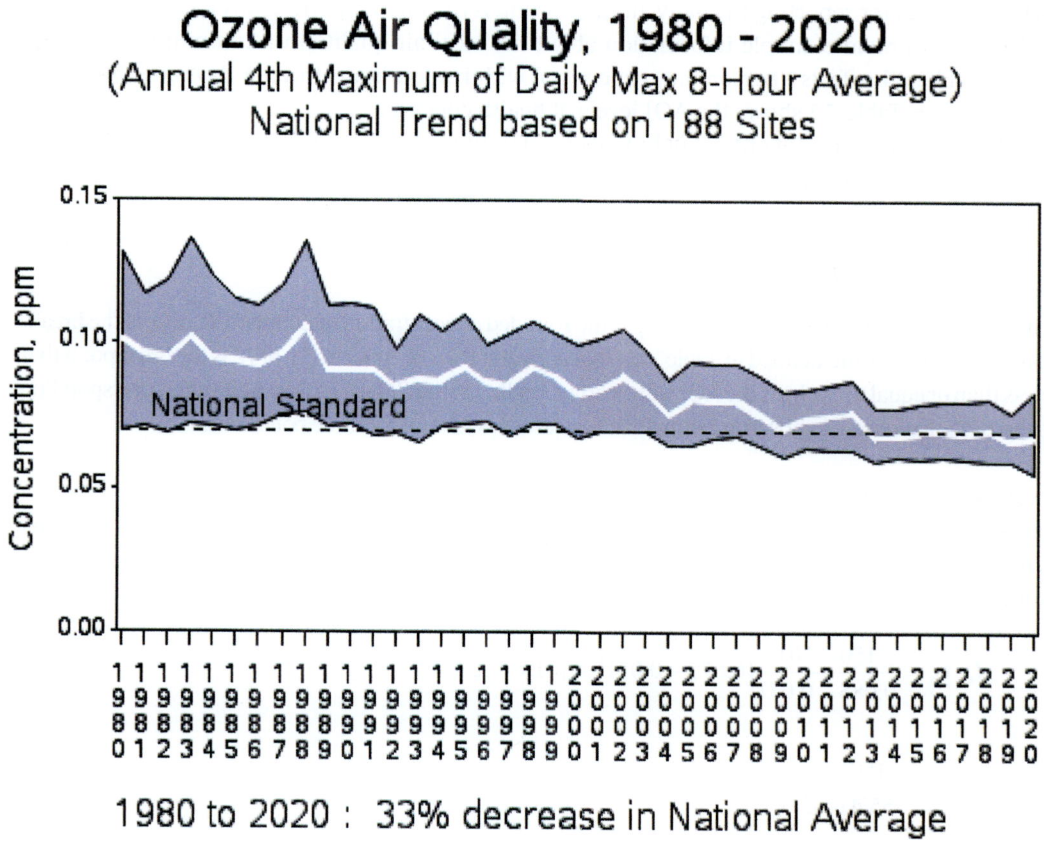

Figure 7.9 Ozone air quality between 1980 and 2020; reproduced from the US EPA [5].

7.5 Clean Air Act

The Clean Air Act passed in 1970 and amended in 1990 provides the framework to regulate outdoor pollutants, known as the NAAQS. The Act requires that the US EPA set NAAQS (40 CFR part 50) for the main pollutants considered harmful to public health and the environment. The Clean Air Act was incorporated into the US Code as title 42, chapter 85. The pollutants regulated in the Clean Air Act are the primary and secondary pollutants, and the EPA has set NAAQS for six principal pollutants called "criteria" air pollutants, which are CO, Pb, NO_2, O_3, PM, and SO_2. The regulated concentration levels of these pollutants are shown in Table 7.2.

The key elements of the Act are:

- To reduce outdoor, or ambient, concentrations of air pollutants that cause smog, haze, acid rain, and other problems.
- To reduce emissions of toxic air pollutants known to, or suspected of, causing cancer or other serious health effects.
- To phase out production and use of chemicals that destroy stratospheric ozone.

Since the implementation and the amendment of the Clean Air Act the critical air pollutants have been reduced in every category (Figure 7.10.)

The Clean Air Act is the main framework for the adoption of the air quality index (AQI). The AQI is an index for reporting the daily air quality and is the main reporting mechanism for critical pollutants. It provides simple information about the local air quality in a community. The AQI focuses on health effects a person may experience within a few hours to a few days after breathing unhealthy air. Table 7.3 shows the AQI levels of health concern.

To translate measures of critical pollutants to AQI, the US EPA provides Table 7.4. Basically, to estimate the AQI, Equation 7.1 is used.

$$I_p = \frac{I_{Hi} - I_{Lo}}{BP_{HI} - BP_{Lp}}\left(C_P - BP_{Lo}\right) + I_{Lo} \tag{7.1}$$

where I_P is the index for the pollutant P; C_P is the rounded concentration of pollutant P; BP_{HI} is the breakpoint – as indicated and defined in Table 7.4 – that is greater than or equal to C_P; BP_{LO} is the breakpoint that is less than or equal to C_P; I_{HI} is the AQI value corresponding to the BP_{HI}; I_{LO} is the AQI value corresponding to the BP_{LO}. Exercise 7.1 shows an example of how to estimate AQI for three pollutants.

Exercise 7.1 Estimation of AQI

Suppose you have an 8-h ozone value of 0.077 ppm, a $PM_{2.5}$ value of 54, and a CO value of 8.4 ppm. What is the AQI?

Solution:

$$I_{O_3\text{-8h}} = \frac{150 - 101}{0.085 - 0.071}\left(0.077 - 0.071\right) + 101 = 122$$

$$I_{PM2.5} = \frac{150 - 101}{55.4 - 35.5}\left(54 - 35.5\right) + 101 = 146$$

$$I_{CO} = \frac{100 - 51}{9.4 - 4.5}\left(9.4 - 8.4\right) + 51 = 61$$

Answer: AQI = 146 due to $PM_{2.5}$, color orange, unhealthy for sensitive groups.

Table 7.2 National Ambient Quality Standards for six critical pollutants; reproduced from the US EPA [11].

Pollutants [links to historical tables of NAAQS reviews]		Primary/ Secondary	Averaging Time	Level	Form
Carbon monoxide (CO)		Primary	8 h	9 ppm	Not to be exceeded more than once per year
			1 h	35 ppm	
Lead (Pb)		Primary and secondary	Rolling 3-month average	0.15 μgm^{-3} (1)	Not to be exceeded
Nitrogen dioxide (NO_2)		Primary	1 h	100 ppb	98th percentile of 1-h daily maximum concentration, averaged over 3 years
		Primary and secondary	1 year	53 ppb (2)	Annual mean
Ozone (O_3)		Primary and secondary	8 h	0.070 ppm (3)	Annual fourth-highest daily maximum 8-h concentration, averaged over 3 years
Particle pollution (PM)	$PM_{2.5}$	Primary	1 year	12 μgm^{-3}	Annual mean, averaged over 3 years
		Secondary	1 year	15 μgm^{-3}	Annual mean, averaged over 3 years
		Primary and secondary	24 h	35 μgm^{-3}	98th percentile, averaged over 3 years
	PM_{10}	Primary and secondary	24 h	150 μgm^{-3}	Not to be exceeded more than once per year on average over 3 years
Sulfur dioxide (SO_2)		Primary	1 h	75 ppb (4)	99th percentile of 1-h daily maximum concentrations, averaged over 3 years
		Secondary	3 h	0.5 ppm	Not to be exceeded more than once per year

1 In areas designated nonattainment for the Pb standards prior to the promulgation of the current (2008) standards, and for which implementation plans to attain or maintain the current (2008) standards have not been submitted and approved, the previous standards (1.5 μgm^{-3} as a calendar quarter average) also remain in effect.
2 The level of the annual NO_2 standard is 0.053 ppm. It is shown here in terms of ppb for the purposes of clearer comparison to the 1-h standard level.
3 Final rule signed October 1,j12 2015, and effective December 28, 2015. The previous (2008) O3 standards are not revoked and remain in effect for designated areas. Additionally, some areas may have certain continuing implementation obligations under the prior revoked 1-h (1979) and 8-h (1997) O_3 standards.
4 The previous SO_2 standards (0.14 ppm 24-h and 0.03 ppm. annual) will additionally remain in effect in certain areas: (1) any area for which it is not yet 1 year since the effective date of designation under the current (2010) standards, and (2) any area for which an implementation plan providing for attainment of the current (2010) standard has not been submitted and approved and which is designated nonattainment under the previous SO_2 standards or is not meeting the requirements of a SIP call under the previous SO_2 standards (40 CFR 50.4(3)). A SIP call is an EPA action requiring a state to resubmit all or part of its State Implementation Plan to demonstrate attainment of the required NAAQS.

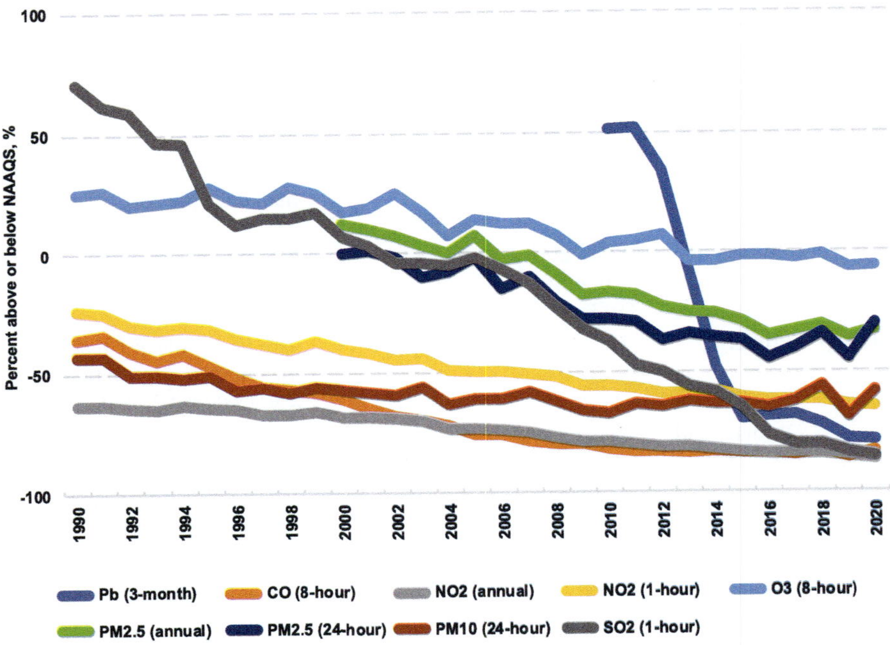

Figure 7.10 Reduction of critical air pollutants between 1990 and 2020 as reported by the US EPA.

The states are required to enforce the Clean Air Act. AirNow.gov is the main source for AQI reporting for all zip codes in the US. Besides the air critical pollutants, the US EPA has identified 188 hazardous air pollutants (HAPs) or urban air toxics (UATs), which need to be reported.

Table 7.3 Air quality index levels of health concern and color codes; adapted from [12].

Air Quality Index (AQI), Level of Health Concern	Numerical Value	Meaning
Good	0 to 50	Air quality is considered satisfactory, and air pollution poses little or no risk.
Moderate	51–100	Air quality is acceptable; however, for some pollutants there may be a moderate health concern for a very small number of people who are unusually sensitive to air pollution
Unhealthy for sensitive groups	101–150	Members of sensitive groups may experience health effects. The general public is not likely to be affected.
Unhealthy	151–200	Everyone may begin to experience health effects; members of sensitive groups may experience more serious health effects.
Very unhealthy	201–300	Health alert: everyone may experience more serious health effects.
Hazardous	301–500	Health warnings of emergency conditions. The entire population is more likely to be affected.

Note: Values above 500 are considered beyond the AQI. Follow recommendations for the "Hazardous category."

Table 7.4 AQI breakpoints for calculation of AQI; reproduced from [13].

These Breakpoints...							... Equal This AQI	... And This Category
O_3 (ppm) 8 h	O_3 (ppm) 1 h[1]	$PM_{2.5}$ (μgm^{-3}) 24 h	PM_{10} (μgm^{-3}) 24 h	CO (ppm) 8 h	SO_2 (ppb) 1 h	NO_2 (ppb) 1 h	AQI	Category
0.000–0.054	-	0.0–12.0	0–54	0.0–4.4	0–35	0–53	0–50	Good
0.055–0.070	-	12.1–35.4	55–154	4.5–9.4	36–75	54–100	51–100	Moderate
0.071–0.085	0.125–0.164	35.5–55.4	155–254	9.5–12.4	76–185	101–360	101–150	Unhealthy for sensitive groups
0.086–0.105	0.165–0.204	(55.5–150.4)[3]	255–354	12.5–15.4	(186–304)[4]	361–649	151–200	Unhealthy
0.106–0.200	0.205–0.404	(150.5–250.4)[3]	355–424	15.5–30.4	(305–604)[4]	650–1249	201–300	Very unhealthy
0.201– (2)	0.405–0.504	(250.5–350.4)[3]	425–504	30.5–40.4	(605–804)[4]	1,250–1649	301–400	hazardous
(2)	0.505–0.604	(350.5–500.4)[3]	505–604	40.5–50.4	(805–1,004)[4]	1,650–2,049	401–500	

[1]Areas are generally required to report the AQI based on 8-h ozone values. However, there are a small number of areas where an AQI based on 1-h ozone values would be more precautionary. In these cases, in addition to calculating the 8-h ozone index value, the 1-h ozone value may be calculated, and the maximum of the two values reported.
[2]8-h O_3 values do not define higher AQI values (≥301). AQI values of 301 or higher are calculated with 1-h O_3 concentrations.
[3]If a different Significant Harm Level (SHL) for $PM_{2.5}$ is promulgated, these numbers will change accordingly.
[4]1-h SO_2 values do not define higher AQI values (≥ 200). AQI values of 200 or greater are calculated with 24-h SO_2 concentrations.

7.6 Case Study 7.1. Air Quality in Delhi, India, in Winter

Every year in winter months, especially from October to December, New Delhi, India, air quality has been the subject of international news. Citizens complain about stinging eyes, sore throat, and breathlessness. The New Delhi overall air quality index, which captures pollutants such as $PM_{2.5}$, rose to 477 on a scale of 500. Values of $PM_{2.5}$ have been reported as high as around 300 μgm^{-3}/24 h in 2010 and 240 μgm^{-3}/24 h in 2018, and with values as high as 1,250 μgm^{-3}/24 h in 2016. The highest pollution values coincide with the winter months during Diwali and crop harvest weeks [14].

The leading causes of pollution in Delhi have been attributed to:

a) vehicular emissions due to cars with engines of questionable quality putting out high volumes of smoke and haze due to the improper combustion of fossil fuels – the automotive sector is responsible for around 40% of the pollution in Delhi;

b) combustion of leftover plant matter or stubble to clear land in agricultural areas after the September harvest;
c) seasonal changes such as decreased wind speed and reduced rain that would otherwise help reduce pollution;
d) the Delhi topography, where winds blow in, can bring with them large amounts of smoke and haze from other parts that are trapped when winds decrease in winter. Furthermore, when the air temperature drops during the winter months, the pollutants in the air remain closer to the ground, forming the mixed cocktail of contaminants responsible for the haze that blankets the city (see Figure 7.11);
e) the last leading cause of pollution is the industrial sector, responsible for around 19% of the pollution, with many factories and production plants of varying sorts located around the city playing a considerable role in the release of smoke into the atmosphere. During periods of high pollution, the government has ordered fuel and coal-based factories to close to halt the shattering rise of pollution levels [15].

High air pollution values are responsible for the health effects of breathing pollutants such as PM_{10} and $PM_{2.5}$. Air pollutants have a significant health risk since they can cause stroke, heart disease, lung cancer, and chronic and acute respiratory diseases. Of the top 10 most polluted cities reported by the World Health Organization (WHO), 9 are in India, with Delhi, Ghaziabad, and Noida among the top ranking. The WHO target level for $PM_{2.5}$ is $10\ \mu\ gm^{-3}/24\ h$ [15].

Delhi's Environment Pollution (Prevention and Control) Authority (EPCA), created in 1998 by the India Supreme Court, has declared air pollution a public health emergency several times each year. In November 2019, the EPCA reported 48 h of levels exceeding the most severe tier in the air quality index: 5 times the standard considered acceptable in India and nearly 10 times the standard deemed safe by the WHO [16]. Several immediate remediation measurements were put in place, such as stopping specific dirty industries, restricting outdoor activities, and ending all construction activities. The journey to improve the air quality of New Delhi will take years or even decades.

Figure 7.11 Air quality in New Delhi, India, and surrounding areas; (left) picture of afternoon smoke near a country road; (right) picture of Taj Mahal on November 23, 2017 with high AQI values.

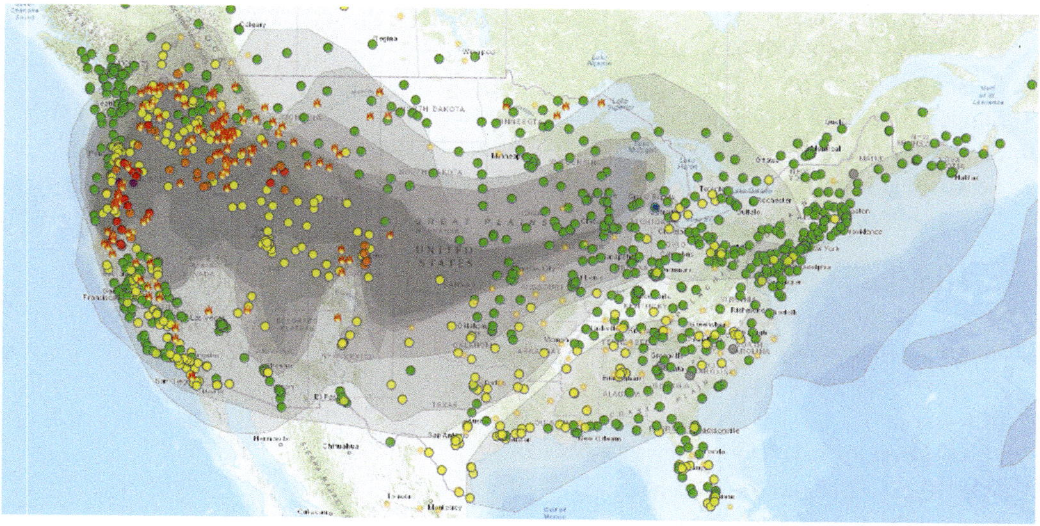

Figure 7.12 Fire and smoke map in September 8, 2020, in the US; reproduced from [19].

7.7 Case Study 7.2. Air Quality in the US in Summer

Every year during summer, the US is experiencing an increasing number of fires due to extensive heat waves and droughts in the west and mountain regions of the country, with Idaho, Montana, Washington, California, Oregon, Wyoming, Nevada, Colorado, and Utah reporting the largest numbers of fires, Figure 7.12. In the last 10-year period (2011–2020), on average more than 40,000 fires and more than 5 million burned acres have been reported [17]. Due to these large fires, air quality in those regions and almost in every state in the US during summer have been affected, with AQI higher than 150 in cities in the west region. Figure 7.12 shows that a large plume of smoke (gray color) is created due to the presence of fires. Recent research has shown that wildfires have accounted for up to 25% of $PM_{2.5}$ in recent years across the US, and up to half in some of the Western regions [18]. As this book was written, there was increasing concern about climate-induced wildfires.

7.8 Study Questions

1. What are the main sources and factors affecting the production and movement of air pollutants?
2. What are the main primary air pollutants and their main sources?
3. What are secondary air pollutants?
4. What is photochemical smog?
5. What does the Clean Air Act regulate?
6. What are the key elements defined in the Clean Air Act?
7. Define AQI; which are the main critical pollutants regulated by the Clean Air Act?
8. What elements are regulated by the Clean Air Act as key elements?
   A CO, HCs, $PM_{2.5}$, PM_{10}, SO_2, NO_2, and Pb
   B O_3, $PM_{2.5}$, PM_{10}, CO, SO_2, and NO_2

C O_3, PM, and Pb
D O_3, $PM_{2.5}$, PM_{10}, CO, NO_2, and Pb
E None of the above

9 Suppose you had a 1-h value of ozone of 0.130, and an 8 h value of 0.110 ppm. What is the AQI?

10 Suppose you had a 1-h value of ozone of 0.140, and an 8-h value of 0.100 ppm. What is the AQI?

11 Lead is regulated as a key critical pollutant by the Clean Air Act. T F

7.9 Additional Resources

1. Criteria Air Pollutants
https://www.epa.gov/criteria-air-pollutants
2. AirNow.gov
https://airnow.gov/index.cfm?action=aqibasics.aqi
3. EARTHDATA – Open access for open science
https://earthdata.nasa.gov/learn/getting-started
4. NOOA – Teaching Climate
https://www.climate.gov/climate-and-energy-topics/causes-climate-change
5. Air Emissions Inventories
https://www.epa.gov/air-emissions-inventories/air-pollutant-emissions-trends-data
6. Fire and Smoke Map
https://fire.airnow.gov#
7. Arunabha Ghosh: Five steps for clean air in India
https://www.ted.com/talks/arunabha_ghosh_5_steps_for_clean_air_in_india
8. Paul Hessburg: Why wildfires have gotten worse – and what we can do about it
https://www.ted.com/talks/paul_hessburg_why_wildfires_have_gotten_worse_and_what_we_can_do_about_it

References

1. Brusseau, M.L., Pepper, I.L., and Gerba, C.P. (2019). *Environmental and Pollution Science*, 2nd Edition. London, UK: Academic Press Publications.
2. US Environmental Protection Agency. (2014). 2014 National Emissions Inventory. https://gispub.epa.gov/neireport/2014/# (accessed May 22, 2019).
3. US Environmental Protection Agency. (n.d.). *National Emissions Inventory (NEI)*. https://www.epa.gov/air-emissions-inventories/national-emissions-inventory-nei. (accessed September 7, 2021).
4. US Environmental Protection Agency. (n.d.). Our Nation's Air 2021. https://gispub.epa.gov/air/trendsreport/2021/#home (accessed September 7, 2021).
5. US Environmental Protection Agency. (n.d.). *National Air Quality: Status and Trends of Key Air Pollutants*. https://www.epa.gov/air-trends (accessed September 7, 2021).
6. US Electronic Code of Federal Regulations (eCFR). (n.d.). https://www.ecfr.gov/current/title-40/chapter-I/subchapter-C/part-50 (accessed February 8, 2021).
7. Terzano, C., Di Stefano, F., Conti, V., Graziani, E., and Petroianni, A. (2010). Air pollution ultrafine particles: Toxicity beyond the lung. *European Review for Medical and Pharmacological Sciences* 14: 809–821.

8 Chen, R., Hu, B., Liu, Y., Xu, J., Yang, G., Xu, D. et al. (2016). Beyond PM2.5: The role of ultrafine particles on adverse health effects of air pollution. *Biochimica Et Biophysica Acta – General Subjects* 1860: 2844–2855. DOI: 10.1016/J.BBAGEN.2016.03.019.

9 US Environmental Protection Agency. (2014). *Particulate matter concentrations*. 12. https://www.epa.gov/roe (accessed October 11, 2019).

10 A rising problem | Earthdata. (n.d.). https://earthdata.nasa.gov/learn/sensing-our-planet/a-rising-problem (accessed February 8, 2021).

11 US Environmental Protection Agency. (n.d.). NAAQS Table | Criteria Air Pollutants | US EPA. https://www.epa.gov/criteria-air-pollutants/naaqs-table (accessed February 8, 2021).

12 US Environmental Protection Agency. (2006). *Guidelines for the reporting of daily air quality – The Air Quality Index (AQI). US environ prot agency*. 1–31. https://nepis.epa.gov/Exe/ZyNET.exe/P1006KOQ.TXT?ZyActionD=ZyDocument&Client=EPA&Index=2006+Thru+2010&Docs=&Query=&Time=&EndTime=&SearchMethod=1&TocRestrict=n&Toc=&TocEntry=&QField=&QFieldYear=&QFieldMonth=&QFieldDay=&IntQFieldOp=0&ExtQFieldOp=0&XmlQuery= (accessed February 26, 2021).

13 Mintz, D. (2013). *Technical assistance document for the reporting of daily air quality – The Air Quality Index (AQI)*.

14 Delhi Air Quality Index (AQI) and India Air Pollution | AirVisual. (n.d.). https://www.iqair.com/us/india/delhi (accessed February 26, 2021).

15 IQAir. (2019). *World air quality report*. 2019 World Air Qual Rep. 1–35. https://www.iqair.com/world-most-polluted-cities/world-air-quality-report-2019-en.pdf (accessed February 26, 2021).

16 Delhi In A Chokehold: Air Pollution As A Public Health Emergency | Health Affairs. (n.d). https://www.healthaffairs.org/do/10.1377/hblog20200130.710866/full (accessed February 26, 2021).

17 National Fire News | National Interagency Fire Center. (n.d.). https://www.nifc.gov/fire-information/nfn (accessed September 8, 2021).

18 Burke, M., Driscoll, A., Heft-Neal, S., Xue, J., Burney, J., and Wara, M. (2021). The changing risk and burden of wildfire in the United States. *Proceedings of the National Academy of Sciences of the United States of America* 118: DOI: 10.1073/PNAS.2011048118.

19 Fire and Smoke Map. (n.d.). https://fire.airnow.gov/# (accessed September 8, 2021).

8

Global Climate Change

8.1 Learning Objectives

After learning from this chapter, you are expected to:

- Understand the concept of radiative forcing.
- Recognize the main pollutants affecting global climate change.
- Understand which main pollutants impact the cooling and warming of the planet.
- Understand the main impacts of global climate change.
- Describe the main climate change agreements.
- Explain the main role of the IPCC regarding increasing awareness about global climate change.

8.2 Introduction

Global climate is determined by the balance between energy entering the earth system from the sun, and energy lost from the earth to space. Several pollutants have a direct effect on this balance by strongly interacting with the radiation present in the atmosphere, affecting how much is absorbed or reflected at various wavelengths. Others have an indirect effect by impacting the amount of ozone and other substances present within the atmosphere that can interact with radiative energy. The energy transferred by electromagnetic radiation is proportional to the wavelength of the light waves. So, shorter wavelength radiation, like visible light produced by the sun, has higher energy than longer wavelength radiation, such as near-infrared radiation emitted by the earth.

The radiation environment on earth is a function of three main radiative processes: the flux of radiant energy reaching the earth's surface from the sun; the redistribution of the radiant energy between the ground and the atmosphere; and the loss of radiant energy to space, as shown in Figure 8.1. The overall radiation balance of the earth-atmosphere involves absorption, scattering, transmission, and emission of energy. Mostly, two wavelength intervals are important: short wave (solar) and longwave radiation. Shortwave radiation consists of wavelengths from about 0.15 to 3.0 μm (in this range are ultraviolet (0.15–0.36 μm), visible (0.36–0.75 μm), and near-infrared (0.75–3.0 μm) light). Longwave radiation is between 3.0 and 100 μm. Near-infrared, ultraviolet, and visible light are the most important radiation types for the global environment and climate

Life Cycle of Sustainable Packaging: From Design to End of Life, First Edition. Rafael A. Auras and Susan E. M. Selke.

Companion Website: www.wiley.com/go/Auras/lifecycleofsustainablepackaging

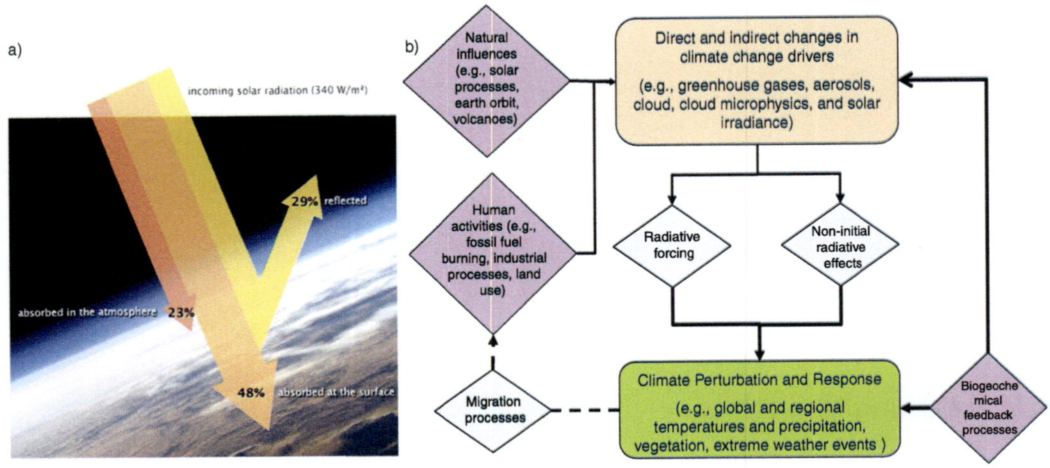

Figure 8.1 (a) Radiative forcing (RF) concept, and (b) components of the climate change process; adapted from [1].

since they carry most of the energy. Around 43% of the energy in the form of light that is emitted by the sun is in the visible range. Of the remainder, 49% is in the near-infrared and 7% in the UV range. Less than 1% is emitted as X-rays, gamma waves, and radio waves. UV-C and UV-B radiation contain a lot of energy, but they are mostly filtered out by the earth's upper atmosphere, so they have relatively little effect on the climate.

Once the sun's energy reaches earth, it is absorbed by gases such as ozone, oxygen, and water vapor. Some of the sun's energy is reflected back to space. As radiation is absorbed by chemical compounds, the atoms in the chemical compound move faster. This absorbed energy is transformed into heat energy, which plays a crucial role in regulating the temperature of the earth's crust, surface waters, and lower atmosphere. The balance between absorption and reflection is quantified using the concept of radiative forcing (RF, defined as the difference between insolation (sunlight) absorbed by the Earth and energy radiated back to space). Typically, "RF is quantified at the tropopause in units of watts per square meter of the Earth's surface. A positive forcing (more incoming energy) warms the system, while negative forcing (more outgoing energy) cools it. Causes of radiative forcing include changes in insolation and the concentration of radiatively active gases, commonly known as greenhouse gases and aerosols" [1].

As energy from the sun passes through clear air, some energy is scattered by atmospheric molecules. Visible light reaching the earth's surface is absorbed and energy heats the soil and air and heats or evaporates water. The earth's atmosphere is mostly opaque to most longwave radiation, but a window exists between 8 and 11 μm, which permits energy to escape. Absorption and re-emission of longwave energy within the atmosphere occurs within the vibrational energy mode of various molecular species such as water vapor, carbon dioxide, nitrous oxide, and methane, which are known as greenhouse gases. These gases, by reabsorbing the longwave energy that would otherwise escape from the earth, create the warming known as the greenhouse effect, much as the glass in a greenhouse keeps the inside of the greenhouse warmer, by interfering with heat loss.

Without the warming of the greenhouse effect, the earth's temperature would be much lower. Currently, the average temperature is 15°C. However, as greenhouse gases accumulate, there is concern that the average global temperature will increase by 1.4–5.8°C, with associated impacts on precipitation.

8.3 Greenhouse Gases

The primary greenhouse gas, meaning the one present in the largest quantities and having the greatest total effect, is carbon dioxide (CO_2). The next most significant greenhouse gas is methane (CH_4). Methane is a greenhouse gas itself, and when it is decomposed, it becomes CO_2, so it is still a greenhouse gas. Other greenhouse gases include nitrous oxide (N_2O), chlorofluorocarbons (CFCs), and hydrofluorocarbons (HFCs). Human influences have resulted in significant increases in a variety of greenhouse gases, leading to concerns about global climate change.

Water vapor is a very potent greenhouse gas and is present in the atmosphere in relatively high amounts. However, since it remains in the atmosphere for a very short time compared to other gases, and further because it is so variable with weather and temperature, its overall impact on climate change is not as large as might be expected. For these reasons, water vapor is seldom listed explicitly as a greenhouse gas. Rather, its impact on climate is built into climate models that include related measures such as the impact of clouds of different types.

CO_2 concentrations and N_2O emissions have risen primarily as a consequence of industrial activities involving the burning of fossil fuels. CH_4 concentrations have risen due to a variety of human activities ranging from drilling to natural gas to cattle raising to the cultivation of rice to municipal waste landfills. Other potent greenhouse gases such as CFCs and HFCs are almost entirely due to human activity. Table 8.1 shows the change in concentration of some greenhouse gases compared to preindustrial levels.

To quantify the overall impact of various greenhouse gases, the common approach is to convert them to equivalent amounts of CO_2 (CO_2-eq), defined as the amount of CO_2 that would have the same impact on global warming. Another useful measure is the Annual Greenhouse Gas Index (AAGI), defined as the ratio of total direct radiative forcing due to long-lived greenhouse gases to the 1990 base year. Figure 8.2 shows changes in total greenhouse gases, CO_2, and the AAGI.

Figure 8.3 shows the changes in the proportion of radiative forcing due to various greenhouse gases over time, compared to 1750 as the baseline year.

There are other influences on climate, in addition to greenhouse gases. The major positive and negative influences, including natural factors such as changes in solar irradiance as well as anthropogenic factors (associated with human activity), are summarized in Figure 8.4.

Sources of greenhouse gases

The primary source of greenhouse gases, globally, is energy production, followed by agriculture, as shown in Figure 8.5.

Table 8.1 Selected greenhouse gas concentrations. Data obtained from the NOAA Global Monitoring Laboratory [2].

Greenhouse Gas	Concentration			
	Preindustrial	1990	2010	2021
CO_2, ppm	280	354	368.9	415.5
CH_4, ppb	700	1,714.5	1,773.2	1,891.6
N_2O, ppb	270	308.7	315.5	334.1
CFC-11, ppt	0	259.5	259.4	223.6
HFC-23, ppt	0	7.4	12.6	30
CF_4, ppt	40	2.33	4.58	9.96

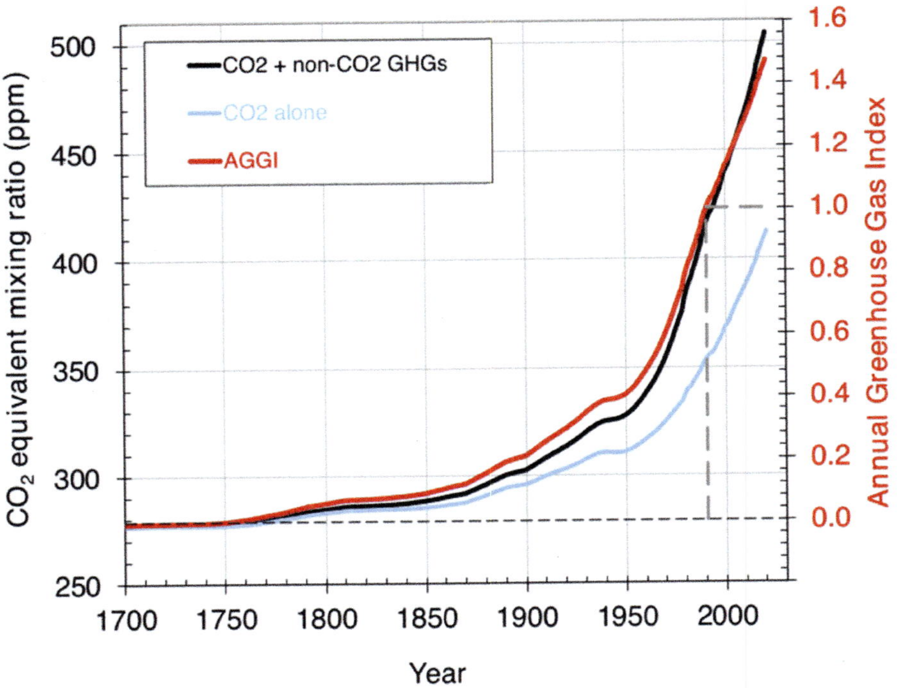

Figure 8.2 Changes in greenhouse gas concentrations and Annual Greenhouse Gas Index; reproduced from the National Oceanic and Atmospheric Administration [2].

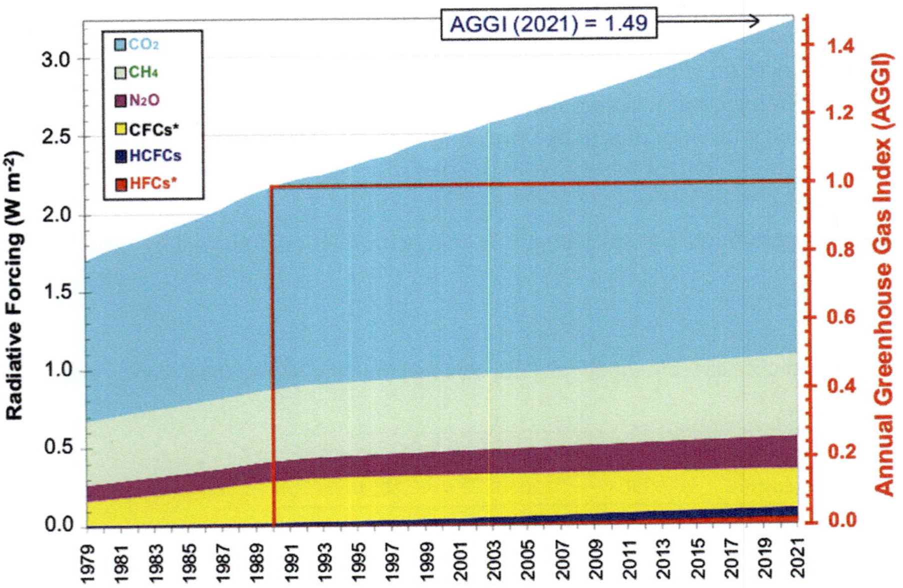

Figure 8.3 Radiative forcing from greenhouse gases, compared to 1750 as the baseline; reproduced from the National Oceanic and Atmospheric Administration, US Department of Commerce [2].

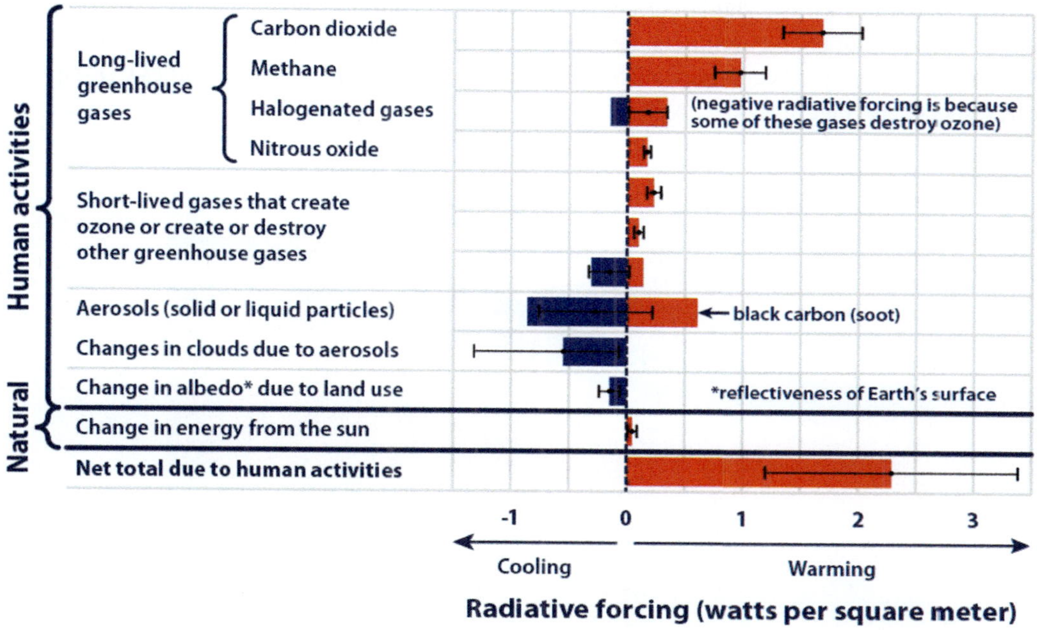

Figure 8.4 Climate forcing factors, for 2011 compared to 1750; reproduced from the US EPA [3]. *Albedo refers to the proportion of the incident radiation that is reflected by the planet.

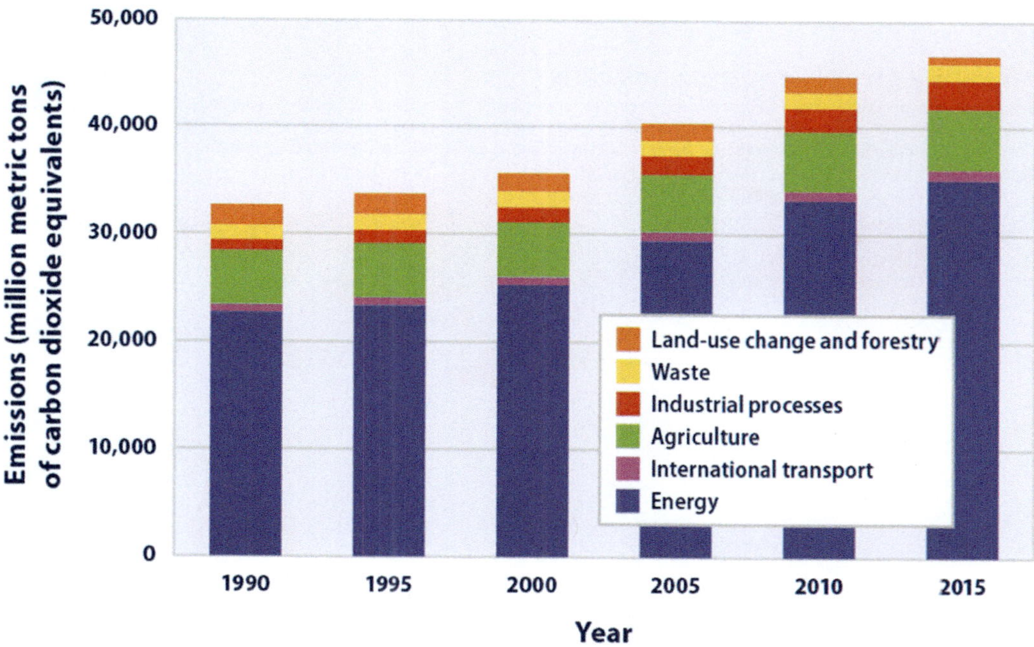

Figure 8.5 Global greenhouse gas emissions by sector, 1990–2015; reproduced from the US EPA [3].

Figure 8.6 shows global greenhouse gas emissions by region. The rapid increase in emissions from Asia is clear; much of this increase is from China, which is now the country with the largest annual emissions of greenhouse gases.

However, this number changes when we account for the population of each country (i.e., CO_2 emissions per capita) with Saudi Arabia being the first followed by Kazakhstan, Australia, US,

Canada, and South Korea. The larger emissions for small countries is because they are oil producers. Of the major country emitters, for example, in the US the average person emits more than 18 tonnes of CO_2-eq per year while in China and India it is less than 8 and 2.5 tonnes CO_2-eq per year, respectively (Figure 8.7).

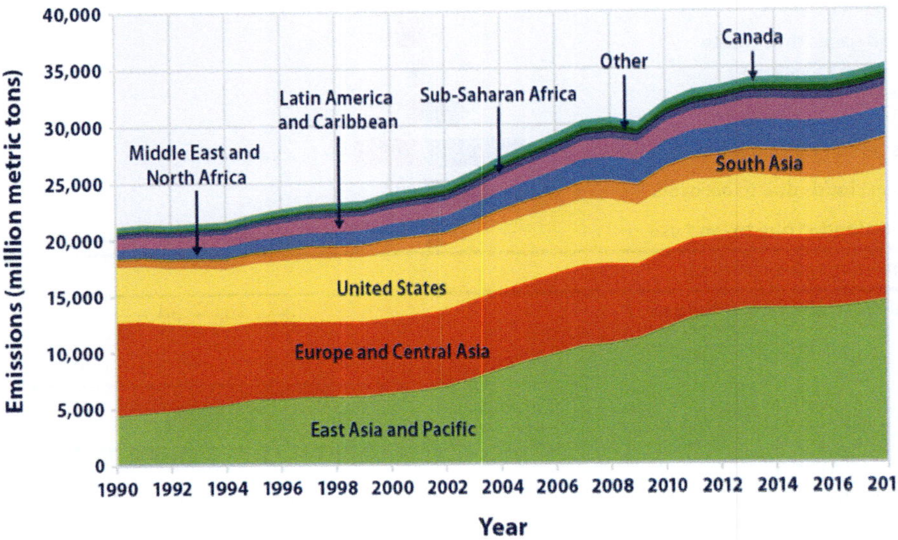

Figure 8.6 Global greenhouse gas emissions by region, 1990–2018; reproduced from the US EPA [3].

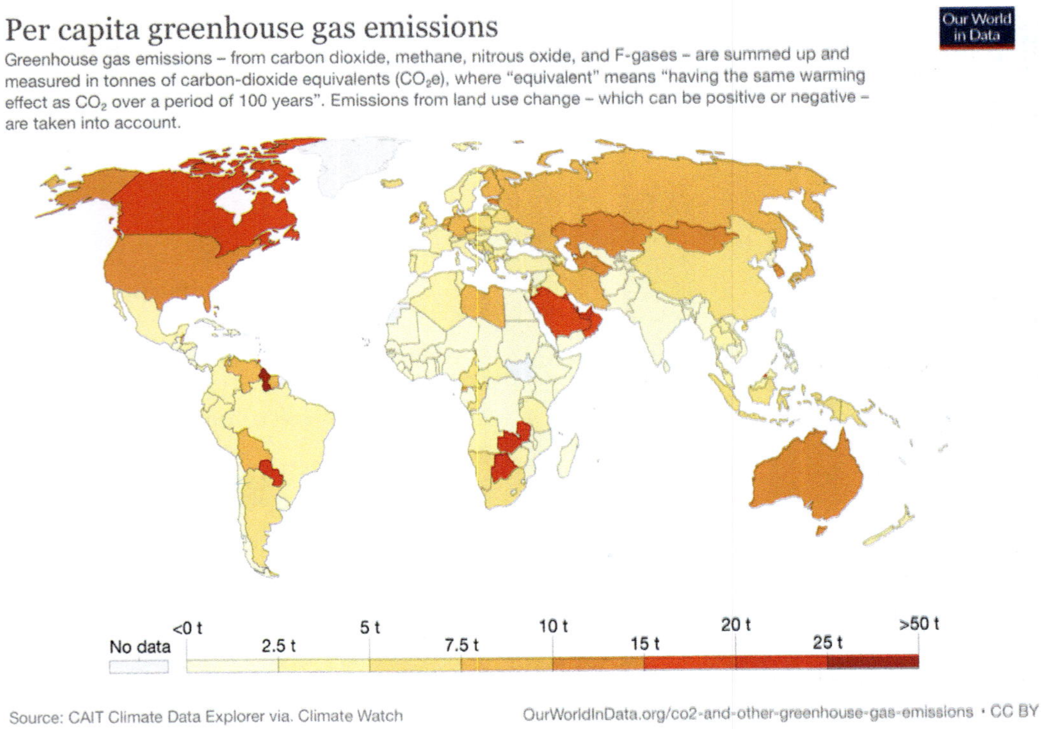

Figure 8.7 Greenhouse gas emissions per capita for 2016; reproduced from [4] under the Creative Commons License CC BY 4.0.

8.4 Impacts on Global Climate

One of the main effects of increasing concentrations of greenhouse gases and the associated increases in global radiative forcing is an increase in average global temperature. Figure 8.8 shows average global temperatures for 1880–2020. The pattern of increasing temperatures, superimposed on year-to-year fluctuations, is clear. Several sources list 2020 as tying with 2016 for the warmest year on record; others have it as a close second. All analyses agree that the 2010–2019 decade was the warmest decade on record, surpassing 2000–2010, which in turn broke the 1990–2000 record. Average global temperature in 2020, according to NOAA, was 0.98°C (1.76°F) above the average for the twentieth century.

Increased global temperatures have a variety of impacts. They contribute to increased heat waves in urban areas, lengthen growing seasons, affect productivity of agricultural land, contribute to melting of ice and snow, etc. Thorough coverage is beyond the scope of this book, but we will briefly mention some of the major consequences. It should also be noted that temperature changes are not uniform around the world; polar regions have warmed much more than lower latitudes.

One of the major impacts of global climate change is an increase in the variability of weather. Global warming leads to more intense precipitation events, both droughts and floods. While there is disagreement about whether hurricanes are becoming more frequent, there is general agreement that they are, on average, stronger. Melting of glaciers and polar ice caps, along with thermal expansion of water, leads to an increase in global sea level (Figure 8.9). Increased amounts of CO_2 dissolved in the oceans results in an increase in the acidity of oceans, and in turn impacts the ability of organisms such as crustaceans to build strong shells. Changes in ocean temperature and salinity can lead to changes in ocean currents.

A particular concern is the long-term nature of many of these impacts. Some of this is due to the long atmospheric residence time of many greenhouse gases, so emissions now can impact global temperatures for decades or even longer. Feedback cycles add another complication. For example, increasing global temperature leads to increased melting of permafrost, releasing methane, which itself is a strong greenhouse gas and thus causes further warming. A major concern is that there are likely "tipping points" that, once passed, make it impossible to easily reverse the impacts of climate change.

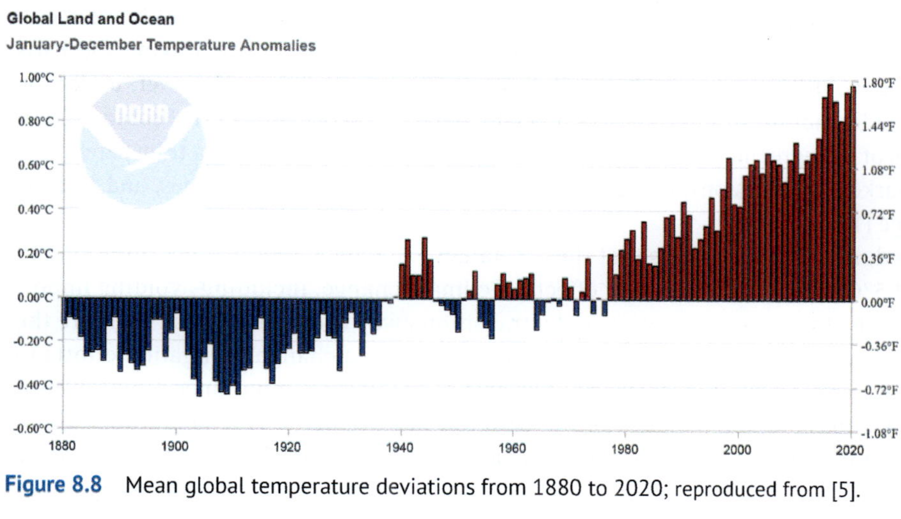

Figure 8.8 Mean global temperature deviations from 1880 to 2020; reproduced from [5].

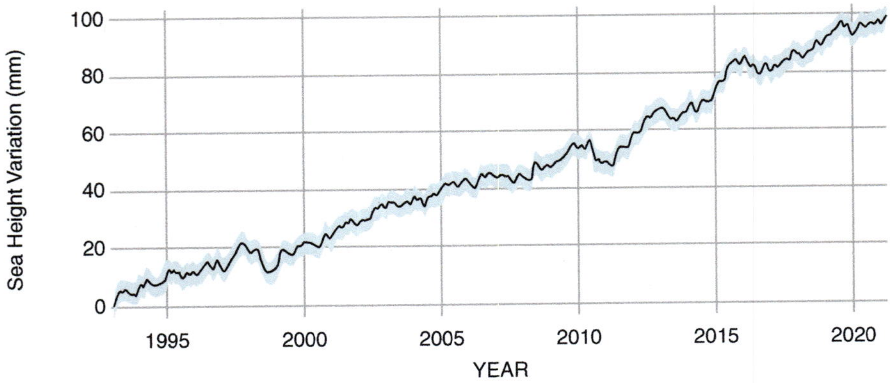

Figure 8.9 Global sea level rise; figure reproduced from [6].

Analyses of the impacts of global climate change are typically based on climate models. A number of different models exist, and they are continually being refined to more accurately represent the past history and expected future changes. These models incorporate both natural and anthropogenic influences on climate. The IPCC is the major international body that issue reports on effects of climate change and recommends policy actions. The major IPCC Sixth Assessment Report was released in 2021 and has been released in segments in 2021 and 2022. A 2019 IPCC special report recommended limiting global temperature increase to 1.5°C as necessary to prevent major deleterious impacts of climate change.

8.5 Climate Change Agreements

Because global climate change is, obviously, a global issue that requires action by countries around the world, a number of international agreements have been instituted to try to deal with the issue. These include the 1982 Rio Summit [7], the 1997 Kyoto protocol [8], and the 2015 Paris Agreement [9].

The Paris Agreement sets out a "a global action plan to put the world on track to avoid dangerous climate change by limiting global warming to well below 2°C. The agreement went into force in 2020" [10]. In 2017, the Trump administration announced that the US was leaving the Paris Agreement effective in November 2020, the earliest possible effective date. However, under the newly elected Biden administration, the US officially rejoined the agreement in February 2021. The main goals of the agreement are to work on mitigating emissions, creating a transparent and global trade market for emissions, to create adaptative infrastructure and societies, and to avert loss and damage [10].

As mentioned, the IPPC now feels that the 2°C target is too high, and that warming should be limited to 1.5°C to stave off the worst effects of climate change, including avoiding important tipping points. Energy conservation and a transition away from use of fossil fuels are the most significant actions that can be taken to minimize climate change. Mitigation efforts to limit the damage caused by climate change are also of increasing importance, since even if greenhouse gas emissions immediately fell to zero (an impossible task) temperatures would continue to rise for a number of years due to the long-term effects of the emissions that have already occurred.

8.6 Case Study 8.1. History of the Intergovernmental Panel on Climate Change (IPCC)

The IPCC was created in 1988 and endorsed by the United Nations General Assembly under Resolution 43/63 of December 6, 1988. The main intention of the creation of the IPCC was to form an institution able to prepare comprehensive reviews and recommendations regarding "the state of knowledge of the science of climate change; the social and economic impact of climate change, and potential response strategies and elements for inclusion in a possible future international convention on climate" [11]. Since the creation of the institution, the IPCC has delivered sixth Assessment Reports (Table 8.2) in addition to several methodological, special, and technical reports in response to particular scientific and technical issues.

The IPCC is led by an executive committee and composed of several working groups: working group I – physical science basis; working group II – impacts, adaptations, and vulnerability; working group III – mitigation of climate change, and the task force on national greenhouse gas

Table 8.2 Examples of main IPCC assessment and special reports.

Year	Report	Content
1990	First IPCC Assessment Report (FAR)	It highlighted the importance of climate change as a challenge with global consequences and requiring international cooperation.
1995	The Second Assessment Report (SAR)	It provided important material for governments to draw from in the run-up to adoption of the Kyoto Protocol in 1997.
2001	Third Assessment Report (TAR)	It focused attention on the impacts of climate change and the need for adaptation.
2007	Fourth Assessment Report (AR4)	It laid the groundwork for a post-Kyoto agreement, focusing on limiting warming to 2°C.
2013	Fifth Assessment Report (AR5)	It provided the scientific input into the Paris Agreement.
2019	Special report, Global Warming of 1.5°C (SR15)	Drafted to address the Paris Agreement.
2019	2019 Refinement	It updated the 2006 IPCC Guidelines on National Greenhouse Gas Inventories.
2019	Special Report on Climate Change and Land (SRCCL)	It provides the best understanding on the impact of climate change on issues such as desertification, land degradation, sustainable land management, food security, and greenhouse gas fluxes in terrestrial ecosystems.
2019	Special Report on the Ocean and Cryosphere in a Changing Climate (SROCC)	It addressed the impacts of climate change on ocean, coastal, polar and mountain ecosystems, consequences for human communities that depend on them, and options for adaptation and climate resilient development.
2022	Sixth Assessment Report (AR6)	It will provide the latest up-to-date physical understanding of the climate system and climate change.

inventories. Hundreds of leading experts are part of the IPCC reports, volunteering their time and expertise. In 2007, the IPCC together with the US Vice-President Al Gore were awarded the Nobel Peace Prize "for their efforts to build up and disseminate greater knowledge about man-made climate change, and to lay the foundations for the measures that are needed to counteract such change" [12].

As climate change keeps impacting society, the scientific information provided by the IPCC is becoming an essential tool for deciding on applying mitigation technologies to reduce global climate change.

8.7 Study Questions

1. Define radiative forcing (RF). What is the current RF value?
2. What are the main sources of energy relevant to global environment and climate change?
3. What are the main components affecting climate change?
4. Which are the main greenhouse gases?
5. What is the common approach to quantify the overall impact of the main greenhouse gases?
6. Which human activities have a cooling and warming effect on the environment?
7. Which human activities are the primary source of greenhouse gases?
8. Which regions of the world emitted the most greenhouse gases per capita?
9. List some of the main impacts of climate change.
10. Mention some of the main climate change agreements.
11. Is the US part of the Paris Agreement enforced in 2020?
12. What is the common target goal of the Paris Agreement?

8.8 Additional Resources

1. Methodology Reports on Short lived Climate Forcers
   https://www.ipcc.ch/report/methodology-report-on-short-lived-climate-forcers
2. Special report of the IPPC titled "Global warming of 1.5°C"
   https://www.ipcc.ch/sr15
3. IPCC, Sixth Assessment Report,
   https://www.ipcc.ch/report/sixth-assessment-report-cycle
4. US Global Change Research Program, Fourth National Climate Assessment,
   https://nca2018.globalchange.gov
5. NOAA, Global Monitoring Laboratory,
   https://gml.noaa.gov
6. NASA, Global Climate Change,
   https://climate.nasa.gov
7. Gavin Schmidt: The emergent patterns of climate change
   https://www.ted.com/talks/gavin_schmidt_the_emergent_patterns_of_climate_change#t-9399

References

1 IPPC. Climate Change (2014). *Synthesis Report. Contribution of Working Groups I, II and III to the Fifth Assessment Report of the Intergovernmental Panel on Climate Change.* Geneva, Switzerland: IPCC. 2014.

2 National Oceanic and Atmospheric Administration Global Monitoring Laboratory – THE NOAA ANNUAL GREENHOUSE GAS INDEX (AGGI). (2021). https://gml.noaa.gov/aggi/aggi.html (accessed September 9, 2021).

3 US Environmental Protection Agency. (2021). *Climate change indicators: climate forcing* https://www.epa.gov/climate-indicators/climate-change-indicators-climate-forcing. (accessed September 10, 2021).

4 Greenhouse gas emissions – Our World in Data. (2021). https://ourworldindata.org/greenhouse-gas-emissions. (accessed September 10, 2021).

5 National Oceanic and Atmospheric Administration. (2021). *Climate at a glance | Centers for Environmental Information (NCEI)* https://www.ncdc.noaa.gov/cag/global/time-series/globe/land_ocean/ytd/12/1880-2020. (accessed September 9, 2021).

6 Sea Level | Vital Signs – Climate Change: Vital Signs of the Planet. (2021). https://climate.nasa.gov/vital-signs/sea-level. (accessed September 10, 2021).

7 Brunland, G.H., Khalid, M., Agnelli, S., Al-Athel, S.A., Chidzero, B., Fadika, L.M. et al. (1987). *Report of the World Commission on Environment and Development.* New York, NY.: Our Common Future.

8 Kyoto Protocol to the United Nations Framework on Climate Change. (1998). https://unfccc.int/sites/default/files/kpeng.pdf. (accessed May 24, 2019).

9 United Nations. Adoption of the Paris Agreement. (2021). https://unfccc.int/sites/default/files/english_paris_agreement.pdf (accessed May 24, 2019).

10 United Nations. Paris agreement. (2015). https://unfccc.int/process-and-meetings/the-paris-agreement/the-paris-agreement. (accessed October 11, 2019).

11 IPCC. (2021). *History* – IPCC https://www.ipcc.ch/about/history (accessed September 10, 2021).

12 IPCC Statement about the 2007 Nobel Peace Prize. (2021). https://www.ipcc.ch/site/assets/uploads/2018/02/Nobel_statement_final.pdf. (accessed September 10, 2021).

9

Life Cycle Assessment

9.1 Learning Objectives

After learning from this chapter, you are expected to:

- Understand the life cycle thinking approach.
- Define consequential and attributional LCA.
- Describe the basic principle of LCA.
- Describe the main steps to run an LCA and the main standards.
- State the goal, scope, function unit, reference flows, LCI, and LCIA of an LCA.
- Conduct the interpretation of an LCA.
- Identify trade-offs and provide key perspectives of the environmental footprint of product(s) or system(s) using LCA.

9.2 Introduction

Let us imagine that we want to understand what is involved in making a cup of tea from a life cycle thinking perspective. If we are going to drink a specific tea, such as an Earl Grey Tea from Lipton® or Twinings®, both of them are similarly prepared. First, the tea must be grown. The tea leaves used for the matrix of the black tea are grown mostly in South America and the bergamot citrus fruit in the southern Mediterranean region or India. During the growing period, the tea shrub requires water and nutrients. During the harvesting, we need to use resources such as packaging materials and energy in the form of electricity or fuel to remove the leaves. After harvesting, the leaves must also be dried using heat and then packaged in bulk for shipping to the London, UK Twinings® plant. Bulk packaged leaves are transported first by truck until they reach the local ports in South America; then by small and large ships on rivers and oceans; and again, by truck from the ports to the plant in the UK. Once the leaves reach the plant, they are mixed with the bergamot zest coming from the Mediterranean region, which follows similar transportation routes. Then, the amount to be introduced in an individual bag is weighed and sealed into the bag. The production of a single bag includes attaching a string and a label to the bag so that an individual serving is created, which is packed in a polypropylene (PP) film bag. Twelve bags are packed in a paperboard box; produced locally in the UK and wrapped with a PP film. Twenty-four boxes are placed into a secondary container and then palletized for distribution. If you are buying your tea from a local US store, the tea

Life Cycle of Sustainable Packaging: From Design to End of Life, First Edition. Rafael A. Auras and Susan E. M. Selke.

Companion Website: www.wiley.com/go/Auras/lifecycleofsustainablepackaging

bag will travel by truck from the Twinings® plant to the UK port and by ship to the US port, and from the US port to the local distribution center (DC) by truck. From the DC, it is shipped to your local supermarket store where you could buy it and bring it home by car. After you boil the water, you can get a cup and prepare a cup of Earl Grey tea. Let us consider that you prefer the tea straight without sugar, so we do not need to include the life cycle of producing two spoonsful of sugar. After you are done drinking your tea, you will dispose of the tea bag in the landfill or organic recycling bin, wash the cup with detergent and water, and dry it. As we can realize, it takes a lot of steps to produce a cup of tea, as shown in Figure 9.1. We call this way to analyze the production of goods or products a life cycle of a product/package where we include all the inputs and outputs to produce it. This method of evaluating the input and outputs of a product(s) or system(s), life cycle thinking, and the practice of quantification of the input and outputs and the environmental impact of this process is known as LCA, which was briefly introduced in Chapter 1 section 1.8. LCA must be conducted under particular conditions and is the subject of this chapter.

Environmental LCA is the analytical framework for quantifying the resources and the impact on the environment and human health of a product(s), service(s), or system(s) over its entire life cycle [1]. LCA of a product or system comprises the appraisal of the environmental effects created during its entire life cycle, from its beginning as a raw material extracted from the ecosphere, during processing, distribution to the user, potential upgrading, and until its end, usually as a waste. It accounts for the "upstream" and "downstream" phases, and it provides full documentation and evaluation of materials and energy as they flow through the system. LCA typically evaluates the environmental footprint (EFP) and impact in different stages. Potential life cycle stages could include extraction of raw materials from nature, refining of raw materials into technical materials and products, assembling into components, using the products, and their end-of-life. LCA studies can cover different stages of the life of a product.

A large number of disciplines such as engineering, product design, environmental chemistry, policymaking, materials science, packaging, and project management use LCA to support the process of identifying and improving the ecological performance or reducing the EFP of products or systems.

The American Center for Life Cycle Assessment (ACLCA), the main organization publishing and aggregating LCA expertise in the US, defines environmental LCA as:

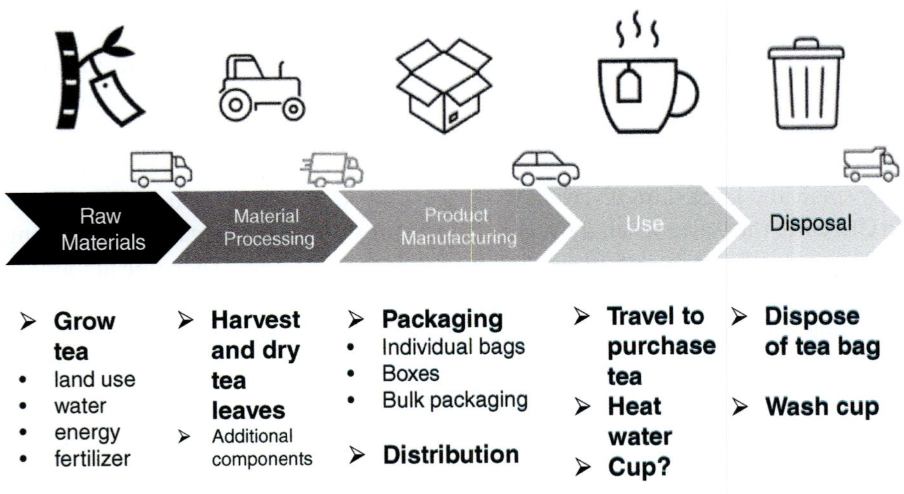

Figure 9.1 Life cycle for producing a cup of tea.

A comprehensive yardstick of the environmental performance of goods and services. LCA is a powerful tool to influence human behavior and environmental outcomes through management and engineering decisions, public policy, and purchasing decisions. The majority of those using the output of an LCA are not in a position to verify every element of the LCA. LCA professionals, those performing, providing life cycle data, and developing life cycle tools and models, therefore, have a great responsibility to provide unbiased, accurate, and transparent data and analyses to the greatest extent possible. In doing so, they use the work of their hands and minds to build a more sustainable world, and they support the strength and quality of the LCA profession.

Table 9.1 shows several reasons to run an LCA study. Typical examples are understanding the environmental impact of a product, trying to promote innovation, understand the EFP of an entire sector, comparing the EFP of products or options, and making claims about the EFP of a product.

The terms ecosphere and technosphere must be defined since they are relevant to LCA modeling.

Ecosphere is defined as the environment. However, this can also refer to the manmade environment such as buildings and dams.

Technosphere is defined as the subject of the study and it is manmade.

The boundary between the technosphere and ecosphere can more easily be recognized and defined as the elementary flow or "single substance or energy entering the system being studied that has been drawn from the ecosphere without previous human transformation, or single substance or energy leaving the system being studied that is released into the ecosphere without subsequent human transformation" [2].

9.3 Provisions of LCA Study

The European Commission – Joint Research Centre – Institute for Environment and Sustainability (JRC-IES) developed a detailed handbook explaining the steps to conduct an LCA study according to the requirements of ISO14040 and 14044:2006 – the two main standards regarding LCA, which provide a framework of best practice, reliability, efficiency, flexibility, fairness and acceptance, transparency and reproducibility, and assured quality [2]. In the handbook, they differentiate between the main goal situations encountered in LCA practice, which are shown in Table 9.2.

Table 9.1 Some reasons to run an LCA.

	Reasons to Perform the Study
Global exploration of options	Obtain a first impression of the environmental effects of certain options.
Company-internal innovation	Assess the environmental impact of a company need to improve a product, product development, or technical innovation.
Sector-driven innovation	Evaluate a sector-oriented industry such as the aluminum industry.
Comparison	Evaluate whether a product or system meets environmental standards, or whether it is environmentally preferred over another product.
Comparative assertion disclosed to the public	Provide an environmental claim regarding superiority or equivalence of one product versus a competing product which performs the same function.

Table 9.2 Decision orientation and support and the kind of consequences in the background and foreground systems when conducting an LCA; reproduced from [3].

		Kind of process-changes in the background system and other systems	
Decision support?	Yes	None or small-scale	Large-scale
		Situation A Microlevel decision support	**Situation B** Meso-/macrolevel decision support
	No	**Situation C** Accounting **C1:** including interactions with other systems, **C2:** excluding interactions with other systems	

Situation A ("Micro-level decision support"): This situation of running an LCA provides support at the microlevel, and it is commonly conducted for product-related queries. The "Micro-level decision support" is presumed to be limited, and no structural consequences happen outside the decision-context. Therefore, a change in capacity or infrastructure is not expected. Thresholds are too small to cause a large-scale consequence or change in the background system or part of the technosphere (i.e., anthroposphere, the part of the environment that is used and modified by human activities and habitats). Examples of situation A are the identification of the critical environmental footprint of a product group (streamlined LCA), development of ecolabels, benchmarking of specific products against an average product, green procurement, comparison of particular goods or services, and evaluation of carbon footprint and primary energy consumption. Situation A mostly supports decisions directly or indirectly to consider products or services that are already in the market. Since the products or services have a limited share of the total production of a single product in an industrial sector, its production, use, and end-of-life can be expected to cause very small or no disturbance in the background system or overall system.

Situation B ("Meso/macro-level decision support"): This situation covers decision support at strategic levels (e.g., raw material strategies, technology scenarios, policy decisions). Meso/macro-level decision support may have consequences outside the decision context by, for example, changing the available production capacity. The decision has significant implications for the background system and parts of the technosphere. Examples of situation B are policy development and information. In situation B, the life cycle decision supports consequences that are so extensive that they impact the background data and parts of the technosphere, such as increasing bio-ethanol content in gasoline in the US from 0% to 20% per gallon. In this situation, due to the large demand for ethanol by the transportation industry, it will create a ripple effect on the supply chain of corn, impacting areas such as human and animal feed and starch production. So, the EFP of increasing bioethanol in gasoline in the entire US should be conducted as a Situation B LCA study.

Situation C ("Accounting"): This situation is a purely descriptive assessment of the past, present, or forecasted future system (e.g., product or system, and sector or country), and does not consider any potential consequences for other parts of the economy. Two subtypes can be recognized in this situation. Situation C1 includes benefits outside the system considered in the analysis, for example, due to recycling benefits, and situation C2 does not consider benefits. Situation C1 examples are monitoring the EFP impact of a nation, industry sector, product group, or category, policy information for classification of products, identifying products with the broad category, certification of a supply chain. Situation C2 examples are accounting studies that do not consider interactions with other systems. Situation C will not support decision making since the entire life cycle is decided before the analysis takes place, and the study is retrospective so that it cannot be used for informing purchase decisions or answer "what if" scenarios. For example, situation C can be used to study the EFP of postconsumer plastic recycling technologies in the past, but it cannot be used to stipulate how new technologies will perform in the future. In this case, situation A or B should be applied.

Overall, conducting an LCA is an iterative process after the goal of the study has been identified and defined (Figure 9.2). After the initial LCA starts, only small revisions of the goal are permitted. As further iterations are conducted to obtain the maximum precision with the minimum necessary effort, a point of diminishing returns should be set in any LCA study to optimize the use of time and effort. For new technologies and complex systems when previous data is not available, the first generation of an LCA study may use generic or average data from the background system. Combined with expert judgment to identify essential processes and flows, the practitioner can and should decide when to stop the study for final review or publication. It is highly recommended to keep a track record and documentation of each stage of the study and all their assumptions so that the study can be perfected in a subsequent iteration. Iterations are included in the study (i.e., the key processes, parameters, and elementary flow quantifications), so the requirements of the goal and scope of the LCA study are finally met.

9.4 Different Approaches to Conduct LCI Studies

There are two main approaches to conduct LCA studies: attributional and consequential LCA. ISO 14040:2006(E) defines these two approaches as:

"a) one which assigns elementary flows and potential environmental impacts to a specific product system typically as an account of the history of the product, and

b) one which studies the environmental consequences of possible (future) changes between alternative product systems."

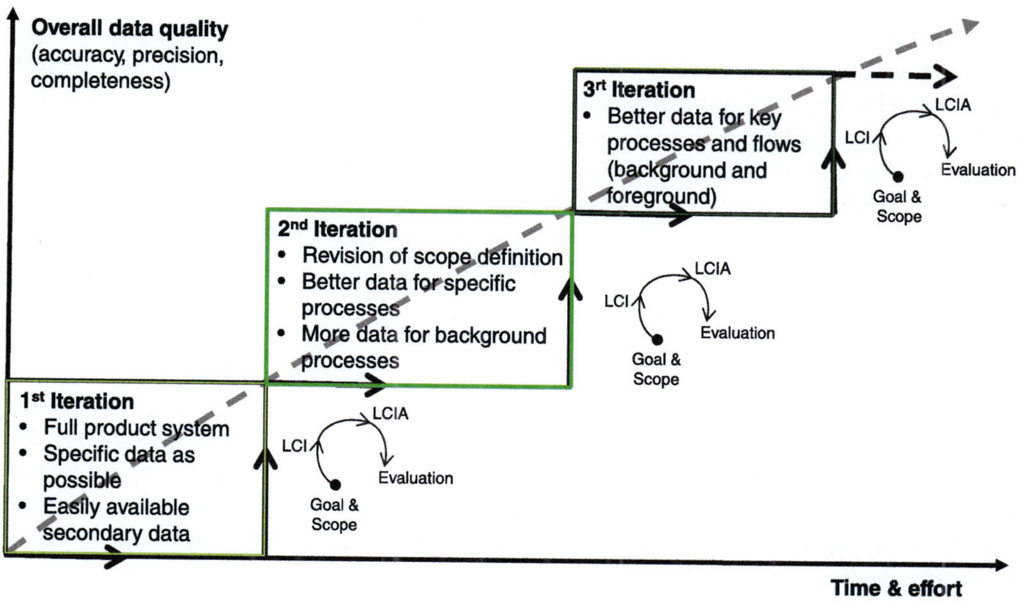

Figure 9.2 Iterative nature of LCA. LCAs are generally performed in iterative loops of goal and scope definition, inventory data collection and modeling, impact assessment, and interpretation analysis; adapted from [2].

The two approaches are recognized as Attributional LCAs (ALCA) and Consequential LCAs (CLCA).

ALCA is of the accountancy type of LCA (typically applied for hot-spot-identification, product declarations, and for general consumer information). ALCA is a system modeling approach in which inputs and outputs are attributed to the functional unit of the study by connecting and/or partitioning the reference flows of the system according to a normative rule (i.e., cut-off and allocation) to isolate the investigated product or system from the rest of the world. So, it ignores many of the physical and economic causalities related to using or implementing this product or system. ALCA can be used to estimate, for example, the EFP of delivering milk in different containers. ALCA can be conducted under situations A and C. ALCA determines the EFP that can be allocated to a system over its life cycle considering the upstream along the supply chain and downstream considering the system and the end of life. ALCA modeling uses historical, factual, and measurable data of known uncertainty, including all the flows and processes relevant to the system under study. In an ALCA study, the system is modeled as it is or was, and in general, average data is used.

CLCA studies the environmental consequences of possible (future) changes between alternative product systems, and it is typically applied in product development and public policymaking. CLCA defines changes in production systems. A definition can be stated as: "CLCA is a system modeling approach in which activities in a product system are linked so that activities are included in the product system to the extent that they are expected to change as a consequence of a demand for the functional unit" [4]. The primary purpose of consequential modeling is decision support. So, the consequences of a decision are examined over time. An example of consequential modeling can be estimating the consequence of increasing the amount of bioethanol in gasoline produced from corn, as previously discussed. In this case, as the demand for gasoline increases, a large amount of corn is used to produce bioethanol, but less is available for human and animal feed. Therefore, a substitution must be done using another product due to the lack of corn for human or animal feed. In this case, the study should be run as situation B if the background change of the market is more than 5%. CLCA modeling is labeled as "change-oriented, effect-oriented, decision-based, or market-based." CLCA concentrates on estimating the consequences that a decision made in the foreground system has for other processes and parts of the economy. CLCA does not reflect the actual or present system, but a hypothetical generic supply-chain, using a prognosis of a possible supply-chain considering interactions and consumer changes. A crucial step in consequential modeling is to identify the marginal flows and processes, since several mechanisms can affect processes and products and their actions to follow – for example, the need to build a new production system due to the outlined consequence of the study. Figure 9.3 shows a symbolic view of the two types of LCA.

9.5 Steps of an LCA Study

The main standards describing how to conduct an LCA study are ISO 14040 and 14044:2006 [1, 5]. Basically, the main steps for conducting an LCA are goal and scope definition including functional unit and reference flow, defining how to collect all the needed inventory data, aggregate this data into an impact assessment methodology, and conduct an interpretation of the results. This process is an iterative process among all the stages. A conceptual diagram is shown in Figure 9.4.

LCA projects can be more than a study or a simple report. They can be used for decision making by industry, governments, and NGOs. However, LCA studies can only be considered for decision-making support if most stakeholders are involved in the analysis of the process, and the

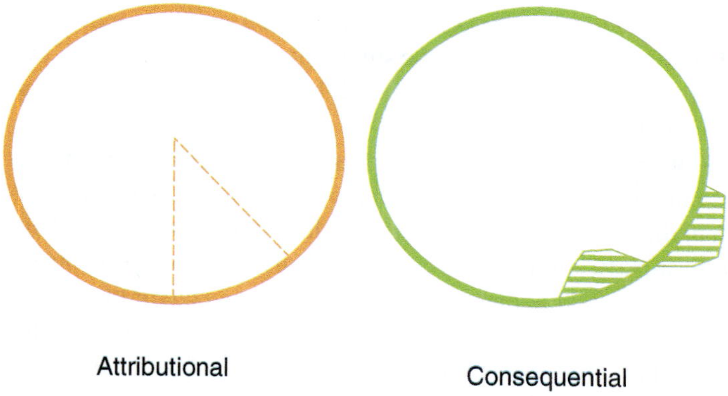

Figure 9.3 **Symbolic view of attributional and consequential LCA.** The circles represent the total global environmental exchanges. In the left circle, ALCA seeks to cut out the piece with dotted lines that belongs to a specific human activity. In the right circle, CLCA seeks to capture the change in environmental exchanges that occur as a consequence of adding or removing a specific human activity; adapted from [4].

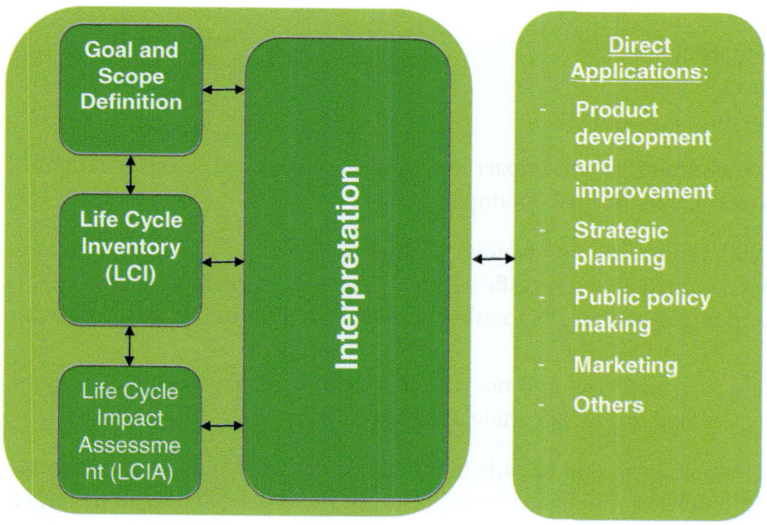

Figure 9.4 **Framework for life cycle assessment,** including the main steps of goal and scope, LCI, LCIA and interpretation; adapted from [1].

individuals involved: (a) take responsibility for the tasks, (b) influence decisions made during the study, (c) evaluate and deal with bottlenecks of the study, (d) participate in the management process, and (e) conduct a transparent and robust study. A short description of each step is provided below.

9.5.1 Goal and Scope Definition of an LCA

The *goal* definition of an LCA is the step where the decision, context(s), and intended application(s) of the study are identified and the intended targeted audience(s) are specified. The goal of the study must be formulated in terms of the (a) exact question, (b) target audience, and (c) intended application.

The right definition of the goal of the study is pertinent to conduct the full study since the goal definition: (a) guides all the detailed aspects of the scope of the study, (b) informs the frame of the LCI and LCIA, (c) provides the quality control for the study, and (d) informs how the interpretation and recommendations are evaluated.

During the goal step, the goal should be unambiguously and transparently defined, not only in terms of what is to be done but also in terms of the reasons for executing the study. The intended application(s) and decision(s) plan of the study must be supported by the study. If the study will be used for comparative assertion, a third-party will need to review the study. The parties involved, such as researchers, LCA commissioners, target audience, members of steering committees, and expert reviewers, must also be identified. Additional questions to answer during formulation of the goal of a study are: Is LCA the most appropriate tool? What are the starting points for the LCA study? What are the limitations of the study? Are different products being compared?

When defining the intended application(s) of the LCA results in a precise and unambiguous way, some of the following are considered:

- Weak point analysis of a product
- Identification of the EFP of a product
- Comparison of products
- Intention to "green" the supply chain
- Policy information and development
- Monitoring environmental impact of systems

When a particular method, assumption, and impact limitations are included in the goal definition, these should be fully described in the goal. Examples are:

- Carbon footprint studies with limited impact coverage of the emissions
- Limitations such as studies conducted for a specific site with uncommon characteristics
- Limitations about transferability of the results to other similar LCA studies due to the characteristics of the study
- Niche markets relevant to a particular user and not transferable to general consumers. Factors can be due to price, life style, social image, fair-trade, longevity, etc.

The reasons for carrying out the study should be fully identified, along with the decision context, since these can affect the LCI modeling framework and type of study, ALCA or CLCA. The target audience should be properly identified. It is not the same to conduct a study for an internal vs. external or a technical vs. nontechnical audience. Furthermore, when the study is intended to be disclosed to the public, including any comparative assertions, ISO 14040 and 14044 require a third-party review. If the study is commissioned by a third-party or organization, which has influence in the study, this should be declared. In some cases, the goal definition should contain a provision that the goal of the study may be extended to include additional applications of interest, such as environmental product declaration (EPD) or product improvement and design. The goal of the study should identify whether the study is an ALCA or CLCA study.

Before addressing aspects of scope of an LCA, two requirements should be taken care of that crosscut across all the assumptions and conditions, which are: (a) consistency of methods, assumptions, and data; and (b) reproducibility.

Consistency of methods, assumptions, and data: To guarantee that the LCA study is conducted with the highest standards possible, all the assumptions should be made consistently to the extent necessary to ensure the overall accuracy, completeness, and precision of the study. The data should be consistent regarding accuracy, precision, and completeness aligning with the goal of the study. Furthermore, any inconsistencies should be thoroughly documented.

Reproducibility is a qualitative assessment of how the methods, assumptions, and data and data sources are fully documented so that an independent practitioner could fully reproduce the results of the LCI/LCA study or any conclusions or recommendations. Therefore, the data should be fully supported by technical documentation for informing experts and generating information for the general public. In cases where all the data cannot be released due to confidentiality or proprietary data, a separate report should be included for reviewers and experts. For comparative studies, the LCI and LCIA results should always be made public.

The *scope* definition of an LCA study establishes the main characteristics of a planned LCA study. The following considerations should be included when deriving the scope from the goal of the study:

- The type(s) of deliverable(s)
- The product, system, or process to be studied
- The LCI modeling framework
- System boundary, completeness requirements, allocation procedures, and related cut-off rules
- LCIA impact categories
- LCI data quality requirements regarding temporal, technological coverage, geographical, time-related appropriateness, and representativeness
- Type, quality, and sources of required data and information considering precision and maximum uncertainty
- Special consideration for comparison between products, systems, or processes
- Impact categories
- Level of sophistication
- Type and format of the report and results

Overall, the initial scope definition of an LCA could be fine-tuned or even revised during the study if there are unforeseen limitations or constraints such as lack of data, information not sufficient for the target levels of interpretation (i.e., completeness, precision, accuracy, and intended applications).

Thus, the *goal and scope* of an LCA should be properly stated with an equivalent functional unit if systems or products will be compared (same functional unit and equivalent methodological considerations). If the study is intended to be used for a comparative assertion intended to be disclosed to the public, an independent panel critical review must be conducted. Examples of the goal and scope of several published LCA studies are provided below.

- "The aim of this study was twofold (sic): to identify the most critical life cycle stages of an aged Spanish wine from the point of view of the associated environmental impacts and to compare its environmental performance with that of other wines and beers for which comparable information could be found in the scientific literature. All the product's life cycle stages were accounted for, namely: grapes cultivation (viticulture), wine making and bottling, distribution and sales, and disposal of empty bottles" [6].
- "The goal of this study was to compare the environmental impact of PLA, PET and PS thermoformed clamshell containers used for the packaging of strawberries. The scope of the study ranged from the extraction of the raw material for the three polymers and the processes of their resin production, through container formation, followed by their end-of-life disposal. The scope includes consideration of global warming, acidification, ozone depletion, aquatic eutrophication, non-renewable energy, land occupation, respiratory organics, respiratory inorganics and

aquatic ecotoxicity as impact categories. The functional unit was chosen as 1000 containers of capacity 0.4536 kg (1 lb) each for the packaging of strawberries" [7].

- "The main goal of this work was to equip milk delivery industry stakeholders (milk processors, packaging material manufacturers, and retailers) with timely, science-based information in order to incorporate environmental performance into decision-making and drive innovative new products, processes, and services. LCA provides quantitative, confirmable, and manageable models to evaluate production processes, analyze options for innovation, and improve understanding of the complexity in systems. The results of this study will allow fluid milk manufacturers in the US to benchmark their performance against the 2009 industry average. However, this study does NOT support comparative assertions to be made public" [8].

9.5.2 Function, Functional Unit, and Reference Flow

The function, functional unit, and reference flow are central structures of an LCA study since, without them, a meaningful comparison or description cannot be done. Any LCA study is based on the basic quantitative description provided by the functional unit to analyze the system. The function provides a quantitative and qualitative analysis of how to conduct and compare the systems under study, and it is performed by defining the functional unit.

The functional unit describes the primary function(s) fulfilled by a product system and indicates how much of this function is to be considered in the intended LCA study. It is used as a basis for selecting one or more alternative product systems that might provide these function(s). The functional unit allows diverse systems to be treated as functionally equivalent and permits the reference flows to be determined for each of the systems. Based on the functional unit, many substitute product systems can be declared functionally equivalent, and reference flows can be determined for these systems. The functional unit is the basic unit of the LCA study, and it is common to all alternatives and forms being compared.

The reference flow is defined as a measure of the outputs from processes in each product system required to fulfill the function expressed by the functional unit. The purpose of the functional unit is to provide a reference to which the input and output data are standardized. The reference flow should be stated and measurable. Table 9.3 shows the function, functional unit, and reference flows of an LCA study to determine the EFP of delivering a soft drink to the consumer in different packages. Even though the beverage comes in different sizes and containers, the functional unit normalizes the

Table 9.3 Function, functional unit, and reference flows of an LCA study about beverage packaging.

	Al Can	PET Bottle	Glass Bottle

Volume, fl oz (mL)	12 (355)	16 (473)	8 (236)
Function	Volume of beverage	Volume of beverage	Volume of beverage
Functional unit, fl oz (mL)	48	48	48
Reference flows, units	4	3	6

volume across all the containers so that the study can be framed as the amount of beverage delivered to the consumer. However, to deliver the same amount of beverage to the consumer, different numbers of containers are needed for each package. In this case, four Al cans, three PET bottles, and six glass bottles are required. Therefore, the study will need to consider the different ratios of material to product needed to deliver the beverage using each container. In this case, a direct comparison of material amounts for each container is meaningless and misleading. The functional unit must be defined in the context that the product is going to be used, the volume of beverage delivered to the consumer.

Table 9.4 provides another example of the functional unit and reference flows for an LCI/LCA study about lighting. Several considerations besides the primary function were taken in this study since cost, material amount, and aesthetics could impact the decisions of the study. So, the relevant quantifiable properties and the technical function and performance of the study must be distinguished. In this case, for quantifying the functional unit of the bulbs, we need to take into consideration the duration of use and the extent/quantity to be used to deliver the same function (i.e., 60,000 h at 600 lumens of light intensity.) Secondary factors such as aesthetics and perception should be considered if they are essential to ensure a fair comparison among products. For example, an incandescent bulb releases more heat than a fluorescent or LED bulb. If the released heat is a crucial factor, this should be considered in the functional unit.

Overall, we need to make sure that the function, functional unit, and reference flow identifying the system or process are in line with the goal of the study. We may provide pictures and specifications of the functional unit. The functional unit may provide one or more functions and quantity.

Table 9.4 Function, functional unit, and reference flows of an LCI/LCA study for lighting.

Bulb Type	Incandescent	Fluorescent	LED

Power, Watt	60	11	8.5
Lifetime, h	1,000	6,000	10,000
Weight, g	45	226	136
Primary function	Lighting	Lighting	Lighting
Secondary function	Aesthetics, heating, nice atmosphere	Aesthetics, heating, nice atmosphere	Aesthetics, heating, nice atmosphere
Functional unit	60,000 h @ 600 lumens of light intensity	60,000 h @ 600 lumens of light intensity	60,000 h @ 600 lumens of light intensity
Reference flow	60,000/1,000 = 60	60,000/6,000 = 10	60,000/10,000 = 6
Energy used	60 W × 60,000 h/FU = 3,600 kWh/FU	13 W × 60,000 h/FU = 780 kWh/FU	8.5 W × 60,000 h/FU = 510 kWh/FU
Weight, g	60 bulbs × 45 g = 2,700 g	10 bulbs × 226 g = 2,260 g	6 bulbs × 136 g = 816 g
Cost electricity, U$D	3,600 kWh × 0.1$/KWh = 360 $/FU	780 kWh × 0.1$/KWh = 78 $/FU	510 kWh × 0.1$/KWh = 51 $/FU
Cost bulbs,	60 b × 2.185$ = 131$/FU	10 b × 2.49$ = 24.9$/FU	6 b × 4.02$ = 24.12$/FU
Total cost	491 $/FU	102.9 $/FU	75.12 $/FU

The obligatory properties (i.e., features that the product or system must have), and the positioning properties (i.e., features that are used to position the product in the market as more attractive than another product) must be defined. Standard methods and techniques should be used to measure the reference flow, such as thickness, weight, and volume. The functional unit should be properly defined and identified. When many processes or products are produced by a unit flow, which is identified as the functional unit, each reference flow and functional unit should be adequately defined. If a product or system also provides alternative functions, the most important alternative functions must be described.

9.5.3 Life Cycle Inventory Modeling Framework

Early in the goal and scope definition of the study, decisions must be made about the type of LCI modeling framework to be used in modeling the system/s such as ALCA or CLCA and allocation or system expansion/substitution since these decisions have implications for the modeling of the LCI and the data collection.

During the LCI of the product system (or product systems if there is more than one alternative) the following items must be considered:

- System boundaries (economic, environmental, and cut-off)
- Designing the flow diagrams with unit processes
- Collecting the data for each process
- Allocation

The main result of an LCI is an inventory table listing the quantified inputs from and outputs to the environment associated with the functional unit in terms of the basic units. The LCI defines the range of the system under study, setting the life cycle boundaries: gate-to-gate, cradle-to-gate, cradle-to-grave, or cradle-to-cradle as shown in Figure 9.5. Together with the boundaries, the geographical region or boundaries must be defined since this information has consequences for the background and foreground systems.

Additional consideration is taken in LCI to determine where the data is obtained. Foreground and background systems are defined according to the specificity and management perspective of the study. Figure 9.6 shows a graphical representation of the ecosphere, technosphere, and foreground and background systems.

Foreground system: The foreground system is defined as the processes of the system that are unique to the study. Data for the specific supplier for the product or process must be considered. The foreground processes are those that are directly under the control of the producer of the goods, or processes of the service or user of the goods over which the user has some decision-making power.

Background system: The background system is the processes where, due to the averaging effect across suppliers, a homogenous market with average values can be assumed as representing the system. The background system consists of the processes used or that need to be operated as part of the system. They are not under the direct control of the user, or the user has no decisive influence on them.

9.5.3.1 Flows and Multifunctionality

To conduct the LCI of an LCA study, the flows and processes to be included in the LCI must be defined. For example, Figure 9.7a shows that for producing 1 poly(ethylene terephthalate) (PET) bottle, we need 0.032 kg of PET resin and 0.0891 MJ of electricity, and we will emit 0.044 kg of CO_2 and 1.97E-3 kg SO_2. If we want to account for the production of the resin, we need to include a flow

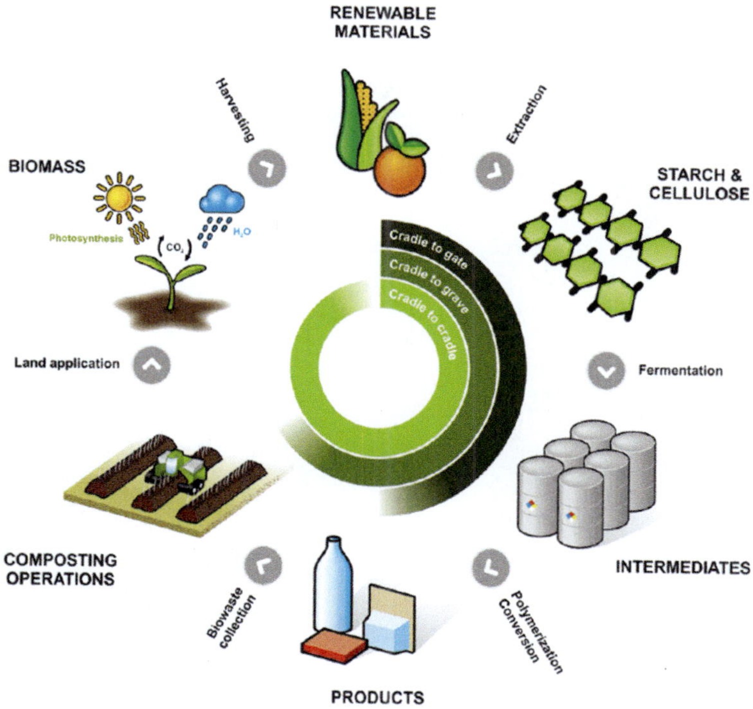

Figure 9.5 Life cycle inventory boundaries, cradle-to-gate, cradle-to-grave, cradle-to-cradle; reprinted from [9], Copyright 2008, with permission of John Wiley & Sons, LLC.

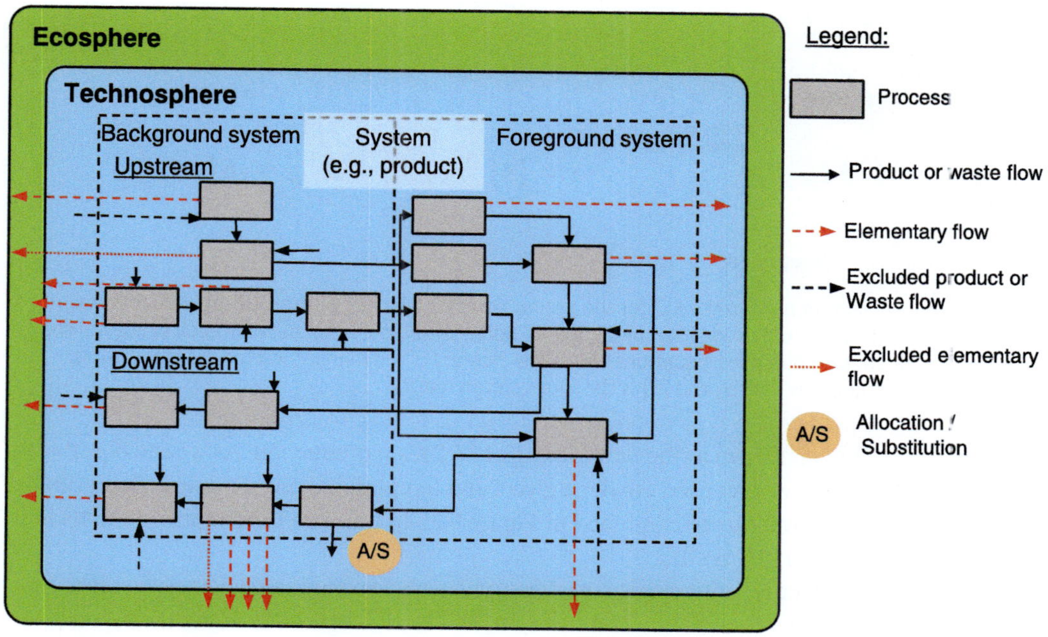

Figure 9.6 Foreground and background system in a specific LCA study. Dashed lines separate the foreground and background system from the technosphere and ecosphere. The system can be divided as the foreground system, which are the processes to be studied, and the background system, which are the average flows; adapted from [2].

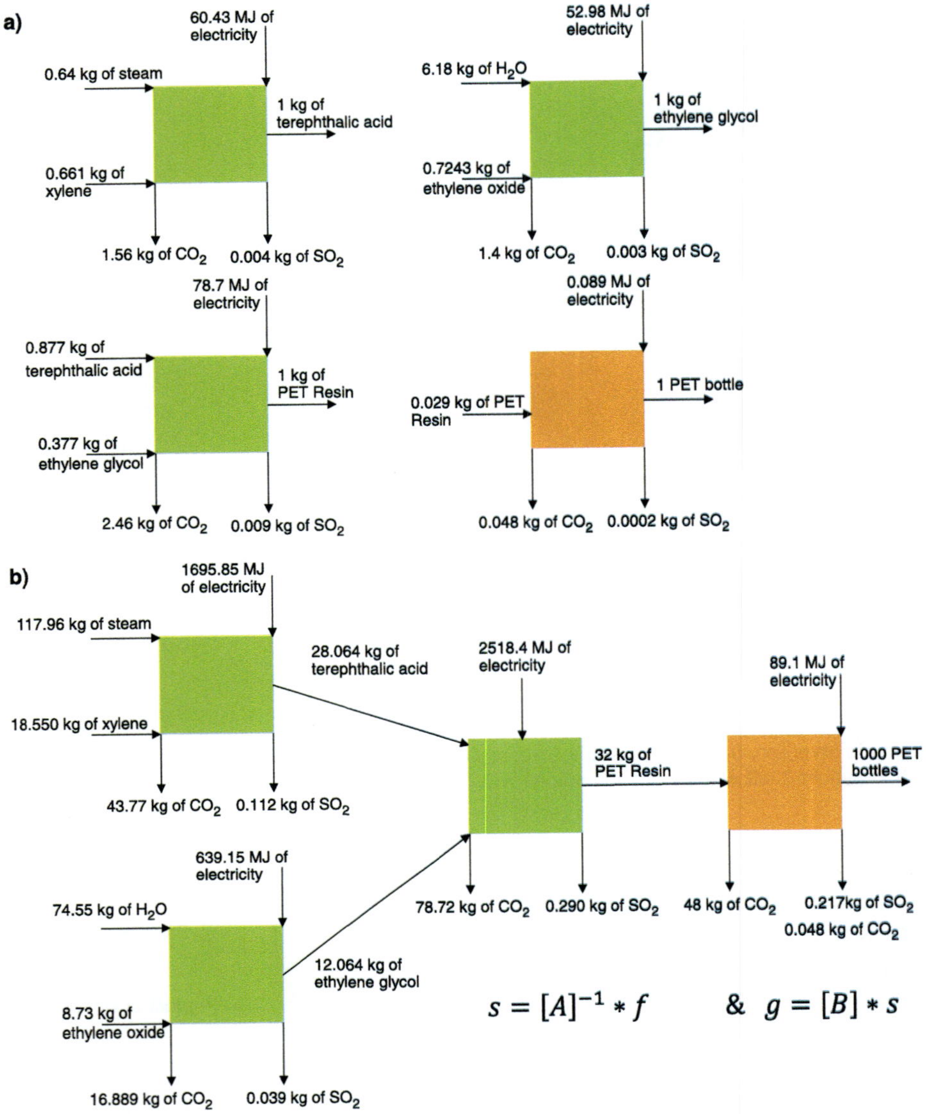

Figure 9.7 Flow diagram used in the LCI for the production of 1,000 PET bottles: (a) individual flows, (b) connection of flows for the production of 1,000 PET bottles. Inputs and outputs for each process were simplified for illustration purposes. Equations inside the figure: *f* is the demand vector, *s* is the scaling factor, [A] is the technology matrix, and [B] is the intervention matrix.

for this process, which is shown as the flow for 1 kg of PET resin production. As we can observe, to produce the PET resin, we also need ethylene glycol and terephthalic acid. So, these flows can be included, and all the flows can be combined in Figure 9.7b to estimate the simplified inputs and outputs to produce 1,000 PET bottles.

These basic flow diagrams and tree networks are used by most commercial LCA software to represent the foreground and background systems of an LCI study. Figure 9.8 shows the representation of the kg of CO_2-eq emitted for the production of 1 PET bottle as depicted by Simapro™.

A tabular representation of all the unit flows for the production of 1 PET bottle is shown in Table 9.5. Commercial software such as Simapro™, Gabi TS, and Humberto base their commercialization and user appeal on compiling the data for these processes and simplifying the LCI calculations.

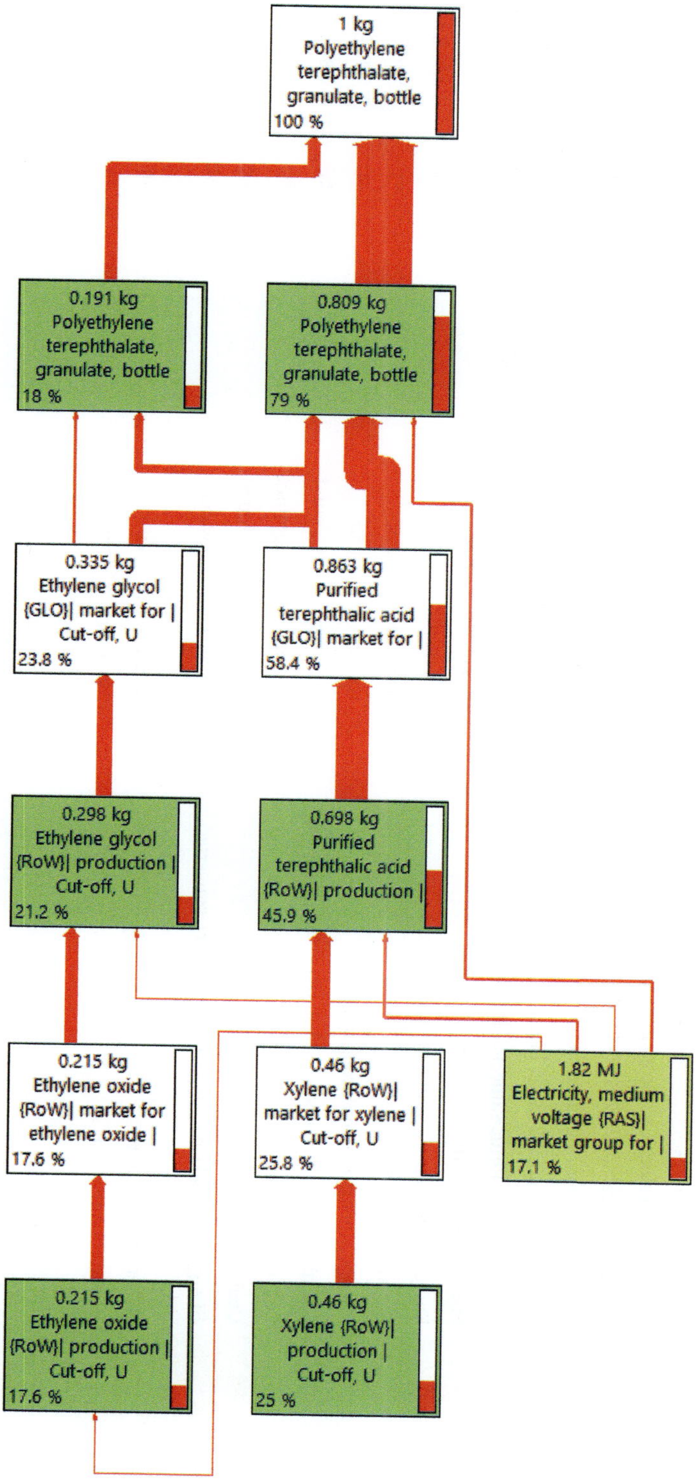

Figure 9.8 Unit process representation to produce 1 kg of PET, granulate, bottle grade {GLO} | market for | Cut-off, U; showing the kg CO_2-eq emitted as estimated by Simapro™ 9.3.0.2, and report according to ReCiPe 2016 v1.1 Midpoint (E).

Table 9.5 Selected unit process for the production of 1 kg of PET bottle grade resin, at plant as reported by PlasticsEurope's Eco-profiles [10]. Data is provided just as an example; selected unit processes with small contributions were removed due to space constraints.

Explanations	Input-group	Output-group	Name	Location	Category	Sub-category	Infrastructure-Process	Unit	PET, bottle grade, at plant
			Location						RER
			InfrastructureProcess						0
			Unit						Kg
information, primary energy demand	4		Energy, geothermal		information	primary energy demand		MJ	2.31 E-3
	4		Energy. gross calorific value, in biomass		information	primary energy demand		MJ	8.91 E-1
	4		Energy, gross calorific value, in hard coal		information	primary energy demand		MJ	1.66E+0
	4		Energy, gross calorific value, in lignite		information	primary energy demand		MJ	1.08E+0
	4		Energy, gross calorific value, in natural gas		Information	primary energy demand		MJ	1.67E+1
	4		Energy, gross calorific value, in oil		information	primary energy demand		MJ	4.71 E+1
resource, biotic	4		Energy, gross calorific value, in biomass		resource	biotic		MJ	4.81 E-1
	4		Energy, gross calorific value, in biomass, primary forest		resource	biotic		MJ	1.73E-6
Resource, in air	4		Carbon dioxide, in air		resource	in air		kg	9.63E-2
	4		Energy, kinetic (in wind), converted		resource	in air		MJ	2.48E-1
	4		Aluminium, 24% in bauxite, 11% in crude ore, in ground		resource	in ground		kg	2.84E-5
Resource, in ground	4		Calcite, in ground		resource	in ground		kg	1.05E-2
	4		Coal, brown, in ground		resource	in ground		kg	1.01 E-1
Resource, in water	4		Energy, potential (in hydropower reservoir), converted		resource	in water		MJ	4.46E-1
	4		Water, cooling, lake		resource	in water		m3	3.45E-4
	4		Water, cooling, river		resource	in water		m3	2.09E-3

	4	Water, cooling, unspecified natural origin	resource	in water	m3	5.64E-2
	4	Water, cooling, well, in ground	resource	in water	m3	3.12E-4
	4	Water, process, lake	resource	in water	m3	1.37E-4
	4	Water, salt, sole	resource	in water	m3	8.12E-4
	4	Water, unspecified natural origin	resource	in water	m3	2.29E-3
	4	Water, well, in ground	resource	in water	m3	2.54E-4
Resource, land	4	Occupation, agriculture + forest	resource	land	m2*year	1.23E-2
	4	Occupation, mineral extraction site	resource	land	m2*year	6.55E-5
	4	Occupation, sealed soil	resource	land	m2*year	9.65E-4
	4	Occupation, unknown	resource	land	m2*year	6.73E-6
	4	Transformation, from agriculture + forest	resource	land	m2	8.03E-6
	4	Transformation, from sealed soil	resource	land	m2	2.05E-6
	4	Transformation, from unknown	resource	land	m2	6.08E-6
	4	Transformation, to agriculture + forest	resource	land	m2	8.07E-6
	4	Transformation, to sealed soil	resource	land	m2	5.98E-6
air, unspecified	4	1,4-Butanediol	air	unspecified	kg	3.11E-15
	4	2-Methyl-1,3-dioxolane	air	unspecified	kg	1.11E-6
	4	2-Propanol	air	unspecified	kg	7.58E-8
	4	Acetaldehyde	air	unspecified	kg	1.28E-5
	4	Acetic acid	air	unspecified	kg	2.62E-4
	4	Acetone	air	unspecified	kg	7.39E-8
	4	Water	air	unspecified	kg	1.71E+0
	4	Xenon-133	air	unspecified	kBq	6.66E-2
	4	Xylene	air	unspecified	kg	3.89E-5
	4	Zinc	air	unspecified	kg	1.58E-7

(Continued)

Table 9.5 (Continued)

Soil, unspecified	4	Antimony		soil	unspecified		kg	6.50E-16
	4	Arsenic		soil	unspecified		kg	1.69E-11
	4	Carbon		soil	unspecified		kg	2.89E-7
	4	Lead		soil	unspecified		kg	3.19E-10
	4	Mercury		soil	unspecified		kg	2.38E-12
	4	Molybdenum		soil	unspecified		kg	1.48E-11
	4	Phosphorus		soil	unspecified		kg	2.05E-8
	4	Vanadium		soil	unspecified		kg	8.25E-11
	4	Zinc		soil	unspecified		kg	9.31 E-9
Water, unspecified	4	1,3-butadiene, hexachloro-		water	unspecified		kg	7.60E-13
	4	Acetaldehyde		water	unspecified		kg	3.00E-6
	4	Acidity, unspecified		water	unspecified		kg	1.37E-7
	4	Cobalt		water	unspecified		kBq	5.21 E-6
	4	Cobalt-60		water	unspecified		kg	2.15E-4
	4	COD, Chemical Oxygen Demand		water	unspecified		kg	2.31 E-2
	4	Copper, ion		water	unspecified		kg	8.36E-6
	4	Dissolved solids		water	unspecified		kg	3.23E-5
	4	DOC, Dissolved Organic Carbon		water	unspecified		kg	6.62E-3
	4	Heat, waste		water	unspecified		kg	9.40E-1
	4	Hydrocarbons, haloginated		water	unspecified		kg	5.53E-4
	4	Hydrogen-3, Tritium		water	unspecified		kBq	5.12E+2
	4	Iodine-129		water	unspecified		kBq	2.50E-4
	4	Suspended solids, unspecified		water	unspecified		kg	1.23E-3
Output	0	PET, bottle grade, at plant	RER			0	kg	1.00E+0

As should be evident, the data quality in the LCI stage of an LCA is crucial since data quality has a significant influence on the results. One of the primary considerations for including data is the data quality requirements. Data should be reliable, robust, and must be validated. LCI modeling works as does any model: if wrong information is introduced, incorrect results and misleading conclusions can result. In short, "garbage in = garbage out."

Special considerations are needed when unit flows or processes output more than one product as shown in Figure 9.9. In these cases, ISO 14040 and 14044 consider this a problem of multifunctionality, and provide some modeling provisions and ranking according to the LCA situation that is being modeled [1, 5].

The first approach to solve multifunctionality according to ISO is a subdivision of the multifunctional process. Figure 9.10 shows an example where ISO recommends collecting additional information for those monofunctional processes that are related to the studied system. In this case, the actual required process is removed from the multifunctionality problem and data is individually collected for the flow. This situation can be used if the separated process is not also a multifunctional process.

The second option according to ISO is system expansion or substitution. In this case, it is required to add or to subtract a comparable process as shown in Figure 9.11. In this case, the system is enlarged by selecting an equivalent process. This methodology is also known as "system expansion, substitution, crediting, or avoided burden approach." In the case of expanding the system boundaries, we substitute a not required function with an alternative way of providing the equivalent production of the product. Another situation is when several units will be produced in a comparative study. In this case, we can expand the system and boundaries, adding the missing functions and inventories of a respective monofunctional inventory.

An example of this is if we compare a combined printer, copier, scanner, and fax machine with a combined copier, scanner, and fax machine. The missing function printer can be added to the latter product system to create equivalent systems. Special consideration must be taken in these studies since misleading interpretations can arise.

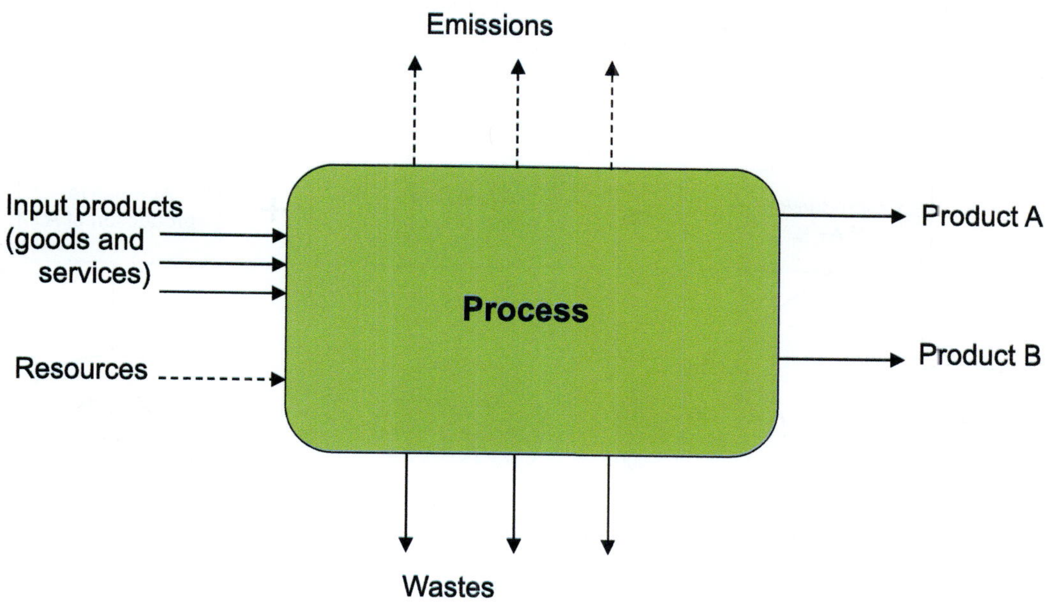

Figure 9.9 Multifunctional process with several inputs products and resources consumed and various wastes and emissions generated to produce two coproducts A and B; adapted from [2].

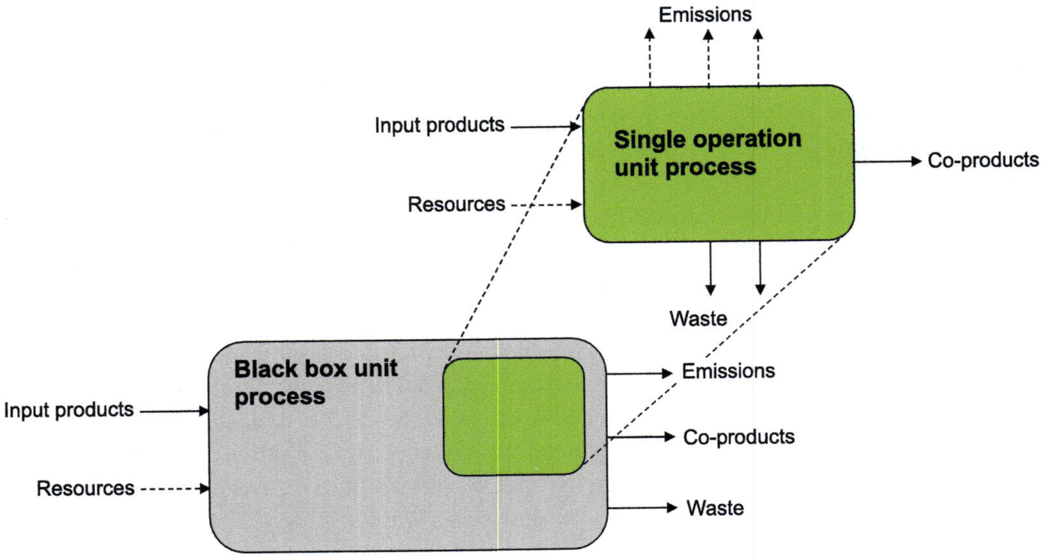

Figure 9.10 Black box unit process and single unit process; adapted from [2].

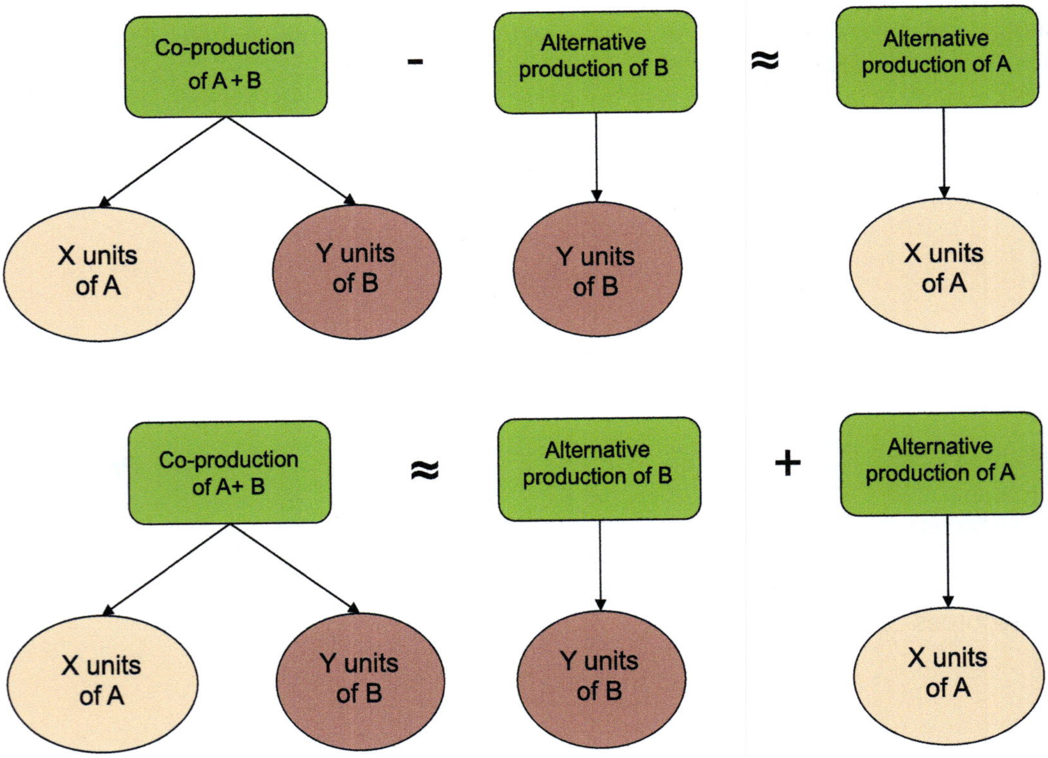

Figure 9.11 Addition or subtraction of unit process in the system expansion process. Functional equivalence is obtained by adding or subtracting an alternative process; adapted from [2].

The third and final method considered in the ISO hierarchy is called allocation. In the case of allocation, we are partitioning the flow according to a property such as element content, mass, market price, or energy content as shown in Figure 9.12.

A classic example of allocation is an oil refinery, which produces fuel oil, naphtha, kerosene, residue, and waste. Several approaches can be used to allocate the EFP to each product. For example, we could allocate based on sales = quantity × price of the products. Table 9.6 shows the proceeds of the six products as produced by the installation. The share of each product in the total proceeds is the allocation factor for that product. According to ISO, allocation is the least desirable method to deal with multifunctionality; however, market share allocation is often found to be practical since it does reflect the market for a product.

9.5.3.2 Completeness/Cut-off and Loops

Even for simple LCI studies, all economic activities can be part of an LCA study. However, the number of processes that quantitively and directly impact a product is somewhat limited. So, in practice, "all quantitatively not relevant nonreference product flows, waste flows, and elementary flows can be ignored" [2], so they are cut off. Special care must be taken so that all the essential flows needed to deliver the goal and scope of the study are not cut off.

Endless loops can exist in modeling. For example, the production of aluminum requires energy, and the creation of energy requires equipment made from aluminum, the production of which requires energy again. These loops can be complicated to resolve, but with new computer capability and new numerical methods, the loops can be resolved in LCA software.

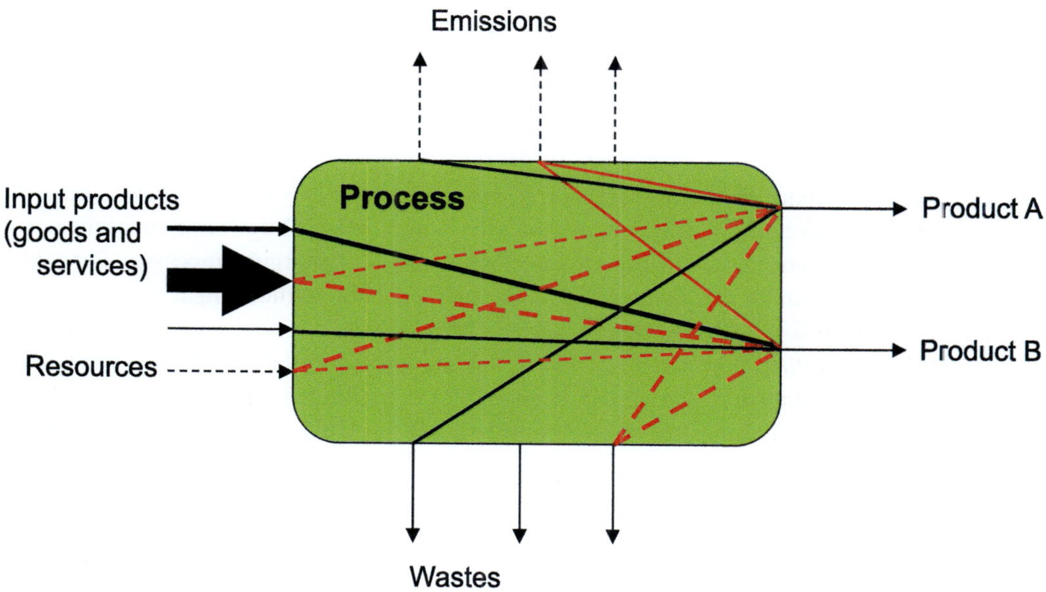

Figure 9.12 Solving multifunctionality by using allocation as the main criteria. In this case the line thickness indicates the proportion of input, resources, wastes, and emissions that are allocated to each product according to a previously defined criterion such as mass, energy, or market price; adapted from [2].

Table 9.6 Allocation of flows using marker share price for the production of products for a refinery per year.

	Product	Quantity, 10^6 kg	Price, €/kg	Proceeds, 10^3 €	Allocation Factor
1	Fuel oil	10	0.15	1,500	0.25
2	Naphtha	9	0.2	1,800	0.3
3	Kerosene	3	0.1	300	0.05
4	Long residue "waste" good	4	0.05	200	0.03
5	Food waste (service)	4	0.15	600	0.1
6	Wood waste (service)	2	0.8	1,600	0.27
	Total	32			1

9.5.3.3 Provisions for LCI according to Situations A, B, and C of LCA

In general, the inventory is the same regardless the situation of the LCA study; however, some considerations are applied to each situation.

Situation A – microlevel decision support – considerations: In this case, the LCI model should be conducted as an attributional model. Multifunctionality should be addressed by subdividing the multifunctional process. If the procedure only affects the system under study, system expansion should be performed using substitution. When a multifunctional process is for a general background process, it should be substituted by an average market consumption mix of the process. If the market alternative is an extensive process and it is being substituted, this should be conducted as a situation B LCA. If the process information does not exist, allocation could be conducted as long as it does not favor the system under study. For waste flow consideration, system expansion should be conducted. Further descriptions of factors can be found elsewhere [2].

Situation B – meso-/macrolevel decision support – considerations: The provisions used for situation A also apply for situation B. Two differences should be considered: for large-scale consequences, the process should be modeled as the expected mix of the long-term marginal processes. For comparative studies, uncertainty calculations should be conducted due to the comparative nature of the study. This can be done by using Monte Carlo methodologies.

Situation C – Accounting support considerations: The same provisions as situation A apply, but two special situations should be considered. In situation C1, multifunctionality should be solved by substitution via system expansion without considering the absolute amount to be substituted. In situation C2, the general cases of multifunctionality should be resolved by allocation.

9.5.4 Impact Assessment

The LCIA aggregates the LCI data to facilitate and support the interpretation. LCIA aims at making the results of the LCI more understandable and more manageable concerning human health, the availability of resources, and the natural environment. During the LCIA, the LCI is converted to impact measures such as global warming potential, eutrophication, and ozone depletion. Figure 9.13 shows a basic example of conversion of the unit process flow emissions through classification and characterization factors to estimate the final impact of each indicator.

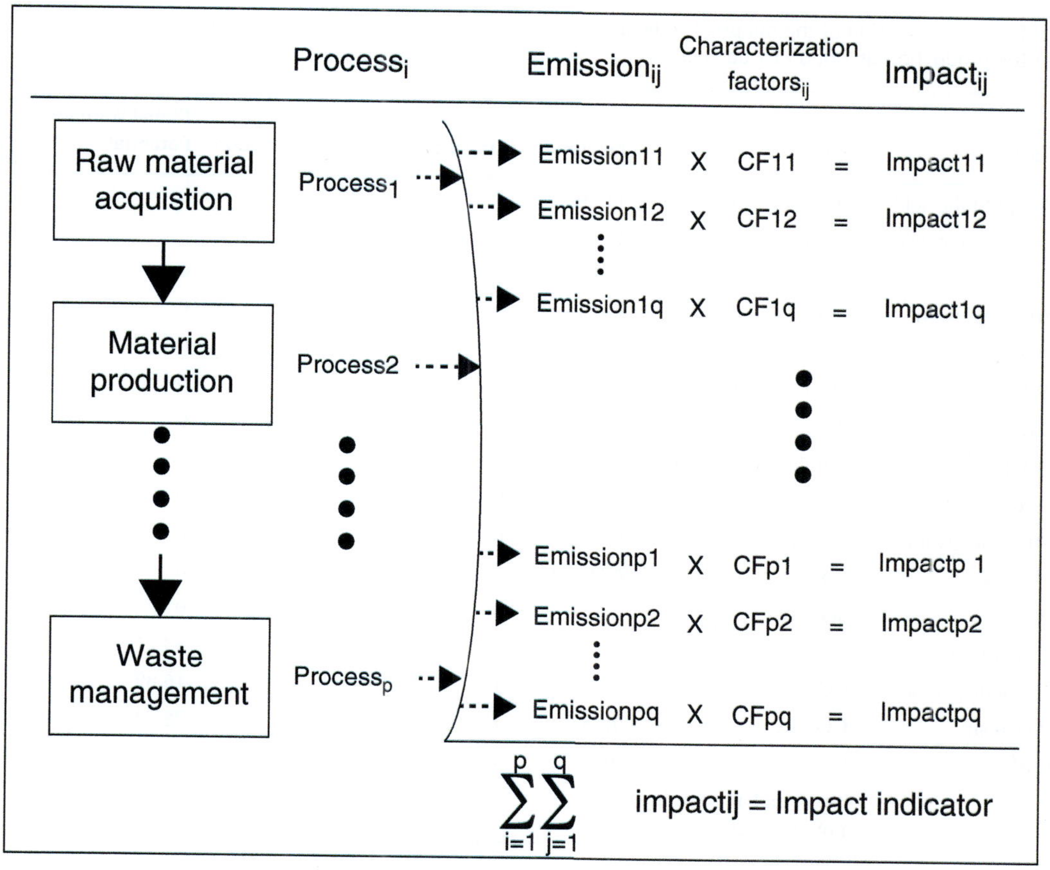

Figure 9.13 Example of LCI conversion to LCIA.

Table 9.7 provides the classification and characterization factors for the production of 1,000 PET bottles as shown in Figure 9.8. The characterization factors are provided according to Impact 2002 + v2.1. To estimate the total global warming potential, we need to multiply the amount of CO_2 emitted by each process times the classification and the characterization factors. Then, by adding each contribution, we can estimate the total global warming potential. This is a very simplified table showing the global warming potential for 1,000 PET bottles.

As illustrated in the calculation for global warming potential, the same steps are conducted for each midpoint impact indicator such as nonrenewable energy. Figure 9.14 shows a representation of how aggregated LCI data are converted to midpoint indicators and then to endpoint indicators. Midpoint indicators include a variety of impact categories such as climate change, (stratospheric) ozone depletion, human toxicity, respiratory inorganics, ionizing radiation, (ground-level) photochemical ozone formation, (land and water) acidification, (land and water) eutrophication, ecotoxicity, land use, (mineral, fossil, and renewable energy resources, water) resource depletion. Midpoint indicators can be combined into endpoint indicators, most often human health, the natural environment, and natural resources. The selection and development of LCIA methods is guided by ISO 14044:2006 [5], which has widespread acceptance by stakeholders and internationally.

Table 9.7 Estimation of the midpoint indicator of global climate change according to IMPACT 2002 + V2.1, for the production of 1,000 PET bottles.

Flow	Flow	Amount	Classification	Characterization	Global Warming Potential
Terephthalic acid	Steam, kg	117.96	0	0	0
Terephthalic acid	Xylene, kg	18.55	0	0	0
Terephthalic acid	Electricity, MJ	1,695.85	0	0	0
Ethylene glycol	Electricity, MJ	639.15	0	0	0
PET resin	Electricity, MJ	2,518.4	0	0	0
ISBM	Electricity, MJ	89.1	0	0	0
Ethylene glycol	Water, kg	74.55	0	0	0
Ethylene glycol	Ethylene oxide, kg	8.73	0	0	0
Terephthalic acid	SO_2, kg	0.1122	0	0	0
Ethylene glycol	SO_2, kg	0.039	0	0	0
PET resin	SO_2, kg	0.29024	0	0	0
ISBM	SO_2, kg	0.217	0	0	0
Terephthalic acid	CO_2, kg	43.77	1	1	43.77
Ethylene glycol	CO_2, kg	16.89	1	1	16.89
PET resin	CO_2, kg	78.72	1	1	78.72
ISBM	CO_2, kg	48	1	1	48
	Total, global warming potential, CO_2-eq				187.38

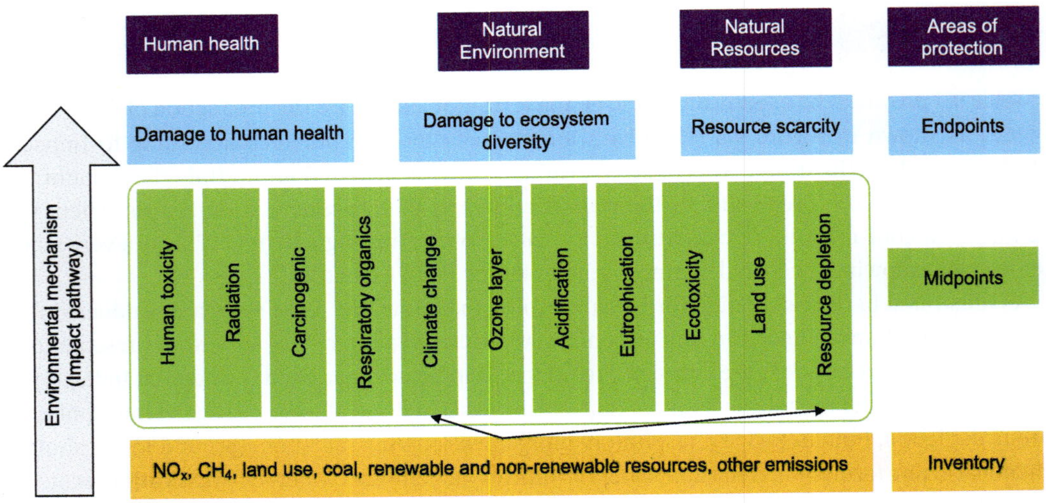

Figure 9.14 Life cycle impact assessment. Schematic steps for estimation from LCI to midpoint to endpoint indicators and their area of protection.

During the LCIA step, several steps must be followed:

- *Selection*: Select and define the impact categories, such as global warming, acidification, and eutrophication, to be used for reporting the data of the study. This should be part of the scope of the study, and properly identified before the LCI and collection of data is commenced to ensure that appropriate and relevant LCI data is collected. The selection of the impact categories, normalization factors, and weighting parameters should agree with the LCI. The choice of the impact categories should be consistent with the goal of the study and cover all the relevant environmental issues related to the analysis system. Generally, commonly accepted LCIA methodologies and models are preferred.
- *Classification*: Classify the aggregated and collected data by assigning the results from the LCI to the relevant impact categories.
- *Characterization*: Characterize the aggregated LCI results in terms of adequate factors to report midpoint indicators and endpoint indicators.

Most studies finish their evaluation and reporting at midpoint indicators. ISO recommends to not complete studies to endpoint indicators unless specifically required for policy purposes. The main objection to the last steps of the LCIA (i.e., normalization and endpoint or damage indicators) is that they need a lot of subjective aggregation of data, with which LCA stakeholders may not agree. Table 9.8 shows examples of single and combined midpoint and endpoint indicators commonly used to assess LCI and report LCIA.

9.5.5 Interpretation

The interpretation step is the phase in which the results of the analysis and all choices and assumptions made during the study are evaluated in terms of soundness and robustness, and overall conclusions are drawn. The main steps are as follows:

9.5.5.1 Evaluation of the Results

- Consistency: during the consistency check, the assumptions, methods, models, and data are checked for consistency with the goal and scope of the study.
- Completeness: during the completeness check, all the relevant information and data needed for the interpretation phase are checked and completed. They should be checked for false assumptions, model choices, and data.

9.5.5.2 Analysis of the Results

- Robustness: during this step, we detect influence on the results of variations in process data, model choices, and other variables.
- Contribution: during this step, the overall contribution of each unit flow or process is estimated and checked against the total emissions for each impact category so that it is possible to estimate the contribution of each process or stage to the total impact.
- Perturbation: during the perturbation analysis, we study the effect that small changes have in the results of LCA. So, we introduce a small perturbation to the input data and determine how the perturbation affects the results. Perturbation analysis can be calculated at many levels. The main difference between contribution and perturbation analysis is that perturbation takes into account not only environmental flows but also economic flows between unit processes.

Table 9.8 Examples of midpoint and endpoint indicators and their area of impact.

Midpoint								
Recipe			Impact 2002 +			Traci 2.0		
Category	Unit	Compartment	Category	Unit	Compartment	Category	Unit	Compartment
Human toxicity (HT)	$kg^{14}DCB$	Urban air	Human toxicity	Chloroethylene into air.eq.kg	Air			
Particulate matter formation (PMF)	kg PM_{10}	Air	Respiratory effects	kg $PM_{2.5}$ into air.eq/kg	Air			
Ionizing radiation (IR)	kg U^{235}	Air	Ionizing radiation	Bq C^{14} into air. eq/Bq	Air			
Ozone depletion	CFC-10^5	Air	Ozone layer depletion	kg CFC^{-11} into air.eq/kg	Air	Ozone depletion	kg CFC^{-11} eq/ kg substance	Air
Photochemical oxidation (POF)	kg NMVOC	Air	Photochemical oxidation	kg Ethylene into air.eq/kg	Air			
			Aquatic ecotoxicity	kg Triethylene glycol into water.eq/kg	Water			
Terrestrial ecotoxicity (TET)	kg 14-DCB	Industrial soil	Terrestrial ecotoxicity	kg Triethylene glycol into soil.eq/kg	Soil			
Freshwater ecotoxicity (FET)	kg 14-DCB	Fresh water						
Marine ecotoxicity (MET)	kg 14-DCB	Marine water						
			Aquatic acidification	kg SO_2 into air-eq/kg	Air			

Freshwater eutrophication (FE)	kg phosphorus	Fresh water	Aquatic eutrophication	kg PO_{43}- into water-eq/kg	Water	Eutrophication	kg N eq/kg substance	Air, water
Marine eutrophication (ME)	kg Nitrogen	Marine water						
Terrestrial acidification (FA)	kg SO_2	Air	Terrestrial acid/nutrification	kg SO_2-eq into air/kg	Air	Acidification	kg SO_2 eq/kg substance	Air
Agricultural land occupation (ALO)	$m^2.y$	Agricultural land						
Urban land occupation (ULO)	$m^2.y$	Urban land	Land occupation	m^2 Organic arable land-eq · y/ $m^2 \cdot y$	Soil			
Natural land transformation	m^2	Natural land						
			Water turbined	m^3 water turbined	Water			
Climate change (CC)	kg CO_2	Air	Global warming[1]	kg CO_2-eq into air/kg	Air	Global climate	kg CO_2-eq/kg substance	Air
Fossil fuel depletion (FD)	kg oil		Nonrenewable energy	MJoule	Soil			
Mineral resource depletion (MRD)	kg Fe		Mineral extraction	MJoule	Soil			
			Water withdrawal	m^3 of water needed	Water			
Water depletion (WD)	m^3	Water	Water consumption	m^3 of water consumed	Water			

(*Continued*)

Table 9.8 (Continued)

Midpoint

Recipe		Impact 2002 +		Traci 2.0		
				Human health criteria	$PM_{2.5}$ eq/kg substance	Air
				Human health cancer	CTUcancer/kg	Freshwater, seawater, natural soil, agricultural soil
				Human health noncancer	CTUnoncancer/ kg	Freshwater, seawater, natural soil, agricultural soil
				Smog formation	kg O_3 eq/kg substance	Air
Endpoint						
Human health (HH)	DALYs	Human health	DALYs			
Ecosystem diversity (ED)	Loss of species during a year	Ecosystem quality	PDF.m_2.y			
		Climate change	kg CO_2-eq into air/kg			
Resource availability (RA)	$ Increased cost	Resources	MJoule			

[1]According to 100-year time horizon
DALY: Disability-adjusted life years = years of life lost (YLL) + years of life disabled (YLD)
PDF: Potentially disappeared fraction of species over a certain amount of m^2 during a certain amount of year
CFC: Chlorofluorocarbon
NMVOC: Non-methane volatile organic carbon compound
14-DCB: 1,4 dichlorobenzene

- Sensitivity and uncertainty analysis: during the sensitivity stage, changes are deliberately introduced to determine the robustness of the results. During the uncertainty analysis, typically Monte Carlo analysis is conducted by randomly iterating input data bounded by the uncertainty ranges to estimate the impact on the final results.

9.5.5.3 Formulation of Conclusions and Recommendations

During the last stage of an LCA, conclusions are drawn, and recommendations are prepared for the intended audience of the study. The conclusions and recommendations are based on the information gathered in the phases of the LCA, and they are combined with the results of the previous steps of the interpretation phase.

Most interpretation results are based on two approaches. First, there are the results of all forms of the evaluation and analysis of the results, such as consistency and uncertainty analysis. These results lead to several judgments relating to the quality and the robustness of the findings of the inventory analysis and impact assessment. Second, the practitioner will provide final conclusions and recommendations, for instance, as to product choice or improvement. As shown in Figure 9.2, LCA is an iterative process, so initial conclusions are used to generate further studies and recommendations.

9.6 LCA Software

We can divide the tools to conduct LCA into full and streamlined LCA tools as described in Chapter 1 section 1.12. Full LCA software that allows running LCAs that are fully compliant with ISO 14040/14044 include SimaPro® developed by PRé Consultants, GaBi ts® developed by PE International, Umberto® NXT developed by ifu™ Hamburg, and Open LCA. These software programs provide complete LCI and LCIA capabilities.

Derived from the full LCA software, streamlined LCA software programs have been developed, often for specific industries such as packaging. For packaging, the main software programs used to evaluate the environmental footprint of packaging systems are COMPASS® (Comparative Packaging Assessment) currently administered by Trayak™, PackageSmart commercialized by EarthShift Global, and PIQET® developed by the Centre for Design at RMIT University (Melbourne, Australia), in association with the SPA.

9.7 Case Study 9.1. LCA Study of Beverage Packaging Systems

This streamlined LCA study concentrated mostly on evaluating the environmental footprint of distributing beverages in three different packages, aluminum cans, PET bottles, and glass bottles. Figure 9.15 shows a graphical representation of the system. The study was conducted using PackageSmart software, a streamlined LCA software used for estimating the EFP of packaging systems.

Goal, scope, and functional unit definition

The goal of the study was to estimate the environmental footprint of delivering carbonated drinks to the consumer. The functional unit and reference flow is 355 mL, which in this case is also the size of one container (all containers are the same size). The system boundary of this study is

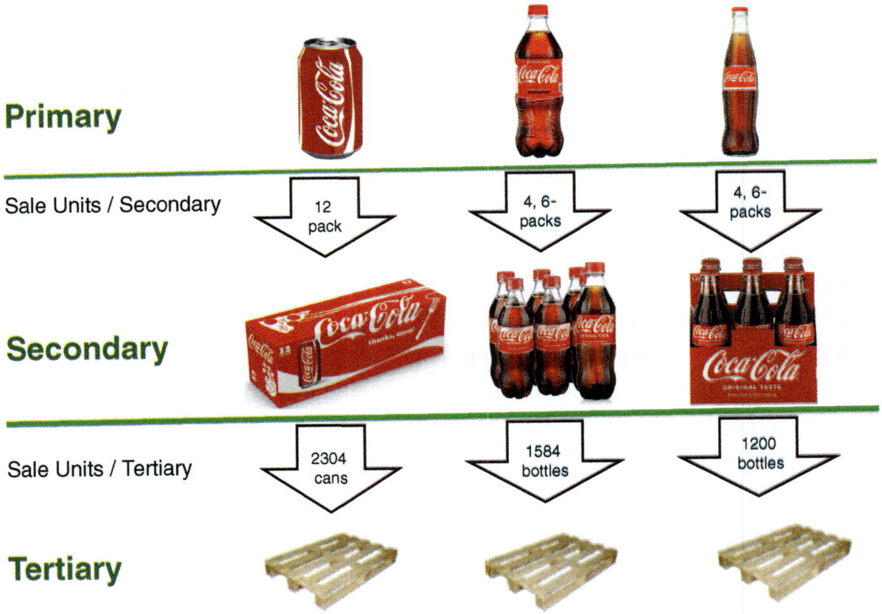

Figure 9.15 Representation of the primary, secondary, and tertiary packaging system to deliver beverage drink.

"cradle-to-grave," and to make a clear interpretation of contribution analysis, it is discretionally divided into 5 main sections: 1. Material production, 2. Al can and PET and glass bottle production, 3. waste management 4. environmental credits, and 5. transportation. These five main sections were established based on the general package production. Several assumptions were made during the study that, for reasons of space, are not provided in this section. A similar study can be checked elsewhere [11].

Life cycle inventory analysis (LCI)

The source of data in this LCA was literature and databases. Table 9.9 shows detailed information about the weight and configuration of each packaging system, and the main parts of each component. This example is for educational purposes, and it should not be used for commercial claims.

Life cycle impact assessment (LCIA)

Figure 9.16 shows the EFP of the plastic bottle, Al can, and glass bottle. Figure 9.16a shows the EFP of each component of the plastic bottle, label, PET bottle, and closure. We can observe that the bottle is the main contributor to the EFP. Figure 9.16b shows the EFP contribution of the materials, conversion, and end of life of the primary plastic bottle. Again, we can observe that the material and conversion are the main contributor to each EFP indicator. Figure 9.16c shows the EFP of the primary, secondary, and tertiary components of the packaging system. The primary plastic bottle is the main contributor to each EFP, with more than 85% contribution in each category. Figure 9.16d shows the EFP of the primary, secondary, and tertiary component for the Al can. For the Al can, the primary and the secondary paperboard container are responsible for the main EFP contributions. Figure 9.16e shows the EFP of the glass bottles. In this case, the glass bottle is responsible for the main EFP. Figure 9.16f shows the comparison of the EFP of the three packaging systems.

Table 9.9 Aluminum can, PET bottle, and glass bottle life cycle inventory of the weight, dimensions, material, processing, and transportation for the primary, secondary, and tertiary. This is an example, and it should not be used for commercial statements.

Package Level	Component	Amount	Unit	Number of Sales Unit	Unit Measure (W × L × H)	Sales Unit Volume	Material	Processing/ Conversion	Raw Material Transportation	Distribution	End of Life	Comments
Aluminum												
Primary	Aluminum can	12.8	g	1 can	2.5″ × 2.5″ × 4.8″	355 mL	62% (7.9 g) Aluminum/ North America and 38% (4.9 g) Aluminum, recycled/North America	Aluminium "average" production process/US	50 km by average truck	500 km by average truck	Packaging waste scenario	
Secondary	SUS board box	86.0	g	12 cans	5″ × 16″ × 5.5″		Solid unbleached board/US	Production of cartons including offset printing/US	50 km by average truck			Build on a per g basis.
Tertiary	Wood pallet	18.1	kg	2,304 cans	40″ × 48″ × 44″		Pallet, wood/US (includes conversion)	Included in material process	50 km by average truck			Build on a per kg basis. Assume 5 uses.
Tertiary	Stretch wrap	300	g				Packaging film, LDPE/US	Extrusion, plastic film/ US	50 km by average truck			Build on a per g basis.

(Continued)

Table 9.9 (Continued)

Package Level	Component	Amount	Unit	Number of Sales Unit	Unit Measure (W × L × H)	Sales Unit Volume	Material	Processing/ Conversion	Raw Material Transportation	Distribution	End of Life	Comments
Plastic Bottle												
Primary	Plastic bottle	24.8	g	1 bottle	2.5″ × 2.5″ × 7″	355 mL	PET, bottle grade/US	Injection molding/US	50 km by average truck	500 km by average truck	Packaging waste scenario	
Primary	Primary closure	3.00	g				Polypropylene/ North America	Injection molding/US	50 km by average truck			Proxy for injection stretch blow molding.
Primary	Label	2.000	g				Polystyrene, expandable/US	Extrustion, plastic film/ US	50 km by average truck			Full wrap label.
Secondary	Ring	6.92	g	24 bottles (4, 6-packs)	11″ × 16″ × 7″		Packaging film, LDPE/US	Extrustion, plastic film/ US	50 km by average truck			Will require 4 rings.
Secondary	Corrugated board box	113.5	g				Corrugated board, 100% recycled, single wall/US	Production of cartons including offset printing/US	50 km by average truck			Build on a per g basis.
Tertiary	Wood pallet	18.1	kg	1,584 bottles	39.5″ × 44″ × 42″		Use above					
Tertiary	Stretch wrap	300	g				Use above					

Glass Bottle

Primary	Glass bottle	282	g	1 bottle	6 cm × 6 cm × 25 cm	355 mL	Packaging glass, white (includes conversion)	Included in material process	50 km by average truck	500 km by average truck	Packaging waste scenario	
Primary	Steel Closure	2.12	g				Iron and steel/ US	Sheet rolling, steel/US	50 km by average truck			
Secondary	Corrugate board box	113.5	g	24 bottles (4, 6-packs)	10.5″ × 15″ × 8.75″		Use above					
Secondary	SUS board box	53.89	g				Use above					Will require 4 SUS boards.
Tertiary	Wood pallet	18.1	kg	1,200 bottles	36″ × 45″ × 43.8″		Use above					
Tertiary	Stretch wrap	300	g				Use above					

Key

(dark shading)	Company or supplier's data responses
(light shading)	Your modeling choices

Note: This is a fictitiuous study that should not be used to derive any judgments. It is just provided as a training exercise.

The glass bottle has a larger EFP for all the six EFP indicators followed by the PET bottle and the Al can. For climate change, ecosystems, and resources, the glass bottle presents 45, 50, and 70% more EFP impact than the Al can, respectively. We must remember that these results are for this

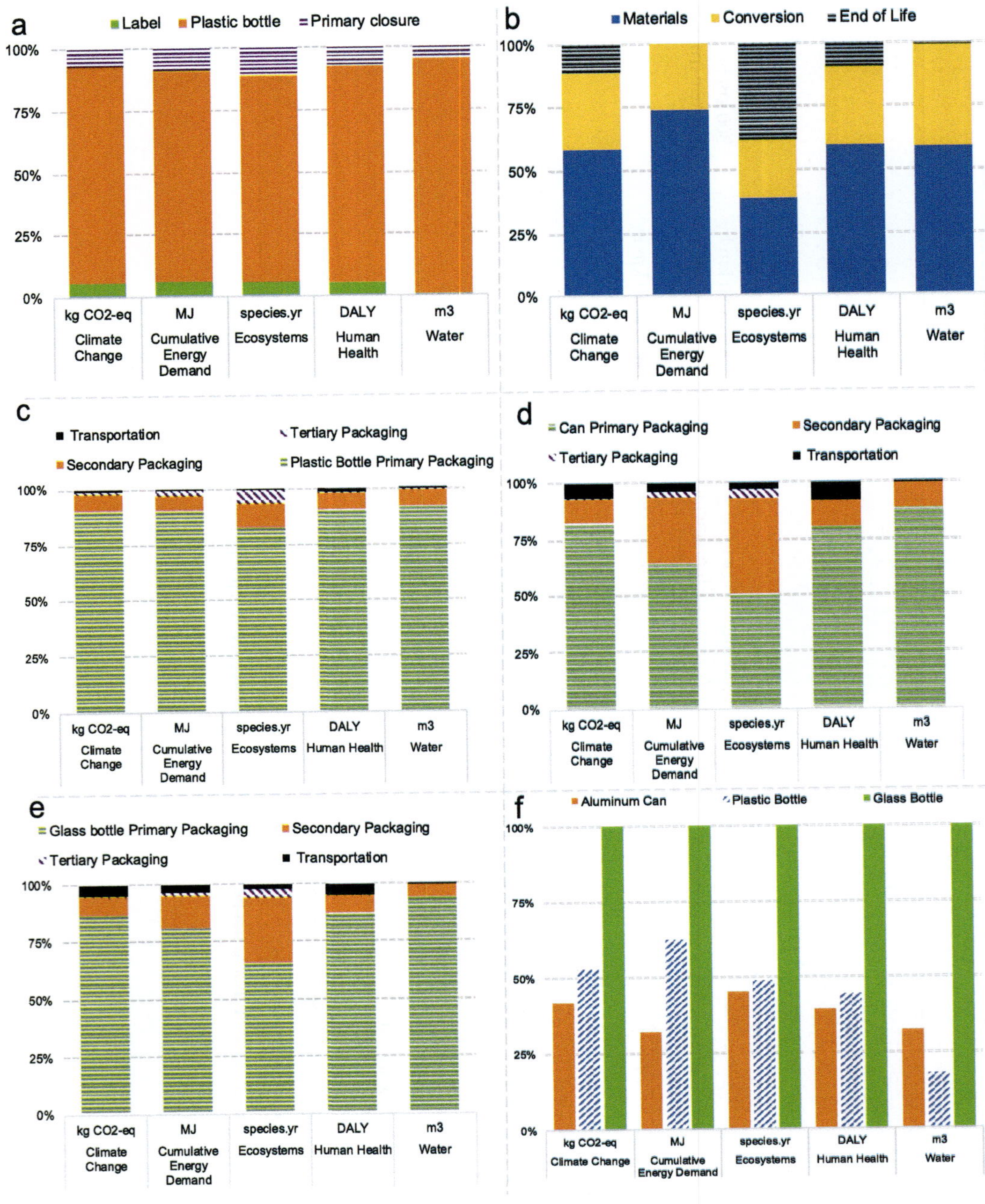

Figure 9.16 Environmental footprint of PET bottle, Al can, and glass bottle: (a) EFP of primary PET bottle components; (b) process contribution of PET bottle; (c) EFP of primary, secondary, and tertiary for the PET bottle; (d) EFP of primary, secondary, and tertiary for the Al can; (e) EFP of primary, secondary, and tertiary for the glass bottle; (f) EFP comparison of the PET bottle, Al can, and glass bottle.

model with boundary conditions, so extrapolation to general scenarios should not be performed, and is not recommended.

9.8 Study Questions

1 List the main ISO standards describing and providing a framework to conduct an LCA study.
2 The main areas of protection in an LCA are:
 A Human health, climate change, and fossil resources
 B Respiratory organics, noise, and land use
 C Human health, natural resources, and natural environment
 D Human health, natural resources, and climate change
 E None of the above
3 Provide two names each for full and streamlined LCA software.
4 Why are cut-off rules used in LCA?
5 Using the sequential or aggregation method, calculate the total amounts of PET resin, electricity, terephthalic acid, and ethylene glycol that are needed to produce 200 PET bottles. Also, estimate the amount of CO_2 and SO_2 release.

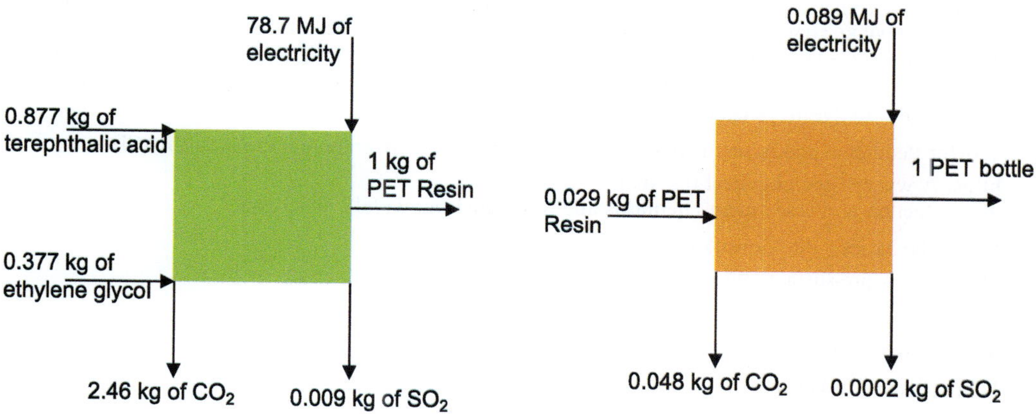

6 Define the main steps of the interpretation section of an LCA study.
7 Define life cycle assessment (LCA) in your own words. What is the main objective of an LCA study? Draw a diagram indicating and labeling the main steps to conduct an LCA as defined by ISO 14040/14044 2006?
8 What is the main difference between attributional and consequential LCA?
9 Define midpoint and endpoint indicators. Describe the process of classification and characterization in LCI. Provide two examples of midpoint indicators.
10 Define the goal (exact question, target audience, and intended application), scope (temporal, geographical coverage, technological coverage, technological coverage, mode of analysis, overall level of sophistication), functional unit, and reference flows for an LCA study to evaluate the environmental footprint of delivering to consumers Domino sugar in a 1 lb paper bag, 1.25 lb paperboard tube, and 4 lb plastic container.
11 Define the goal, scope, functional unit, and reference flows for an LCA study to evaluate the environmental footprint of delivering orange juice in 0.5-, 1-, and 2.5-liter PET containers to consumers.

9.9 Additional Resources

1. LCA youtube video several tutorials
https://www.youtube.com/watch?v=4wQ2Jm6i9F0
https://www.youtube.com/watch?v=fGhoInz-VUs
https://www.youtube.com/watch?v=KrJUpSiCOoU
https://www.youtube.com/watch?v=cYOC8_jJcII
https://www.youtube.com/watch?v=zFaG4QZpzIs
2. Consequential vs. attributional LCA
https://lca-net.com/blog/consequential-vs-attributional-lca
3. The life cycle initiative – United Nations
https://www.unenvironment.org/explore-topics/resource-efficiency/what-we-do/life-cycle-initiative
https://www.lifecycleinitiative.org/starting-life-cycle-thinking/life-cycle-approaches/environmental-lca
4. US EPA Life cycle assessment: Principles and practice
https://nepis.epa.gov/Exe/ZyNET.exe/P1000L86.TXT?ZyActionD=ZyDocument&Client=EPA&Index=2006+Thru+2010&Docs=&Query=&Time=&EndTime=&SearchMethod=1&TocRestrict=n&Toc=&TocEntry=&QField=&QFieldYear=&QFieldMonth=&QFieldDay=&IntQFieldOp=0&ExtQFieldOp=0&XmlQuery=&File=D%3A%5Czyfiles%5CIndex%20Data%5C06thru10%5CTxt%5C00000002%5CP1000L86.txt&User=ANONYMOUS&Password=anonymous&SortMethod=h%7C-&MaximumDocuments=1&FuzzyDegree=0&ImageQuality=r75g8/r75g8/x150y150g16/i425&Display=hpfr&DefSeekPage=x&SearchBack=ZyActionL&Back=ZyActionS&BackDesc=Results%20page&MaximumPages=1&ZyEntry=1&SeekPage=x&ZyPURL
5. Tool for Reduction and Assessment of Chemicals and Other Environmental Impacts (TRACI)
https://www.epa.gov/chemical-research/tool-reduction-and-assessment-chemicals-and-other-environmental-impacts-traci
6. LCA – the basics – PRé Consultants
https://www.pre-sustainability.com/sustainability-consulting/lca-methodology-basics
7. What is LCA? Thinkstep
https://www.thinkstep.com/life-cycle-assessment
8. LCA software Umberto®
https://www.ifu.com/en/umberto/lca-software
9. PackageSmart
https://www.earthshiftglobal.com/software/packagesmart
10. Compass®
https://trayak.com/compass

References

1. ISO (2006). ISO 14040:2006 – Environmental management – Life cycle assessment – Principles and framework.
2. European Commission – Joint Research Centre. ILCD Handbook – General guide for Life Cycle Assessment – Detailed guidance (2010). Ispra, Italy. DOI: 10.2788/38479.
3. EUILCD-handbook general guide LCA Provisions and action steps, EUR 24378 en – 2010. (2010). DOI: 10.2788/94987.

4 Weidema, B. (2006). Market information in life cycle assessment. 2.-0 LCA consultants. https://lca-net.com/files/Market-information-in-life-cycle-assessment.pdf (accessed August 22, 2019).
5 ISO (2006). ISO 14044:2006 – Environmental management – Life cycle assessment – Requirements and guidelines.
6 Gazulla, C., Raugei, M., and Fullana-i-palmer, P. (2010). Taking a life cycle look at crianza wine production in Spain: Where are the bottlenecks? *The International Journal of Life Cycle Assessment* 15: 330–337. DOI: 10.1007/s11367-010-0173-6.
7 Madival, S., Auras, R., Singh, S.P., and Narayan, R. (2009). Assessment of the environmental profile of PLA, PET and PS clamshell containers using LCA methodology. *Journal of Cleaner Production* 17: 1183–1194. DOI: 10.1016/J.JCLEPRO.2009.03.015.
8 Burek, J., Kim, D., Nutter, D., Selke, S., Auras, R., Cashman, S. et al. (2018). Environmental sustainability of fluid milk delivery systems in the United States. *Journal of Industrial Ecology* 22: 180–195. DOI: 10.1111/jiec.12531.
9 Kijchavengkul, T. and Auras, R. (2008). Compostability of polymers. *Polymer International* 57: 793–804. DOI: 10.1002/pi.2420.
10 Plastics Europe (2002). Eco-profiles. https://www.plasticseurope.org/en/resources/eco-profiles.
11 Kang, D., Auras, R., and Singh, J. (2017). Life cycle assessment of non-alcoholic single-serve polyethylene terephthalate beverage bottles in the state of California. *Resources, Conservation & Recycling* 116: 45–52. DOI: 10.1016/J.RESCONREC.2016.09.011.

10

Municipal Solid Waste

10.1 Learning Objectives

After learning from this chapter, you are expected to:

- Define municipal solid waste (MSW).
- Understand the effect of MSW for low-, middle-, and high-income economies.
- Understand the composition of MSW in the US.
- Understand the end-of-life scenario of MSW in the US and its trends.
- Define the 5R concept and its main implications for packaging.

10.2 Introduction

Solid waste, commonly known as garbage, is an increasing issue affecting societies, environments, and economies around the world, but it most negatively impacts the most vulnerable individuals and communities. Poorly managed solid waste is contaminating oceans, clogging fluvial discharges, causing flooding, incubating diseases, increasing respiratory issues, and affecting economies. As countries develop from low-income to middle- and high-income economies, the generation of solid waste is expected to more than double by 2050 [1]. If proper solid waste management practices are not globally implemented, further damage to society, economies, and the environment will occur. However, interest in controlling solid waste has flourished as the new push for a circular economy gains traction.

Solid waste management can be the largest budget item of many cities and counties, accounting for 4% to 20% of their budgets [1]. In 2018, the world generated around 2 billion tonnes of MSW annually. For the purposes of this book, MSW will be defined as described by the US EPA: "MSW is commonly known as trash or garbage discarded by consumers – consists of everyday items such as product packaging, grass clippings, furniture, clothing, bottles, food scraps, newspapers, appliances, paint, and batteries." By 2050, it is expected that the world will produce 3.4 billion tonnes of MSW[1] annually. Solid waste contributed 1.6 billion tonnes of CO_2-eq emissions in 2016, and it is estimated that it will produce 2.6 billion tonnes of CO_2-eq in 2050, accounting for around 5% of the total global GHG emissions [1]. Although it is challenging and systems may be overwhelmed, several routes and solutions have been devised to divert MSW from landfill to other waste management

1 1 tonne = 1,000 kg = 2204.62 lb = 1.1023 US ton.

Life Cycle of Sustainable Packaging: From Design to End of Life, First Edition. Rafael A. Auras and Susan E. M. Selke.

Companion Website: www.wiley.com/go/Auras/lifecycleofsustainablepackaging

options such as reduce, reuse, recycling, and recovering, and to work on creating robust solid waste recovery programs rather than only relying on mitigation. In this chapter, we will first provide a worldwide perspective on MSW, and then we will discuss the US situation so that in later chapters, we can concentrate on specific waste management techniques.

10.3 World Picture of Municipal Solid Waste

MSW is a product of urbanization, economic development, and population growth. Data about global MSW has been hard to obtain, so recently the World Bank Group decided to provide a more systematic report on solid waste management around the world [1]. To report the data, the World Bank divided the world into regions: East Asia and Pacific, Europe and Central Asia, Latin America and the Caribbean, Middle East and North Africa, North America, South Asia, and Sub-Saharan Africa. Figure 10.1 shows how MSW varies around the world and is highly dependent on the income of each region and country. Generally, high-income economies generate more than 1.5 kg (3.3 lb) of MSW per day per person, and in total upper-middle and high-income economies generated around 66% of the MSW in the world. The North America region produced the highest average amount of waste of 2.2 kg (4.9 lb) per capita per day. The global range is mostly between 0.11 and 4.54 kg (0.24–10 lb) per person per day. The global average for 2016 was 0.74 kg (1.6 lb) of MSW per capita per day. MSW generation is positively correlated to per capita GDP and percentage urbanization. Of global regions, South Asia and Sub-Saharan Africa are expected to have the highest increase in waste generation by 2050, although the per capita MSW generation is expected to be only one-third of the amount generated in North America. MSW generation is tightly connected to income. As economies shift from low income to middle and high income, MSW generation will increase around the world. A similar increase is not expected for high-income economies since they may have already reached a point where the GDP is no longer tightly linked to material consumption. MSW is also tightly connected to urbanization. Urbanized areas generated more MSW than rural areas.

10.4 Environmental Kuznets Curve (EKC)

A hypothesized relation between economic development and environmental pollution is known as the classic EKC, which suggests that economic development initially leads to a deterioration in the environment, but after a certain level of economic growth, the society commences to improve its relationship with the environment and levels of environmental degradation decrease (Figure 10.2). From a simplistic point of view, it indicates that economic growth is good for the environment. A few items that reinforce this view are the empirical evidence of declining pollution by cars in the US with economic growth due to stricter regulation; with higher income, consumers are more amenable to paying higher prices in return for more energy-efficient appliances; improved technology provides new ways to derive energy from renewable resources; and the diminishing marginal utility of income (i.e., we as humans give higher importance to living standards when our basic needs are covered.)

Several critics have argued that there is no guarantee about these claims since empirical evidence is mixed; pollution is not only a function of income; global pollution is still increasing; growth leads to greater resource use; and countries with the largest GDP have the highest level of

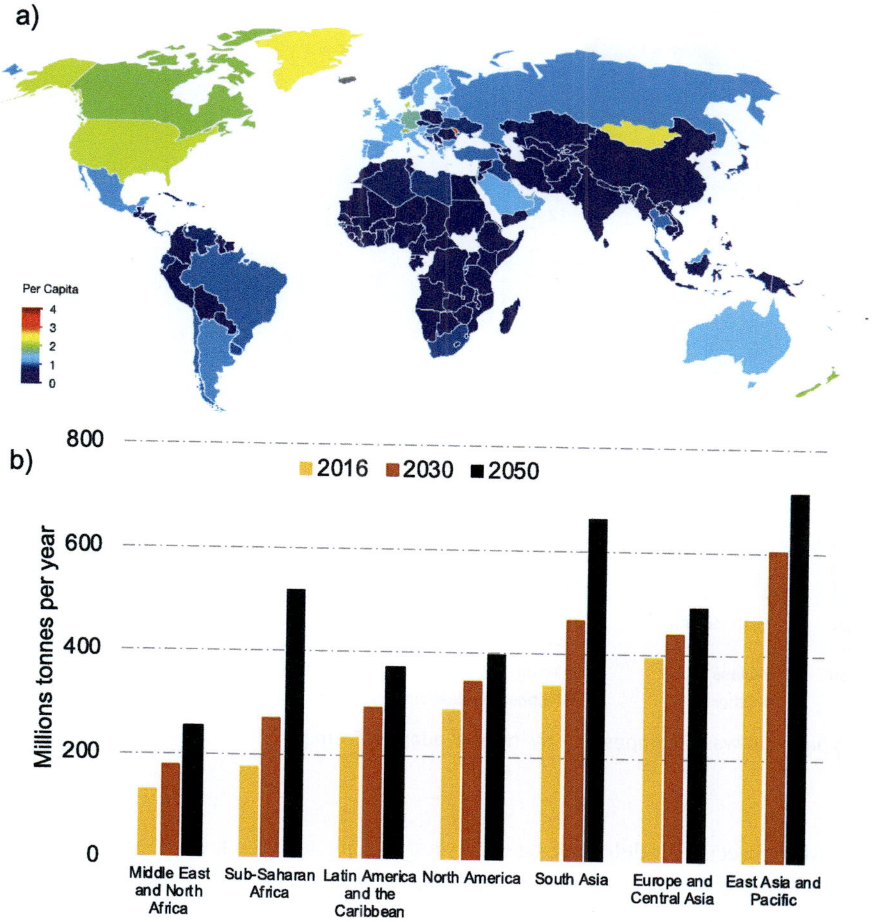

Figure 10.1 Municipal solid waste (MSW) around the world, (a) per capita generation per day in kg for 2016 and (b) amount generated per year by region; adapted from [1].

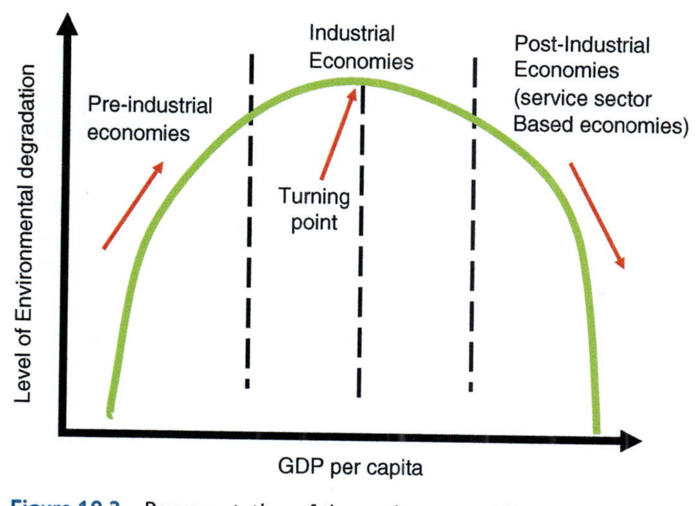

Figure 10.2 Representation of the environmental Kuznets Curve (EKC).

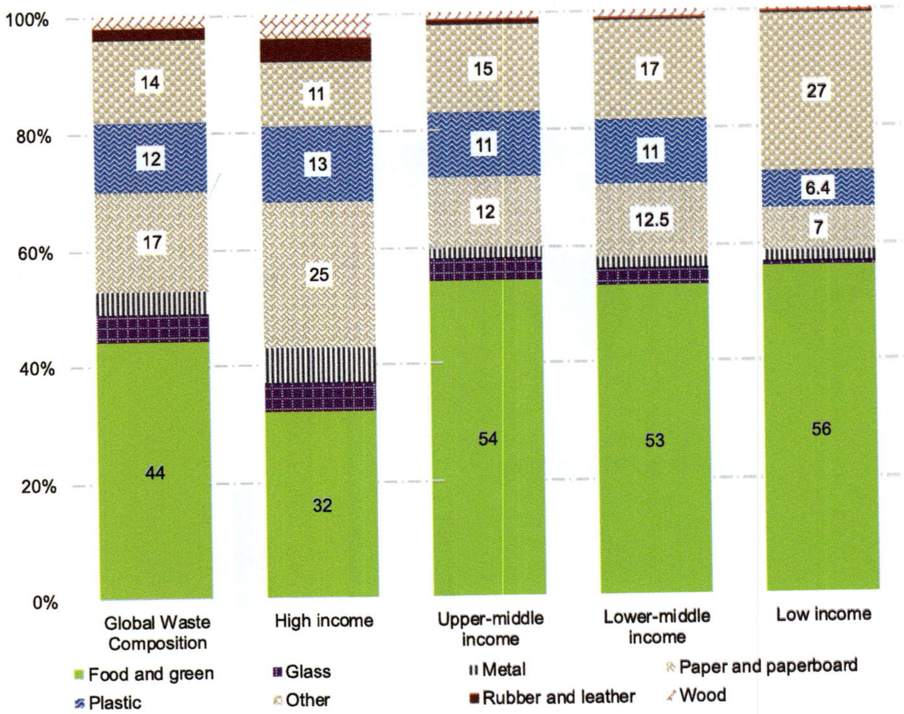

Figure 10.3 Municipal solid waste composition by income; adapted from [1].

CO_2-eq emissions. More precise models have been created to represent the level of environmental degradation and economic growth [2–4] as discussed in Case Study 1.2.

The economy of a country profoundly influences its MSW composition. Figure 10.3 shows that in low-, lower-middle-, and upper-middle-income countries the main MSW component is food and green waste, comprising 44% of the total, followed by paper and paperboard and plastics (29%). Food loss and waste (FLW) is a significant issue for all economies. Studies have shown that, on average, the world wastes 30% of all produced food. The places where FLW happen vary. In low-income countries, FLW mostly occurs during the production, postharvest, storage, and processing phases. In middle- and high-income countries, the most significant amount of food waste is at the consumer level, due mostly to consumer behavior. Materials such as glass, metal, paper, and plastic are important resources in all economies, and if adequately collected and processed, they are resources for establishing a circular economy.

The collection of MSW is one of the most common services provided by a municipality to a community. In general, a truck operator drives weekly or biweekly through the neighborhoods and collects MSW as single or multiple collection streams. In low-income countries or regions, the collection could be a nonformal system or provided by a drop-off facility. In some communities, a signal such as a ringing sound or siren is used to announce that the collection truck is in the neighborhood. The collection is an essential service to be able to handle MSW and to avoid trash reaching the environment. However, consumer engagement is critical to be able to recover MSW. Figure 10.4 shows the collection rate of MSW by income level around the world and in urban and rural areas. High-income countries have almost a 100% collection rate in urban and rural areas. However, upper-middle-, lower-middle-, and low-income countries have a large gap between the collection of MSW in urban and in rural areas. In general, urban area collection requires transportation between city centers and disposal sites between 3 and 50 km, but this varies widely by geographical region. Rural MSW collection requires longer distances.

Of the collected MSW around the world, lower-middle, and low-income regions of the world send most of their waste to open dumps (Figure 10.4) (i.e., piles of waste materials or garbage accumulated or left at a location). In many countries, open dumps are illegal, and individuals are fined

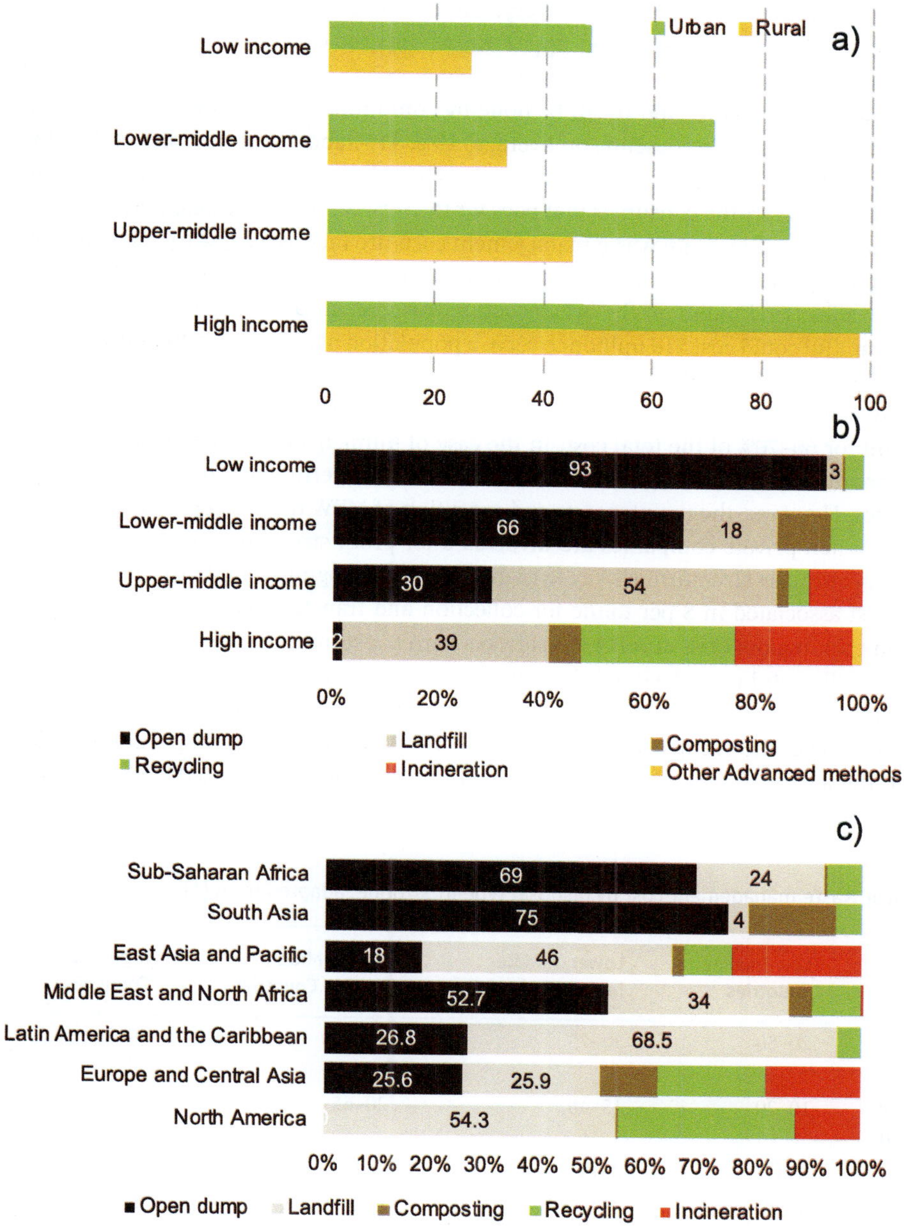

Figure 10.4 Municipal solid waste collection across the world; (a) urban and rural areas; (b) disposal methods by income; (c) disposal methods by regions; adapted from [1].

for illegally dumping waste if they are caught. Open dumps are a breeding ground for mosquitos, provide a reservoir for diseases, invite animals such as rodents to freely reproduce, and emit nuisance odors. In high- and upper-middle-income regions of the world, the most significant disposal method is sanitary landfill (i.e., a safer and organized way to dispose of waste in a regulated area where a region is designated for managing waste in compacted layers). Recycling is a crucial method to divert MSW from reaching the environment, mostly favored in North America, Europe, and Central Asia. The recovery of disposed of materials such as plastic, glass, metal, and paper is increasing and is the main approach in many regions around the world to reduce MSW. Organized incineration or incineration with energy recovery, on the other hand, due to the required investment, is mostly practiced in high- and upper-middle-income countries. However, uncontrolled burning of MSW is widespread in lower-income economies.

Waste management is expensive, and it can be done by communities or by private companies. Community waste management represents, on average, 19%, 11%, and 4% of the total budget for low-, middle-, and high-income economies, respectively. When the waste management is done by private companies, it is mostly the consumer that pays for the service. In some cases, states or communities provide subsidies. Two main waste management costs are incurred for waste managed by municipalities and companies: (a) capital expenditures to set up the infrastructure, and (b) operational expenses to collect and maintain the infrastructure. As an example of capital expenditures, construction of a landfill could cost $10 million to serve a population of around 1 million people [1]. However, the largest financial challenge for municipalities and companies is to cover the ongoing operations of MSW collection and management such as labor, fuel, and maintenance of equipment. These can account for 60–70% of the total cost. In the case of municipalities, capital expenditures are regularly supported by grants, subsidies, or donations from governments, NGOs, or international communities. However, the maintenance and execution of MSW management programs are left to the towns. When private companies are involved a for-profit structure must be accommodated. Therefore, proper cost structuring is key to managing MSW programs. Table 10.1 provides an example of the cost associated in $ per tonne for collection and transfer and the typical disposal routes of MSW in each region. Cost numbers are always hard to estimate and to obtain, but a general framework is still useful to understand the implications of MSW management programs and how to deploy them. The average collection fee varies by income group and category. For example, for high-income countries the average fee per year is $168 per household and $314 per commercial entity. However, the fee for commercial entities tends to vary by tonnage produced. These fees for low-income countries average $37 and $155 for households and commercial entities, respectively.

Table 10.1 Typical waste management cost by disposal type in $/tonne; adapted from [1].

	Low-Income Countries	Lower- Middle-Income Countries	Upper- Middle-Income Countries	High- Income Countries
Collection and transfer	20–50	30–75	50–100	90–200
Controlled landfill to sanitary landfill	10–20	15–40	20–65	40–100
Open dumping	2–8	3–10	N/P	N/P
Recycling	0–25	5–30	5–50	30–80
Composting	5–30	10–40	20–75	35–90

N/A: not available; N/P: not permitted.

10.5 Municipal Solid Waste in the US

In the US, MSW is typically categorized into durable goods, nondurable goods, containers and packaging, food scraps, yard trimmings, and other wastes. Table 10.2 shows the main constituents for each category. Durable goods are mostly things that are not single use. Nondurable goods are "stuff" mostly made from paper fibers. We could consider that some of them, such as paper plates and cups, could be counted as packaging, but they are not. Containers and packaging comprise several materials, such as glass, metal, paper and paperboard, and plastic. Finally, the category of organic wastes includes food, yard, and miscellaneous mostly natural materials that can be biodegraded.

Two methods are mostly used to estimate MSW for a country or a region: (a) site-specific (i.e., collecting data from each site) and (b) material flow (i.e., evaluating inputs and outputs of a system). The US EPA uses the latter. The materials flow technique estimates the waste streams of the entire country by considering inputs and outputs in a region. The methodology was developed in the early 1970s by the EPA's Office of Solid Waste. It considers production data by weight for the materials and products accounted for in MSW. Production data is adjusted for imports and exports and diversions from MSW for materials. The data is also adjusted for the lifetime of different products. Some minor data for food scraps, yard trimmings, and other minor amounts of miscellaneous inorganic wastes are obtained by collecting information from selected sites. One problem with the material flow methodology is that residues left in containers, such as food, detergent, and paint, are not accounted for.

Figure 10.5 shows the total MSW and MSW generation per capita from 1960 to 2018. From 1960 to 1980, the amount of MSW steadily increased. Therefore, in 1980, the US EPA was concerned about this increase and firmly pushed for an MSW reduction program trying to reduce MSW at the source, known as the reduce and reuse program. Since 1990, the amount of MSW generated per person per day stayed almost constant; however, the total MSW increased due to the increase in

Table 10.2 Municipal solid waste categories and composition in the US as defined by the US EPA.

Durable Goods	Nondurable Goods	Containers and Packaging	Other Wastes
• Major appliances • Small appliances • Furniture and furnishings • Carpet and rugs • Rubber tires • Batteries, lead acids • Miscellaneous durables (consumer electronic and others)	• Newspapers • Books and magazines • Office-type papers • Directories • Standard mail • Commercial printing • Tissue paper and towels • Paper plates and cups • Plastic plates and cups • Trash bags • Disposable papers • Other non-PKG paper • Clothing and footwear • Towels, sheets, and pillowcases • Miscellaneous	• Glass PKG • Steel PKG • Al PKG • Paper and paperboard PKG • Plastic PKG • Wood PKG	• Food scraps • Yard trimmings • Miscellaneous

population. US consumers in 2018 generated approximately 292 million tons of MSW, corresponding to 4.90 lb per person per day (2.2 kg per person per day), showing that US consumers are in the upper range of waste generation in the world, as previously discussed.

Of the amount generated by US consumers in 2018, containers and packaging represented the largest category (~37%), as shown in Figure 10.6. Organic wastes, including food, yard, and miscellaneous, were the second-largest category, at 29.9%. In the case of yard trimmings, many are recovered through composting, but disposal of food waste is mostly in landfills since very few cities have a collection of organics. Nondurable and durable goods represent 40.4% (see Figure 10.6).

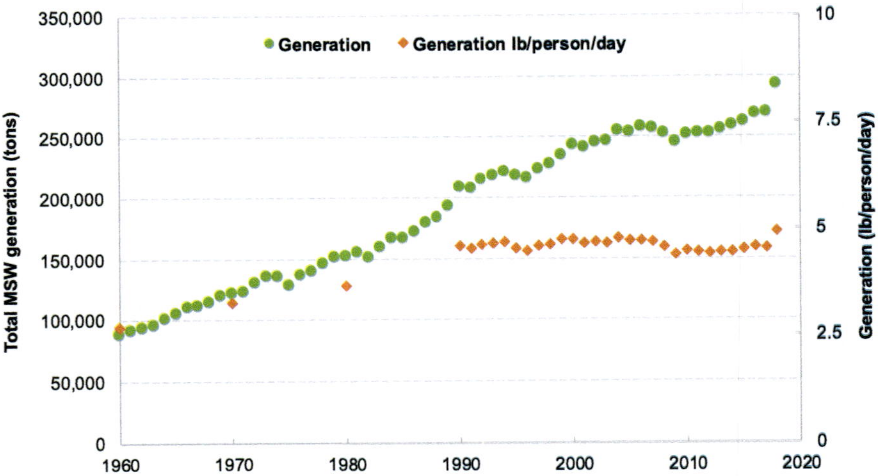

Figure 10.5 Municipal solid waste generation between 1960 and 2018 including per capita generation; original data obtained from [5]. Note: MSW generation rose considerably from 2017 to 2018 due to EPA's enhanced method to measure food waste.

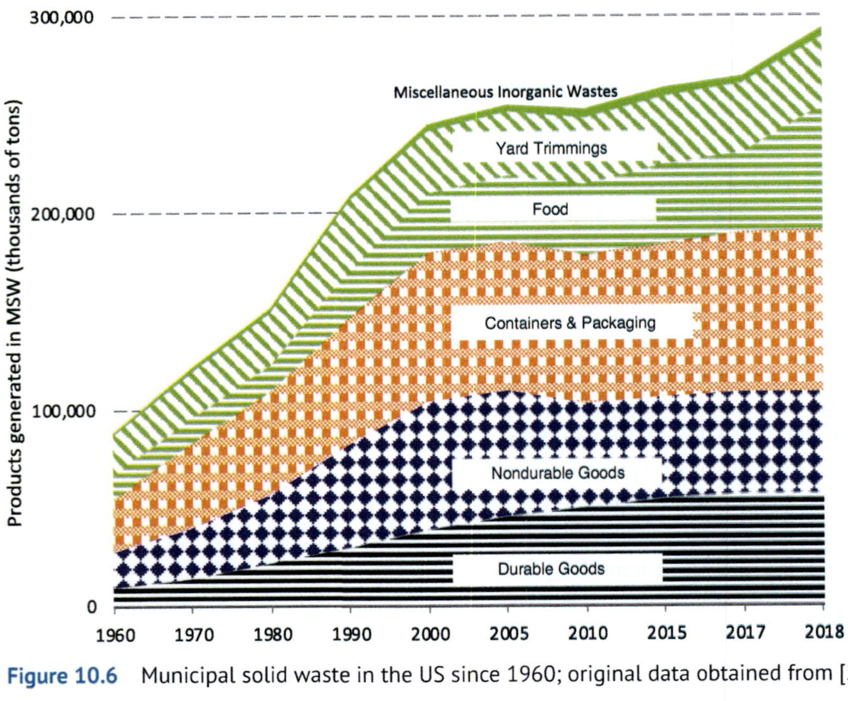

Figure 10.6 Municipal solid waste in the US since 1960; original data obtained from [5].

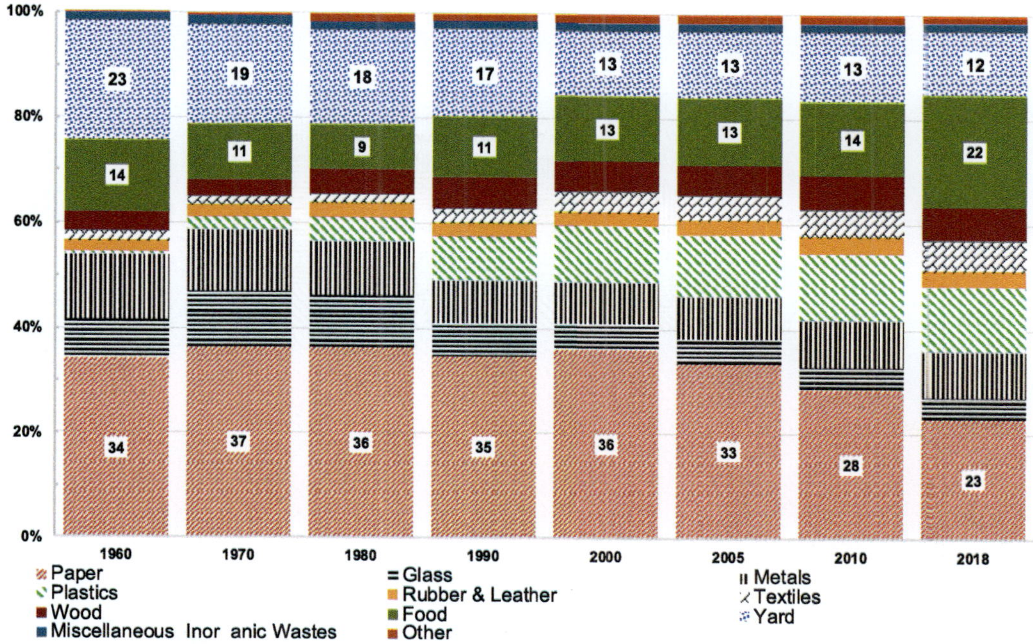

Figure 10.7 The municipal solid waste material composition between 1960 and 2018; original data obtained from [5].

The material composition of MSW since 1960 has been dominated by paper, food, and yard waste, as shown in Figure 10.7; however, plastic has been increasing. Organics (i.e., food and yard waste) represent around one-third of MSW.

Most of the MSW generated between 1960 and 1980 in the US was sent to landfills. However, by 1980, there was concern about sufficient landfill carrying capacity to divert all the trash as the population grew. Additionally, there was general concern that landfills were disappearing, due to the combination of landfill sites closing due to normal end of life (reaching maximum capacity) or to issues of groundwater pollution. Fewer new landfills were opening although the available landfill capacity was not significantly reduced. Consequently, there was increased interest in composting (mostly for yard trimmings), incineration of MSW with energy recovery, and recycling (see Figure 10.8). Landfill, composting, and combustion with energy recovery have not changed in total amount much since 2005, while recycling has increased modestly, as shown in Figure 10.9. In 2018, around 50% of MSW was sent to landfill, with only 32% of MSW recovered through recycling (including 8.5% through composting).

10.6 Municipal Solid Waste in Different US States

MSW management and collection across the US vary since the collection of MSW is left to the states and cities. Different states have different policy and approaches toward MSW management. More environmentally concerned states such as those on the east and west coasts tend to have stronger policies to control MSW disposal. So, MSW numbers vary around the US and data collected by the different states differ widely.

The US EPA divides the country into 10 regions (see Figure 10.10) and collects data provided by states. However, data submission is not mandatory; so data, reports, and information are not consistent [5], making it difficult to make national policy regarding MSW. Most states report selected metrics, but not on their entire MSW composition and recovery, tending to report data that reflect

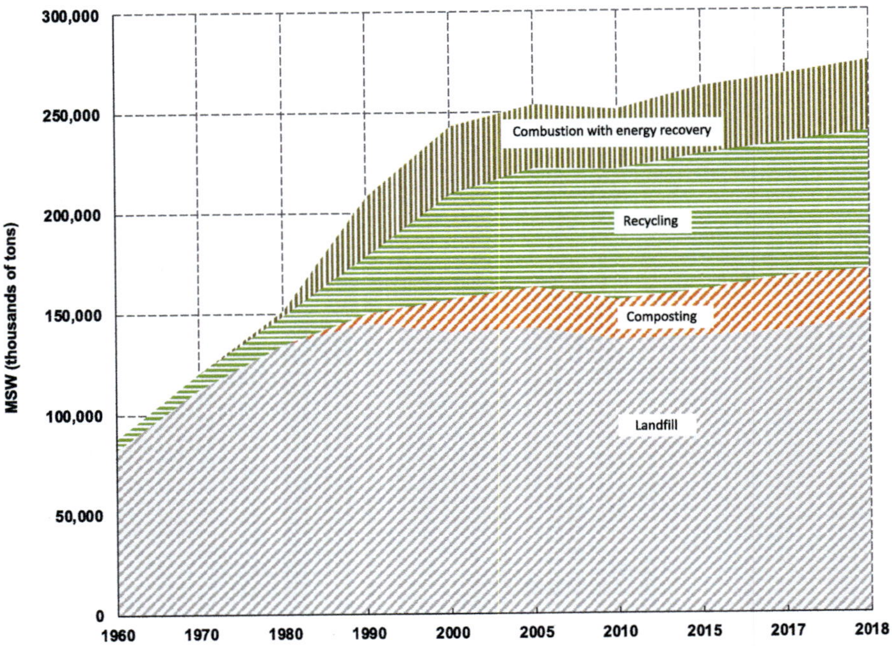

Figure 10.8 Generation, recycling, composting, combustion with energy recovery, and landfilling of municipal solid waste; original data obtained from [5].

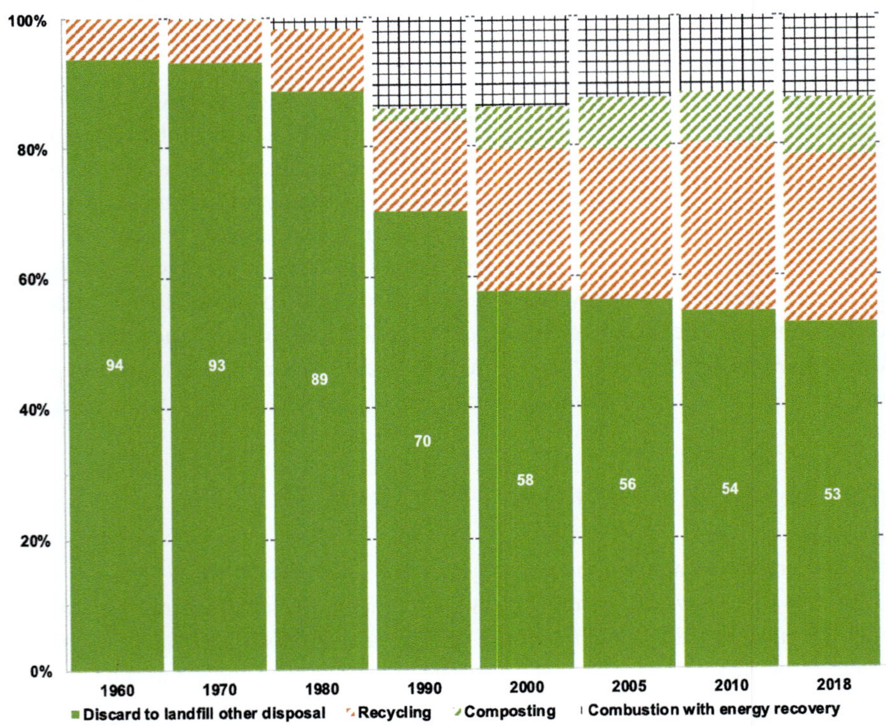

Figure 10.9 Municipal solid waste between 1960 and 2018 by disposal category in percentage; original data obtained from [5].

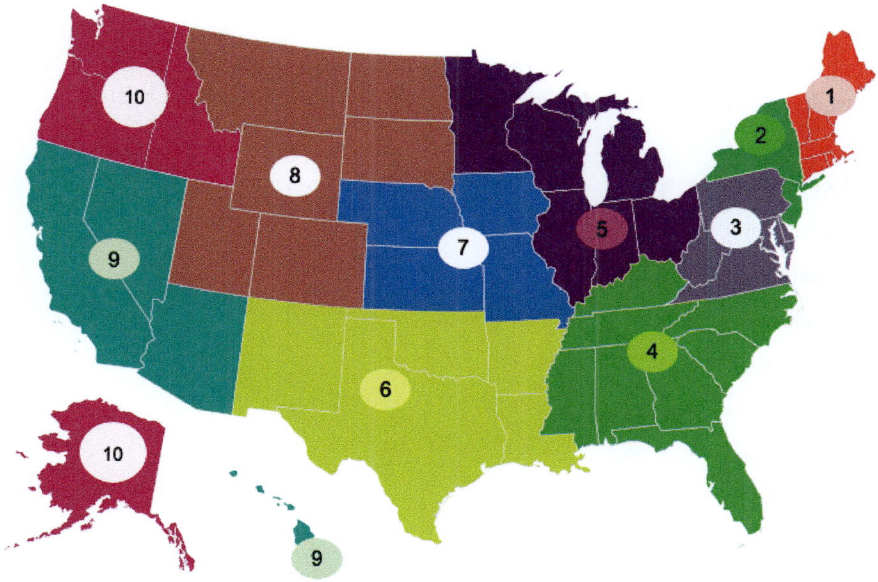

Figure 10.10 US EPA regions in the US; adapted from [5]. Puerto Rico is in region 2.

favorably on them. Several efforts have been made to report the amount of garbage produced in the US, collecting or aggregating individual data from each state. One of the most well-known reports was The State of Garbage in America, reported by the *Biocycle* magazine. States reported the MSW generated, recycled, composted, send to waste-to-energy, and landfilled. However, due to the lack of continuous data reporting, some data was extrapolated from previous years and the report is no longer being updated [5]. Furthermore, numbers from these reports do not fully match the US EPA data for the same time frame since they differ in methodology and definitions. So, it is a continuous struggle to obtain reliable data of the MSW for individual US states.

Different US states have different approaches to MSW management. The California Environmental Protection Agency through CalRecycle provides an annual report about disposal and recycling. In 2019, CalRecycle reported that 55% of the MSW generated in CA was sent to landfill, 19% was exported recyclables, and 12% was composted, sent to anaerobic digestion or used as mulch (Figure 10.11).

The Oregon Department of Environmental Quality (DEQ) reported that Oregon generated more than 5.5 million tons of waste in 2018, a 2.9% increase from 2017. Around 3.3 million tons were disposed of in landfills and incinerators, which represented around 60% of the generation, and the remaining 40% was recovered through recycling and composting. The recycling rate was 2% lower than in the 2017 [7]. The majority of the other US states reported lower recycling rates than California and Oregon. Although California and Oregon report larger recovery numbers than other US states, these rates are still lower than in many European countries, which will be further discussed in Chapter 13.

10.7 Municipal Solid Waste Management Approaches

MSW has become a significant concern for communities and municipalities. Many different approaches are being used to reduce the amount of MSW reaching landfills. The main recognized approach to manage MSW follows the order of priority provided by the US EPA: *reduce*, *reuse*, and *recycle*, recognized as the "3R waste management system." In this book and in the following chapters, we will recognize two more categories, *recovery* and *residual management*, making it a 5R system. We will call this system the "5R MSW management method," as shown in Figure 10.12. The benefits and trade-offs of each methodology will be explored in the following chapters.

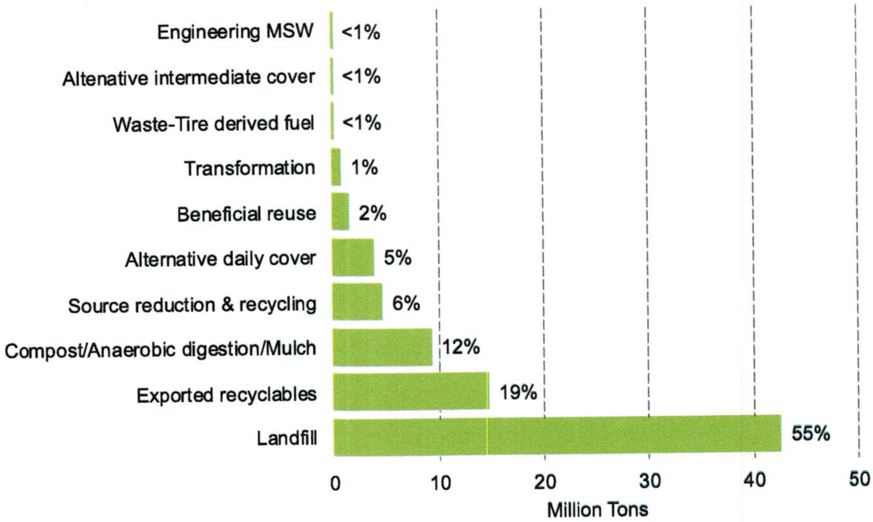

Figure 10.11 Amount of MSW managed in California in 2019 – total 77.5 million tons; adapted from [6].

Figure 10.12 Municipal solid waste prevention and reduction methods.

The EPA defines these terms as follows:

Source reduction is "consuming and throwing away fewer goods by reducing the amount of material used and/or the toxicity of the material used to accomplish any task, reusing a product in its original form, or using repairable, refillable, or durable products that result in a longer useful life. Manufacturing or using materials or products (including shipping containers and packaging) to reduce their amount or toxicity before they become waste is also considered a source reduction method" [8]. This concept will be explored in Chapter 11.

Reuse is the use of materials or items several times. Reuse of materials and products, and the creation of a reuse system, may conserve valuable natural resources and reduce the amount of soil, water, and air pollution. Items can be reused by repairing, donating them to charity and community groups, or selling them for a second life. Chapter 12 will cover reuse.

Recycling is defined by the US EPA as a "series of activities that includes collecting recyclable materials that would otherwise be considered waste, sorting, and processing recyclables into raw materials such as fibers, and manufacturing raw materials into new products." A detailed discussion of recycling will be provided in Chapter 13. Organics recycling – composting – will be discussed in Chapter 14.

Recovery, energy recovery from the combustion of MSW, is a key component of the nonhazardous waste management hierarchy as shown in Figure 10.12, which ranks from most to least environmentally preferred. Energy recovery ranks below source reduction and recycling/reuse but above residuals management. Confined and controlled burning with energy recovery, known as combustion, can decrease the volume of solid waste destined for landfills and recover energy from the waste-burning process. The combustion of MSW can generate renewable energy and reduce carbon emissions by offsetting energy from fossil sources and reducing methane generation from landfills. We will discuss in detail combustion with energy recovery in Chapter 15.

Residuals management, disposal in landfill, is the least preferred option of the US EPA waste management hierarchy. The US EPA allows the construction of modern landfills, which are well-engineered facilities designed for particular types of waste, including MSW, construction and demolition debris (C&D), and hazardous waste. Modern landfills must not be confused with dumps or open dumps, since these facilities are designed to protect the environment from contaminants, which may be present in the solid waste. Chapter 16 will cover landfill disposal and the type of landfills in the US.

10.8 Case Study 10.1 – Environmental Footprint of PET Bottles Managed According to the US EPA Waste Management Hierarchy

Using the example from Case Study 9.1, we can estimate the EFP of 1,000 PET bottles disposed according to the 2018 US EPA MSW waste management scenario (i.e., 29.9% recycling for PET at EoL), the 2050 goal of MSW by most countries (i.e., 50% recycling for PET at EoL), and a hypothetical 100% recycling for PET at EoL. Table 10.3 shows that recovering PET bottles at end-of-life reduces the environmental footprint indicators. Climate change reduces by 0.78% and 2.72%, by increasing recycling at EoL to 50% and 100%, respectively. There is an important point to consider in these three cases. We estimate that the PET bottle avoids incineration and landfill at the EoL, changing the percentage from 71%, 50%, to 0%, but this does not indicate that the recycled bottles will be used as new recycled material. So, we are simply not penalizing them for disposal in the landfill and incineration with energy recovery. Chapter 13 will evaluate the benefits of using these postconsumer PET bottles as recycled content to produce new bottles, which will result in a distinct reduction in the EFP. It is also important to note that if we do not recover

Table 10.3 Environmental footprint indicators for different end-of-life scenarios for 1 PET ISBM bottles. Data estimated using PackageSmart using Packaging Impacts midpoint indicators, and Waste Scenario as the 2018 US EPA.

		Recycling at End-of-Life*		
Indicator	**Units**	**29%**	**50%**	**100%**
Climate change	kg CO_2-eq	290	288	282
Cumulative energy demand	MJ	5,345	5,344	5,340
Ecosystems	species.yr	2.74E-06	2.72E-06	2.67E-06
Human health	DALY	5.66E-04	5.62E-04	5.54E-04

*The amount not recovered through recycling is disposed of as 19% incineration with energy recovery and 81% landfill.

the PET bottle at EoL, they cannot be used as postconsumer recyclate (PCR) content to produce new bottles.

10.9 Study Questions

1 What are the main consequences of poorly managed solid waste?
2 MSW is most strongly affected by:
 A Demographics
 B Geographical boundary
 C Human health
 D Country's economy
 E None of the above
3 What is the range amount of MSW per capita per day around the world in 2016?
4 Provide examples of high-income countries disposing the large amount of MSW per capita per day.
5 What is the percentage of containers and packaging in US MSW in 2018?
6 What is the largest category of waste composition for all types of income levels around the world?
7 Explain the main difference between urban collection and rural collection rates by income level around the world.
8 What are the main categories of MSW in the US according to the EPA? What method does the US EPA use to estimate the data reported in their yearly report now titled "Advancing Sustainable Materials Management"?
9 Considering the different materials reaching MSW in the US since 1960, describe the main variation in the material composition over time.
10 How many regions are included in the US EPA division of the US?

10.10 Additional Resources

1 Advancing Sustainable Materials Management: Facts and Figures Report
https://www.epa.gov/facts-and-figures-about-materials-waste-and-recycling/advancing-sustainable-materials-management
2 What a Waste 2.0: A Global Snapshot of Solid Waste Management to 2050
https://openknowledge.worldbank.org/handle/10986/30317

3 Reduce, Reuse, Recycle
https://www.epa.gov/recycle

4 Solid waste management
https://www.unenvironment.org/explore-topics/resource-efficiency/what-we-do/cities/solid-waste-management

5 CalRecycle
https://www.calrecycle.ca.gov

6 Oregon Department of Environmental Quality
https://www.oregon.gov/deq/Pages/index.aspx

7 EGLE – Michigan Department of Environment, Great Lakes, and Energy
https://www.michigan.gov/egle

8 New York State – Department of Environmental Conservation
https://www.dec.ny.gov

9 Florida Department of Environmental Protection
https://floridadep.gov

10 Texas Commission on Environmental Quality
https://www.tceq.texas.gov

References

1 Kaza, S., Yao, L., Bhada-Tata, P., and Van Woerden, F. (2018). What a waste 2.0: A global snapshot of solid waste management to 2050. The World Bank. https://doi.org/10.1596/978-1-4648-1329-0.

2 Stern, D.I. (2004). The rise and fall of the environmental kuznets curve. *World Development* 32 (8): 1419–1439.

3 Dogan, E. and Inglesi-Lotz, R. (2020). The impact of economic structure to the environmental kuznets curve (EKC) hypothesis: Evidence from European countries. *Environmental Science and Pollution Research* 27: 12717–12724. https://doi.org/10.1007/s11356-020-07878-2.

4 Churchill, S.A., Inekwe, J., Ivanovski, K., and Smyth, R. (2018). The Environmental Kuznets Curve in the OECD: 1870–2014. *Energy Economics* 75: 389–399. https://doi.org/10.1016/j.eneco.2018.09.004.

5 US Environmental Protection Agency. Advancing Sustainable Materials Management. (2018). Tables and figures assessing trends in materials generation and management in the United States. U.S. EPA: 2020.

6 CalRecycle. (2017). State of disposal and recycling in California: for calendar year. Drrr-2019-1643 2019:29. https://www2.calrecycle.ca.gov/Publications.

7 Oregon Department of Environmental Quality. (2019). Materials management program 2019-20 report to the legislature 33. www.oregon.gov/DEQ.

8 Brown, K. and Andre, P. (1993). Waste prevention: Source reduction now. How to implement a source reduction program in your organization. https://www.leg.mn.gov/docs/2021/Other/210738.pdf (accessed October 29, 2019).

11

Reduction

11.1 Learning Objectives

After learning from this chapter, you are expected to:

- Define *reduce* and its relation to packaging.
- Describe the main benefits of reduction in packaging systems.
- Provide examples of packaging systems where reduction provides a decrease in the environmental footprint of a packaging system.
- Describe the main trade-offs in reduction.
- Understand how to implement lightweighting in a packaging system without impacting overall damage.

11.2 Introduction

Of the MSW management strategies discussed in the previous chapters, the first two recommended approaches making the most impact on MSW are *reduction* and *reuse*, according to the US EPA. These two techniques have been, for a long time, the preferred methods to manage pollution and MSW from its inception. The central rationale is that if the goods are not created in the first place, energy will not be used, and emissions will not be released. With the increasing use of LCA, detailed and thorough characterization of the inputs and outputs needed to produce consumer and nonconsumer goods demonstrates that as a general rule the two methods (i.e., reduction and reuse) could be encouraged, but they may not always result in the lowest EFP. Therefore, caution must be taken when generalized statements considering MSW management prevention and EFP reduction are the main objectives since the overall sustainability goal must be to reduce the total EFP to produce the goods, including the products and the packaging systems.

Reduction, especially, weight reduction or minimization, also known as "lightweighting," has been a long tradition in the production of durable and nondurable goods, since products with lower weight require less use of raw material, as a consequence using less energy and emitting fewer pollutants. A classic example is the car industry, which has conducted intensive research to reduce automobile weight to minimize gasoline consumption per mile. Although weight reduction is initially appealing for producers of goods and waste program managers, we must be aware that

Life Cycle of Sustainable Packaging: From Design to End of Life, First Edition. Rafael A. Auras and Susan E. M. Selke.

Companion Website: www.wiley.com/go/Auras/lifecycleofsustainablepackaging

lightweighting has been used for many decades due to the accompanying direct cost reduction. In this chapter, a detailed discussion is provided of how *reduction* can be used to create systems with lower EFP, especially considering each packaging material, metal, glass, paper, and plastic.

11.3 Reduction

The EPA defines *source reduction*, also known as waste prevention, as "reducing waste at the source. It can take many different forms, including reusing or donating items, buying in bulk, reducing packaging, redesigning products, and reducing toxicity"[1].

Source reduction is a useful tool for institutions, industries, commercial establishments, and consumers to reduce the generation of MSW, and it can be used at any step in the creation of a product. Reduction can work during industrial operations such as mining, forestry, agriculture, processing of raw material, manufacturing, shipping, packaging, or final use of a product.

Many approaches have been used as source reduction techniques in business, such as avoiding unnecessary printing, printing on both sides of paper, eliminating heavy metals in pigment for inks, removal of solvents, replacement of fluorescent tubes by light emitting diode (LED) lamps, and removal of toxic or carcinogenic compounds from original formulations. Source reduction, especially lightweighting, has been a crucial issue for the automotive industry. Auto manufacturers are increasingly adopting lightweight materials for the body, structures, interior, and parts. The long-term goal is to reduce overall vehicle weight to improve fuel economy, meeting future demands and regulatory requirements. The search for lightweight materials is creating switching from heavier materials such as carbon-steel to thinner high-strength steel and aluminum and plastic composites. Several success stories have been shared by the automotive industry, such as the all-aluminum F-150 developed by Ford Motor Co., and the reduction of more than 350 pounds per vehicle in 14 of the GM vehicles launched since 2016 [2]. Another example of source reduction is the banning on January 1, 1996, of tetraethyl lead (TEL) used to improve octane rankings. California banned TEL in 1992. By early 2000, most industrialized countries had banned TEL because of concerns over air and soil lead pollution levels (see Chapter 7).

11.4 Reduction in Packaging

Packaging is a commodity material for delivering goods and providing a service, so cost is an essential factor. Material reduction in packaging usually directly translates to cost reduction if the overall structure and product are not compromised by the lightweighting. Packaging professionals refer to source reduction as *packaging lightweighting*, and it has been ongoing for glass, metal, paper, and plastic packaging since the creation of packaging. The greatest benefit from lightweighting is obtained when lightweighting is conducted for high density and energy intensive materials and structures, or when a high concentration of toxic compounds is present.

The EFP of overpackaging (use of excessive amounts of packaging) has been extensively discussed and is a common public concern. Remember how many times you have complained about the amount of packaging. However, when considering a packaging system, we need to balance over- and underpackaging to ensure that the product is protected, and the amount of packaging material is optimal. Figure 11.1 shows that if we have minimal packaging and maximal damage, the environmental impact is the largest. As more packaging is introduced to protect the product, less of the product is damaged and the EFP decreases until we arrive at an optimal packaging level.

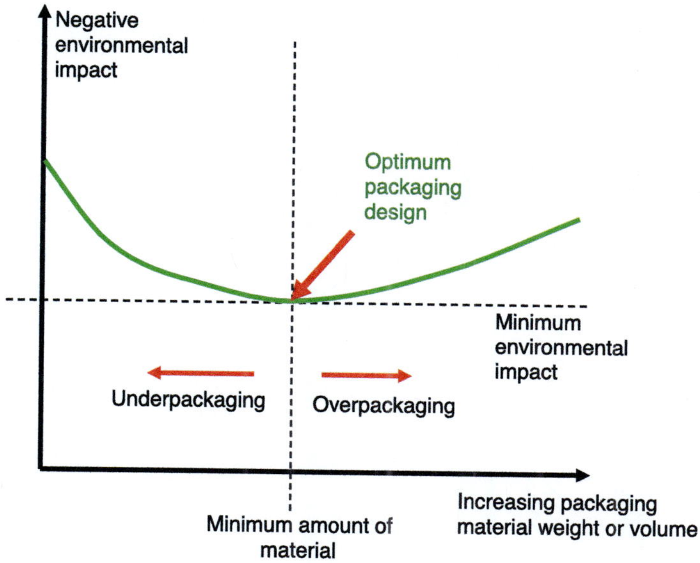

Figure 11.1 Optimum packaging design considering trade-offs between under- and overpackaging and environmental impact.

After this optimum, overpackaging will again produce an increase of the EFP of the product/package system.

11.4.1 Glass

Glass is an energy-intensive material to produce and has a density of around 2.5 gcm^{-3} [3]. So, any reduction in weight can be translated to a decrease in the EFP. However, one of the main issues with achieving significant weight reduction in glass packaging is its fragility, which tends to be inversely correlated to weight (i.e., lighter weight glass bottles are more fragile.) Several companies have successfully reduced the weight of glass bottles, reducing, among other EFP indicators, the greenhouse warming potential.

Coors Brewers Ltd., the owner of Grolsch beer, a premium Dutch beer brand, together with Owens Illinois and Rockware, worked on reducing the weight of its 300 mL beer bottle by launching a 10% lightweighted bottle, reducing 20 tonnes of glass and 14 tonnes of CO_2-eq for each million bottles [4]. Similar effects can be observed using Case Study 9.1 as an example. If we reduce the weight of the glass bottles by 10% or 20%, all the EFP indicators are reduced by similar amounts (Figure 11.2.) However, caution must be used to avoid excessive extrapolation of lightweighting of bottles since when we reduce the weight of the bottles, we also increase the risk of having more damage due to increased fragility, which may translate to an increase in the overall EFP of the system.

11.4.2 Metal

Steel and aluminum are also energy-intensive materials to produce, with a density of around 7.8 gcm^{-3} and 2.7 gcm^{-3}, respectively [3]. So, there is a lot of opportunity for lightweighting. The weight of metal cans has been reduced since their creation. Some of this was due to the transition from three-piece to two-piece cans, and the weight reduction from the elimination of the side seam. Most of these reductions were obtained due to the production of stronger metal alloys and

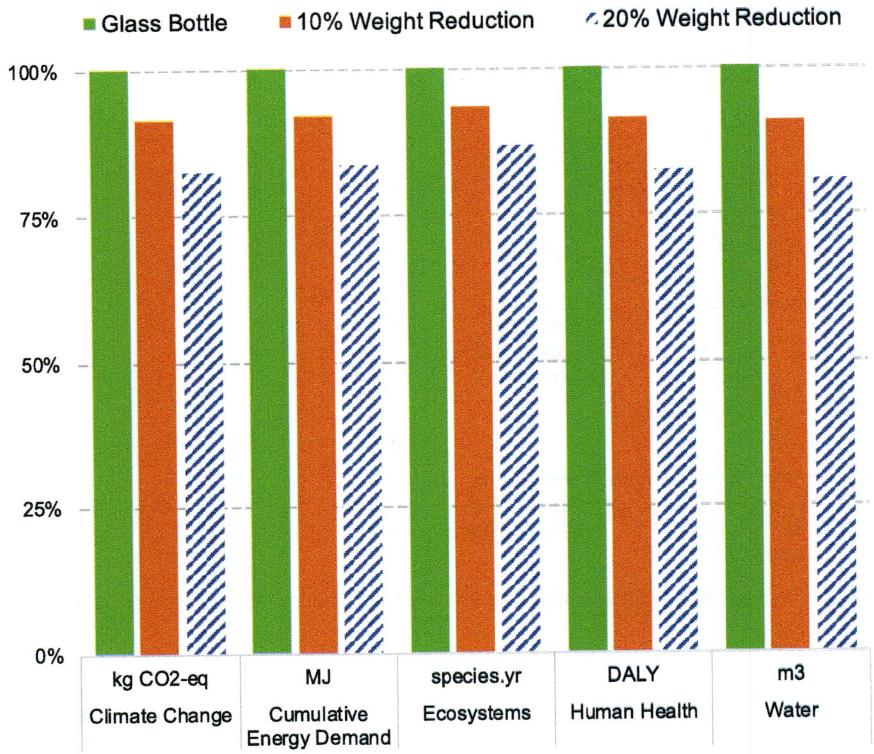

Figure 11.2 Environmental footprint of glass beverage drink and alternative scenarios with 10% and 20% weight reduction of the glass bottle.

better control of the alloy structure (see Figure 11.3) [5]. In the last two decades, the weight of steel cans decreased by around 33% while the average weight of the 12 oz aluminum beverage can and end in the US market decreased by around 41% between 1970 and 2016 [5].

For the aluminum can, a large reduction was obtained from reducing the weight of the aluminum by improving the alloy and by introducing nitrogen gas in noncarbonated drinks, increasing the compression strength. Figure 11.4 shows the weight reduction of the aluminum can since 1972, showing that more than 35% weight reduction was achieved. In 2018, nearly 34 cans were produced from 1 lb of aluminum.

At present, the average thickness for aluminum beverage cans is 0.083 mm and 0.14 mm for the steel two-piece food cans [5]. Reduction of the weight of the aluminum can, similarly to the glass bottle, reduces the overall EFP of the can, mostly in a direct proportion (i.e., larger weight reduction implies larger EFP reduction). However, further weight reduction of aluminum cans is challenging without modifying the aluminum alloy to increase the can compression strength.

11.4.3 Paper, Paperboard, and Corrugated Board

The average weight of corrugated board – as indicated by the corrugated board basis weight (CBW) as a proxy – has been reduced since 1970 in Japan, Europe, US, and most parts of the world (Figure 11.5). Similar trends have been observed for paper and paperboard. A large difference can be seen in the average grammage between Europe, Brazil and the US, Japan, and Thailand, mainly due to the type of fibers used to produce the corrugated board.

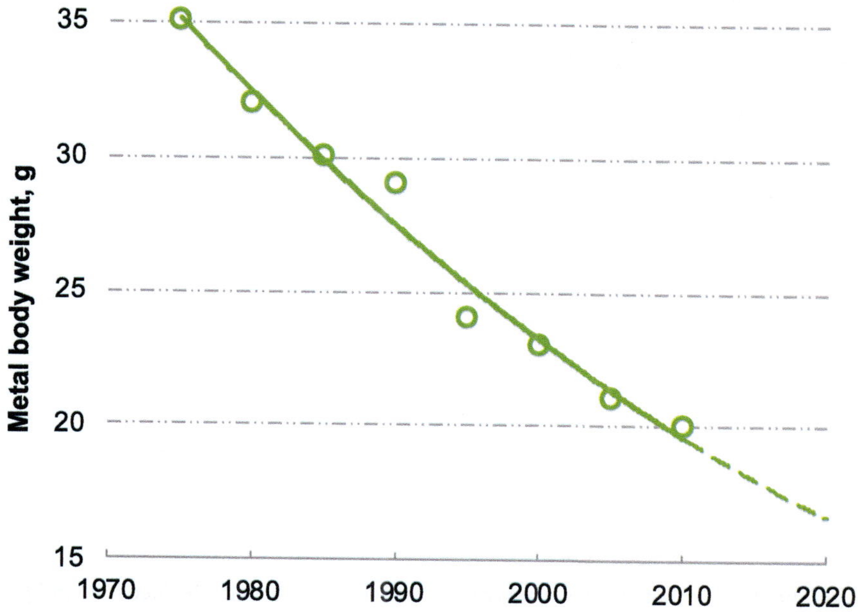

Figure 11.3 Weight of steel beverage can 33 cL between 1973 and 2015; line indicates average values; original data obtained from [6]. Data from 2010 to 2020 were extrapolated. Eventually, it should plateau.

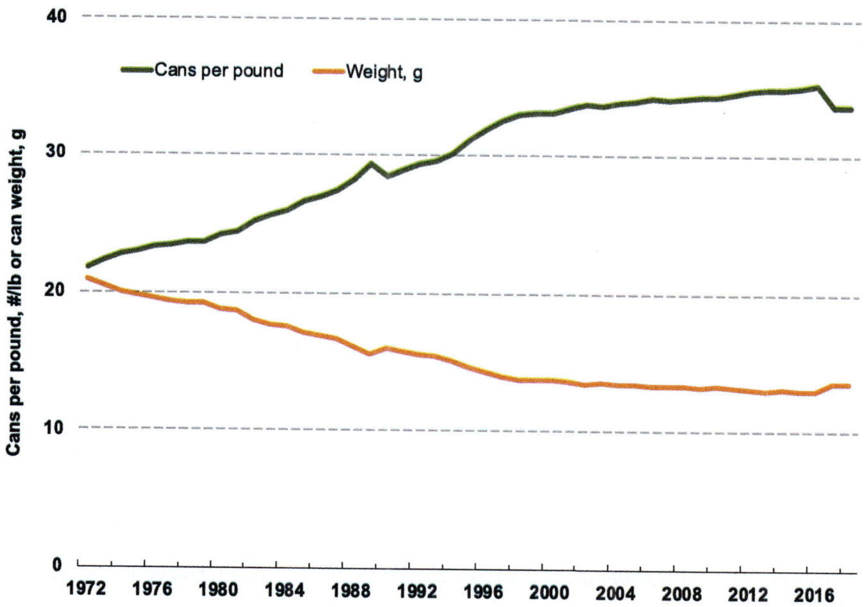

Figure 11.4 Aluminum can weight reduction since 1975, number of cans per pound and individual weight of an aluminum can since 1972, data obtained from [7].

11.4.4 Plastic

As with other packaging materials, single plastic packages can be made with reduced weight to save cost and lessen the EFP. Figure 11.6 shows an example of weight reduction of PET bottles since 1983. Although some studies reported a large reduction of PET bottle weight for individual companies, others show an almost constant weight for some categories such as isotonics.

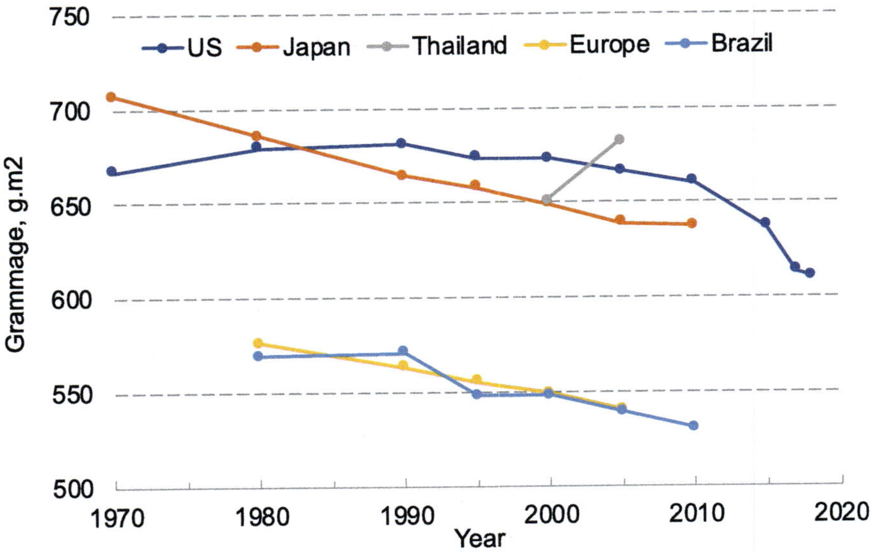

Figure 11.5 Average weight of corrugated board in grams per square meter (GSM); data obtained from [8, 9].

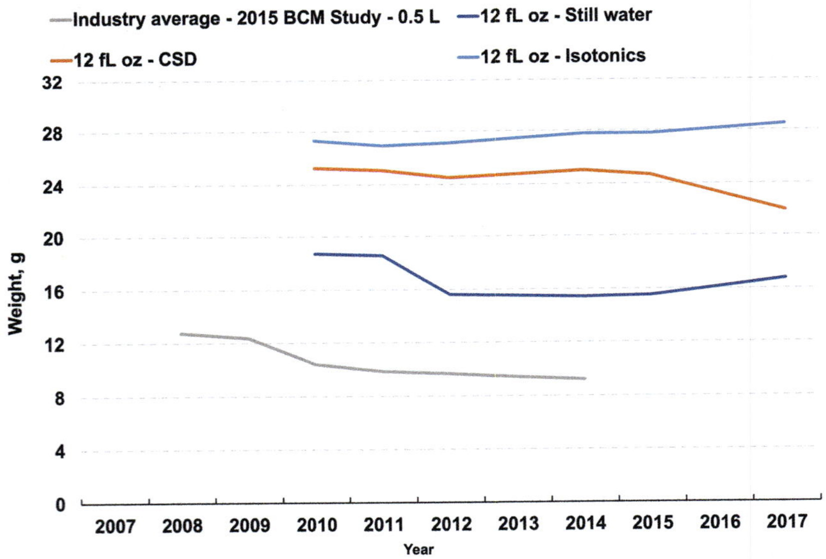

Figure 11.6 Industry average data for the weight of 0.5 l PET bottles vs. time commercialized in the US, data obtained from [10, 11].

Other types of plastic bottles have also been lightweighted. Walmart™ and Costco® introduced a new HDPE milk container design intended to reduce the weight of the entire packaging system, reducing distribution cost as well as lowering the EFP [12]. However, another study reported that the use of returnable crates with the original milk container had a lower EFP [13]. Furthermore, although some companies benefitted from this design change, consumers had mixed opinions since they reported more spilling, and that the new design was hard for kids to pour. This example is clear evidence that EFP reduction without fully understanding the consumer requirements cannot achieve the ultimate goal of improving the sustainability of the packaging system.

Besides the weight reduction of bottles, a transition from rigid to flexible packaging has occurred for many products packaged initially in rigid plastic or paperboard containers. Since 2016 the market for flexible plastic packaging has increased from $130 billion to $150 billion, and it is expected to grow to more than $180 billion by 2024 due to consumer demands for single-serve packaging and lightweighting. Global consumption of flexible packaging reached 30 million tonnes in 2019, with 89% flexible plastic, 6% flexible paper, and 5% flexible foil packaging [14]. The use of flexible packaging can minimize transport costs and the EFP of the product-packaging systems due to the lower packaging weight and consequent decreased use of primary resources. Flexible packaging can also reduce the use of additional labeling through printing directly on the package. Investment for producing flexible plastic packaging is lower, reducing the initial investment for delivering safe products. These attributes make it ideal for low-income economies, where significant growth in this packaging format has occurred, with the Asia-Pacific region being the largest consumer. North America is the second-largest market, and South and Central America are the smallest markets [14].

However, most of the new flexible plastic packaging structures are made from several plastic layers to deliver similar performance as rigid plastic or glass or metal structures, so recovering them is challenging. The primary material in terms of market share is polyethylene (PE), followed by biaxially-oriented polypropylene, biaxially-oriented poly(ethylene terephthalate), and then paper and foil. Food packaging is the primary driver of flexible packaging [14]. Overall, in the next five years, the flexible packaging industry is expected to grow. Still, consumers, governments, and associations are pressing for single-use flexible packaging to be reduced through regulations, to prevent them from damaging wildlife and ecosystems.

11.5 Case Study 11.1. Bacon Packaging

This study was conducted to estimate the EFP of weight reduction and different material composition of packages using LCA of a traditional bacon package (L-board bacon package – composed of PE/wax-coated paper/PE) compared to a new lightweighted bacon package (OLB bacon package – composed of reverse printed oriented polypropylene (OPP)/expanded polystyrene (EPS) with adhesive); see Figure 11.7 for an example of the bacon insert. The entire study is described in further detail elsewhere [15]. In the following paragraphs, we provide a short overview of this study.

"The goal of this study was to compare the environmental burden of the L-board and OLB-board using LCA. The target audience of this study was brand owners and stakeholders. The life cycle of this study was mainly cradle-to-grave, except for corrugated boxes and PE bags used in the packing process that had cradle-to-cradle boundaries. In other words, raw material extraction, intermediate product production, assembly, transportation, waste treatment and environmental benefits generated by recycling were included in this study. Since postconsumer bacon packages are contaminated by the fat of the bacon, it was reasonable to rule out recycling as an end-of-life option for the bacon packages. For this reason, the end-of-life for L-board and OLB-board bacon packages was modeled as 81.5% landfill and 18.5% incineration, the US average in the year of the study (see Figure 11.8 for additional details). Energy recovery from incineration of waste and scrap was not included. The end of life scenarios of corrugated boxes and PE bags were modeled as avoided burden after they were used to deliver the L-board and OLB-board bacon package to the assembly plants. To handle the environmental benefit from the reuse of corrugated box and PE bags, the system expansion or avoided burden method was used.

Figure 11.7 Example of bacon insert.

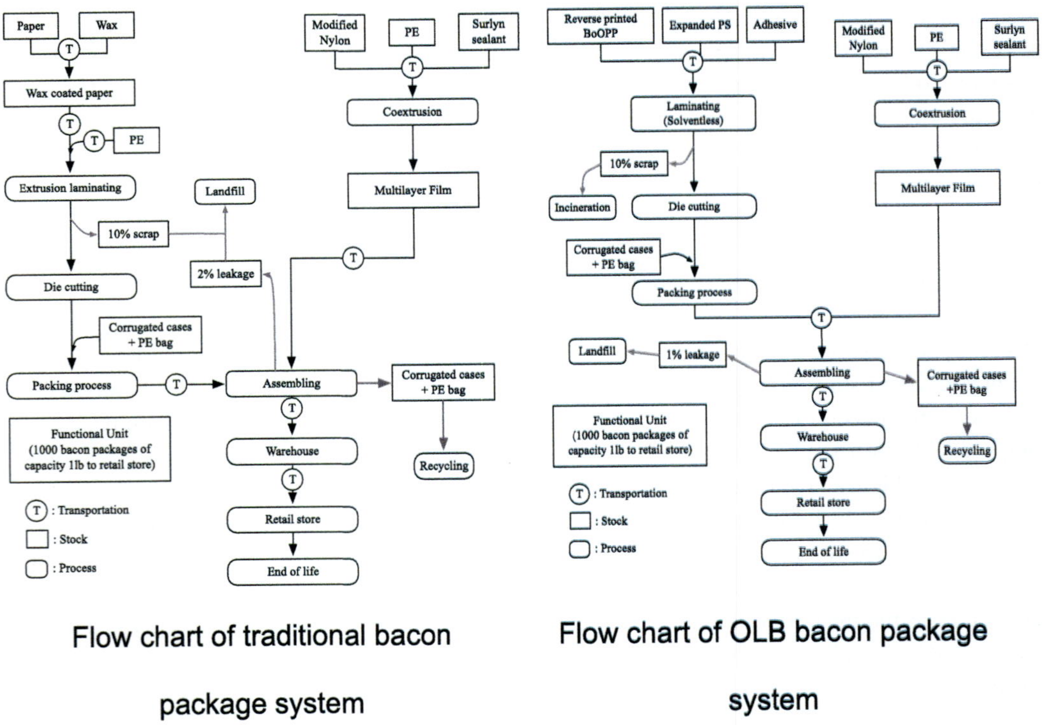

Figure 11.8 Flowcharts of traditional and OLB bacon package system; reproduced from [15].

The functional unit was determined as the needed packages to deliver 1000 lb of bacon to the retail consumer (1,000 bacon packages of capacity 1 lb each) in the US" [15].

To evaluate the environmental impact of lightweighting and changes of material, several hypothetical scenarios were established.

- OLB first (O1) was defined as the new lightweight bacon package. EPS resin for bacon board was produced in Dusseldorf, Germany, and delivered to Wisconsin for making film, via Spain and New York.

- OLB second (O2) was defined as an alternative new lightweight bacon package. EPS resin for bacon board is produced in Houston, Texas, and delivered to Wisconsin for making film.
- L-Board was defined as the traditional bacon board insert.
- L-Board 25 or 50 were defined as the traditional bacon package with 25% or 50% less material used (mass basis), respectively."

"The only difference between OLB first and OLB second is the origin of the EPS resin and film. L-Board 25 and L-Board 50 were hypothetical scenarios with reductions of 25% and 50% across all the materials used for the L-Board" [15]. For example, L-Board 25 is composed of 25% less wax, paper, and PE. Table 11.1 shows the weight of the OLB 1 and 2 and L-Board, L-Board 25, and L-Board 50.

"For the transportation phase, there were two assembling facilities for the L-Board, each accounting for roughly 50% of traditional bacon board production, so 50% of the L-Board production was modeled as delivered to each of the two facilities. The same rule was used for the pouches. For the OLB bacon board, there were also two assembling facilities, so modeling was the same. However, for the OLB bacon pouch, there were also two co-extrusion facilities. Therefore, the model assumed 25% of the OLB bacon pouches were delivered to each of the two assembling facilities from the two co-extrusion facilities. This may have somewhat increased the transportation related impacts, as it did not allow for preferential shipment to minimize distance" [15].

Figure 11.9 "shows the contribution analysis of the O1, O2 and L bacon package systems. The largest contribution to global warming for the OLB first, second and L board bacon package systems were the material production phase, specifically the production of nylon 6. The O1 and O2 bacon package systems had approximately 30% lower global warming value than the L-board bacon package system. The primary reason for the difference in global warming between these two

Table 11.1 Weight of packaging components per functional unit for OLB, L-Board, L-Board 25, and L-Board 50.

	OLB*	lb (kg)	%	L-Board*	lb (kg)	%	L-Board 25*, lb (kg)	%	L-Board 50*, lb (kg)	%
Board	EPS	11.61 (5.27)	40.0	Paper	19.17 (8.69)	35.3	14.38 (6.52)	31.4	9.585 (4.37)	31.3
	BOPP	0.7245 (0.33)	2.5	Wax	9.7 (4.40)	17.9	7.275 (3.30)	15.9	4.85 (2.21)	15.9
	Adhesive	0.4572 (0.21)	1.6	Polyethylene	4.989 (2.26)	9.2	3.741 (1.70)	8.2	2.494 (1.13)	8.2
	Ink	0.2286 (0.10)	0.8	Ink	0.231 (0.10)	0.4	0.231 (0.10)	0.5	0.231 (0.10)	0.8
Pouch	Nylon 6	5.29 (2.40)	18.2	Nylon 6	6.68 (3.03)	12.3	6.68 (3.03)	14.6	6.68 (3.03)	21.8
	Polyethylene	10.69 (4.85)	36.8	Polyethylene	13.49 (6.12)	24.9	13.49 (6.12)	29.5	13.49 (6.12)	22.1
Total		29 (13.15)	100.0		54.26 (24.61)	100.0	45.80 (20.77)	100.0	37.33 (16.93)	100.0

*All weights for the L-Board and OLB bacon packages include 10% scrap, as well as 2% leakers for the L-Board package and 1% leakers for the OLB bacon package.

bacon package systems is the differing amounts of nylon used; 3.03 kg of nylon 6 for L-board is used, whereas 2.4 kg is used in OLB board. The second reason is due to changes in the type of material. Unlike the OLB board bacon package systems, Kraft paper is one of the main components for the L board bacon system and approximately 9 kg CO_2-eq is produced during landfill of this packaging paper mainly due to the production of methane at end of life" [15].

"For the acidification and respiratory effects impact indicator (Figure 11.9), material production was also the main contributor for the O1, O2, and L-Board bacon package systems. Specifically, the production of natural gas for LDPE and the production of nylon contributed the most for O1, O2, and L-board bacon package systems. The OLB first and second bacon package systems had 41 to 44% lower acidification values and 46 to 47% lower respiratory effect values than the L-board bacon package system. The major reason for the differences in the acidification and respiratory effect values is the different amounts of (the same) material used (i.e., 8.38 kg of LDPE is used for the board and pouch of the L-board bacon package system, whereas only 2.4 kg of LDPE is used for the OLB bacon board).

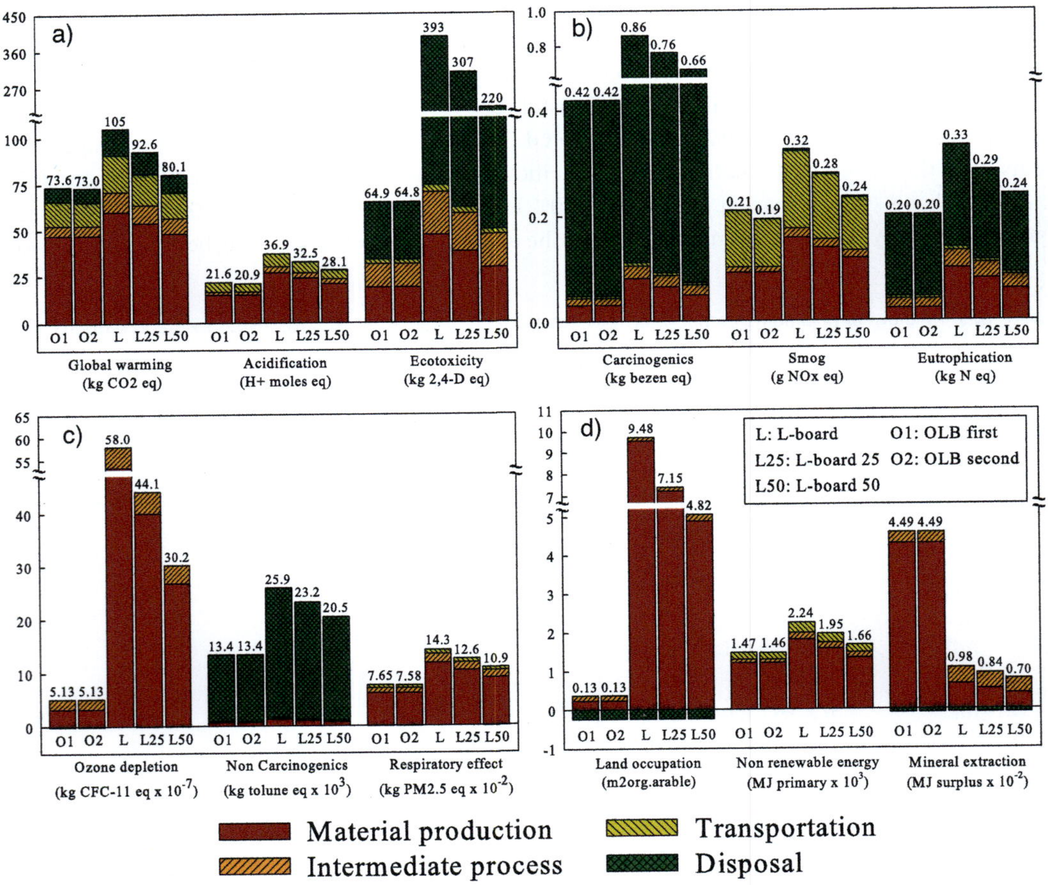

Figure 11.9 Results of LCIA using TRACI 2 (a, b, and c) and Impact 2002+ (d); numbers at the top of each impact indicator column represent the average value of the impact indicator. Letter at the bottom of each impact indicator column describes the different scenarios (O1: OLB first scenario, O2: OLB second scenario, L: L-board, L25: L-board 25% weight reduction, L50: L-board 50% weight reduction); reproduced from [15].

The largest contribution to the carcinogenic and non-carcinogenic impact assessment indicators for the O1, O2, and L-board bacon package systems was the disposal phase, specifically the landfill of plastics (Figure 11.9). The OLB first and second bacon package systems had about 51% lower carcinogenic and about 48% lower non-carcinogenic values than the L-board bacon package system. The major reason was the different amount of plastic landfilled. Even though the total amount of plastics in the OLB bacon package system was higher than in the L-board bacon package system, more of the plastics in the L-board bacon package system ended in landfill than in the OLB bacon package system.

Material production was found to be the main contributor to ozone depletion for the O1, O2, and L-board bacon package systems (Figure 11.9). Production of EPS was the largest contributor for ozone depletion for the OLB first and second bacon package systems whereas production of crude oil for naphtha was the largest contributor for the L-board bacon package system.

In the case of eutrophication and ecotoxicity indicators, disposal was the largest contributor in the OLB first, second and L-Board bacon package systems (Figure 11.9.) For O1 and O2, the main process for eutrophication was landfill of EPS and LDPE, while landfill of EPS and plastics was the main process for ecotoxicity. For the L-Board package, landfill of Kraft paper and LDPE was the main process for eutrophication, and landfill and incineration of Kraft paper for ecotoxicity. In eutrophication, two factors, type of materials landfilled, and differing amounts of the same material landfilled, account for the differences in the impact value, whereas different materials landfilled is the reason for the differing ecotoxicity effects.

Transportation was the main contributor in the O1, O2, and L-board bacon package systems for smog, specifically diesel-powered single unit trucks, Figure 11.9. It was also found that the OLB first and second scenarios had 35 to 40% lower smog values than the L-Board package. The main reason is differing amounts of material delivered during the life cycle of each bacon package system. A total 83.7 kg of material is transported in the life cycle of the L-board bacon package system whereas only 38.5 kg of material is delivered in the life cycle of the OLB bacon package system.

Material production was the main contributor for non-renewable energy in the O1, O2, and L-Board package systems, specifically production of nylon and production of LDPE (Figure 11.9). The OLB first and second bacon package systems had about 35% lower non-renewable energy than the L-board bacon package system due to the differing amounts of material used (2.4 kg of nylon 6 and 4.85 kg of LDPE for the OLB bacon package system, 3.03 kg of nylon 6 and 8.38 kg of LDPE for the L-board bacon package system).

In the case of mineral extraction, the O1, O2 scenarios and L-Board packages have material production as the main contributor (Figure 11.9). Specifically, production of EPS was the largest contributor to mineral extraction for the OLB first and second scenarios and production of ferronickel and bauxite was the main process for the L-Board package. In order to bleach the pulp, bauxite is used to produce hydrogen peroxide for sulfate pulp, which is used for producing Kraft bleached paper. Ferronickel was used for power sawing machines for hardwood pulp. It was found that the OLB first and second bacon package systems had higher mineral extraction values than the L-board bacon package system. This is because, among all the raw materials used in the material production phase, copper generates the highest burden in mineral extraction, resulting in 78% higher mineral extraction value than L-Board. Copper is used to produce the aluminum that is used in blasting charges for explosives used in coal mining" [15].

The LCA results showed that a 25% to 50% weight reduction of the wax coated paper and polyethylene layer of the L-board will overall reduce the environmental burden of the L- Board package as expected; however, this reduction is not enough to outweigh the benefit of changing the main component from paper to EPS. In addition, weight reduction of the wax coated paper structure may not guarantee acceptable performance of the package (e.g., barrier and mechanical properties) compared to what the EPS (expanded polystyrene) with BOPP (biaxially oriented polypropylene) structure can provide. Thus, the environmental burden difference between OLB-Board and L-Board is mainly from changing the material for bacon board from wax coated paper to the EPS with BOPP structure, which also leads to reduced package weight.

"The results presented in this study emphasize that designing the optimal relationship between the type and weight of the packaging material during the product development stage can optimize not only the performance of the packaging systems, but also reduce the environmental burden of the entire life cycle of the product/package system" [15].

11.6 Study Questions

1. Define source reduction according to the US EPA.
2. Reduction in packaging can be done in:
   - **A** Metal packaging
   - **B** Glass packaging
   - **C** Paper, paperboard, and corrugated board packaging
   - **D** Plastic packaging
   - **E** All of the above
3. Provide an example of source reduction.
4. Why is source reduction especially advantageous for glass and metal packaging?
5. Provide examples of source reduction for metal. Discuss the reduction of basis weight for corrugated board in the US and around the world.
6. Explain lightweighting in plastic packaging.
7. What are the main findings of the bacon packaging paper?
8. Provide an example of source reduction in any packaging category. Considering the main differences in packaging materials and uses, which main opportunities do you see for lightweighting in packaging?

11.7 Additional Resources

1. Pollution Prevention (P2)
   https://www.epa.gov/p2
2. Pollution Prevention Law and Policies
   https://www.epa.gov/p2/pollution-prevention-law-and-policies
3. WRAP UK
   http://www.wrap.org.uk
4. How to Prevent Wasted Food through Source Reduction
   https://www.epa.gov/sustainable-management-food/how-prevent-wasted-food-through-source-reduction
5. Stop the waste: UN food agencies call for action to reduce global hunger
   https://news.un.org/en/story/2019/10/1049181

References

1 Brown, K. and Andre, P. (1993). Waste prevention: Source reduction now. *How to implement a source reduction program in your organization.* https://www.leg.mn.gov/docs/2021/Other/210738.pdf (accessed October 29, 2019).

2 Lightweighting Is Top Priority for Automotive Industry | 2018-06-07| Assembly Magazine (n.d.) https://www.assemblymag.com/articles/94341-lightweighting-is-top-priority-for-automotive-industry (accessed October 21, 2019).

3 Engineering toolbox. Density of Various Solids (2019). https://www.engineeringtoolbox.com/density-solids-d_1265.html (accessed October 22, 2019).

4 WRAP (2014). *Lightweight beer bottles : Setting a bold new standard.*

5 Thomas, J. (2019). *The future of metal packaging and coatings to 2023. 2019.*

6 Birat, J.P. (2015). Steel and humanity's grand challenges. *Materials and Technologies* 103: DOI: 10.1051/mattech/2015015.

7 The Aluminum Can Advantage | The Aluminum Association (2019). https://www.aluminum.org/aluminum-can-advantage (accessed October 21, 2019).

8 Fire Box Association (2019). *Shipped average basis weight for the U.S.*

9 Popil, R. (2019). *What's important in liner and medium physical properties.* http://www.rbi.gatech.edu/sites/default/files/documents/newsletter_0911.pdf (accessed October 22, 2019).

10 NAPCOR (2019). *PET bottle weight history between 2010 and 2017.*

11 BCM (2015). *Report of average weight bottle by year.*

12 Rosenbloom, S. (2008). Solution, or mess? A milk jug for a green Earth. *New York Times* A1.

13 Singh, J., Krasowski, A., and Singh, S.P. (2011). Life cycle inventory of HDPE bottle-based liquid milk packaging systems. *Packaging Technology and Science* 24: 49–60. DOI: 10.1002/pts.909.

14 Cameron, G. (2019). *The future of global flexible packaging to 2020.* Smithers Pira:. Surrey, UK, 108.

15 Kang, D., Sgriccia, N., Selke, S., and Auras, R. (2013). Comparison of bacon packaging on a life cycle basis: A case study. *Journal of Cleaner Production* 54: 142–149. DOI: 10.1016/J.JCLEPRO.2013.05.008.

12

Reuse

12.1 Learning Objectives

After learning from this chapter, you are expected to:

- Define *reuse* and describe the main benefits of reuse systems.
- Explain the trade-offs of implementing reusable packaging systems.
- Provide examples of packaging systems where reuse provides a reduction of the environmental footprint of a packaging system.
- Describe EFP trade-offs for reuse systems for metal, glass, paper, paperboard, corrugated board, and plastic packaging systems.

12.2 Introduction

Reuse is considered one of the two MSW management strategies to reduce waste at the source making the most impact on the reduction of MSW and helping achieve a zero-waste society. Reuse programs have been created in communities, cities, and countries around the world to redistribute durable and nondurable goods to new users [1]. Reuse programs, for example, for resources such as water are being encouraged by the US EPA [2].

As economies are developing and transitioning from low- to middle- or high-income economies, the number of single-use items has been increasing, mostly due to the increasing consumer disposal income, since as income rises society tends to increase disposal until a certain point, as discussed in Chapter 10 Section 10.4 (Environmental Kuznets curve). In 1980, to reduce MSW generation, the US EPA encouraged *reuse* programs for organizations such as businesses, schools, and prisons, and for events. Reusable goods are generally built differently, such as with higher weight so that they can last longer. Reuse is often seen as a preferred method to reduce the EFP of goods and is often considered as similar to reduction. However, LCA studies that fully characterize product systems, especially those that include a broad range of EFP indicators, have shown that reuse does not always result in reduction of EFP. Over the last few decades, *reuse* of primary packaging systems has decreased in favor of single-use packaging. On the other hand, *reuse* systems for selected secondary and tertiary packaging systems have increased, facilitated by the better control and digitalization of the supply chain.

In this chapter, a detailed discussion of how *reuse* can be used to create systems with lower EFP is offered for all the major packaging materials: metal, glass, paper, and plastic.

Life Cycle of Sustainable Packaging: From Design to End of Life, First Edition. Rafael A. Auras and Susan E M. Selke.

Companion Website: www.wiley.com/go/Auras/lifecycleofsustainablepackaging

12.3 Reuse

The US EPA classifies *reuse* as the use of materials or items several times. Reuse of materials and products and the creation of a reuse system may conserve valuable natural resources and reduce the amount of soil, water, and air pollution. Items can be reused by repairing, donating them to charity and community groups, or selling them for a second life. As the old saying goes, "One person's trash is another person's treasure." So, instead of discarding unwanted tools, furniture, books, working electronics or appliances, they can be sold or donated. Flea markets, thrift stores, estate sales, and garage sales are conventional methods to encourage the reuse of products, benefiting all parties involved. Reusing products, when possible, can be better than recycling because the item does not need to be reprocessed before it can be used again. Many organizations, such as churches, community centers, schools, and nonprofits, accept a variety of donated items.

Reuse systems can be categorized by two main approaches, *closed* and *open loop* (see Figure 12.1.) A *closed-loop* system is defined when a product is reused for the same or a similar purpose. A closed-loop product system requires a well-defined circuit of shipping and delivery points. Returnables are defined as products, packaging, or containers designed to be reused in several systems such as manufacturing and distribution. They eliminate the need to purchase new products, packaging, or containers, or disposing of them. Returnable products, packages, or containers can be designed to be more durable than expendables for added protection. Examples of reuse systems include refillable milk bottles, reusable pallets for delivery of goods to warehouses, and reusable crates for delivery of auto parts.

An *open-loop product* system refers to a vaguely defined circuit of initial shipping of a product in which after its first use, it is repurposed into a different application. In this case, eliminating the need to purchase a new product or replacement may happen. For example, used tea cans may be reused to store miscellaneous items.

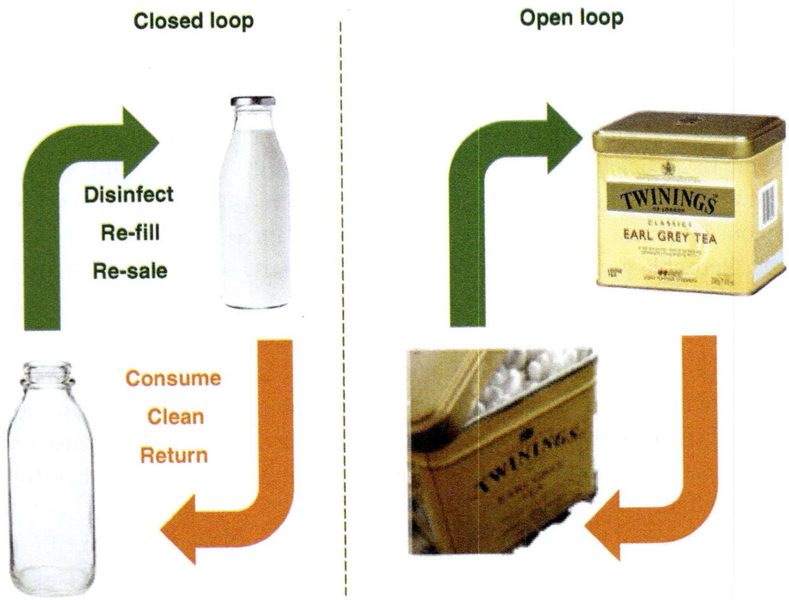

Figure 12.1 Example of closed and open reuse loop systems.

Table 12.1 Source reduction methodology to determine the weight and volume reduction between two product options.

Product A	Product B
Volume of discarded product	Volume of discarded product
Weight of discarded product	Weight of discarded product
Cost of product	Cost of product
Number of uses per year	Number of uses per year
Volume, weight, and cost per year	Volume, weight, and cost per year

In the early 1980s, the US EPA incentivized a methodology for source reduction. The Minnesota Office of Waste Management published a series of bulletins explaining this methodology [3]. Mostly, the method quantifies mass and volume reduction due to the implementation of reuse systems. Table 12.1 shows an example of a method to quantify source reduction between two products, encouraging a reuse system. Example 12.1 shows a calculation example.

Example 12.1 A school cafeteria would like to change from single-use plates to reusable plates. Several questions are being asked to understand whether the changes will reduce waste. The original study can be found elsewhere [4].

- What is the annual disposal volume of the single-use and reusable plates?

To simulate the disposal volume, 25 plates were thrown into a large box. Then, the box was shaken to help settling and occupied volume was measured by measuring the length, height, and weight, which for the 25 single-use plates was 1,760 in^3. A case has 500 plates, and the disposal volume of 500 single-use plates will be 35,200 in^3. On average, 64 cases are used per year, so the disposal volume for one year is 2,252,800 in^3/y.

The reusable plate measures 8.5 in. diameter × .375 in. in thickness, with a volume of 21.26 in^3.

- How many reusable plates are needed?

The cafeteria needs 72 plates, which is a volume of 1,531 in^3.

- How long will the reusable plates last?

The reusable plate has a lifespan of 3 years, according to the manufacturer, so 510 in^3 of waste are generated per year.

- What is the difference in the waste volume amount for these two plates?

The difference in volume of waste is $2{,}252{,}800 - 510 = 2{,}252{,}290$ in^3 or 39.8 yd^3/y or a 99.9% volume reduction.

An additional amount of solid waste is generated by the soap used for cleaning of the dishes, but this will not be included in this exercise at this time. Furthermore, this calculation not only takes into account the space actually occupied by the reusable plates but also includes a very large amount of void space for the disposable plates. More detailed data can be collected to obtain more precise estimation of the waste volume for these two scenarios. A solid waste compactor can be used to reduce the volume of waste for the single-use plate by using pressure to reduce the volume.

- What is the annual disposal weight of the single-use plates?
  The case of single-use plates weighs 20 lbs, so the 64 cases weigh 1,280 lb/y.
- What is the annual disposal weight of the reusable plates?
  The reusable plate weighs 5.625 oz, so 5.625 oz × 72 plates/3 y = 135 oz/y or 8.4 lb/y.
- What is the difference of weight between the two plates?
  1,280 − 8.4 = 1,271.6 lb/y or 99% weight reduction.
- What is the cost of the single-use plates?
  The cost is $36/case × 64 cases = $2,304/y

Additional costs are incurred due to the use of single-use plates, such as labor required to order, stock, deliver, and dispose. Hauling waste costs are also incurred based on volume. For example, we can assume that it will cost $260 to empty a 15 yd^3/dumpster, so, $17.33/yd. If we reduce 40 yd^3/y, we are reducing cost by $693/y.

What is the cost of the reusable plates?
The cost of each plate is $2.08 × 72 = $150/3 y life = $50/y.

- What is the cost of cleaning?
  The washing cost for each plate can be estimated as $14 water + $1.50 electricity + $80 soap = $95.50. The total cost per year: $50 + $95.50 = $145.5/y.
- What is the difference in the cost of the two options?

Cost of single use $2,304/y – cost of reusable plate $145.5/y = $2,158 savings/y. This simple calculation can be seen as fundamentally unfair; costing for the reusable plates does not include the labor costs for the reuse system, though similar costs mentioned for the disposables can provide an initial estimation. Therefore, the cost calculation could include additional components to get a more precise estimation.

Table 12.2 shows a simplified summary of the source reduction evaluation using the procedure outlined above. Although these initial studies of source reduction were prevalent until 2000, they did not fully incorporate the EFP of the single and reusable plate systems, sometimes hiding the true EFP of the system. Further examples will be provided in the following section when we look at different reuse systems in packaging.

Table 12.2 Summary of the source reduction of the single-use and the reusable plate.

	Single-Use Plate	Reusable Plate
Volume, in^3/y	2,282,800	510
Weight, lb/y	1,280	8.4
Number used, unit/y	32,000 in 1 year	72 in 3 years
Cost, $/y	2,304	146

12.4 Reuse in Packaging

An early example of reusable packaging is the use of amphora, made of ceramic, between 1500 BCE and 500 CE to ship wine and other products throughout the Mediterranean. Several shapes of amphora were used for this purpose. Ships and secondary structures were built around the design of the containers for their transportation [5]. Casks also appeared in the same period, evolving from open-top wooden tubs. Early use of wooden barrels made of hollow logs and ends covered with dried animal skins was found during the Armenian period for shipping wine in the Euphrates River. Wooden barrels have been used since then through history to ship many liquid products [6]. Barrels and casks are currently made of several materials, such as wood, steel, and plastic, and they have become standardized in size and specifications [7].

The automobile and furniture industries have also used reusable containers for many years. The Ford Motor Company implemented returnable containers in the Rouge plant as early as 1930. Parts were received in crates, and the wood from the crates was reused to make bumpers and running boards for cars [8]. During the push for single-use packaging and a lack of understanding of the entire EFP of a packaging system, many companies shifted to single-use packaging systems, mostly for their primary packaging. However, with the implementation of LCA studies, a better understanding of the entire EFP of packaging systems has been obtained, and many companies, such as the automotive, beverage, medical device, military, fresh produce, milk, and furniture industries, shifted again to reusable packaging systems as an integral component of the new Lean Manufacturing Process (i.e., creation of customer value of goods and quality systems using, for example, a six-sigma quality process for high standards of product delivery) [8].

New companies have been created such as Loop® for developing reusable systems based on the delivery of the product in reusable packages [9]. Since consumers are really concerned about single-use packaging and the increase of MSW, the implementation of these approaches is reaching consumers' core values. Companies such as Nestle®, Procter & Gamble, and Clorox™ are participating with selected products in this early distribution system mimicking old models, such as the milk bottle distribution system in the UK.

Recently, Mahmoudi and Parvizionran reviewed the literature on reusable packages and proposed a flow of reusable containers between a sender and a recipient (Figure 12.2) [10]. For companies willing to adopt reusable containers, managing the reusable containers is the main concern due to the high volumes and the frequency of the deliveries/pickups, further complicated with hyperlocal economies. The main concerns are (a) how to manage the inventory of reusable containers at the warehouse (i.e., understanding how to plan for future supply and demand), (b) scheduling and routing of the reusable packages (i.e., estimating the delivery and pick-up of empty containers to optimize zero lead time), (c) the optimal repairing, replacement, and disposal schedule, and (d) understanding and measuring the performance of the reusable packaging systems.

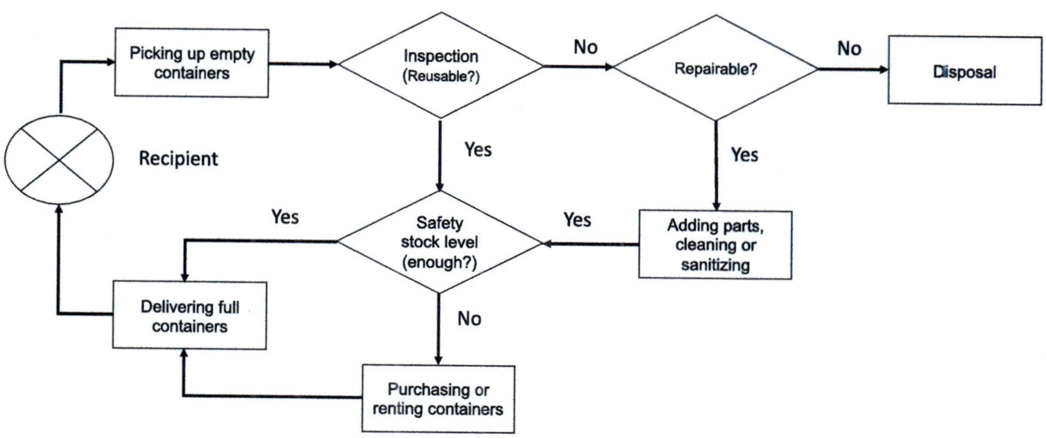

Figure 12.2 Flow of reusable containers between a sender and a recipient, adapted from [10].

12.4.1 Metal

Metal is an ideal material for the creation of reusable packages due to its durability, weight, and high strength. Several reuse systems are commonly presented in metal packaging such as drums, tin cans, and pallets, as shown in Figure 12.3. Metal pallets are mostly used for products where static electricity is not an issue. Drums are commonly used for business to business (B2B) operations to supply ingredients in the form of powders and liquids for industrial production. Raugei et al. evaluated the EFP of single-use fiber drums vs. reusable steel drums where the single-use disposable fiber drums are landfilled at the end of each one-way trip, and a reusable steel drum is returned to the chemical production plant and used for 200 times before final disposal. The weight of the steel drum is 18 kg, and the fiber drum 5 kg. The authors evaluated the global warming potential, acidification potential, overall gross energy requirements, and total waste, and concluded that the steel drum had lower EFP in all the categories [11]. However, caution must be taken since the number of reuses and the distance travelled for the return of the steel drums are critical factors affecting the EFP. When the reusable containers have a much larger weight than the single-use containers, the number of reuses directly impacts the initial EFP of producing the container since the initial EFP is allocated to all the uses (i.e., initial EFP: production EFP/number of uses), and the distance to distribute these heavy containers directly affects the EFP. Reuse systems implemented in smaller geographical regions or countries have lower transport EFP than reuse systems implemented in larger geographic areas like the US.

Other examples of reusable metal packaging systems are reusable water bottles, lunch boxes, and military containers. Generally, if the supply chain is tightly controlled and almost 100% of the

Figure 12.3 Examples of returnable metal packaging: (a) returnable pallet, (b) metal drum, (c) refillable body lotion used by the Loop® company, (d) metal box, (e) aluminum ice cream container used by Loop®, (f) food canisters – Indian Tiffin – commonly used in several countries for homemade cooked food.

metal containers can be returned, metal containers can be an attractive option, depending on the number of reuses and the transportation distance.

12.4.2 Glass

Glass is another excellent material for the creation of reusable packaging systems. Glass, being an inert material, is ideal for reusable jars, bottles, and containers since after returning, cleaning can be easily done without interaction of the cleaning agents with the glass. It was used for centuries to distribute milk, wine, and beverages in closed-loop systems. Home delivery of milk in refillable glass bottles is still available in a few states such as Michigan, Illinois, California, and Florida, although with very minimal market share. Selected food stores also offer milk in reusable glass bottles. However, as single-use packaging has become more convenient and industry consolidates, distribution distances become larger, so the high weight of reusable glass containers has prohibited extensive reuse systems for glass in countries such as the US and Australia where transportation distances are considerable. The main drawback is the trade-offs between the initial EFP of producing the glass container, the number of reuses, and the total distribution distance for each glass bottle. For example, if we look at the Case Study 9.1, if we reuse the glass bottle 50 times, the initial environmental footprint will be divided by these 50 uses. The additional EFP for cleaning the bottles between applications and the backhaul transportation must also be considered. Although for selected companies, reusable glass systems work, the global tendency around the world is moving to single-use containers that are recycled. Figure 12.4 shows the decline of refillable bottles, mostly glass in the soft and beer market, to mostly one-way containers between 1947 and 1998 in the US. Newer data were not available. However, numbers have not changed

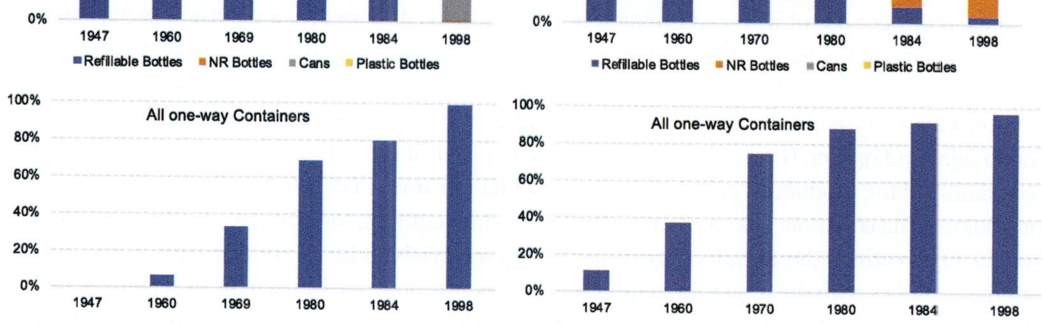

Figure 12.4 Soft and drink beer containers in the US from 1947 to 1998; data obtained from [12]. Source Soft Drink: Can and Bottle Bills, CalPIRG, 1981 for 1947–60 data;1986 Statistical Profile, National Soft Drink Association, 1986 for 1969–84 data; Beverage World, June 1999 for1998 data. Source Beer Containers: Can and Bottle Bills, CalPIRG, 1981 for 1947–80 data; Beer Institute, 1998 for 1984 data, 1999 for 1998 data. Note: Figures may not add up to 100% due to rounding.

much [12]. The US and many parts of the world have mostly transitioned from refillable to one-way recyclable containers.

12.4.3 Paper, Paperboard, and Corrugated Board

Most reuse systems in this category are reusable corrugated boxes. Some companies and businesses can reuse corrugated boxes that their vendors have used to ship supplies or parts, reducing the need and costs of new corrugated boxes. Corrugated boxes designed for reuse need to be free of address labels and other markings that could confuse shipping. Nike developed a sticker for its box reuse program, which can cover all addresses. Some companies using reusable boxes have common areas used to stack unfolded boxes to facilitate their reuse. Some businesses differentiate between reused corrugated boxes for shipment to customers vs. boxes reused internally, with the higher quality, "cleaner" boxes used for customer orders. One caution is that boxes should not be reused for shipment if they are printed with labels such as "hazardous," "flammable," etc. The Portland order fulfillment center of OfficeMax® reuses corrugated boxes from vendors and large consumers such as universities and schools, which use a large amount of supplies [13]. Several companies stock their own products in stores, so they have the opportunity to reuse their own corrugated boxes. However, this is not widespread.

Not many reports are available about reusable corrugated boxes (CBs). However, Battini et al. proposed the reuse of display ready corrugated (DRC) containers for traditional and short fresh food supply chains by introducing two or three layers of plastic film in the bottom of the DRCs. They demonstrated that this solution can provide lower EFP than regular reusable plastic containers (RPCs) for shorter distribution supply chains [14].

12.4.4 Plastic

Plastic containers have been used for several reusable applications from replacing DRC for RPC to distribute fresh produce and milk to storing liquids and food at home. An entire industry has been created around reusable plastic containers. Figure 12.5 shows several examples of RPC used for several industries and applications. Generally, most of these containers need to be cleaned before the next use.

In the case of food packaging, special consideration needs to be taken to reuse RPC. First, they must be cleaned and disinfected before reusing for another food application to avoid off-odors and cross-contamination. Second, if the RPCs are used in the microwave, they must be approved for this use since improper use of RPC can threaten health. Plastic is different than glass and metal since it has low density and is permeable to small molecular weight compounds. So, the RPC used in food packaging must be specially designed for several uses. When companies create a packaging system, they consider the type of food (high-acid vs. low-acid, alcoholic or not alcoholic), the sensitivity of the food to light and oxygen (fat, milk, juices), the type of physical protection (tomatoes, eggs, etc.), and the visibility of the products. They also need to consider the amount of heat required for sterilization and cooking, the container size, and the home processing required, for example microwave or oven. The US FDA verifies the initial safety of packaging materials to be in contact with foods. If the material complies with the migration requirements, it is authorized to be used for food contact applications [15, 16]. So, some basic guidelines can be provided for reusable food packaging containers:

1) Packages that are not designed for food should not be used in contact with food.
2) Reuse food containers with food with similar acidity and sugar, fat, or alcohol content.
3) Reuse food containers for food that is subject to similar processing.

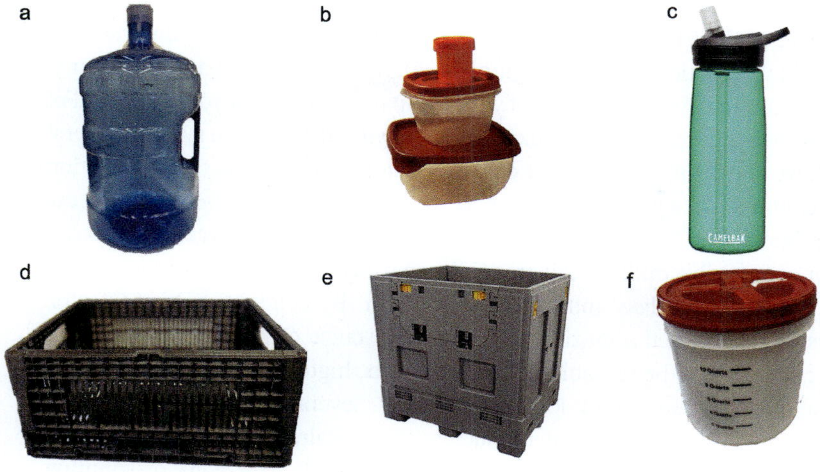

Figure 12.5 Examples of reusable plastic containers, (a) 5-gallon bottle jug, (b) plastic containers, (c) water bottle, (d) reusable plastic container, (e) reusable plastic pallet, (f) bucket.

4) Do not reuse porous packages.
5) Do not reuse microwave packages containing "heat susceptors," such as popcorn bags.
6) Avoid storing foods with strong odors or flavors in RPC.
7) RPC that are safe to use in microwave ovens are distinctly labeled.
8) Do not expose the printed area to the food.
9) If a container was used to store a nonfood item such as gasoline or solvents, do not reuse for food storage even when cleaned.
10) Use only food grade approved materials.
11) If in doubt about whether the RPC is approved for food contact, transfer the food to an inert container.

An entire research area is dedicated to the migration of chemical compounds such as additives for processing, stabilizers, and colorants from containers, and plastic is a large portion of these studies. Several researchers have demonstrated that RPC in contact with several chemical compounds can sorb chemical compounds, and then transfer them when in contact with a new product, or compounds present in the plastic can be released to a different product [17–24]. Hahladakis et al. conducted a comprehensive review of the chemical additives present in plastics and able to migrate to food during their useful life due to food–additive–plastic interactions and enter the environment through recycling or disposing to landfill or littering to the environment [20].

Example 12.2 considers the reuse of pallets for the distribution of glass bottles as previously discussed in Case Study 9.1.

Example 12.2 Considering Case Study 9.1, we can think about using a reusable pallet for the distribution of glass drink bottles. Wood pallets are used fewer times than reusable plastic pallets (RPP). In this example, we assume that the wood pallet is used 5 times, and the RPP is used 100 times. We can observe that this change does not make a significant difference in the EFP of the packaging system (Figure 12.6).

However, if we evaluate the EFP of the tertiary packaging system (pallet), we can observe a significant reduction in this portion of the packaging system. If you are the soft drink company, this may not be as relevant in reducing the EFP of your company; however, if you are the pallet pooling company, this is a great case to minimize the EFP of your distribution system. Several additional studies have shown the same effect of reuse systems. As long as the distribution system can be kept contained in a geographical region, the reuse system can compete with the single-use system. When the EFP of returning the reusable container is too large, the trade-off is not as evident. For a returnable milk container, a German study determined that the average transport distance for refillable glass bottles was just 90 km. By conducting an LCA, the European Commission determined a maximum transport distance of 241 km; at more than this distance, some of the EFP of the reusable containers may be higher. So, reusable milk containers are beneficial or better designed for localized products. Several factors should be taken into consideration when designing these systems such as the container weight, transport distance, backloading of empty crates (reverse distribution), end-of-life treatment, distribution environment, turnover, filling speeds, space utilization, transit damage, and shelf life [25].

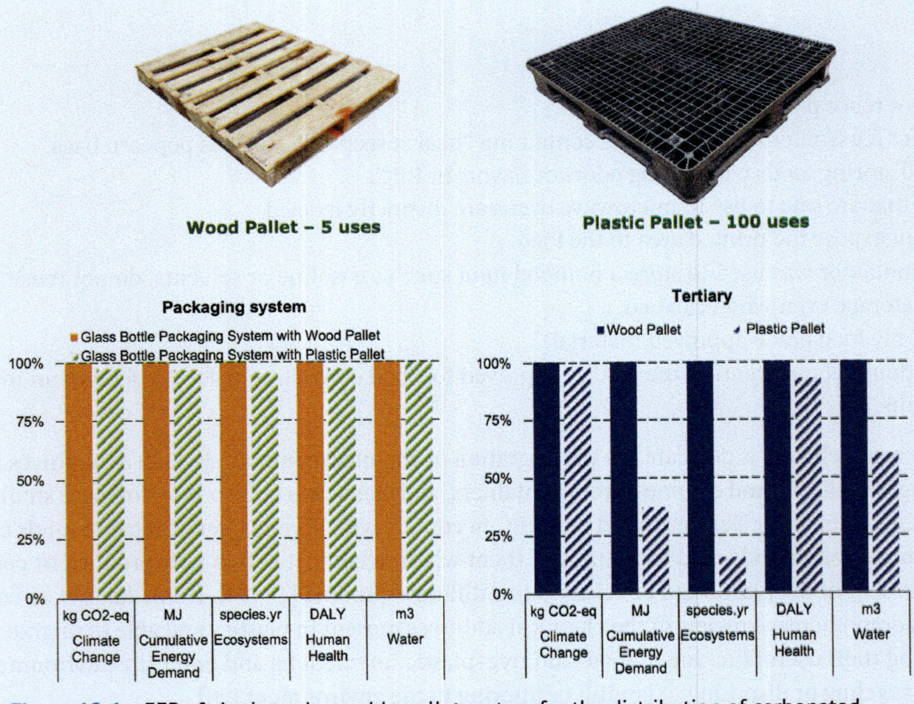

Figure 12.6 EFP of single and reusable pallet system for the distribution of carbonated drinks in glass containers based on Case Study 9.1.

12.5 Case Study 12.1. Reusable Cups

An LCA study was conducted to determine the environmental effects of using reusable cups during a major event that took place in Barcelona in 2004 during the Olympic games. In this study, the authors compared single-use and reusable cups made of polypropylene, but with different physical characteristics such as mass, shape, and capacity [26]. Table 12.3 shows the main information for

Table 12.3 Physical characteristics of single-use and reusable cups; data reproduced from [26].

Characteristics	Single-Use Cup	Reusable Cup
Capacity, mL	200	300
Height, cm	9	15
Weight of a cup, g	3.20	44.89
Number of cups needed to serve 1,000 L of draught beverages	5,000	3,333
Weight of cups kg equivalent to 1,000 L beverage distribution	16.00	149.63

the single-use and reusable cups. The authors decided to study four main scenarios: scenario 1: 2 uses (1 cleaning), scenario 2: 9 uses (8 cleanings), scenario 3: 10 uses (9 cleanings), scenario 4: 14 uses (13 cleanings).

Figure 12.7 shows the EFP of the single-use cup and the reusable cup with four different scenarios. Observing all the EFP indicators, the first thing that we can state is that after 10 uses the EFP of the reusable cup in most indicators is lower than the single-use cup. However, the ozone, carcinogenic, and solid waste indicators never approached lower values than the single-use cup. This study provides several insights. First, weight reduction is not always achieved straight away. Many uses must be obtained from reusable cups before the solid waste production is much lower than the single cup. Second, use conditions for the products determine a lot of the trade-offs among the EFP indicators. Additionally, how many event cups can we take home before they become redundant?

12.6 Case Study 12.2. Reusable Plastic Containers (RPC)

Display ready corrugated (DRC) containers are being used for the distribution of perishable and sensitive products such as tomatoes, avocados, and grapes. However, an increase in RPC is happening also in this category; adopted by several larger companies such as

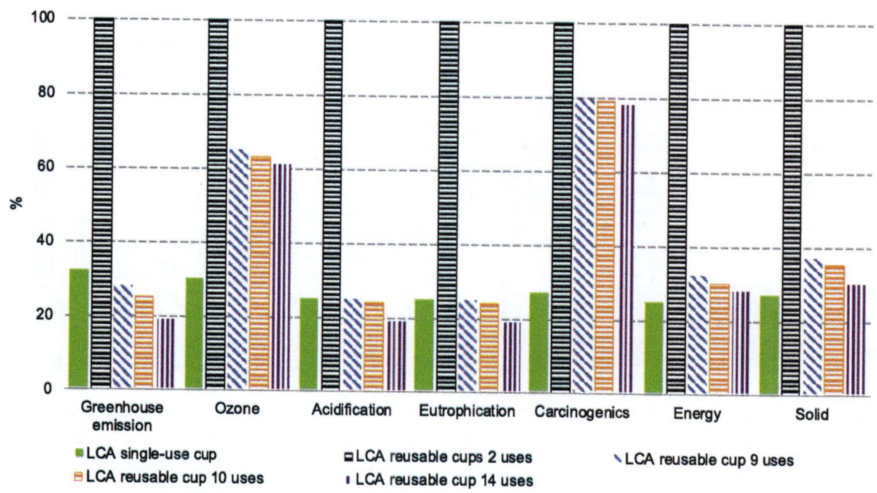

Figure 12.7 The environmental footprint of single-use and reusable cups; original data obtained from [26].

Walmart®. Singh et al. reported an early study initially conducted by Franklin Associates – an LCA consulting company – to determine how the energy, solid waste, and GHG of DRC compared with RPC. Figure 12.8a and Figure 12.8b show a picture of a DRC and a RPC, respectively, and Figure 12.8c, Figure 12.8d, and Figure 12.8e show the materials flow from "cradle-to-grave" and the life cycle of DRC and RPC, respectively. The authors considered two scenarios for the RPC containers, one with an average backhauling operation where the pooling systems for RPC allow the growers to obtain RPC from the nearest RPC pooling location – average scenario defined as RPC with average use/loss rates, and another scenario with 80% backhaul for the RPC. They compared these two scenarios with the standard DRC scenario with a recovery rate of 95% overall in the US for grocery stores. The functional unit of the study was 1,000 tons of produce distributed to the distribution center and consumed. The authors estimated that for 10 different produce items, such as oranges, tomatoes, onions, apples, carrots, grapes, and strawberries, the RPC requires 39% less total energy, produces 95% less total solid waste, and generates 29% less total GHG in the North American market [27].

An expansion study of the RPC and DRC was conducted by Battini et al., considering that the DRC can be reused if two or three layers of removable plastic films are used at the bottom of the containers [14]. After reviewing different supply chains, they concluded that the DRC with two or three removable plastic films was the best choice with the lowest manufacturing costs and environmental impact. They also recommended that the new DRC with removable plastic films would only be a solution to implement for a short lifespan product. Similar findings regarding the implementation of RPC have been reported by other authors [28].

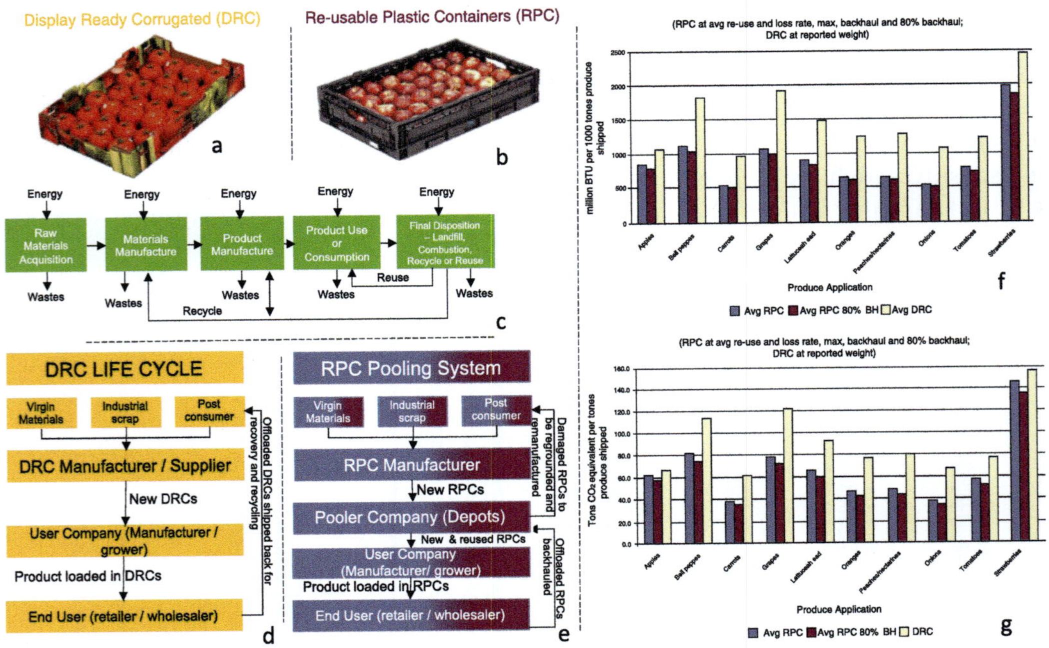

Figure 12.8 Evaluation of the environmental footprint of distribution of fresh produce: (a) display-ready corrugated (DRC), (b) and reusable plastic containers (RPC), (c) general materials flowchart from "cradle-to-grave," (d) DRC life cycle, (e) RPC life cycle, (f) average scenario energy comparison, (g) average scenario greenhouse gases. Figures a, b, f, and g are reproduced from [28]. Copyright from John Wiley & Sons.

12.7 Study Questions

1 Define *reuse* as described by the US EPA.
2 What are the characteristics to consider when designing a reusable packaging system?
  A Container weight
  B Transportation distance
  C Reverse distribution
  D End of life
  E All of the above
3 Provide two examples of closed- and open-loop reuse systems.
4 A school cafeteria would like to change from using single-use cups to reusable cups. Several questions are being asked to understand if the changes will reduce waste. Assuming that 600 children eat lunch every day and that each child uses one cup per lunch, estimate the change in weight and volume waste for one year of consumption. Data: Assume that the single-use cup and the reusable cup have the same characteristics as the cups described in Case Study 12.1. The reusable cups are used for 5 years, replacing 10% of the cups per year due to losses, damage, and miscellaneous encounters.
5 What is the main difference between the source reduction methodology introduced by the US EPA in the 1980 and LCA methodology?
6 Provide a short summary of the history of reusable packaging.
7 Provide examples of reusable metal containers.
8 Provide examples of reusable glass containers.
9 Provide examples of paper, paperboard, or corrugated board reusable containers.
10 Provide examples of reusable plastic containers.
11 Explain the trade-offs between initial EFP for production, weight, and transportation distance for reusable containers.
12 What special consideration should be taken when reusing plastic containers for food?

12.8 Additional Resources

1 Reducing and reusing basics
https://www.epa.gov/recycle/reducing-and-reusing-basics
2 Reduce, reuse, recycle
https://www.epa.gov/recycle
3 Water Reuse Action Plan
https://www.epa.gov/waterreuse/water-reuse-action-plan
4 Reusable Packaging Association
https://reusables.org
5 Loop® store
https://loopstore.com
6 Reusable transport packaging
https://reusabletranspack.com/product-category/plastic-pallets
7 Reusable pallets
https://www.supplychainquarterly.com/media/videos/play/297-making-the-switch-to-reusable-pallets-in-your-supply-chain

References

1 Cogan, S. (1979). *Reuse as a key element in waste reduction.*

2 US Environmental Protection Agency. (2019). *National reuse water action plan.* https://www.epa.gov/waterreuse/water-reuse-action-plan (accessed November 16, 2021).

3 Minnesota Office of Environmental Assistance. (1996). Reduction.pdf Source Reduction Now. https://www.leg.mn.gov/docs/2021/Other/210738.pdf (accessed October 4, 2019).

4 Brown, K. and Andre, P. (1993). Waste prevention: Source reduction now. *How to implement a source reduction program in your organization.* https://www.leg.mn.gov/docs/2021/Other/210738.pdf (accessed October 29, 2019).

5 Twede, D. (2002). The packaging technology and science of ancient transport amphoras. *Packaging Technology & Science* 15: 181–195. https://doi.org/10.1002/pts.597.

6 Twede, D. (2005). The cask age: The technology and history of wooden barrels. *Packaging Technology & Science* 18: 253–264. https://doi.org/10.1002/pts.696.

7 Lewandowski, K. (2016). Growth in the size of unit loads and shipping containers from antique to WWI. *Packaging Technology & Science* 29: 451–478. https://doi.org/10.1002/pts.2231.

8 Chism, C. (2010). *Optimizing and benchmarking returnable container processes within an automotive distribution system.* http://scholarworks.rit.edu/theses (accessed October 24, 2019).

9 Casper, R. and Sundin, E. (2018). Reverse logistic transportation and packaging concepts in automotive remanufacturing. *Procedia Manufacturing* 25: 154–160. https://doi.org/10.1016/J.PROMFG.2018.06.069.

10 Mahmoudi, M. and Parviziomran, I. (2020). Reusable packaging in supply chains: A review of environmental and economic impacts, logistics system designs, and operations management. *The International Journal of Production Economics* 228: 107730. https://doi.org/10.1016/j.ijpe.2020.107730.

11 Raugei, M., Fullana-i-Palmer, P., Puig, R., and Torres, A. (2009). A comparative life cycle assessment of single-use fibre drums versus reusable steel drums. *Packaging Technology & Science* 22: 443–450. https://doi.org/10.1002/pts.865.

12 Container Recycling Institute. (2019). *The decline of refillable beverage bottles in the U.S.* http://www.container-recycling.org/index.php/refillable-glass-bottles/53-facts-a-statistics/glass/428-the-decline-of-refillable-beverage-bottles-in-the-us (accessed October 25, 2019).

13 US Environmental Protection Agency. (n.d.). *Reducing and reusing basics.* https://www.epa.gov/recycle/reducing-and-reusing-basics (accessed October 21, 2019).

14 Battini, D., Calzavara, M., Persona, A., and Sgarbossa, F. (2016). Sustainable packaging development for fresh food supply chains. *Packaging Technology & Science* 29: 25–43. https://doi.org/10.1002/pts.2185.

15 Wolfe, H.B. (2009). Reusing food packaging... Is it safe? *WIT Trans Built Environ.* 108: 389–397. https://doi.org/10.2495/SAFE090361.

16 US Food & Drug Administration. (n.d.). *Guidance for industry: Preparation of premarket submissions for food contact substances (chemistry recommendations) | FDA* https://www.fda.gov/regulatory-information/search-fda-guidance-documents/guidance-industry-preparation-premarket-submissions-food-contact-substances-chemistry (accessed November 7, 2019).

17 Safa, H.L. and Bourelle, F. (1999). Sorption-desorption of aromas on multi-use PET bottles A Test Procedure. *Packaging Technology Science* 12: 37–44. https://doi.org/10.1002/(SICI)1099-1522(199901/02)12:13.0.CO;2-F.

18 Komolprasert, V. and Lawson, A.R. (1997). Considerations for reuse of poly(ethylene terephthalate) bottles in food packaging: Migration study. *Journal of Agricultural and Food Chemistry* 45: 444–448. https://doi.org/10.1021/jf9603801.

19 Sonchaeng, U., Iñiguez-Franco, F., Auras, R., Selke, S., Rubino, M., and Lim, L.-T. (2018). Poly(lactic acid) mass transfer properties. *Progress in Polymer Science* 86: 85–121. https://doi.org/10.1016/J.PROGPOLYMSCI.2018.06.008.

20 Hahladakis, J.N., Velis, C.A., Weber, R., Iacovidou, E., and Purnell, P. (2018). An overview of chemical additives present in plastics: Migration, release, fate and environmental impact during their use, disposal and recycling. *Journal of Hazardous Materials* 344: 179–199. https://doi.org/10.1016/j.jhazmat.2017.10.014.

21 Seltenrich, N. (2015). A hard nut to crack: Reducing chemical migration in food-contact materials. *Environmental Health Perspectives* 123: A174–9. https://doi.org/10.1289/ehp.123-A174.

22 Simoneau, C. (2010). *Applicability of generally recognised diffusion models for the estimation of specific migration in support of EU Directive 2002/72/EC*. https://doi.org/10.2788/85958.

23 Samsudin, H., Auras, R., Mishra, D., Dolan, K., Burgess, G., Rubino, M. et al. (2018). Migration of antioxidants from polylactic acid films: A parameter estimation approach and an overview of the current mass transfer models. *Food Research International (Ottawa, Ont.)* 103: 515–528. https://doi.org/10.1016/J.FOODRES.2017.09.021.

24 Bhunia, K., Sablani, S.S., Tang, J., and Rasco, B. (2013). Migration of chemical compounds from packaging polymers during microwave, conventional heat treatment, and storage. *Comprehensive Reviews in Food Science and Food Safety* 12: 523–545. https://doi.org/10.1111/1541-4337.12028.

25 Golding, A. (1999). Reuse of primary packaging. *Abfallberatung Müllvermeidung Recycl* 5: 103.

26 Garrido, N. (2007). Alvarez del Castillo MD. *Environmental Evaluation of Single-use and Reusable Cups. Int J Life Cycle Assess* 12: 252–256. https://doi.org/10.1065/lca2007.05.334.

27 Singh, S.P., Chonhenchob, V., and Singh, J. (2006). Life cycle inventory and analysis of re-usable plastic containers and display-ready corrugated containers used for packaging fresh fruits and vegetables. *Packaging Technology & Science* 19: 279–293. https://doi.org/10.1002/pts.731.

28 Tua, C., Biganzoli, L., Grosso, M., and Rigamonti, L. (2019). Life cycle assessment of reusable plastic crates (RPCs). *Resources* 8. https://doi.org/10.3390/resources8020110.

13

Recycling

13.1 Learning Objectives

After learning from this chapter, you are expected to:

- Define recycling and the requirements for successful recycling.
- Describe the primary recycled materials in the US and the US states and around the world.
- Explain the main systems for collection of industrial and household recycling items and the main differences between them.
- Discuss the differences between motivation and incentives for recycling.
- Explain the operation of a material recovery facility (MRF).
- Explain the main elements of the cost of recycling.
- Explain the main differences among glass, metal, paper, paperboard, corrugated board, and plastic recycling.
- Discuss the impact of adding recycling content and recycling at the end of life on the environmental footprint of packaging systems.

13.2 Introduction

The US EPA considers recycling as the preferred method for disposal reduction, in contrast to reduction and reuse, which are considered methods of source reduction and are preferred on the overall waste management ladder. Recycling is defined by the US EPA as a "series of activities that includes collecting recyclable materials that would otherwise be considered waste, sorting, and processing recyclables into raw materials such as fibers, and manufacturing raw materials into new products." The concept of recycling cannot be discussed without talking also about recyclable, recovered, and recycled materials. Recycling systems vary across the globe, and implementation of successful recycling is not only a function of collection, infrastructure, and markets, but also consumer engagement and customs. Figure 13.1 shows the recycling and composting rates as a share of the total MSW in the OECD countries in 2018. Germany leads the effort on recovering material through recycling and composting followed by South Korea, and then other European countries such as Austria, Belgium, and Switzerland. The US ranked 18th in this effort. Several reasons can account for the lower recycling rates in the US. Recycling schemes that work excellently in countries such as Germany, Japan, and Switzerland may not be easily translated to

Life Cycle of Sustainable Packaging: From Design to End of Life, First Edition. Rafael A. Auras and Susan E. M. Selke.

Companion Website: www.wiley.com/go/Auras/lifecycleofsustainablepackaging

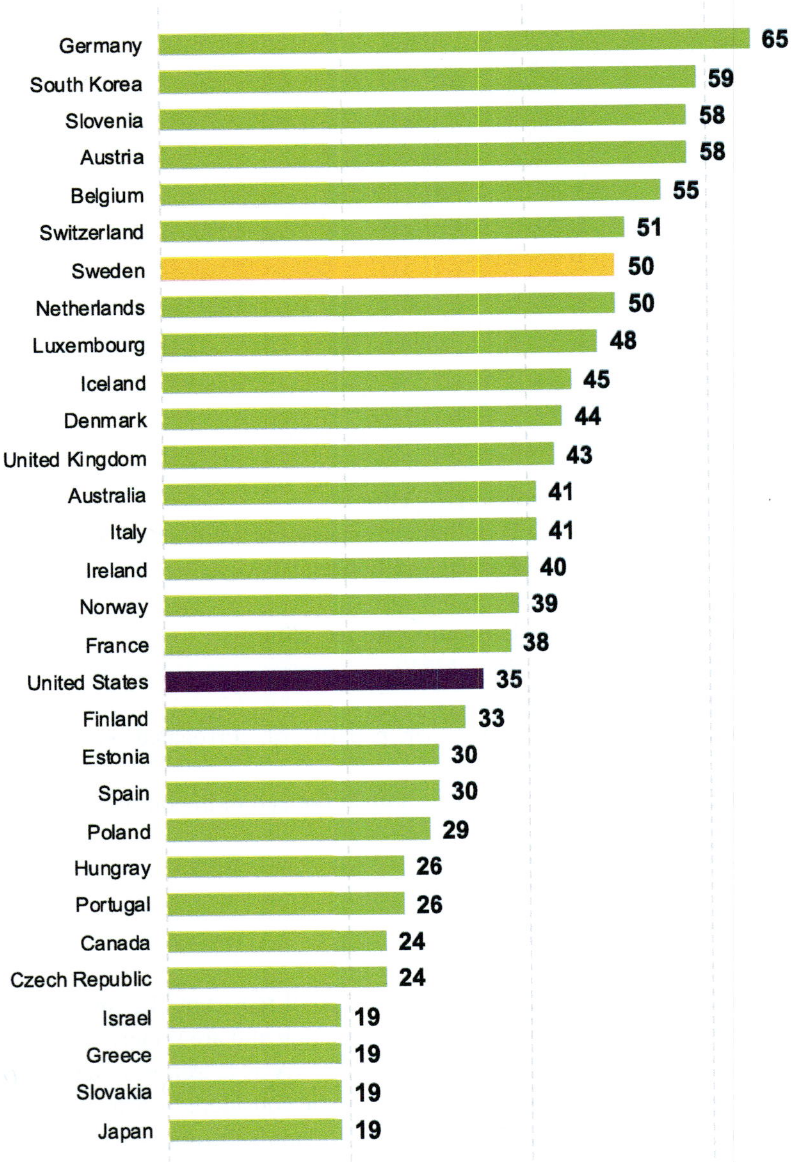

Figure 13.1 Recycling and composting waste rates as a share of the total MSW in the OECD countries in 2018.

the US or other countries. Therefore, several scientific areas of knowledge such as environmental engineering, waste management, consumer behavior, and economics must be engaged to create consistent and successful collection of recyclables and their effective use; these will be discussed in this chapter.

13.3 Requirements for Successful Recycling

Picturing recycling as a simple task of collection of MSW has created reiterative failures in efficient recovery of materials. Recycling happens for three main reasons: (a) altruistic reasons – a general need for protecting the environment and conserving nonrenewable resources; (b) economic imperatives – the cost of environmentally acceptable disposal has been rising, so recycling is becoming favorable, and customers may give preference to companies seen as environmentally responsible; (c) legal considerations – both public demand and governmental requirements due to lack of alternative waste disposal methods, leading to economic and civic penalties and incentives being instituted to encourage recycling. So, increasing support has been given to recycling of materials in general. However, for recycling systems to work, several steps must successfully be implemented and work together. Figure 13.2 shows the basic requirements for recycling systems to be successful: (a) consumer engagement, (b) collection, (c) sortation, (d) reprocessing, and (e) end markets – if any of these components are not included or are not working in a recycling system, low efficiency can be expected for material recovery. In the next section, the components and important aspects of each of these requirements will be discussed in detail.

13.3.1 Consumer Engagement

When we think of recycling as a type of human behavior that we would like to encourage, we can recognize that there are two main factors that determine whether someone will engage in recycling. One factor is *motivation*; the higher the participant's degree of motivation, other things being equal, the more recycling they will do. The other factor is *convenience*; other things being equal, the more convenient recycling is (the less people have to deviate from their accustomed ways to participate), the more they will recycle. Of course, we do not usually isolate these factors from each other, but different approaches do differ in the relative balance between them.

13.3.1.1 Motivation

A variety of motivating factors have been investigated to encourage recycling. One type of motivation is to mandate recycling by law. If people want to consider themselves, or be considered, law-abiding citizens, they must participate in recycling to the extent that the law requires. Legally enforced systems have been established in countries such as Switzerland based on the "polluter pays principle" where the polluter is responsible to pay for the trash [2]. In Switzerland, there are trash policy enforcement officers. So, if citizens do not properly participate in recycling or recycle materials incorrectly, they can be fined for wrongdoing with penalties up to 500 CFR [3].

Other motivators are financial. For example, Rockford, Illinois, ran a system in which each week a household was randomly chosen, and their garbage was inspected. The main goal was to encourage the recycling of paper and aluminum cans. They offered a $1,000 prize if the person whose trash was selected did not place paper or aluminum cans in their trash. If there was no winner one week, the prize increased by $1,000 for the next week. The "garbage lottery" was created because the town of 139,000 in 1986 found that it was not enough to rely on citizens' good intentions. The program continued for a while and city officials reported that three times greater amounts of paper and aluminum were recycled than the year before [4]. However, the program was later discontinued.

Figure 13.2 Requirements for successful recycling; adapted from [1].

Another type of financial motivator is widely used in Europe, and has expanded to Asia, and even to Canada. Under this system, companies must pay a fee on all their packaging to pay for its eventual recycling (or disposal). The fee usually depends on both the type and the amount of the material used [5]. This system is better known as "Extended Producer Responsibility (EPR)"; this approach is used for several categories of waste in the state of Michigan, such as electronics. We will discuss more about EPR in Section 13.3.2.4.

A financial motivator that applies directly to individuals is a deposit system. In this system, generally known as the "bottle bill," the consumer pays a deposit on the covered containers, and then gets that money back when the containers are turned in for recycling. Therefore, the individual has a direct monetary interest in returning the containers for recycling; deposit amounts vary. Further discussion of this system will also be provided in Section 13.3.2.4.

13.3.1.2 Convenience

Recycling systems can be set up in ways that provide a great deal of convenience, or very little. One common categorization of recycling systems targeted at households is curbside or drop-off.

In a drop-off program, the consumer must deliver the recyclables to some location. The location of the drop-off center, availability, times, and access can widely vary, impacting the success rate of these sites. For example, locally, MSU, in East Lansing, MI, offers a drop-off recycling site that is open to the public, 24 hours a day, 7 days a week. Several communities around MSU, including East Lansing itself, also operate drop-off recycling sites. It is common for drop-off sites to accept some materials that are not included in the recycling collection program of those communities.

Curbside programs are inherently more convenient than drop-off programs that are otherwise equivalent. In a curbside program, the consumer merely puts their recyclables at the curb, along with their other garbage. Someone then comes and takes the recyclables away. Usually, collection is done on the same days as regular garbage pickup, which further increases convenience, or on alternate weeks. To increase participation, most communities run a commingled recycling program (i.e., all the recyclables are collected in one container). In some large urban cities such as New York, Buenos Aires, and Amsterdam, curbside programs work more frequently such as a couple of times a week.

Programs targeted at collection of recyclables from businesses obey these same principles but are more hybrid. At MSU, we run a multiple stream collection system inside buildings (i.e., recyclables are separated in different bins – see Figure 13.3). These provide essentially the same level of convenience as a curbside collection system – or arguably even more, since the user does not even need to take the recyclables out of the building. This system allows for collection of recyclables with less contamination. However, when run in communities, requiring extensive separation tends to reduce participation. One approach that is of increasing interest is dual collection, in which curbside recyclables are separated into two streams, typically one for containers and one for paper of various types.

The amount of preparation of the recyclables that is required also has a large impact on convenience, and therefore on participation rates. Virtually all programs require the recyclables to

Figure 13.3 Multiple stream collection system for recyclables at MSU.

be reasonably clean. Some, as discussed, require various degrees of separation of different recyclable materials from each other. A few require that labels be removed, or containers be flattened. In general, such demands in some way facilitate or simplify the recycling process, but they also decrease convenience. Programs are always running a fine line between requirements that are needed for efficient management of the program, and the effects of these requirements in reducing participation.

One of the most effective techniques for maximizing participation in curbside recycling is providing a container for recyclables to be placed in. This does several things. It increases convenience by providing a place for the recyclables to be stored, and an easy way of getting them to the curb. It also can increase motivation by bringing peer pressure to bear. If the bright blue (the most common color in the US and Canada) recycling bin is not out in front of your house on collection day, all your neighbors will know that you are not recycling. There is an increasing trend to provide large, wheeled carts to residents, in place of the original small open bins. This provides a larger space to collect more recyclables, easier transport to the curb, a cover to protect against weather, and allows automatic pick up by the collection truck, decreasing labor costs and increasing worker safety (Figure 13.4).

13.3.1.3 Education/Publicity

The most well-designed program, with a good balance of motivation and convenience, will not work if no one knows about it. It is essential that there be education of the community about recycling opportunities or requirements, and how to participate. One time is not sufficient, as people move into and out of communities all the time, so an ongoing strategy is required. It has been found that one of the most effective ways to do this education is in the local school systems, especially at the elementary school level. Elementary school children tend to be very rule-oriented, and to believe what their teachers say. The hope is, also, that if recycling habits are established early, they will remain for a lifetime. The containers mentioned above (see Figures 13.3 and 13.4) also provide an education function. Their presence in a home serves as a reminder of the existence of the recycling program. When a program is initiated, delivery of the container to every targeted

Figure 13.4 Examples of home collection landfill (black lid), recyclables (yellow lid), and yard waste (green lid).

home helps ensure that everyone in the community is aware of the program. It is pretty easy to overlook a mailed notice, or something in the local paper, but pretty hard to ignore the big brightly colored bin left at your front door. In general, the participation rate can be expected to be the largest for commingled collection, which requires the least effort by the generator of the recyclables.

13.3.2 Collection

Before we discuss collection, three main terms must be defined. Recycling programs often quote *diversion rates*, *participation rates*, and *collection rates* as indicators of their success in recovering household materials. These three terms are related but have distinct differences.

The *diversion rate* (or recovery rate) is a measure of the amount of a material "diverted" from the waste stream. A 75% diversion rate for milk bottles, for example, would mean that 75% of the milk bottles used were sent to recycling, and only 25% were sent to disposal.

The *participation rate* is the proportion of households that "participate" in recycling. Participation is usually defined as diverting some materials from garbage to recycling at some time during a designated period, often one month. The argument is made that counting participation for a single week, for example, would underestimate the true participation, as most everyone misses putting out their recyclables at least once in a while. Further, many programs encourage residents to put out their recycling bin only if it is full or nearly so, to increase the collection efficiency. Generally, one month is considered a reasonable period of time. Participating households rarely divert 100% of their available recyclables. As in the example above, most people do put some recyclables into the garbage due to misinformation or inattention. Therefore, a *participation rate* of 80% will not translate into a *diversion rate* of 80%. The diversion rate will be lower – maybe much lower.

The *collection rate for MSW* is the amount of material collected, as a percentage of total waste generated. Several waste collection service models are used in the US and around the world. Different reports can be found regarding the waste collection rate in the US and around the world, and they agree that the waste collection rate in the US is at least 94%, provided by various collection methods. Figure 13.5 shows the waste collection rates for different regions around the world [6]. However, this number does not provide an indication of the collection rate for recyclables.

A report conducted from 2015 to 2016 specifically for the US indicated that 94% of US residents had access to some type of recycling program. Of these, 73% had access to curbside recycling programs,

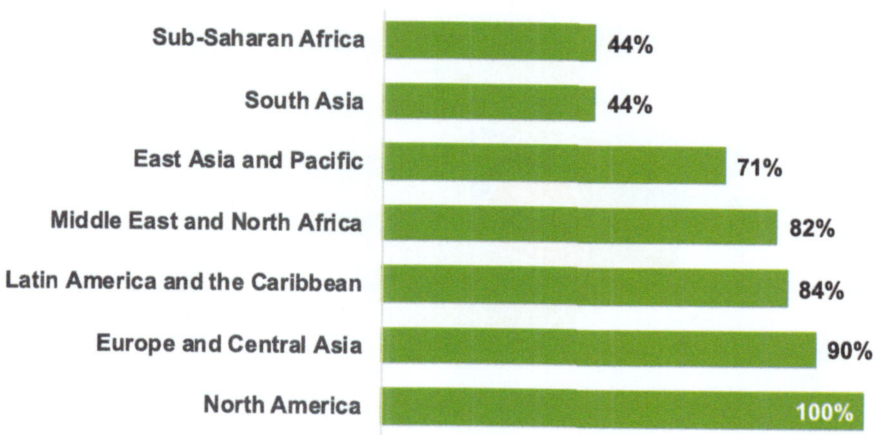

Figure 13.5 Waste collection rates by region; adapted from [6].

and 21% to drop-off programs only. The remaining 6% did not have access to any recycling program [7]. Figure 13.6 shows the availability of recycling programs in the US by geographical region.

Collection of recyclables from household and commercial MSW is a main component of any solid waste management system and is one of the most common services provided by a municipality or county. The four main types of systems for collection of household recyclables in the US and also around the world are: (a) curbside collection; (b) multidwelling collection; (c) drop-off sites; and (d) deposit systems.

13.3.2.1 Curbside Collection

In curbside collection, trucks are used to pick up recyclables outside of households at a predetermined frequency. If street spaces are narrow or difficult to drive, such as in old European and Asian cities, collection is conducted with small vehicles. Some communities without predetermined collection days communicate through a signal, such as a bell or a cellphone app, when the collection truck is coming and available. If containers are provided, as shown in Figure 13.4, higher participation and recovery rates are expected for curbside collection. Curbside collection is most often weekly, in some places the same day as landfill trash collection. Higher recovery rates are obtained when commingled materials are collected.

13.3.2.2 Multidwelling Collection

Multidwelling units and commercial collection are very similar to curbside collection; however, the collected amount and pick up location are different. In this case, the truck must only stop once to collect the recyclables from several dwellings or a commercial facility, reducing distance travelled and fuel consumption, thereby reducing the environmental footprint of the collection

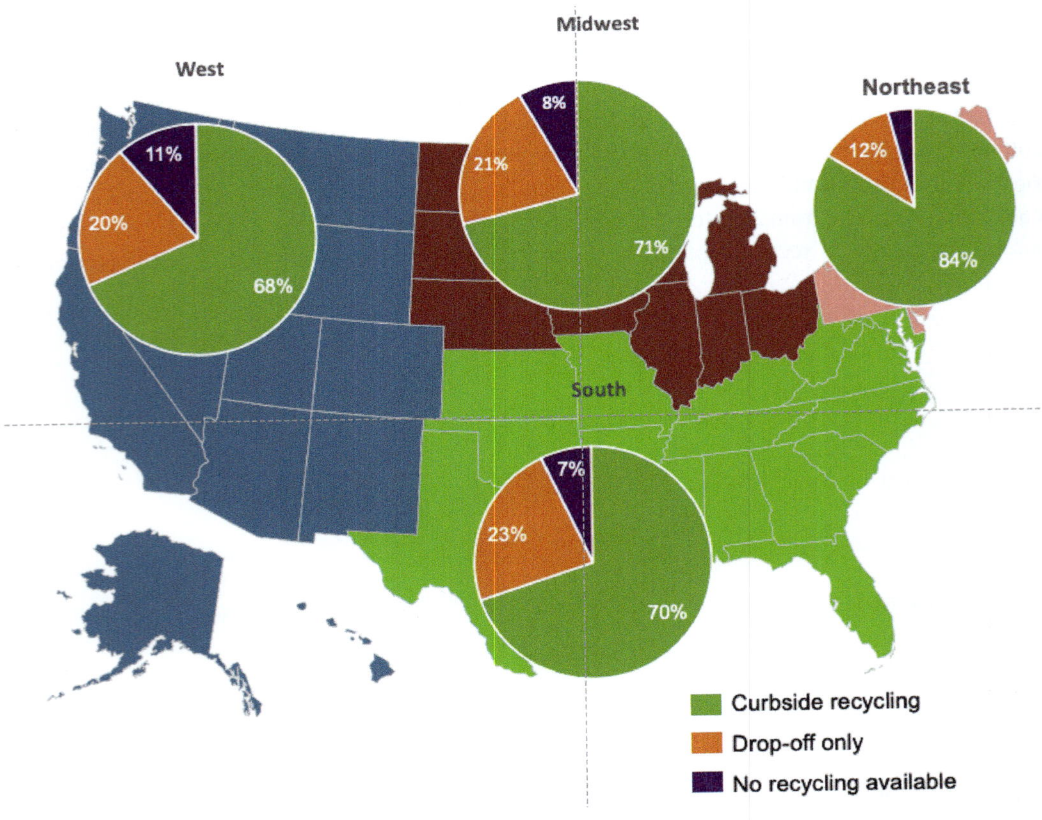

Figure 13.6 Availability of recycling programs by region in the US; adapted from [7].

process. This collection process is well established in major cities around the world, often with daily collection.

13.3.2.3 Drop-off Sites

Drop-off sites are designated locations where the community can leave recyclable materials. Drop-off centers provide a service to a community for cost-effective recycling. Additionally, they can supplement curbside collection, especially for materials that are not accepted through curbside such as, in some locations, glass, wood, and foam. Drop-off collection centers can also provide a recycling opportunity for multifamily dwellings and for those consumers located outside of a curbside collection area or for consumers unwilling or unable to pay for recyclables collection. They can be available 24 hours a day 7 days a week, so they can provide an excellent place to collect recyclables. They are more economical to operate, and they can collect a wider variety of single recyclable materials, which are presorted by consumers from the beginning, so contamination is low (provided consumers follow instructions and do not use the site as a dumping place for garbage). Since they are not as convenient, they do not typically achieve as high participation and/or diversion rates as curbside collection. Figure 13.7 shows some examples of drop-off centers in the US and in Europe.

13.3.2.4 Deposit Systems

A deposit system for recyclables consists of a monetary deposit paid when the product is acquired. Typical examples are tires, automotive batteries, and selected beverage containers and electronics in some states. When these goods are returned or replaced, the deposit is given back to the consumer. However, in many cases the consumer does not see the reimbursement since a new battery or tire is acquired, which also requires payment of the new deposit. Most items covered by deposit systems have a recovery rate of about 90% as long as the deposit fee is meaningful. In the case of automotive batteries, recovery is more than 95%.

Bottle bills: The bottle bill in the US refers to deposits paid by consumers on certain types of beverage bottles, in effect in 10 states as shown in Figure 13.8. The deposit varies between 5 and 15 cents.

Figure 13.9 shows the typical initiation and redemption process in the bottle deposit system. In brief, when the consumer goes to the store and purchases a filled container, they pay a deposit. When the consumer returns the empty container, the money is refunded.

Figure 13.7 Examples of several drop-off sites: (a) East Lansing, MSU; (b and c) Granger facility in Lansing, MI; (d) Denmark; (e) Switzerland.

Figure 13.8 US bottle deposit states.

A number of states have expanded their bottle bills, but no new states have enacted laws since Hawaii passed its law in 2002 (implemented in 2005). However, one US territory, Guam, did enact deposit legislation in 2010. In 2010, Delaware became the first state to rescind statewide deposit legislation, enacting instead a temporary 4 cent recycling fee intended to expand recycling in the state. A report in 2011 said collections of that fee were running more than $220,000 per month less than expected [8]. The fee went into effect on December 1, 2010, and automatically expired on December 1, 2014.

The deposit in most states is 5 cents. Michigan is unusual in having a 10-cent deposit. Several states have raised the deposit amount, or are considering doing so, as that has been found to increase collection rates. The motivational value of 5 cents is not the same now as when these systems were implemented, due to inflation! Oregon's deposit increased to 10 cents in April 2017. Maine has a 15-cent deposit on wine and liquor bottles, and Vermont has a 15-cent deposit on liquor. California has a 10-cent deposit on containers greater than 24 ounces in size. In early 2019, legislation was introduced in Florida that would have instituted a 20 or 30 cent deposit on containers [9], but it did not pass. In most states, the deposit applies to beer and carbonated drinks, but increasingly covers other beverages as well. Maine's deposit applies to all beverages except milk and "unprocessed cider and blueberry juice produced in Maine." Hawaii's deposit applies to all nonalcoholic drinks except milk and dairy products, but only to limited alcoholic drinks (beer, malt beverages, mixed spirits, and mixed wine). Several states include bottled water, but Michigan does not. There are numerous other differences between states, such as some that specifically include liquor (e.g., Iowa), and others that specifically exclude it (e.g., Oregon).

In most states, the consumer can take the deposit container back to any retailer that sells that brand of beverage (actually a bit broader – to any retailer that deals with a distributor who handles that brand), even if they do not carry the particular container involved, or to a redemption center. One drawback is the increasing volume of brand owners' proprietary drinks that can only be

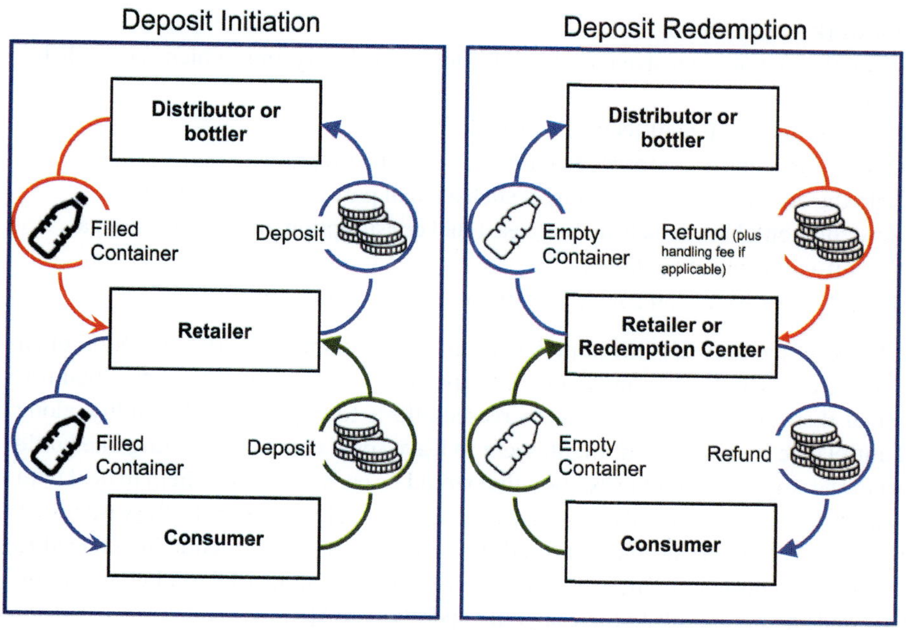

Figure 13.9 Typical bottle deposit system, (left) initiation and (right) redemption; adapted from the Container Recycling Institute.

redeemed at their stores such as Costco (Kirkland™ brand), Walmart (GreatValue™), and Carrefour (Carrefour™). The deposit originates with the distributor. In many states, at least a proportion of unclaimed deposits goes to the state. It is also common for the distributor to pay a handling fee to the retailer, who bears the increased costs associated with handling the returned containers, returning the deposit money, etc.

California's refund value system, which is similar but not quite identical to a deposit, applies to many beverages, although a number have been exempted. In California, retailers are not required to accept returned containers; rather the containers must go to designated redemption centers. In California, the bottler pays the refund value plus a processing fee to the state, which then reimburses the redemption centers for their costs. California's system has encountered significant trouble in recent years, with the closing of many redemption centers. In particular, the sudden August 2019 closing of all of the 284 remaining rePlanet redemption centers (after 191 were closed in January 2016) was a major blow [10].

Many stakeholders such as the aluminum industry and plastic recyclers are supportive of the bottle deposit system since it is claimed that deposits:

a) prevent litter,
b) promote recycling and reduce waste,
c) create jobs,
d) have widespread public support,
e) encourage producer and consumer responsibility,
f) produce high-quality recyclable materials,
g) provide financial incentives for recycling,
h) create additional opportunities for recycling.

On the other hand, several stakeholders, such as the American Beverage Association, and the plastic industry do not favor the deposit system since they state that deposit systems:

a) simply do not work as advertised,
b) cost much more than comprehensive recycling or litter control programs, which accomplish the same goals,
c) penalize and hinder more efficient recycling programs,
d) are inconvenient, particularly compared to curbside recycling programs,
e) do little to help the environment as they target such a small part of the waste stream,
f) create new environmental burdens from increased fuel consumption and greenhouse gas emissions to return and collect containers,
g) impose a hidden, regressive tax on consumers in the form of higher prices.

Historically, the beverage industry, the suppliers of their containers (with the exception of Al can suppliers), and retailers have adamantly opposed deposits [11]. More recently, some organizations have softened their position on deposits. Increased efforts to improve sustainability, along with increased antiplastic packaging sentiment, seem to leave some companies feeling like bottle bills may be part of the way to ward off legislation that would be even harder for them to deal with – or at least they are sometimes talking like they might be in favor [12]. Interestingly, the suspension of collection of deposit containers in many states during the COVID-19 pandemic in 2020 led to statements of support by a number of industry organizations for deposit legislation, in recognition of the value of the collected materials to their ongoing operations [13].

Whether one agrees or disagrees with the deposit fee, it is hard to argue with the fact that containers under the deposit fee achieve greater rates of return for covered containers than any other system. Even though redemption rates have been falling somewhat in recent years, in Michigan, with its 10-cent deposit, redemption rates remain around 90% – an unheard-of rate in most other recycling collection systems. A report on Oregon's bottle deposit system in February 2019 indicated that the increase in deposit amount resulted in a dramatic increase in redemption rates, reaching 90% in 2018 compared to only 64% in 2016 before the deposit was increased [14].

In addition, deposit systems accomplish the primary reason most of them were passed in the first place: reduction of littering of beverage containers. Deposit systems work so well because they provide a powerful motivator – enough to get people to engage in the mild inconvenience of saving up their deposit containers and taking them somewhere that they can be redeemed. Figure 13.10a shows the recycling rate between deposit and nondeposit states for aluminum, PET, and glass, indicating the deposit states recovered larger amounts of material than nondeposit states. A higher number of containers recycled per-capita can be observed for the deposit states (Figure 13.10b), except Delaware.

Many European countries and other countries around the world have deposit systems for bottles, including Australia, Austria, Barbados, Belgium, Canada, Croatia, Denmark, Estonia, Fiji, Finland, Germany, Guam, Iceland, Israel, Kiribati, Micronesia, the Netherlands, Norway, Palau, Sweden, Switzerland, and Turks and Caicos [16].

Extended Producer Responsibility (EPR): EPR is a method in which the producers of a product take financial responsibility, and sometimes operational responsibility, for its end-of-life management. The P can also denote Product, and there are significant differences between these two meanings of EPR. To add to the confusion this creates, the term "product stewardship" is also used, sometimes meaning extended producer responsibility, sometimes extended product responsibility, and sometimes simply the broader concept of minimizing adverse impacts of products and their packaging. However, if any types of EPR programs are effectively implemented, they can radically improve how waste is handled since producers have to finance or manage the end-of-life care of their products, regardless of their disposal stream (e.g., litter, garbage, recycling, etc.). The operational management of the system can either be retained by municipalities or handed off to the producers. Effective EPR programs set recycling targets for producers, which are monitored. The ultimate goal is to incentivize producers to prevent trash and embrace circular economies.

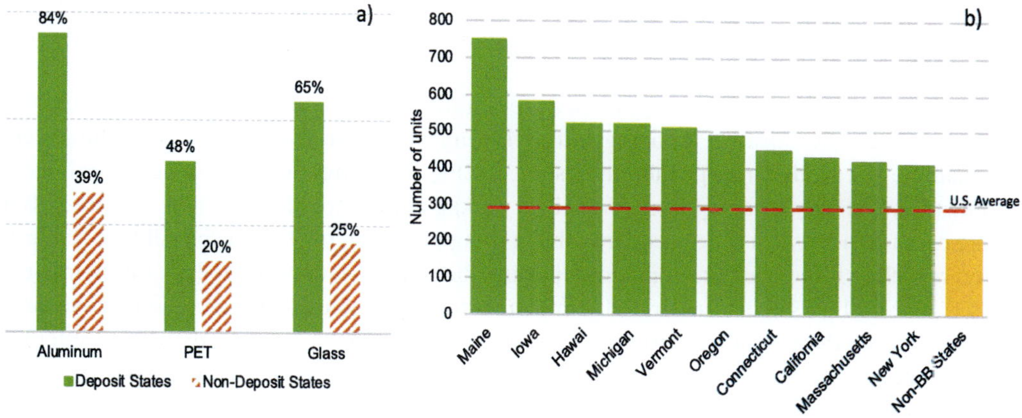

Figure 13.10 (a) Deposit and nondeposit states beverage container recycling rates; (b) per-capita container recycling rates in deposit and nondeposit states – all container types, 2010; adapted from [15].

Well-designed EPR programs transfer cost to producers and encourage them to design out waste. If EPR is implemented at the state or federal level, it can create a level field for competition.

United States: In the US, 118 EPR programs exist in 33 states for products such as electronics, carpets, pharmaceuticals, and paints [5]. The US has not yet adopted any general producer responsibility systems for packaging materials. Information about US EPR and product stewardship initiatives can be obtained from the Product Stewardship Institute [17]. Pesticide containers are currently the only packaging specifically listed on their map of extended producer responsibility state laws [18].

The State of California has adopted a policy on EPR that calls for placing "shared responsibility for end-of-life product management on the producers, and all entities involved in the product chain, instead of the general public; while encouraging product design changes that minimize a negative impact on human health and the environment at every stage of the product's lifecycle" [19]. This law has resulted in the passage of specific stewardship laws for carpet, paint, mercury thermostats, recalled products, auto refrigerants, and pesticide containers. There have been efforts to establish an EPR program for packaging in the state, but nothing has been implemented yet. After a set of bills regarding recycling, recycled content, and EPR failed to pass in 2019 or were vetoed, it is expected that this will be a hot topic for 2021 to 2025 [20].

Oregon became the first US state to pass a paint stewardship law, in 2009 [21]. The states of California, Colorado, Connecticut, Maine, Minnesota, Rhode Island, and Vermont and the District of Columbia now also have paint stewardship laws [22]. Maine passed LD 1631 in 2010 titled "An Act to Provide Leadership Regarding the Responsible Recycling of Consumer Products" [23] that is based on product stewardship. It provided that beginning on January 15, 2021, annual reports were to be delivered by the Dept. of Environmental Protection indicating what products and product categories could be appropriately managed by EPR. Some categories were specifically excluded: motor vehicles, and pulp and paper manufacturers "except conversion facilities for consumer product packaging." The program website does encourage people "to choose products with less packaging" [24]. A number of EPR or related bills were worked on in various states in 2019 [25].

Canada: Ontario and Quebec have adopted a system of fees for packaging and printed material, designed to cover part of the costs of running municipal recycling systems. Ontario's system is run by Stewardship Ontario [26]. The historical pattern of fees charged in Ontario, which vary by material, can be found elsewhere [27]. For 2021, fees range from 38.69 ¢/kg (Canadian) for most plastics other than PET and HDPE bottles to 0.57 ¢/kg for CNA/OCNA newsprint. (These acronyms stand for the Canadian Newspaper Association and the Ontario Community Newspaper Association – newsprint that does not come from these associations is charged a much higher rate, 7.19 ¢/kg.) The next lowest fees are for clear glass at 2.38 ¢/kg. However, Al food and drink cans are given a credit of −6.09 ¢/kg.

Germany: The German Green Dot system, better known as "Der Grüner Punkt" was introduced in 1990 before the European "Packaging and packaging waste directive – 94/62/EC." The system is an example of a producer responsibility system or EPR. It requires and binds all companies and manufacturers using packaging to recover their packaging. Companies that join the Green Dot scheme can use the system to collect recyclable packaging; if not, they are responsible for collecting and recycling it themselves. Companies joining the scheme must use the logo on their packaging (Figure 13.11). Since it was initially implemented in Germany, similar systems have been extended to more than 23 other European countries. The basic concept underlying the system is that consumers who see the logo on the packaging know that the manufacturing company contributes to the recovery and recycling of the packages. The green dot license finances the scheme. Fees vary by material and country.

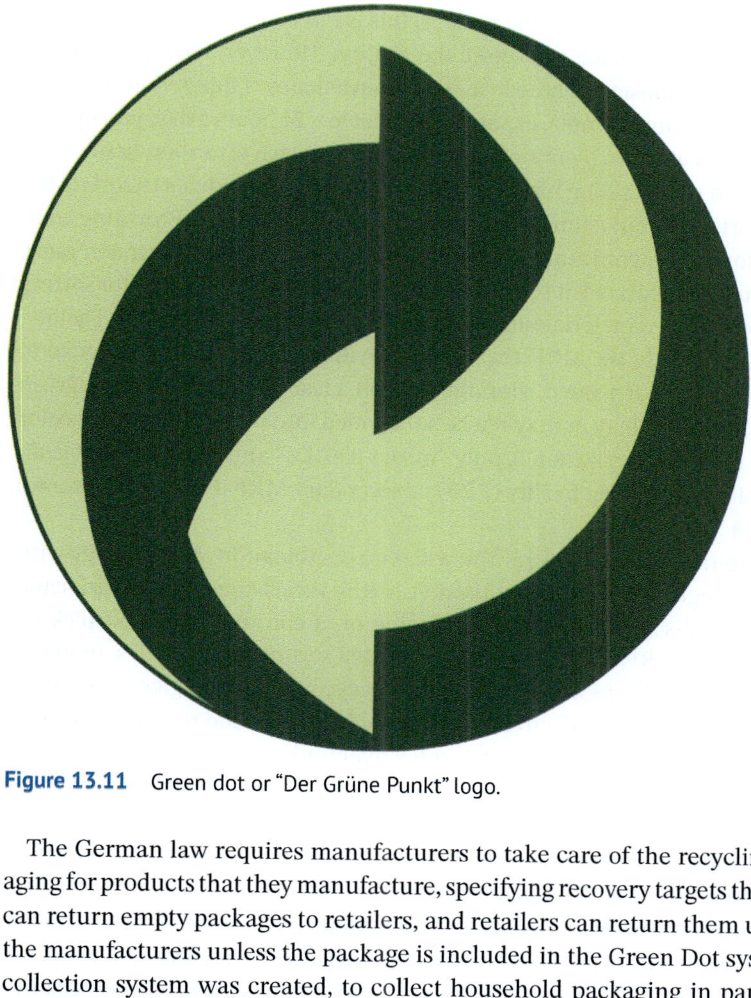

Figure 13.11 Green dot or "Der Grüne Punkt" logo.

The German law requires manufacturers to take care of the recycling or disposal of any packaging for products that they manufacture, specifying recovery targets that must be met. Consumers can return empty packages to retailers, and retailers can return them up the distribution chain to the manufacturers unless the package is included in the Green Dot system. Consequently, a dual collection system was created, to collect household packaging in parallel to the existing MSW. The industry system is operated by the Duales System Deutschland GmbH or DSD. The DSD only collects packaging material from manufacturers that pay the license of the Green Dot Scheme. Consumers separate the packages with the Green Dot symbol into yellow bags or yellow wheelie bins collected by DSD operators. The Green Dot symbol does not indicate that the package the logo is displayed on is necessarily recyclable but that it is collected through the DSD program. The separation, recovery, recycling, and final use depend on the country and market conditions for the materials, as long as mandated targets are met. Retailers often are unwilling to carry products from manufacturers that do not participate in the system, as then they would have to carry the burden of collecting the used packaging and returning it to the next step up the supply chain.

13.3.3 Sortation

For recyclable materials to be useful, they must be separated by the type of material. Obviously, there is not much use for a mixture of paper, glass, aluminum, plastic, etc. In drop-off recycling programs, this is handled relatively simply by having different containers for the different materials. Of course, failure of people to follow instructions and put only the designated materials into the container can be a significant problem. Usually, for understandable reasons, this is more of a concern at unstaffed sites than those where someone is designated to oversee the process. However, having the site be open only when there is staff present considerably decreases the convenience of this option. The local East Lansing drop-off recycling sites, for example, are always available – 24 hours a day, year-round.

When materials are collected, as we have mentioned, a common approach is to allow some or all materials to be mixed together in the recycling bin, rather than requiring individuals to sort all the different materials into categories and put a number of containers out at curbside. Providing commingled collection has been found to significantly increase both participation and diversion rates. Obviously, these materials must be separated at some point. The most common strategy for sorting recyclables is to collect the commingled materials and deliver them to a Material Recovery Facility, known as a MRF (pronounced Merf). In the MRF, the materials will be sorted at least into general categories: mixed paper, newspaper, corrugated, aluminum, steel, clear glass, amber glass, green glass, plastics, etc. The plastics fraction may or may not be subdivided into resin and possibly color categories. Sometimes at the MRF the separation is only "mixed plastics" and separation by resin is done later, at a dedicated plastic recycling facility (PRF) or secondary MRF. Figure 13.12 shows a graphical representation of a MRF.

A video of one MRF can be found in Section 13.13. You will see a combination of technology with considerable hand labor. The degree of technology used at MRFs varies considerably. In some facilities, often small MRFs, hand-sorting of materials passing on a conveyor belt past picking stations is the main method. Other MRFs rely mostly on automated sorting technologies. In many cases, some of both is practiced. Furthermore, innovations in recycling technologies have made the recycling process easier and more cost-effective. Such technologies include reliable detectors and sophisticated decision and recognition software that collectively enhance the productivity and accuracy of automatic sorting of materials, especially plastics. For example, Fourier Transform-Near Infra-Red (FTIR) detectors can run for up to 10,000 hours between faults in the detectors. Furthermore, recently advancements in robotics and artificial intelligence have created impetus for implementation of hand robots to sort comingled materials in MRFs [28].

Recently, there has been some rethinking of the value of commingled collection (single-stream) versus requiring customers to do some separation. Increasing issues with recovery of glass containers from such systems, along with glass contamination of paper streams, have been especially

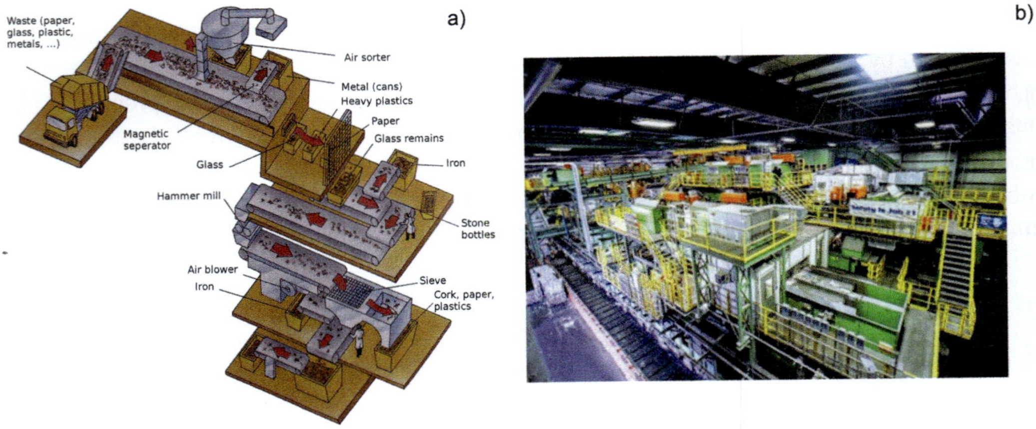

Figure 13.12 **Material recovery facility**: (a) graphical representation by Robin Kerrod reproduced from the public domain, (b) reproduced from Garrett Ziegler, Sunset Park SIMS Material Recovery Facility under CC BY-NC-ND 2.0.

problematic. It may be that we will see return to dual collection to avoid some of the issues associated with single stream – at this point it is too early to tell. Another thing that has happened in some communities is a decision to eliminate recycling of particularly problematic materials altogether – usually meaning glass.

13.3.4 Reprocessing

After comingled or dual stream collection and separation at the MRFs by material type, the materials need to be reprocessed to enter the market. Depending on the type of material that is recovered, several reprocessing steps may be needed. In the case of old corrugated containers (OCC), after collection and removal of contaminants at the MRF, the OCC is delivered to a paper production mill or processor where it is used to produced different paper products such as recycled paperboard, new corrugated boxes, or other paper products. In the case of PET plastic, during reprocessing PET is typically shredded to flakes, washed with water and surfactants, melted, and pelletized. It may also be sent through solid-stating to increase the molecular weight. In the case of metal and glass, after collection and processing by the MRF, they are sent to a metal or glass processing facility for use in new metal or glass products. All these reprocessing facilities are not always close to the communities or MRFs where the materials are recycled. Therefore, networks of primary and secondary MRFs and processing facilities often must be in place to successfully recover these materials. Furthermore, the cost associated with shipping between the collection, MRFs, and reprocessing facilities must not be prohibitive to enable the recovered material to compete with virgin materials. Shipping costs are particularly problematic for low-value high-density materials such as glass.

13.3.5 End Markets

End market demand coupled with supply drives the final price of recycled materials. Generally, low supply and high demand increase price and vice versa. In the case of recycled materials, without strong legislation in place to encourage recycling and incentivize a recycling economy, virgin material directly competes with postconsumer recycled materials.

Not much data is available about the total cost of collection, separation, reprocessing, and final market value of recycled materials. Porter provides some data about the collection of newspaper, glass, aluminum, PET, and ferrous metals (see Table 13.1) [29]. However, these reported costs still do not account for reprocessing and profit before competing with the supply of virgin market materials.

Figure 13.13 shows the price of virgin and postconsumer HDPE and PET resins. We can observe that between 1997 and 2019 at several times the virgin HDPE price was lower than that of postconsumer HDPE due to oil prices and several other factors. Therefore, it is very challenging for postconsumer resins to compete with virgin material since the cost of collecting and cleaning postconsumer plastics do not fluctuate that much.

The information about postconsumer resin provides an example of how several factors must be in place to create a healthy end market for recyclables, among them: (a) create a demand to use recycled materials in end products; (b) have brand owner companies demanding and pulling the recycled materials into new products; (c) put specifications in place for recycled content; (d) remove contaminants in recycled materials, so that end applications are not hindered; (e) provide a robust infrastructure for a consistent supply of recyclables; (f) educate about implementing recyclable materials and their benefits; (g) generate a competitive market where the cost of recycling is not prohibitive [30].

Table 13.1 Estimated costs of recycling various materials (cost in U$D per ton); adapted from [29].

Material	Collection, US$/ton	MRF, US$/ton	Total, US$/ton
Newspaper	72	34	106
Clear glass	60	73	133
Aluminum	581	143	724
Ferrous metals	240	68	308
PET	1,089	184	1,273

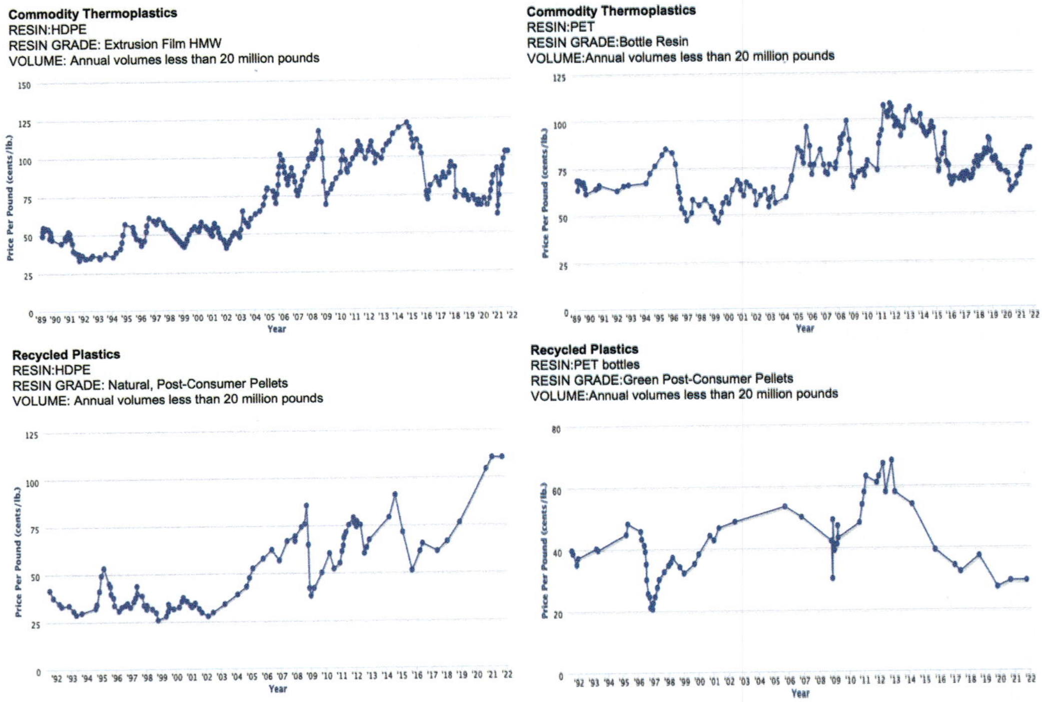

Figure 13.13 Virgin and postconsumer resin prices for HDPE and PET resins; data obtained from PlasticsNews on 10/19/2020.

13.4 Recycling of Packaging Materials

The main materials used in packaging: metal – aluminum and steel, glass, paper and paperboard, wood, and plastics, are recycled at different rates. Let us take a look again at general recycling statistics in the US from the EPA. In the 1980s, the US EPA and consumers were really concerned about the amount of MSW going to landfill. The public was worried that they were running out of landfill space, which was not really the case [31]. At that time, less than 10% of the MSW was recovered through recycling and composting (Figure 13.14). This perception of running out of landfill space incentivized the reduction of MSW reaching landfill by increasing the amount of recovery through recycling. Between 1980 and 2010, the amount of MSW recovered through recycling increased to around 35%. The 2018 recycling rate was 35.2%. There has not been much increase in the recycling rate since 2010; it has remained around 35%. A general consensus has been reached that a substantial increase in recycling will not be achieved by only changes in the system; it will also require an increased consensus and participation by consumers.

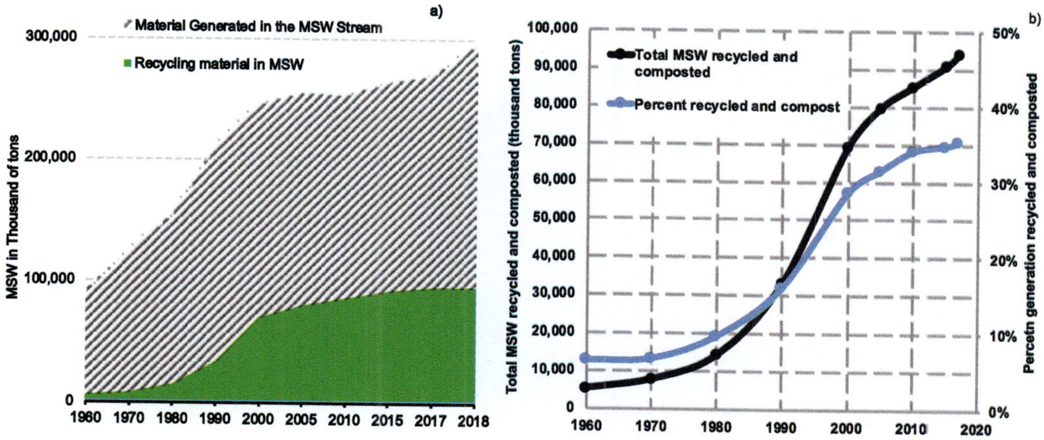

Figure 13.14 (a) Materials generated and recovered through recycling and composting in the MSW since 1960; (b) total and percentage MSW recycled and composted [32].

Figure 13.15b shows the recycling and composting tonnages since 1960 by material. Measured by weight, the most-recycled or composted products and materials in 2017 were corrugated boxes (28.8 million tons), yard trimmings (24.4 million tons), mixed nondurable paper products (9.9 million tons), newspapers/mechanical papers (4.2 million tons), lead-acid batteries (3.2 million tons), major appliances (3.1 million tons), glass containers (3.0 million tons), wood packaging (3.0 million tons), tires (2.6 million tons), food (2.6 million tons), mixed paper containers and packaging (1.3 million tons), and selected consumer electronics (1.0 million tons). These products collectively accounted for 92% of total MSW recycling and composting in 2018 (Figure 13.15a). Since 1960 the largest recycled material has been paper and paperboard (Figure 13.15b). For yard trimmings, several states imposed landfill bans in the 1980s and instituted composting programs, leading to it becoming a significant portion of recovered materials in US MSW.

In 2018, the paper and paperboard recycling rate was 68.2% (45.97 million tons); the yard trimmings composting rate was 63.0% (22.30 million tons); the glass recycling rate was 25.0% (3.06 million tons); the metal recycling rate was 34.1% (8.72 million tons); and the composting rate for food and other organic wastes was 4.1% (2.59 million tons). Table 13.2 shows that recycling (as a percentage of generation) of materials in MSW has mostly increased over the last 47 years. The US EPA in the "America Recycles Summit" in November 2020 announced a new goal to reach 50% recycling by 2030, which will require strengthening markets for recycled materials and providing additional support to regions, states, and municipalities to develop these markets [33]. The main critical action areas are: promote education and outreach, enhance infrastructure, strengthen secondary materials markets, enhance measurement, and cross-cutting actions that support all other areas [33].

13.4.1 Closed- and Open-Loop Recycling

When discussing recycling, regardless of the material, you will often find reference to the terms "open-loop recycling" and "closed-loop recycling." *Closed-loop recycling* (CLR) refers to recycling something into an identical or similar article, such as making new aluminum beverage cans out of old aluminum beverage cans. *Open-loop recycling* (OLR) refers to recycling something into a different type of product, such as recycling steel food cans into a metal beam. Figure 13.16 shows a graphical representation of CLR and OLR. A good example is PET beverage bottles, which can be collected as postconsumer products, reprocessed, and manufactured back into new PET beverage

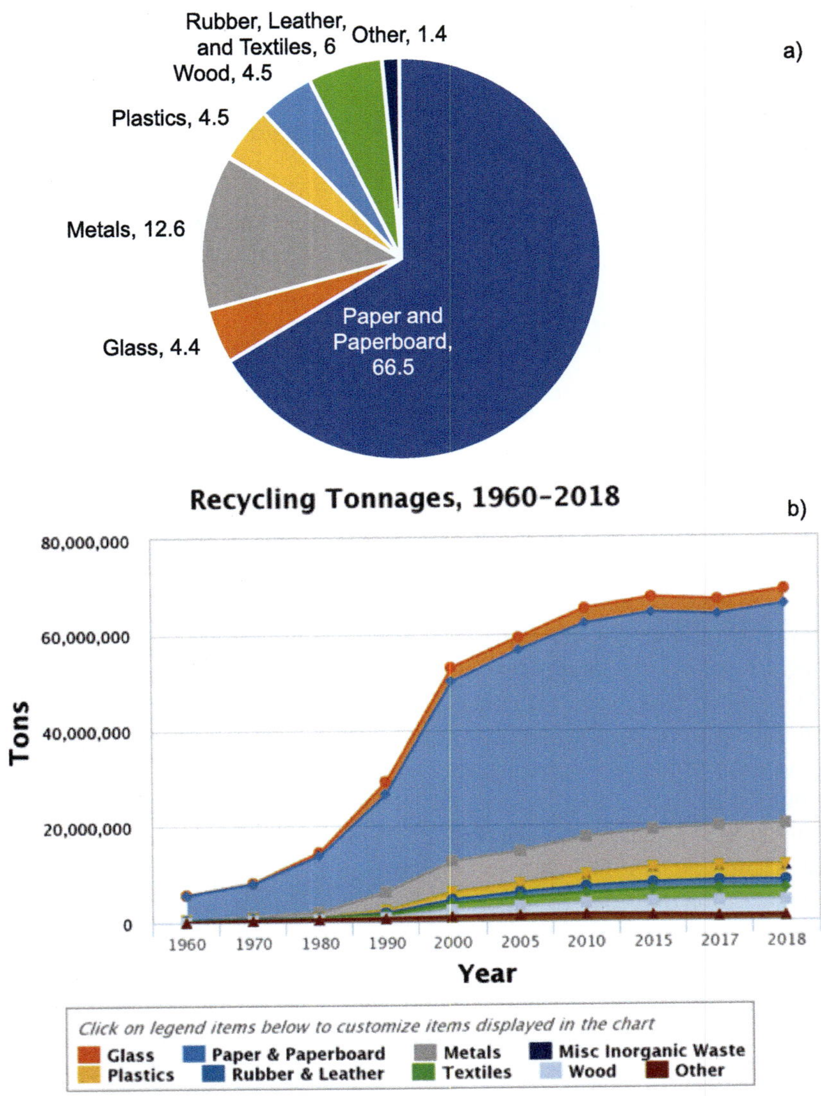

Figure 13.15 Recycled materials in MSW: (a) percentage in 2018; (b) amount since 1960 [32].

bottles (i.e., a bottle-to-bottle recycling process.) Table 13.3 shows common recycling alternatives for packaging materials. The environmental benefit of CLR occurs by using recycled material in place of the same amount of virgin material. OLR, also sometimes called downcycling, involves reprocessing a material into an alternative that is often a lower-value application (Figure 13.16).

Some people feel that CLR is, by its very nature, superior to OLR. They argue that for soft drink bottles the best recycling system is one that continually makes new soft drink bottles from the old ones. However, the position that CLR is inherently better is not really defensible since it depends on many factors such as the environmental footprint of producing similar quality material as the virgin material and also requires that there are adequate markets for the recycled material. To use recycled PET soft drink bottles for making new soft drink bottles and virgin PET for making carpet,

Table 13.2 Recycling as a percentage of generation for the main recycled materials* since 1960.

Materials	Percent of Generation of Each material									
	1960	1970	1980	1990	2000	2005	2010	2015	2017	2018
Recycled										
Paper and paperboard	16.9%	15.3%	21.3%	27.8%	42.8%	49.5%	62.5%	66.6%	65.9%	68.2%
Glass	1.5%	1.3%	5.0%	20.1%	22.6%	20.7%	27.2%	27.8%	25.0%	25.0%
Metals										
Ferrous	0.5%	1.2%	2.9%	17.6%	33.1%	33.0%	34.3%	33.4%	32.7%	33.1%
Aluminum	Neg.	1.3%	17.9%	35.9%	27.0%	20.7%	19.4%	18.3%	15.7%	17.2%
Other Nonferrous	Neg.	47.8%	46.6%	66.4%	66.3%	68.8%	71.3%	64.2%	68.1%	67.3%
Total Metals	*0.5%*	*3.5%*	*7.9%*	*24.0%*	*34.8%*	*34.3%*	*35.3%*	*33.6%*	*33.6%*	*34.1%*
Plastics	Neg.	Neg.	0.3%	2.2%	5.8%	6.1%	8.0%	9.0%	8.5%	8.5%
Rubber and leather	17.9%	8.4%	3.1%	6.4%	12.3%	14.4%	18.6%	18.1%	18.3%	18.2%
Textiles	2.8%	2.9%	6.3%	11.4%	13.9%	15.9%	15.5%	15.3%	15.2%	14.7%
Wood	Neg.	Neg.	Neg.	1.1%	10.1%	12.4%	14.5%	16.3%	16.6%	17.1%
Other**	Neg.	39.0%	19.8%	21.3%	24.5%	28.2%	29.1%	25.2%	21.4%	21.3%
Total materials in products-recycled	**10.3%**	**9.6%**	**13.3%**	**19.8%**	**29.7%**	**32.0%**	**36.6%**	**36.8%**	**35.5%**	**36.4%**
Composted										
Food-composted										
Food^	Neg.	Neg.	Neg.	Neg.	2.2%	2.1%	2.7%	5.3%	6.3%	4.1%
Yard Trimmings-composted										
Yard trimmings	Neg.	Neg.	Neg.	12.0%	51.7%	61.9%	57.5%	61.3%	69.4%	63.0%
Misc. Inorganic Wastes- composted										
Miscellaneous Inorganic Wastes	Neg.	Neg.	Neg.	Neg.	Neg.	Neg.	Neg.	Neg.	Neg.	Neg.
Total – composted	***Neg.***	***Neg.***	***Neg.***	***6.8%***	***25.4%***	***29.9%***	***27.6%***	***29.8%***	***33.8%***	***24.3%***
Other Food Management										
Other Food Management¥										
Food – animal feed										2.9%
Food – bio-based materials/ biochemical processing										2.9%
Food – codigestion/anaerobic digestion										8.3%
Food – donation										7.6%
Food – land application										0.4%

(Continued)

Table 13.2 (Continued)

Food – sewer/wastewater treatment										5.9%
Total Food – other food management										28.1%
Total MSW recycled and Composted-%	***6.4%***	***6.6%***	***9.6%***	***16.0%***	***28.5%***	***31.4%***	***34.0%***	***34.7%***	***35.0%***	***32.1%***
Total MSW recycled and Composted and Other food management-%										***38.2%***

* Recycling of postconsumer wastes; does not include converting/fabrication scrap. Details may not add to total due to rounding.
** Collection of electrolytes in batteries; probably not recycled.
Neg = Less than 5,000 tons or 0.05 percent.
^ Includes collection of other MSW organics for composting.
¥ In 2018, the food waste measurement methodology was expanded to include additional sources of generation and management pathways beyond composting, combustion with energy recovery and landfilling. Please see https://www.epa.gov/facts-and-figures-about-materials-waste-and-recycling/food-material-specific-data.

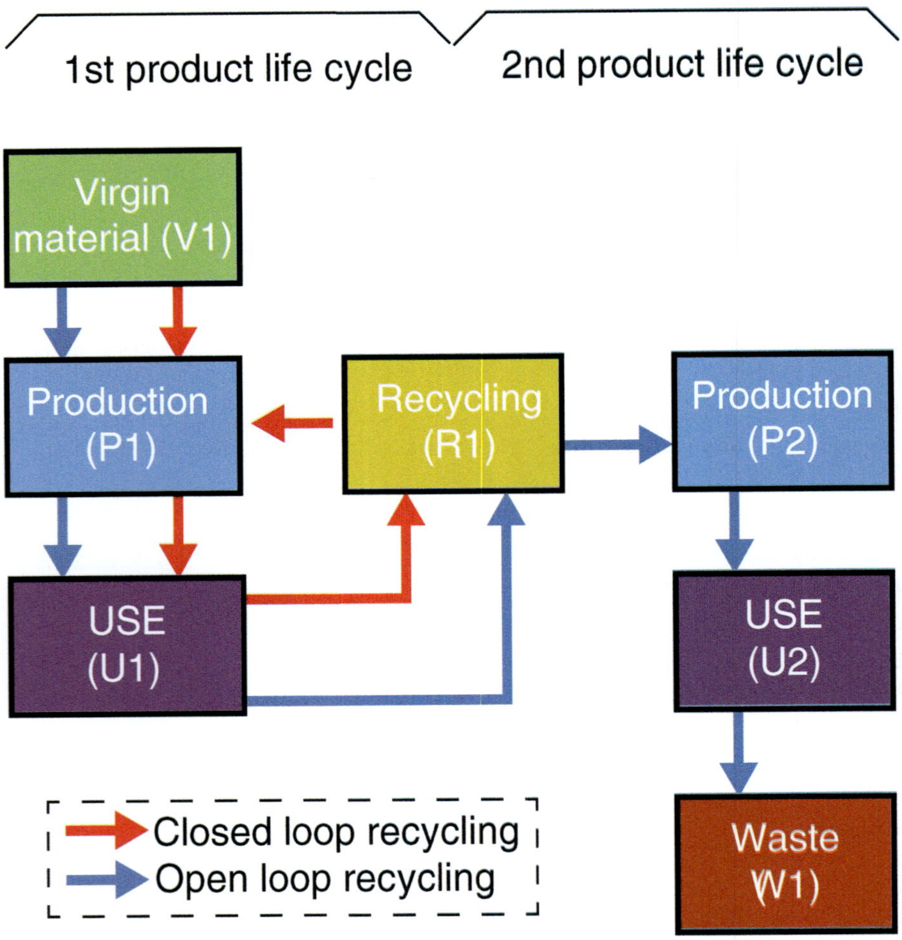

Figure 13.16 Simplified illustration of the processes and material flow in an idealized closed- and open-loop recycling system. Note: the open-loop shows a single second use and then it goes to waste, which is often the case but certainly not always true.

Table 13.3 Common recycling alternatives for standard packaging materials, and barriers for closed-loop recycling of materials.

Material	Closed-Loop Recycling (CLR)	Barriers to CLR	Open-Loop Recycling (OLR)
Polyethylene terephthalate (PET)	Jars or bottles	Quality of recycled PET, contamination with PVC	Sheet, film, or fiber
Glass	Jars or bottles	High cost of transport, mixing of different colored glass	Road base or asphalt
Aluminum	Cans	Minimal	Car component or windows
Steel	Cans	Minimal	Reinforcing rod, pipe, or wire
Paper and paperboard	Boxes, cartons, or bags	Quality of the recycled paper (e.g., length of fiber, color)	Cereal carton, animal litter, insulation, or mulch

rather than using virgin PET for making new soft drink bottles and recycled PET soft drink bottles for making carpet, may not provide any advantages. What makes the real difference is, first, how much virgin material is displaced by the use of recycled material, and second, what are the efficiencies and environmental impacts of the alternative recycling systems.

There are two important qualifiers in the statements that we have made above. First, the statement about adequate markets for the recycled material is crucial. People favoring CLR do have a valid point that unless the material is also recycled from its second use, disposal is simply delayed, not averted. And that brings us to the second crucial statement. We said that the important factor is how much virgin material is displaced by the use of recycled material. Certainly, in the soft drink bottle example, there is much greater potential to reduce use of virgin PET by using recycled content over and over again in the billions and billions of soft drinks bottles we use than there is in reducing use of virgin PET in carpet, perhaps only a single time.

13.5 Metal Recycling

Recycling of metal in the US has been done for decades since the industry recognized that the products that are already produced are a valuable resource for creating more metal products. Metal products present in MSW are mostly in durable goods – ferrous metals (iron and steel in appliances, furniture, tires, and miscellaneous durable), aluminum (in appliances, furniture, and miscellaneous durables), lead in lead-acid batteries, and other nonferrous metals in appliances and miscellaneous durables; nondurable goods – mostly aluminum; and containers and packaging – steel (cans and other), aluminum (beer and soft drinks, other cans, foil, and closures). Metal recycling increased from 1980 to 2000 to around 35% of the generated amount in the MSW; however, since then, the amount of metal recovered has plateaued (Figure 13.17).

Generation and recovery of ferrous materials from MSW collection are the largest percentage of the metal category due to a combination of these metals often being in things like appliances that are source separated, removal from MSW with magnets at landfills and – especially – incinerators, and separation at MRFs. Figure 13.18 shows that ferrous materials represent the largest amount of metal in MSW, followed by aluminum.

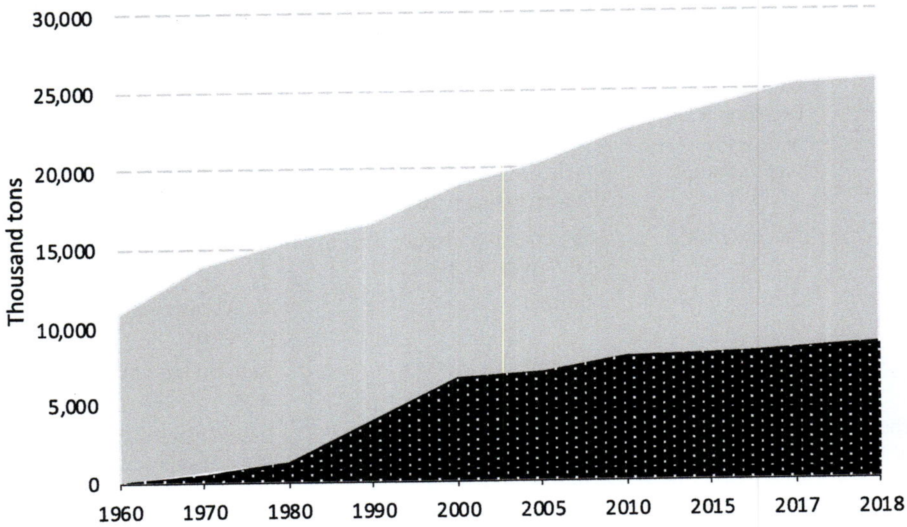

Figure 13.17 Total metals generated and recovered in US MSW since 1960.

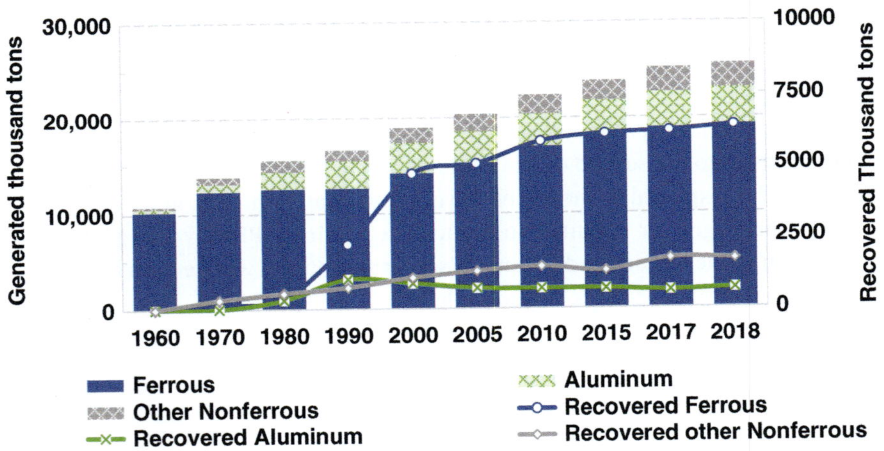

Figure 13.18 Ferrous, aluminum, and other nonferrous material generated and recovered in US MSW since 1960.

13.5.1 Steel Recycling

The US has developed a strong industry for recovering steel scrap. Most of the steel scrap is used in the domestic market, but a large amount is exported. Steel scrap has economic value. It is more economical to make new steel from steel scrap than from virgin materials. The main sources of scrap steel are mills, manufacturers, and postconsumer sources. Current information on steel recycling in the US can be obtained from the Steel Recycling Institute (SRI) [34]. According to SRI numbers as shown in Figure 13.19, the US recycling rate for steel (overall) in 2014 was 86% – down from its record high of 103% in 2009. The rate for steel can recycling was 70%, slightly below the high of 71% in 2011 and 2012. However, it remained below the rates for automotive steel at 85%

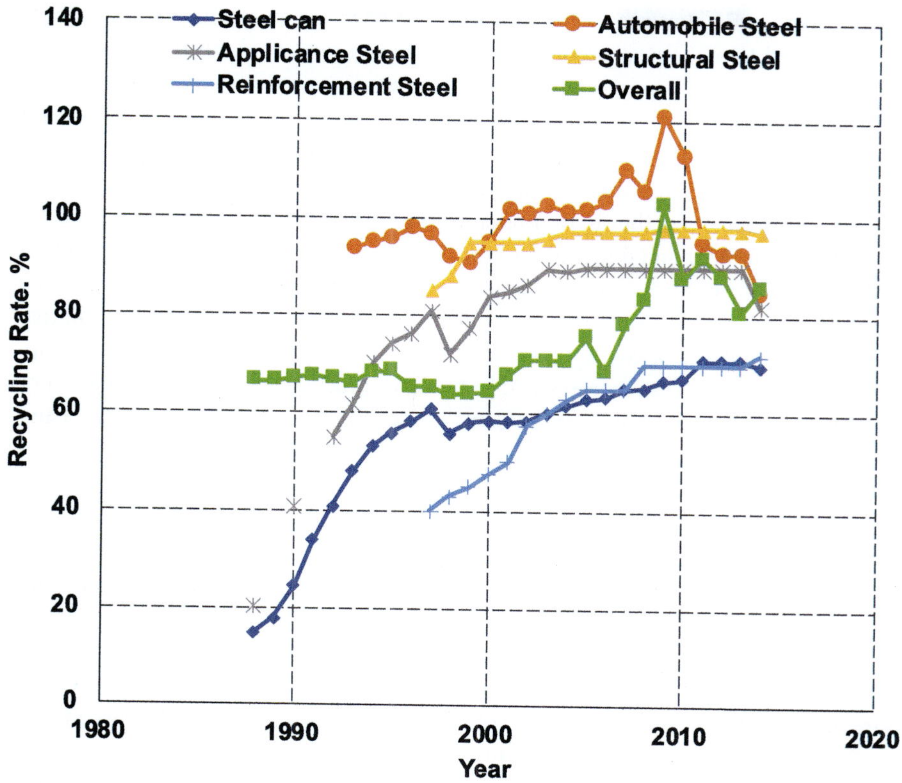

Figure 13.19 Steel recycling rate for can, appliance, reinforcement, automobile, structural, and overall; adapted from [35].

and appliances at 82% [34, 35]. You may be wondering how the overall recycling rate of US steel can possibly be greater than 100% – the answer is that there are backlogs of material that can enter the recycling stream. Since the recycling rate is based on the proportion of materials that enter the solid waste stream, it is possible, by diving into these backlogs, to recycle more steel in one year than entered the waste management stream in that year. Of course, that cannot be sustained indefinitely, as eventually the backlogs will disappear.

Steel is recycled in a predominantly open-loop fashion, with collected steel being directed to the nearest appropriate facility. Steel for can-making is produced in basic oxygen furnaces, which use about 20–35% recycled content.

Steel Recycling Processing and Collection: Steel mills were at one time the best suppliers of steel scrap. Mill scrap or "home scrap," which is leftover pieces from steelmaking and defective or rejected products, was collected and recycled. As new technologies have been developed and steel mills became more efficient in producing steel, this amount significantly reduced. Nowadays, steel mills are highly dependent on outside sources of steel scrap, such as product manufacturers and on postconsumer steel scrap. Manufacturing scrap or "prompt" scrap is steel scrap produced during the production of steel products, such as steel sheet or can-making operations. Postconsumer scrap includes steel collected after consumer use and that has already used for its purpose, such as old appliances and automobiles.

In steel recycling, magnetic separation is the main way to separate steel from nonferrous materials. Scrap merchants sort collected steel into a variety of categories, so that the appropriate "recipe" can be put together for the intended end-use of the recycled material. The material is shredded, melted, and used to make new products. There is waste generated during this process, especially dust and volatiles. In some cases, valuable metals can be recovered from this waste. Steel mills are the largest consumers of out-of-service steel products. The two main steelmaking processes used to recover steel are the basic oxygen furnace (BOF) and the electric arc furnace (EAF) [36]. Modern and integrated steel mills use a blast furnace to reduce iron ore to molten iron; then a BOF consumes scrap steel together with molten iron to produce new steel. In this process, steel scrap is dropped into a hopper and dumped into the furnace. Oxygen is blown into the molten iron which reacts with the carbon to produce steel. The final steel produced by BOF contains around 25% scrap steel. Common products made with BOF steel are appliance shells, automotive parts, aerosol cans, drums, and studs. While BOF uses a combination of iron ore and steel scrap, the EAF process consumes almost 100% scrap steel. In the EAF process, a specific formulation of scrap steel is introduced to the furnace and large electrodes are used to melt the steel scrap using electricity. Products made with EAF steel include structural beams and steel plates.

13.5.2 Aluminum Recycling

Aluminum can be recycled over and over again. According to the Aluminum Association, around 75% of all aluminum ever produced is still in use today. Aluminum can be recycled if it is collected. Figure 13.20 shows the amount of primary production of aluminum has been reduced since 1950 as larger amounts of old scrap have become available. Use of old scrap aluminum is much more cost and environmentally effective than production of virgin aluminum. However, the amount of old scrap is not sufficient to cover the increasing demand for aluminum around the world.

Figure 13.21 shows the main producers of aluminum around the world indicating that China is the main producer of primary aluminum ingots, followed by North America and Europe.

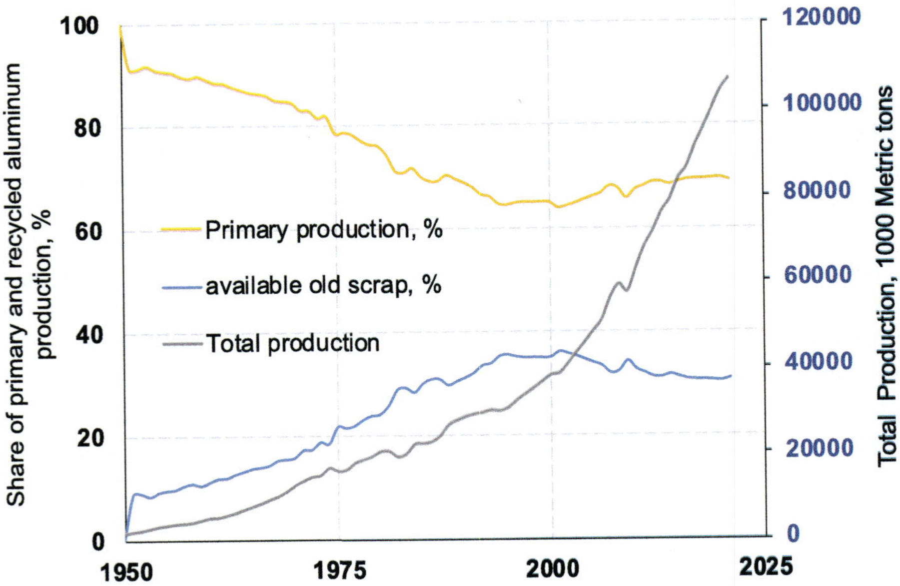

Figure 13.20 Primary and recycled aluminum production share and total aluminum production since 1950 in the world. Values from 2017 to 2021 were extrapolated; data obtained from [37].

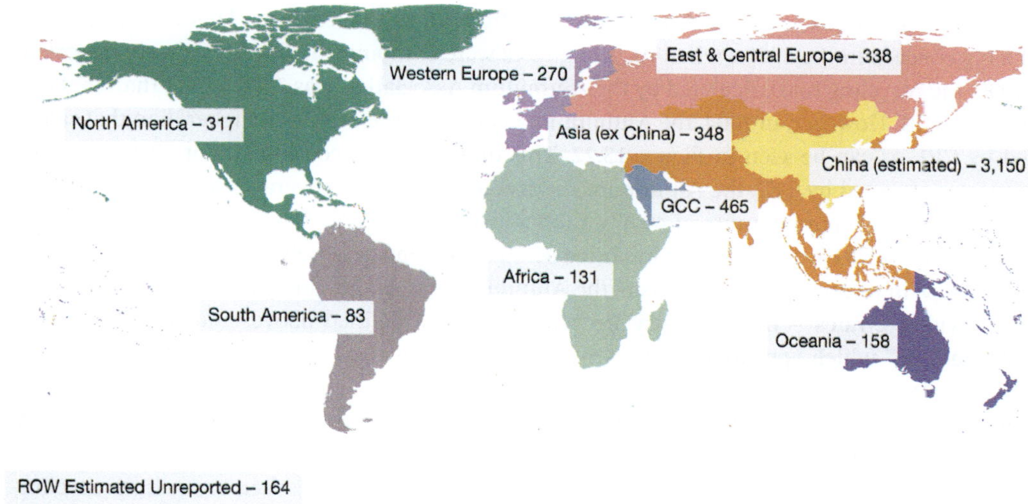

Figure 13.21 Primary aluminum production in different regions of the world; total for September 2020: 5,424 thousand metric tonnes of aluminum; adapted from [37].

Recovering aluminum can be done from postindustrial or postconsumer sources. Postindustrial recycling is scrap aluminum that is left in industrial production. Postconsumer recovery of aluminum can be obtained from aluminum foil, beverage cans, containers, cars, appliances, etc. In recycling aluminum, the material is shredded, primarily to prevent any trapped liquid from entering the furnace. It is then melted. The processing of molten aluminum produces a by-product called "white dross," which is a substance with impurities that contain trapped aluminum. During or prior to the melting process, the lacquers are decomposed by heat and are removed. Proper control over the aluminum dust is necessary to prevent problems such as dust explosions. Between 15% and 80% of the aluminum is present in the white dross and can be effectively recovered by adding the white dross into a bath of molten salt in a rotary barrel furnace. After the removal of aluminum, a residue called saltcake or saltslag remains, containing 3–10% of the original aluminum. The aluminum recovered from the white dross is returned to the aluminum production, reducing the need for primary aluminum.

Aluminum Beverage Cans: Information about aluminum can recycling rates can be obtained from the Aluminum Association [38]. Such information is also available from the Container Recycling Institute (CRI) [39] as well as from the US EPA. The different rates reported have been a point of controversy. In 2005, the Aluminum Association reported, for 2004, the first increase in the beverage can recycling rate since 1997 – 51.2% compared to 50.0% in 2003. In 2006, they reported that the rate for 2005 also rose, to 52.0%. The CRI countered that the rate was only 45.1% in 2004, compared to 44.3% in 2003, and that most of the increase was attributable to the increase in redemption value in California. The difference in these reported values illustrates one of the complexities in calculating recycling rates. The Aluminum Association based its recycling rate on the amount of used beverage cans (UBCs) melted in US facilities, plus exports of can scrap, divided by the number of cans shipped. CRI countered that it is misleading to include imported scrap cans. This is essentially "crediting" the aluminum industry with recovering a higher proportion of cans than it actually does, so CRI excludes these imported UBCs from the rate calculation. This discrepancy was not a big concern in the 1980s, when recycling rates were climbing and few UBCs were imported. In 1990, the recycling rate calculated by the Aluminum Association was 2.7% higher than that calculated by CRI. However, in 2004, the Aluminum Association rate was 6.1% higher than that calculated by CRI, as can be seen in Figure 13.22. This difference has continued. For 2010, the most recent year reported by CRI, the Aluminum Association reported a beverage can recycling rate of 58.1%, while CRI reported a rate of 49.6%, a difference of 8.5%. The EPA determines rates in the same manner as CRI – not including imported scrap cans – and generally reports rates identical to CRI. Perhaps in response to this criticism, the Aluminum Association in its 2015 report for the first time provided 2 rates – the "Industry Recycling Rate" as defined above, and the "Consumer Recycling Rate" which appeared to be essentially the same as the EPA and CRI rates [40]. These rates appear to match the EPA/CRI (Figure 13.22).

Aluminum recycling has a long history because of the impressive energy savings it provides. A widely quoted statistic is that using recycled aluminum saves 95% of the energy that would otherwise be required to make aluminum from ore. This statistic is often misused to say that recycling beverage cans saves 95% of the energy that would otherwise be used – but according to CRI, the savings is about 75%. Where does the difference come from? The savings come from making the aluminum metal; the energy required for the can-making process is the same. So, if you look just at the energy used to make the aluminum sheet, starting with recycled aluminum saves about 95% of the energy, but if you look at the whole can-making process, from raw materials to finished containers, the energy savings is about 75%. In either case, it is substantial, since manufacture of aluminum requires a lot of energy. This creates a definite economic incentive for can-makers to use recycled aluminum as much as is possible.

Other Aluminum Packaging: Aluminum is used in a few food cans, which are likely recycled at about the same rate as steel cans; little direct information is available. Aluminum foil can also be recycled, but its recycling rate is much lower. Thick foil containers can be recycled easily, and

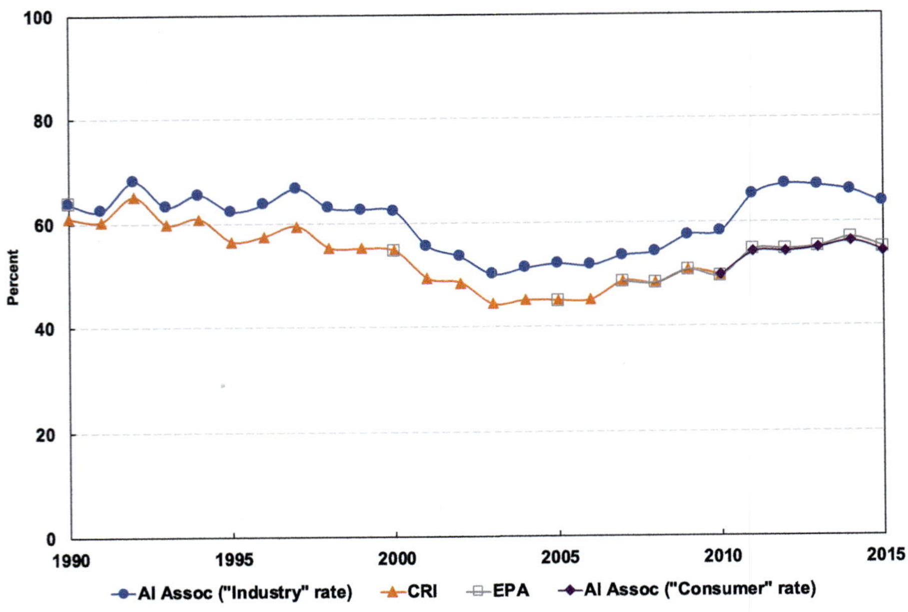

Figure 13.22 Aluminum beverage can recycling rates in the US since 1990.

are accepted in most community recycling programs, mixed in with aluminum cans. Foil contained within multilayer structures is generally not accepted by these programs and has only a very low recycling rate. Aluminum contained in metallized paper or metallized film is not generally regarded as recyclable. Amounts and rates for recycling of nonbeverage-can aluminum packaging materials are contained in the EPA report and are shown in Figure 13.23, along with the rates for beverage cans. Unfortunately, EPA stopped reporting rates for "other [aluminum] cans" or for "foil and closures" beginning with 2010.

For comparison, the UK aluminum drink can recycling rate in 2018 was reported to be 75% and the overall aluminum packaging recycling rate was 52% [41]. For the European Union plus Switzerland, Norway, and Iceland, the beverage can recycling rate in 2017 was reported to be 74.5% [42]. Rates differ by country, as shown in Figure 13.24. A report on global aluminum recycling, reporting data through 2007, can be found elsewhere [43]. The 2017 report stated the global weighted average recycling rate for aluminum beverage cans was 69%.

Aluminum beverage can recycling is a predominantly closed-loop process, as the primary market for UBCs is new beverage cans. In the past, the recovered aluminum went only into the body stock, as the end stock is a slightly different alloy. It is reported that some recycled aluminum does now go into the end stock material, though the proportion is significantly lower than in the body stock.

13.6 Glass Recycling

Glass recycling is almost exclusively limited to recycling of packaging glass. Very little nonpackaging glass is recovered in the US. Figure 13.25 shows the total generated and recovered glass in the US. Since 1980, the amount of generated glass in MSW has decreased from 15,310 thousand tons to 11,380 in 2017. The glass recycling rate was 27% in 2017, with no major increase in glass recovery since 2010.

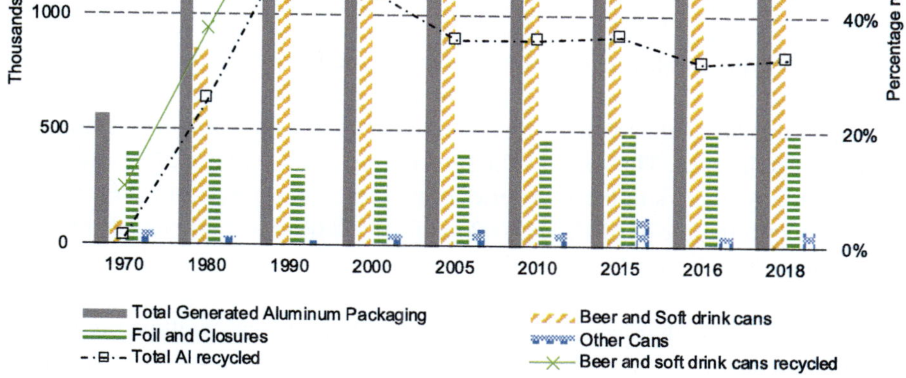

Figure 13.23 Aluminum packaging recycling rates in the US.

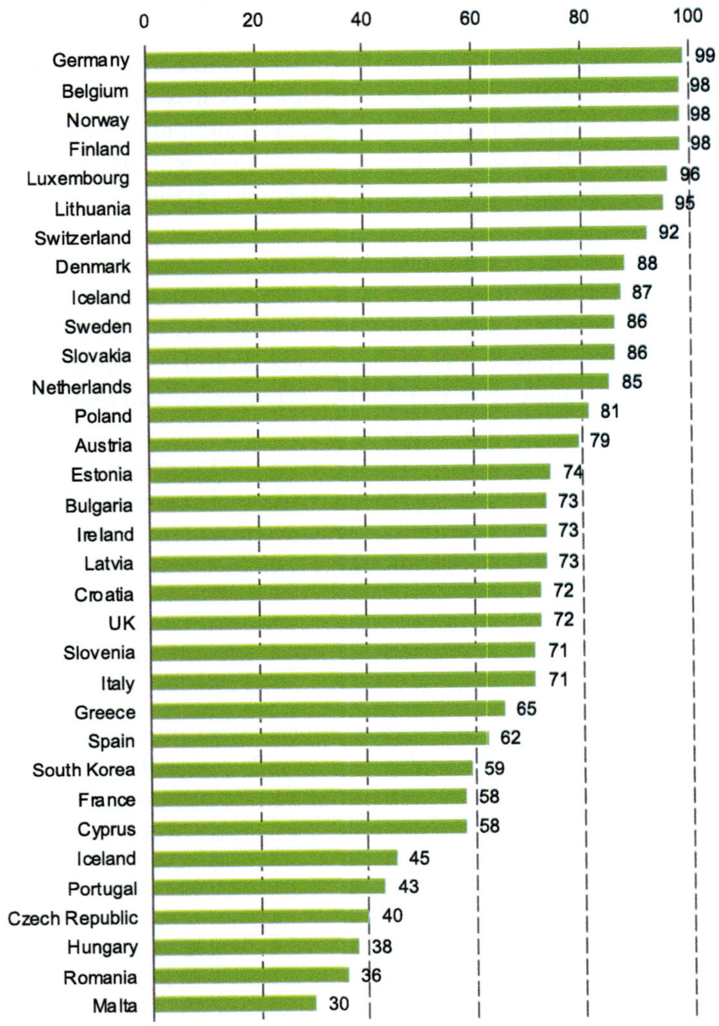

Figure 13.24 Can recycling rates of different European countries (2017, EU28+Iceland, Norway, and Switzerland); adapted from [43].

Figure 13.26 shows the types of glass packaging recycled in the US. The peaks for wine and liquor bottles and for food and other bottles and jars (and the decrease in beer and soft drink bottles) that show up in the 2000 data are likely related more to changes in data sources used by EPA than to a real change in rates. The data for the intervening years between 2000 and 2005 differ dramatically in different years in the EPA series of reports. For example, the 2003 report gives a "food & other" recycling rate of 24.1% for 2002, while the 2006 report gives a rate of only 14.8% for that same category and year. A similar discrepancy occurs in tonnage, with one report of 780 thousand tons and one of 330. The US EPA did not go back and revise the 2000 data like it did for 2001–2004. We suppose that we could have chosen different years to illustrate in the figures, to "hide" this discrepancy, but we think it is useful to illustrate that the "facts" are not always as clear as one might like!

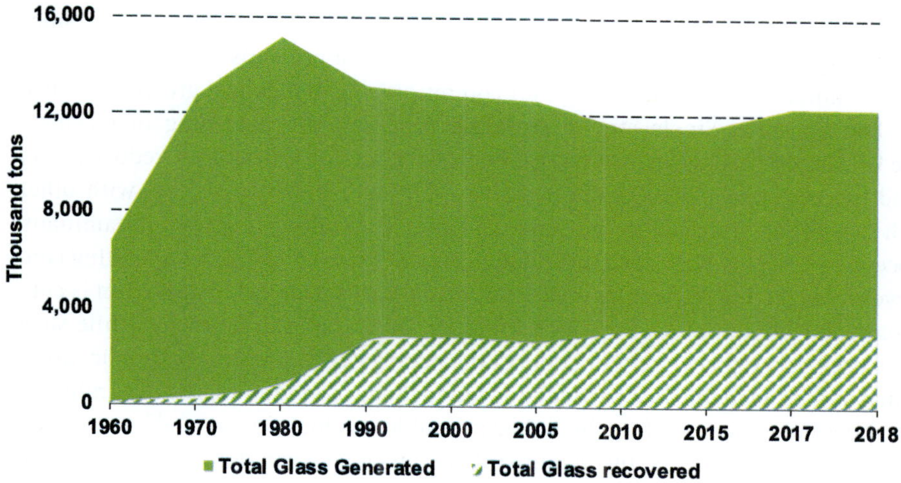

Figure 13.25 Total glass generated and recovered in the US since 1960.

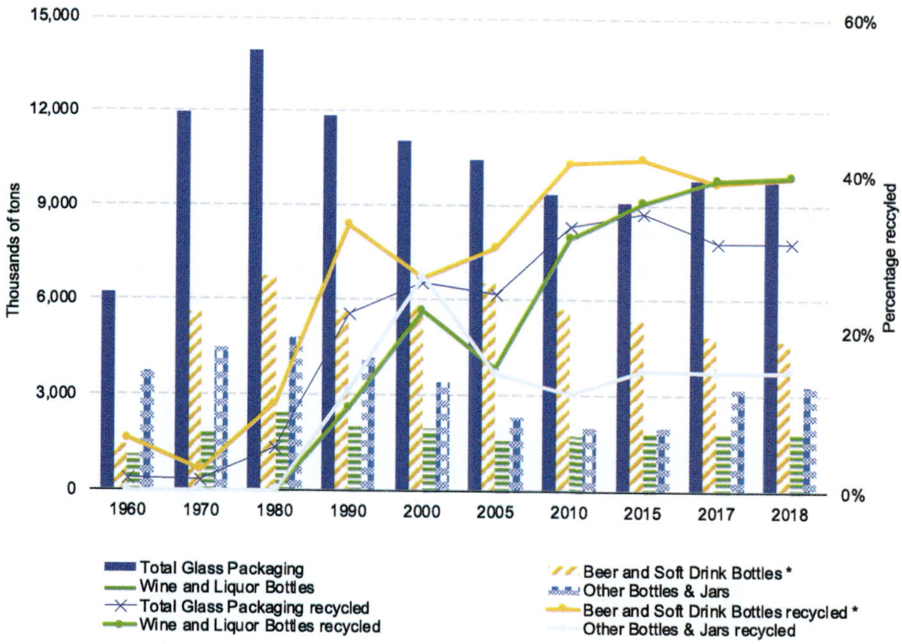

Figure 13.26 Glass packaging recycling amounts and recycling rates since 1960 for different categories. Beer and soft drink bottles are the largest category. No significant amount of glass is recycled from nonpackaging sources. *Beer and soft drink bottles: includes carbonated drinks and noncarbonated water, teas, flavored drinks and ready-to-drink alcoholic coolers and cocktails.

The process for recycling glass back into new containers involves reducing the size of the glass pieces, removing contaminants, and then introducing the cullet into the glass furnace along with virgin raw materials. There is uncertainty about the amount of cullet that can be used in manufacturing new glass containers. The glass industry is much more willing to use postconsumer cullet in recent years than used to be the case 20 + years ago. However, cullet is a relatively low-value material, and the virgin materials it replaces are also low in cost. Since glass is heavy and expensive to transport, opportunities for use of postconsumer glass in making new glass are limited. It is not uncommon for communities to not collect green glass, as it is particularly low in value. Single stream recycling programs are becoming increasingly common in the US. Single stream can lower costs, improve collection efficiencies, reduce injury to workers, and increase diversion. On the negative side, mixing shards of glass with other materials such as paper, introduces an unacceptable, difficult-to-remove contaminant, increasing processing costs for MRFs and mills. Similarly, the glass collected for remelting contains higher levels of objectionable contaminants. In many MRFs a major waste stream is contaminated mixed-color broken glass, for which there are likely no local markets. While such materials can be used as a replacement for some of the aggregate in construction applications, such markets may not be available in a particular location. Some communities have eliminated glass recycling altogether. Because of the high density and low value of glass, and competition with cheap raw material ingredients, the viability of glass recycling can depend quite a lot on where the recycled glass users are located.

Glass Recycling Elsewhere: Glass recycling rates differ dramatically around the world. The glass recycling rate for Europe for 2018 (the most recent available), as reported by FEVE, the European Container Glass Federation, was 76% [44]. Rates per country for 2018 are shown in Figure 13.27.

13.7 Paper, Paperboard, and Corrugated Board Recycling

Paper and paperboard account for by far the largest fraction of materials recycled from municipal solid waste. Figure 13.28 shows the amount of paper and paperboard generated and recovered in the US since 1960. Packaging is both an important source of recycled paper and paperboard, and an important market for these materials, whether originating in packaging or in other paper-based materials.

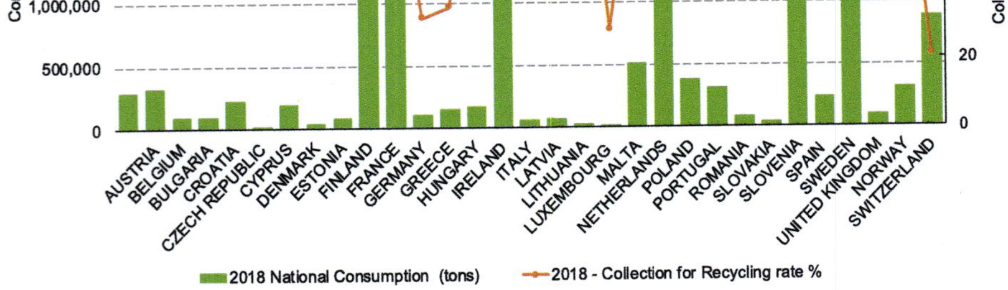

Figure 13.27 Glass container recycling rates in Europe, 2018. Note that these are based on containers collected for recycling, not the actual amount that is recycled [44].

Figure 13.29 shows recycling amounts and rates for paper and paperboard in nondurable goods. Beginning with the 2010 data, EPA no longer separates paper in nondurables into as many categories as previously; rather it lumps newspapers and directories into "Newspapers/Mechanical

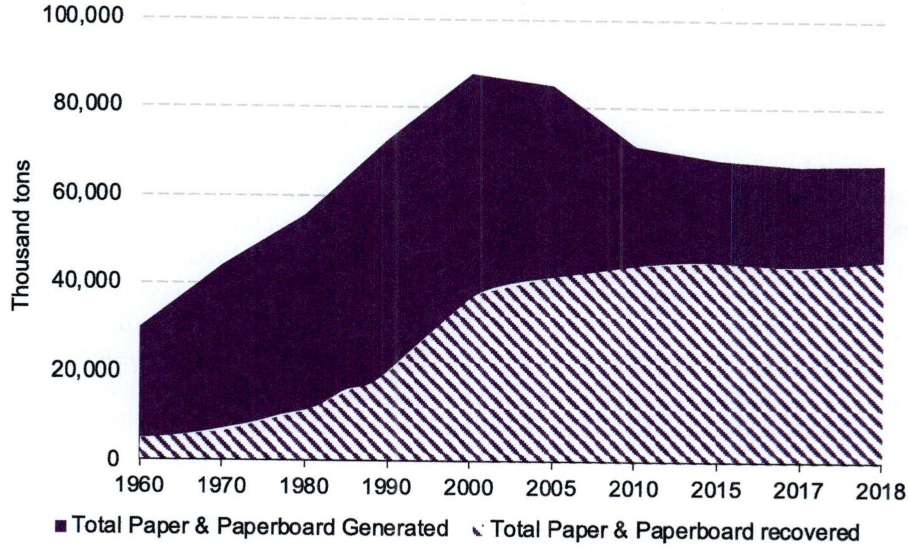

Figure 13.28 Total paper and paperboard generated and recovered since 1960. Note: containers and packaging, which includes corrugated boxes, represented about 50% of the amount recovered in 2018.

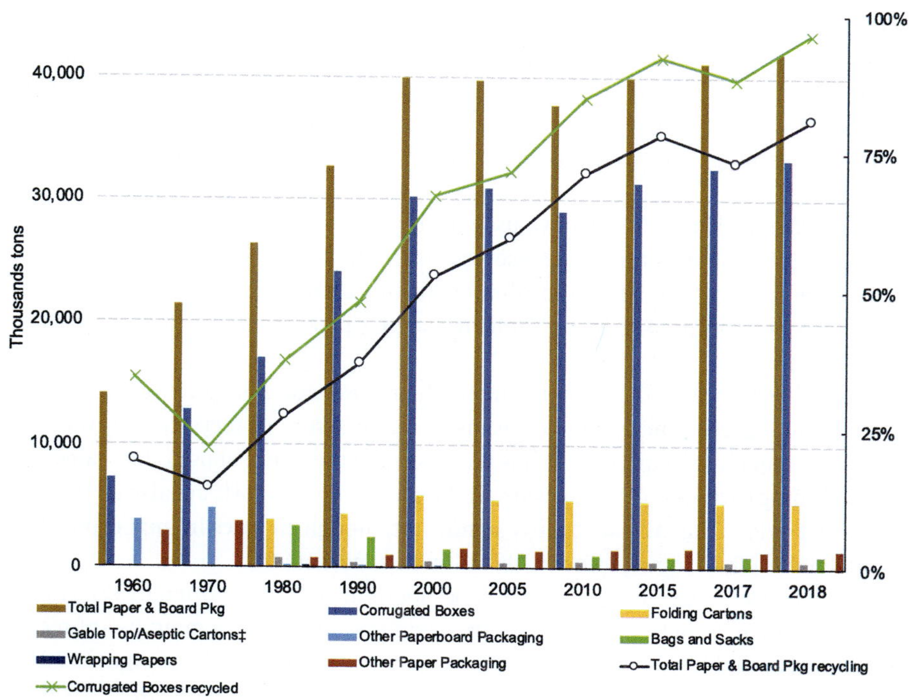

Figure 13.29 Paper and paperboard packaging generated and recycling rates in the US since 1960. Note: Corrugated boxes are the most recycled packaging material.

Papers" and all the others into "Other Paper Nondurable Goods." Also observe the rather steep decline in total amounts recycled and the modest decline in recycling rates (together, a signal of decreasing production) in these categories through 2013.

In the packaging fraction of MSW, corrugated boxes account for the vast majority of recycled paper and paperboard, and also have by far the highest recycling rate – the 2018 rate of 95.5% broke the old record of 92.3% in 2015 (Figure 13.29). The recycling rate for corrugated boxes is higher than that for newspapers and office paper. EPA also changed its categories for paper packaging materials, dividing it into corrugated boxes and "other paper packaging."

The American Forest and Paper Association (AF&PA) provides more recent information about paper than does the US EPA [45]. The AF&PA reported that the overall 2019 paper/paperboard recycling rate set a new high at 66.2%. The AF&PA's goal was to exceed 70% recovery by 2020. Tonnage of paper recycled has not increased to the same extent as the recycling rate, due to falling usage. The tonnage recycled grew substantially in 2019 to 49,176 thousand tons but remained below the record high of 54,325 thousand tons in 2007. AF&PA reported that the recycling rate for corrugated (OCC – old corrugated containers) was 92.0% in 2019, lower than the 2018 rate of 96.2%, which is closest to the EPA values. The tonnage recycled also grew, to 32,195 thousand tons.

For newspapers/mechanical papers, in contrast, AF&PA reported that total recovery in 2019 was 60.9%, a decline in the recycling rate from 77.3% for 2017. Most of this decline was attributed to a drop in exports.

The paper recycling process involves pulping the recovered paper in a hydra pulper, removal of contaminants through various methods, and production of new paper. Various streams of recycled paper may be blended together, or virgin and recycled fibers may be mixed. If printed paper is used for applications that require the paper to be white, deinking is required. Deinking is generally done by flotation, using chemicals to remove the ink from the fiber and cause it to become attached to air bubbles which carry it to the surface of the flotation deinking cell, where it is skimmed off. Loss of fibers during recycling is significant, typically around 10%. Further, the fibers that remain are, on average, shorter. This is the primary reason that recycled paper is generally not as strong as virgin paper. Blending virgin and recycled fiber can enhance the paper strength. Using a higher basis weight of paper can also improve performance. Paper recycling generally operates in an open-loop fashion; downgrading of a portion of the recovered fibers into less demanding applications is commonplace.

Uses of Recycled Paper: Figure 13.30 shows uses of recycled paper in 2018 as reported by AF&PA. The share going to newsprint has decreased substantially – and you may have noticed your local newspaper getting both smaller and thinner. In the 2018 report, AF&PA combined newsprint with other paper (printing, writing, Kraft packaging paper, construction paper and board, molded pulp, etc.).

Old corrugated boxes (OCC) are used predominantly in the manufacture of new boxes (the containerboard shown in the figure). Another major use is in manufacture of recycled paperboard, such as is commonly used in making folding cartons (boxboard, in the figure above). The category "boxboard" includes folding cartons but also gypsum wallboard facings. Exports are also an important market; exports to the Far East are particularly significant, though changes in import policies in China and elsewhere have had some profound effects on this market [46].

Figure 13.30 shows the uses of grades of recycled paperboard in 2007 as reported by the Paper Industry Association Council, which has now been merged into AF&PA. The main use of recycled newspapers used to be making new newsprint. However, that use has declined significantly since the 2007 data. Recycled paperboard is a significant market, and a lot of recycled newsprint is exported. A major export destination is Canada, which is a large manufacturer of both

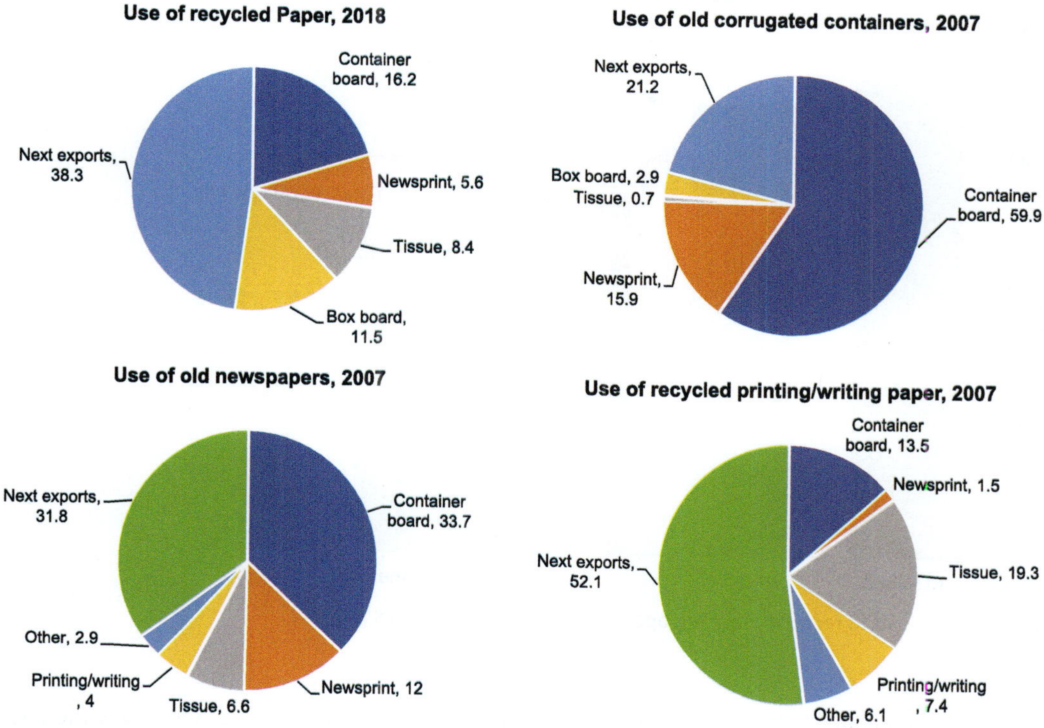

Figure 13.30 Use of recycled paper, old corrugated containers, old newspapers, and printing/writing paper in 2018 in percentage.

recycled and virgin newsprint, a significant amount of which is then imported by the US. Recycled printing and writing paper is also used in manufacture of recycled paperboard, but export is the largest single market. As we can see from the figures above, paper recycling is characterized by a significant mixture of open-loop and closed-loop recycling. It is important to recognize that, in contrast to recycling of metals and glass, the recycling and use history of paper fibers have a significant impact on the strength of the paper made from the fibers. Degradation of the fibers, and hence loss of performance attributes, is to a significant degree unavoidable in recycling of paper.

Paper Recycling in Europe and Other Areas: The European Recovered Paper Council, in their Monitoring Report 2019, reported a 2019 recycling rate for paper in Europe of 72%. The rate for paper and paperboard packaging was 82.1%, compared to 79.5% for steel packaging, 74% for glass packaging, and 73% for aluminum packaging. Fifteen European countries had recycling rates of greater than 70%, while seven countries were under 60%. CEPI (Confederation of European Paper Industries) reports an overall paper recycling rate in Europe of 72 % for 2019 and had a goal of reaching 74% by 2020 [47]. Figure 13.31 also reports the recycling rate of paper and paperboard around the world, showing that Europe is leading the recovery of this material, followed by North America. Villanueva and Wenzel reviewed several LCA studies considering the disposal of paper and reported a broad consensus on recycling paper as the preferred waste management technique with lower EFP [48].

Contamination Issues with Recycled Paper: Another way in which paper recycling differs from that of metal and glass is the issue of contaminants that may migrate into products packaged in recycled materials. For glass and metal, this is not much of a concern. Any organic contaminants in those materials would be destroyed by the very high temperatures involved in melting and

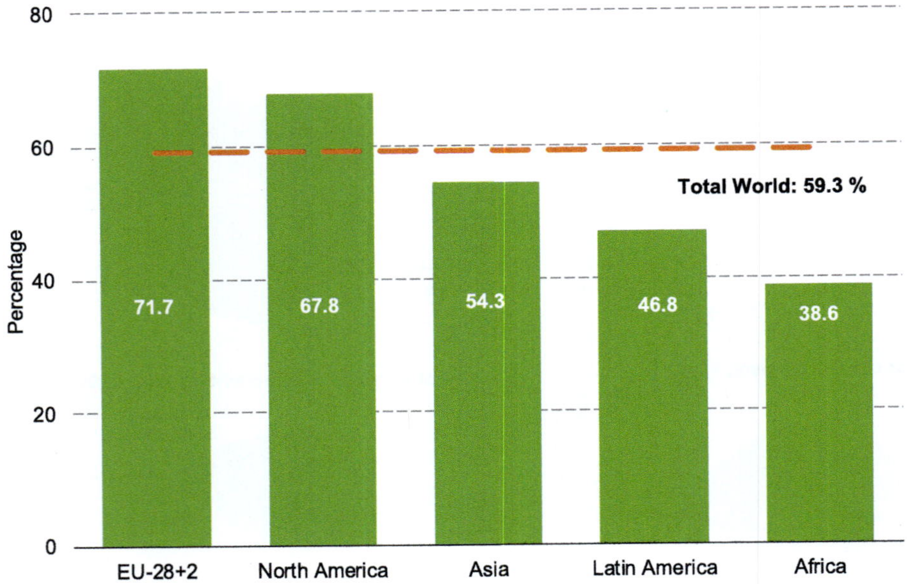

Figure 13.31 World recycling rate of paper around the world in 2018 [48].

reforming the materials. There are potential concerns about contaminants such as heavy metals, but these are typically addressed by controlling the source and composition of the starting materials. For paper, it is not that easy. Microbiological contamination of recycled paper is typically not a concern, as the time/temperature combinations encountered during the drying process will kill most organisms. However, these temperatures are not enough to drive off or destroy various types of other contaminants, and these can potentially migrate into the package contents, including foods. This issue hit the press in 2010, when concerns were raised in Germany about mineral oil in cartons made from recycled paper [49–51]. There are reports that food companies changed some of their practices in use of recycled paper for food packaging, as a result in particular of the mineral oil concerns. Of particular concern is that there is evidence of migration through film liners, in some cases. One major tactic in dealing with this problem is to either change the inks used in printing newspaper, or to remove newspaper from the recycled paper mix. In the US, mineral oil in recycled paper has not been much of an issue. Most US newspapers use vegetable oil-based inks (often soy), rather than mineral oil.

Mineral oils are certainly not the only possible migrant from recycled paper. In 2006, there were reports of contamination of dry foods with trimethyldiphenylmethanes from recycled paper [52] and in 2004 of fluorescent whitening agents. Lately, the presence of per- and polyfluoroalkyl substances (PFAS) have impacted the packaging community since PFAS are highly persistent chemicals, better known as "forever chemicals," that are used to provide water and grease-resistant food packaging [53, 54]. Of course, the FDA also has regulations, as do agencies in a variety of other countries [55, 56]. Generally, dry non-oily foods are less susceptible to contaminant migration than liquids or oily foods. That is the reason that the studies above have concentrated on dry foods – direct contact between recycled paper and food products is generally not allowed for liquids or oily foods, so those situations did not arise. Finally, migration of non-intentionally added substances (NIAS) as a result of reaction by-products, degradation, or oligomers present in recycled paper used for food contact materials (FCM) has been explored, and the main conclusion is that NIAS are mostly derived from printing inks, adhesives, sizing agents and surface coatings [57].

13.8 Plastics Recycling

As is the case for paper, plastics recycling is predominantly an open-loop process, but as with the other materials, recycled plastic can be obtained from primary recycling or secondary recycling. Plastic primary recycling refers to the "in-plant" recycling of scrap material which is a clean uncontaminated single-type waste. Material is used in the same product. Plastic secondary recycling refers to postconsumer plastic, which is rarely suited for primary recycling; mostly coming from handling of mixed plastic streams (commingled plastics). Figure 13.32 shows the total amount of plastic generated and recovered in the US since 1960. Recycled plastics typically go into a variety of end-uses, which are often very different from their original use.

Most recycled plastics comes from packaging, as can be seen in Figure 13.33. Recycling rates for plastics packaging are also higher than recycling rates for most nonpackaging plastics. The exception is PP battery cases, which have a 90 +% recycling rate. Plastics in durable goods had an overall recycling rate of 6.6% in 2015, and plastics in nondurables a rate of 2.2%, while the rate for plastics packaging was 14.6%, and the overall plastics recycling rate was 9.1%. There are significant differences in rates for specific types of packaging plastics, as shown in Figure 13.33.

To a large extent, plastic packaging recycling used to be limited to bottles and jars. Recycling of other plastic containers was limited but growing. A significant fraction of stretch wrap was recycled, but there was little recycling of other packaging films. Plastic bag recycling opportunities had been reasonably plentiful in the early 1990s but had later largely disappeared. This picture has changed quite a bit in the last few years, as we will discuss, but the majority of recycled plastics still comes from bottles.

The main bottle resins are high-density polyethylene – HDPE, used in milk bottles, laundry detergent bottles, etc., and poly(ethylene terephthalate) – PET, used in soft drink bottles, mineral water bottles, juices, etc. Therefore, these two resins account for the majority of recycled plastic, and have the highest recycling rates, as shown in Figure 13.33. The recycling rates for plastics packaging are generally higher than the rates for plastics in MSW, reflecting the preponderance of packaging in the recycled plastics stream. Figure 13.33 shows the recycling percentages by resin reported by EPA.

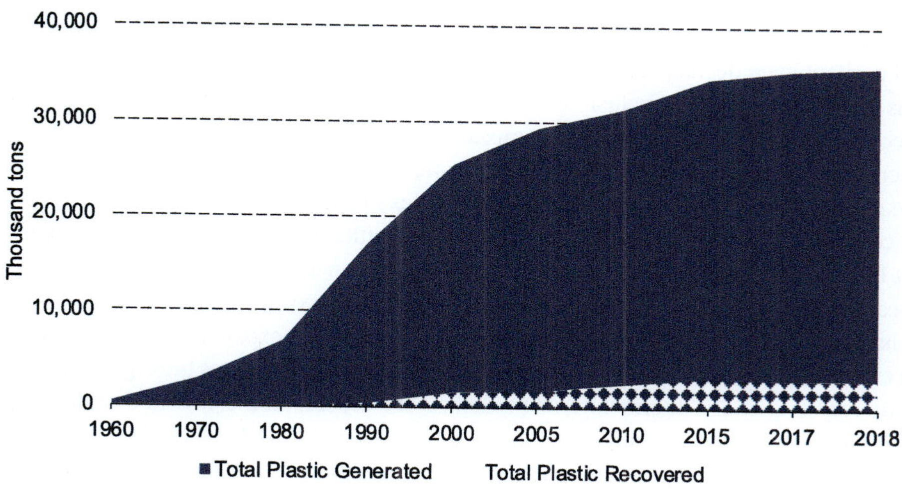

Figure 13.32 Total plastic generated and recovered in the US since 1960.

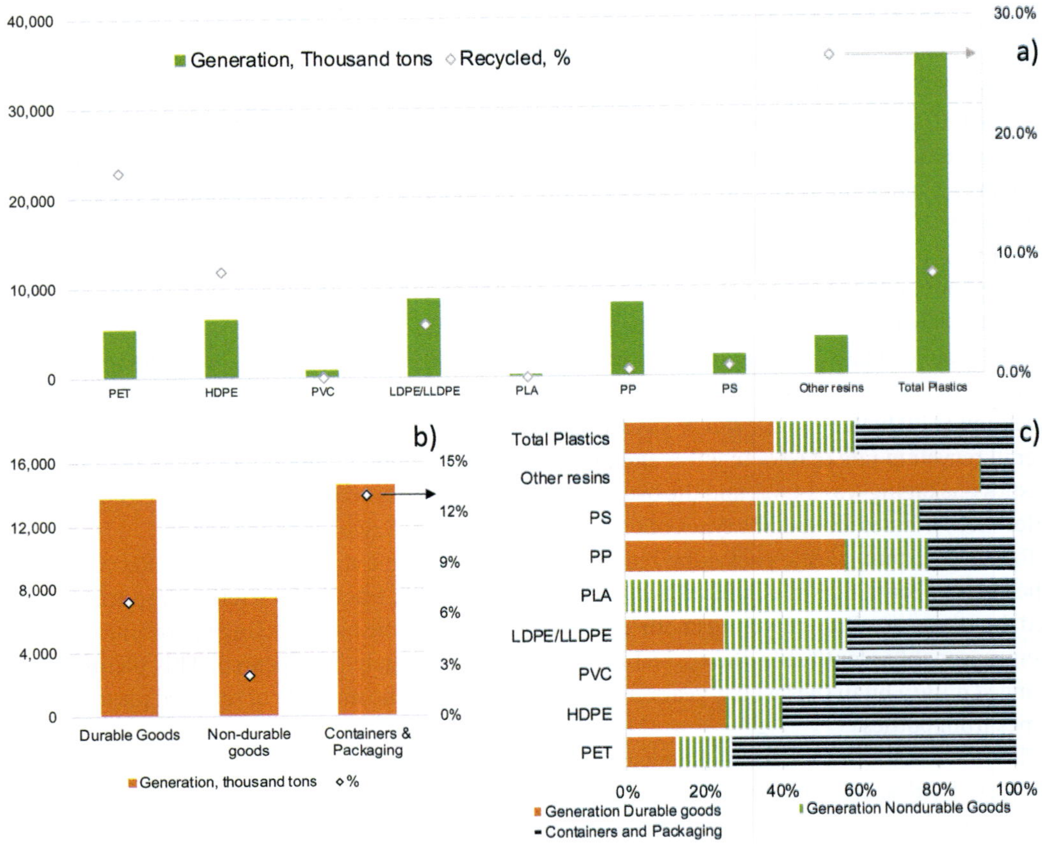

Figure 13.33 Plastics generation in MSW in the US (a) generation and recycling by resin, (b) plastic generation and recycling in MSW by product category, (c) generation of durable goods, nondurable goods, and containers and packaging in 2018.

For a number of years now, the American Chemistry Council (ACC) has been promoting the "all plastic bottle" idea for curbside collection programs, rather than the previously common arrangement of collecting PET and HDPE bottles only. Some studies have found that asking for "all plastic bottles" significantly increases the collection of the desired PET and HDPE bottles, while not increasing, and perhaps even decreasing, the amount of contamination with other types of bottles. Some caution is warranted in these conclusions, however. The parallel movement to single-stream collection of recyclables (all recyclable materials mixed together) has been cited as the probable cause of the increase in contamination levels of baled plastic bottles that has resulted in decreased yields of usable material (see ACC and NAPCOR reports, for example [58, 59]).

Increasingly, communities are going beyond the collection of bottles to accepting a wider variety of types of plastics. However, the types of plastics accepted can fluctuate depending on the market for them. For example, locally MSU collected "all plastics"; however, this changed, becoming more restrictive. The same situation happened in the city of East Lansing. It should be noted that expanded polystyrene was not accepted in either program. The NAPCOR report on PET recycling also discusses the organization's efforts to increase recycling of PET thermoformed packaging and reports an increase in thermoform recycling in the US and Canada to 90.2 million pounds in 2017, compared to 82.4 million in 2016.

ACC did a study on access to recycling in the US, published in May 2011 and updated in 2013 [58]. They reported that 94% of the US population had access to PET and HDPE bottle and cap recycling, either through curbside or through drop-off programs. The population with access to recycling for rigid plastic containers (other than bottles) had grown to 60%.

The most recent ACC report on recycling of other rigid plastics is the 2018 National Postconsumer Non-Bottle Rigid Plastic Recycling Report [59]. It includes nonpackaging as well as packaging. They reported at least 1.35 billion pounds of such plastics recovered in 2017, a decrease of 7% compared to 2016. US purchases increased by 2%, but there was a 33% decrease in exports. ACC reports that the majority of the material collected was polyolefins, especially PP.

A similar ACC report was issued on plastic bag and film recycling for 2018 [60]. It reported a total of about 1 billion pounds of postconsumer bags and film recycled, up by 54% from 2005 – but down substantially from 2016. Another Moore report for ACC on access to film and bag recycling reported that in 2010, 93% of the US population had access to plastic bag recycling within 10 miles, and 74% had access to plastic film recycling within the same distance [60]. With the decrease in the export market, this may have fallen. Several organizations such as Plastic Film Recycling provide information for both consumers and businesses to increase recycling of these materials. Included in the Plastic Film Recycling website is a link where consumers can search for a drop-off location by zip code [61].

The EPS Industry Alliance (EPS-IA) (formerly the Alliance of Foam Packaging Recyclers, AFPR) in 2019 issued a report on EPS recycling for 2018 [62]. They reported 113 million pounds of EPS recycled in 2018, with 39.7 million pounds being "post-use" and 74.1 million pounds postindustrial (compared to their report of 127.3 million pounds of EPS recycled in 2013, including 72.8 million pounds of "post-commercial and post-consumer" packaging).

Figure 13.34 shows EPS recycling amounts as reported by EPS-IA [62]. As can be seen, there was a large decline in 2017, but considerable recovery in 2018 and 2019. The data presented is based on responses from 63 EPS manufacturers and independent recyclers, and EPS-IA therefore considers these minimum amounts, as not all recyclers responded.

There are also systems operated by businesses, independently of waste collection systems, that target specific materials. For example, Terracycle makes products out of hard-to-recycle materials, partnering with individual companies that subsidize the process [63].

Processing of Recycling Plastics: For most applications, plastics must be separated by resin type besides being categorized into rigid and flexible plastics. Mixtures of plastic often tend not to have very good mechanical properties. Like for glass, color sorting is also an issue. Generally,

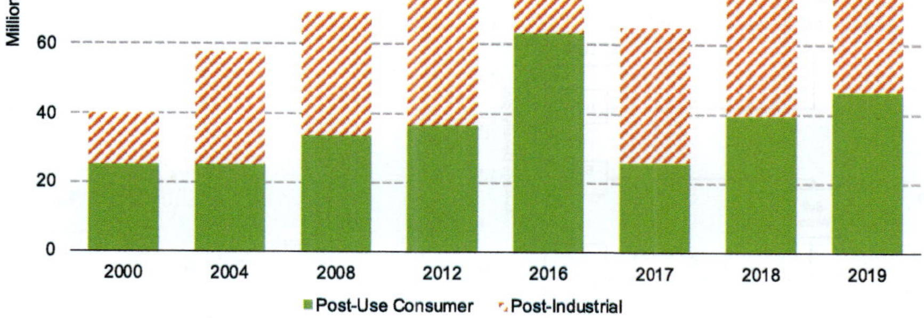

Figure 13.34 EPS-recycling, postconsumer and postindustrial; adapted from [62].

recycling of plastics tends to decrease its weight average molecular weight (M_w). Sophisticated systems have been developed for sorting plastic bottles by resin and by color. The technology works well on whole bottles, but not as well on chipped plastics. Since it is quite expensive, such sorting by resin is generally done only in designated plastics sorting facilities. In many places, hand-sorting is used to separate PET and HDPE bottles, which can easily be differentiated by the high transparency of PET and cloudiness or opacity of HDPE. The plastic resin identification codes are also an aid in doing this sorting but are intended more for telling consumers which bottles to put out for recycling and which to not put out – which is increasingly irrelevant with the growth in "all plastics" collection. There are several methods to recycle polymeric resins after collection, among them mechanical recycling including macrosorting and microsorting, chemical recycling, and enzymatic recycling.

Mechanical Recycling: Mechanical recycling involves shredding, heating, and remolding of the plastics. Particular polymers are separated from other polymers and contaminants. They are reprocessed by conventional melt extrusion. Mechanical recycling is more efficient than making plastic from fossil resources, typically using less than half as much energy [64].

However, mechanical recycling generally degrades, or downcycles, the polymers. The main disadvantage is the deterioration of the product properties in every cycle, such as reduction of M_w due to chain scission caused by heat, stress, H_2O, acidic impurities, etc. Figure 13.35a shows a procedure to separate mixed plastic streams into single materials. The main strategy is to separate the polymers according to their density differences as shown in Figure 13.35b.

Macrosorting requires intact containers. However, if you can have on-truck grinders, you can reduce the volume making collection more efficient. Microsorting uses technologies for sorting mixed plastic chips by resin type. Single resins can also be separated by using infrared and X-ray detectors. Lately, new visual technologies based on artificial intelligence have been developed to distinguish among different plastic packages.

Chemical recycling: Chemical recycling involves deconstruction of polymers by chemical processes to monomers for conversion back into polymers or to molecular intermediates that can be used as fuel or feedstock by the chemical industry. Often the current methods (e.g., pyrolysis – high-temperature decomposition in the absence of oxygen) are energy-intensive and require further processing to make products. Any contaminants present in the process can be removed and new polymers can be polymerized. Common chemical recycling methods have been developed, for example, for PET. Among them are glycolysis, methanolysis, hydrolysis, aminolysis, ammonolysis, and degradation by acid, alkaline, or high-temperature steam [64].

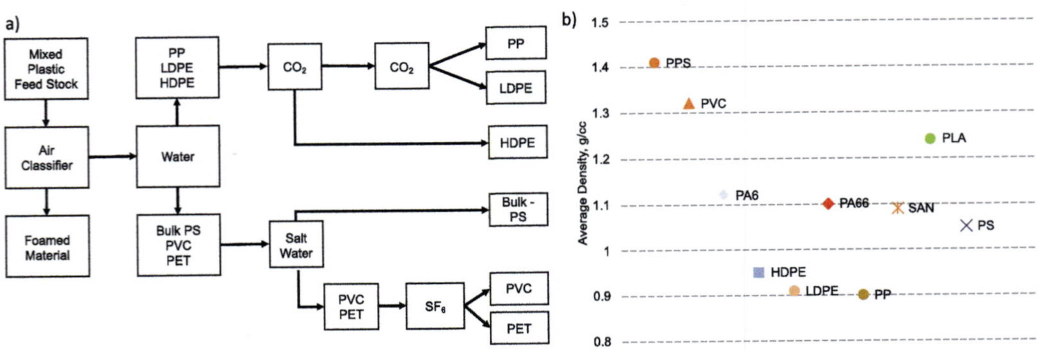

Figure 13.35 (a) Separation of mixed plastic feedstock; (b) main plastic densities.

New attention is being paid to the chemical upcycling of polymers, which is defined as the process of selectively converting discarded plastics into chemicals, fuels, or materials of higher value. These new processes may hold the promise of recovering plastics and converting them into valuable resources. Increasing research and work is going into the "circular lifecycle for plastics, in which the chemical constituents of plastics are reformed into polymers or repurposed to give them another life" [64].

Enzymatic recycling: In this process, particular enzymes are used to degrade polymers at low temperature, using more benign solvents. Several of the enzymatic methods require a preliminary reduction of the polymer's M_w through another method [64]. Lately, a number of articles have reported the selective enzymatic treatment or recycling of PET by PET hydrolases. These enzymes act on PET and break it down into monomeric units. Several factors, such as PET crystallinity, temperature, and pH, affect the enzymatic process. Since for effective hydrolysis of PET high temperature is required, thermostable enzymes are needed, which are mostly produced using enzyme stabilization techniques [65]. Several new enzymatic processes are being researched to recover polymers such as poly(lactic acid), polystyrene, and polyethylene.

Plastic resin coding system: This system was originally developed by SPI, the Society of the Plastics Industry, and enclosed the numbers indicating resin type inside a triangle made up of 3 chasing arrows. The coding system has now been formalized in ASTM D7611/D7611M – 20 [66]. Revisions to the standard were announced in 2010, including a change from the "chasing arrows" triangular logo to a solid triangle, as shown in Table 13.4. ASTM D7611/D7611M – 20 provides two options for the logos, "Option A" and "Option B" marking systems. Generally, the "Option A" symbols are used in the US; these are shown in Table 13.4.

The growing number of collected materials that are multilayer presents special problems for recycling, as most often it is not feasible to separate these materials by resin type. One exception is multilayer PET/EVOH bottles produced without tie layers. These materials (e.g., Heinz ketchup bottle) were designed to be compatible with PET recycling streams. The small amount of EVOH is mostly removed during reprocessing.

Another approach is to build compatibilizers into the structure. One example is Dow Chemical's "RecycleReady Technology" for films, which is designed to permit recycling of multilayer film packaging within the polyethylene film stream [68].

An alternative to separation of plastics by type is applications that can accept mixed ("commingled") plastics. One of the first such applications was "plastic lumber" and substitutes for concrete. These items tend to have large cross-sections, so they are less affected by the poor adhesion between the incompatible plastics in the recycled mix [69]. Often, the processing tends to result in a polyolefin "skin" with the other plastics residing inside the item, out of sight. Not all plastic lumber is made from mixed plastic; the higher quality materials tend to be made from relatively pure streams, or from pure recycled plastic streams blended with wood or paper fibers.

Uses of Recycled Plastics

PET: A major use for PET continues to be fiber, as shown in Figure 13.36. The amount of PET used for food and beverage containers has grown substantially over most of this period, fueled initially in part by pressure from consumer and environmental organizations on Coca Cola, Pepsi, and other beverage companies to use recycled content in their beverage bottles. However, the amount used declined significantly in 2017. Use of recycled PET in the "sheet & film" category grew through 2014 but has declined since then. Much of this is food packaging. Use in nonfood bottles has been fairly flat since 2004. Availability of collected PET for recycling continues to be a major issue for the industry, fueled by a decrease in market share for carbonated soft drinks.

Table 13.4 Plastic packaging resin codes; adapted from [66, 67].

Resin Codes	Description	Applications
1 PETE	Polyethylene terephthalate (PET, PETE). PET is clear, tough, and has good gas and moisture barrier properties. This resin is commonly used in beverage bottles and many injection-molded consumer product containers. Cleaned, recycled PET flakes and pellets are in great demand for spinning fiber for carpet yarns, producing fiberfill and geo- textiles. Nickname: Polyester	Plastic bottles for soft drinks, water, and mouthwash, ketchup, and others.
2 HDPE	High-density polyethylene (HDPE). HDPE is used to make many types of bottles. Unpigmented bottles are translucent, have good barrier properties and stiffness, and are well suited to packaging products with a short shelf life such as milk. Because HDPE has good chemical resistance, it is used for packaging many household and industrial chemicals such as detergents and bleach. Pigmented HDPE bottles have better stress crack resistance than unpigmented HDPE.	Bottle for milk, water, juice, shampoo, and dish and laundry detergents.
3 V	Polyvinyl chloride (PVC, Vinyl). In addition to its stable physical properties, PVC has good chemical resistance, weatherability, flow characteristics, and stable electrical properties. The diverse slate of vinyl products can be broadly divided into rigid and flexible materials.	Rigid packaging applications, blister packs, and clamshells.
4 LDPE	Low-density polyethylene (LDPE). LDPE also is used to manufacture some flexible lids and bottles as well as in wire and cable applications. LDPE is used predominately in film applications due to its toughness, flexibility, and relative transparency, making it popular for use in applications where heat sealing is necessary. Includes linear low-density polyethylene (LLDPE).	Bags for several uses.
5 PP	Polypropylene (PP). PP has good chemical resistance, is strong, and has a high melting point making it good for hot-fill liquids. This resin is found in flexible and rigid packaging, fibers, and large molded parts for automotive and consumer products.	Use for yogurts, margarine, and several thermoform containers. Bottles caps and closures.
6 PS	Polystyrene (PS). PS is a versatile plastic that can be rigid or foamed. General purpose polystyrene is clear, hard, and brittle. It has a relatively low melting point. Typical applications include protective packaging, foodservice packaging, bottles, and food containers. PS is often combined with rubber to make high-impact polystyrene (HIPS) which is used for packaging and durable applications requiring toughness, but not clarity.	Use for take-out foods, and cushioning.
7 OTHER	Other. Use of this code indicates that a package is made with a resin other than the six listed above or is made of more than one resin and used in a multilayer combination.	Several polymers are used for multilayer structures.

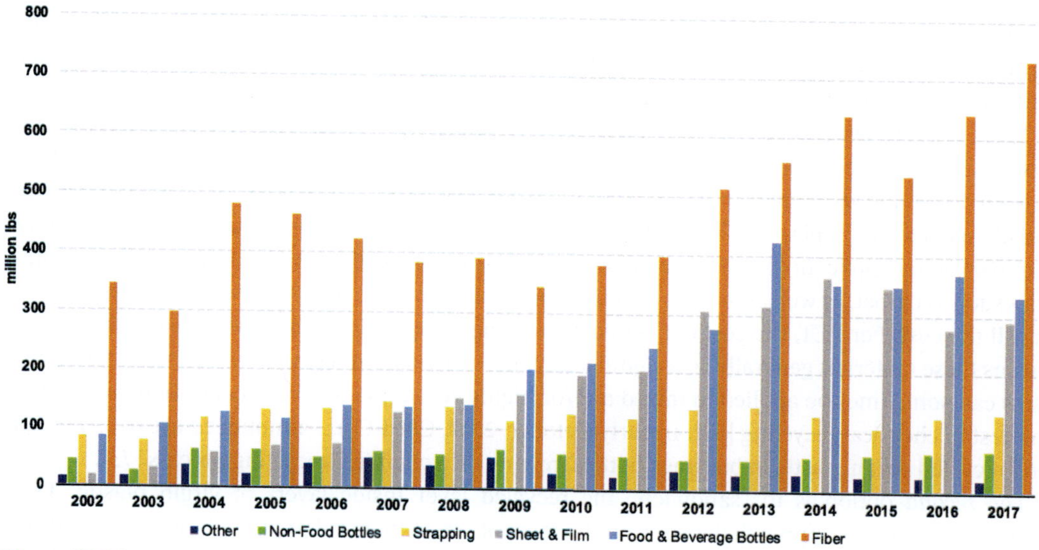

Figure 13.36 Uses of recycled PET from fiber, sheet and film, strapping, food and beverage bottles, and nonfood bottles; adapted from [70].

As stated earlier, the NAPCOR report also discusses PET thermoform recycling, although those amounts are not included in the report data. In a discussion of this recycling stream, they state "While PET thermoforms are technically recyclable with PET bottles, they can be problematic from a performance and yield perspective" – adding that part of the issue is due to concerns about the mingling of PET thermoforms with "look-alikes" of OPS, PVC, PETG (glycol-modified PET), or PLA, along with problems associated with inks, labels, and adhesives that can affect quality.

HDPE: Uses for recycled HDPE are quite different than those for PET. The APR report stated that the largest single uses are in nonfood bottles (37%) and pipe (31%) [59]. Use of recycled HDPE in food containers can be permitted in certain circumstances but will be more limited than use of recycled PET because of the greater potential for migration of contaminants in the recycled material to the product. In 2011, Envision Plastics, which had gotten a "Letter of Non-objection" from the FDA for limited use of recycled HDPE in food containers many years ago but had not commercialized the material, announced the beginning of full-scale production of the resin [71]. Recycled HDPE has been used for milk bottles in the UK [72].

In the original uses of recycled HDPE in bottles, the recycled material was in an inner layer of a multilayer structure. This was the method, for example, in laundry detergent and fabric softener bottles. In applications that are less sensitive to color and odor, such as black-colored motor oil bottles, it was common to blend the recycled HDPE with virgin plastic, in a single-layer structure. With improvements in the technology to "clean up" the recycled plastics stream, use of recycled plastics in single-layer structures has greatly increased.

Polyethylene Film: For film plastics, in 2018 the amount collected fell substantially, due to a large decrease in the export market, much of it related to China's "National Sword" program. For domestic use, 46% went into plastic or composite lumber or decking. Film and sheet accounted for 34%, 12% went into injection molding applications, and the remaining 8% into "other" applications [60].

Polystyrene: Previous EPS Industry Alliance reports have stated that 10–20% recycled content is "common" in EPS packaging, though the last few reports do not address end uses.

Use of Recycled Plastics in Food Packaging

Anyone considering the use of recycled plastics in food packaging, or other sensitive uses, must be concerned about the potential for migration of contaminants from the recycled plastic to the product. As is the case for paper, the processing conditions generally involve sufficiently high temperatures to inactivate microorganisms but will not destroy other types of contaminants. Under suitable conditions, these contaminants may migrate to the package contents.

There are four general approaches to use of recycled plastics in such sensitive applications. The *first* is to use plastic recovered through chemical recycling. In such systems, the recycled plastic is broken down into monomers, purified, and repolymerized as previously explained. Such materials are identical to virgin plastic. This approach is mostly suitable for the condensation polymer family – in packaging applications, this means mostly PET. It has also been applied to nylon, from materials such as carpet, as well as to other condensation polymers. The main drawback of these systems is still the cost. For PET, the cost of chemical recycling, coupled with the cost of initial collection, means these materials generally cost significantly more than virgin resin. However, since these systems can sometimes be applied to mixed and/or highly contaminated feedstocks, there is growing interest in chemical recycling [64]. In early October, 2019, Coca-Cola® announced use of recovered and recycled ocean plastic in beverage bottles, relying on a chemical recycling process [73].

The *second* option is to "sandwich" the recycled layer inside layers of virgin plastic. This "functional barrier" approach depends on the layer of virgin material to protect the contents from migrants originating in the recycled material. It is particularly suited to products with very short shelf lives, as it delays, rather than prevents, migration.

The *third*, and still most common, approach is to process the recycled material in a system that successfully removes most potential migrants. In some cases, this is coupled with control over the source of the material, for instance, using deposit PET bottles rather than curbside bottles since the former is a "cleaner" stream. Increasingly, though, these systems can handle curbside material, relying on intensive cleaning of the recycled stream.

The *fourth* option is simply to restrict use of recycled plastics to less sensitive applications. In such cases, the "normal" recycling process may provide material that is suitable for use without additional clean-up. For example, one of the early uses of recycled PET was for egg cartons and clamshell for products such apples, oranges, and apples. Here, any potential for migration from the package to the product is minimal.

Potential users of recycled plastics in food packaging should consult the FDA document: "Guidance for Industry: Use of Recycled Plastics in Food Packaging: Chemistry Considerations" [74]. FDA also presents a more general discussion of recycled plastics in food packaging [75]. The agency maintains a list of submissions for which it has issued a letter of nonobjection (now referred to as a No Objection Letter, NOL) for food contact; as of this writing, 255 such submissions were listed in the database, including 13 listed in 2021, 19 listed in 2020, 8 listed in 2019, 9 in 2018, and 11 in 2017 [76].

13.9 Labeling

Proper labeling of materials for recycling and recovering has become a widespread concern since proper labeling and identification of the material and the type of recycling methods help consumers to distinguish among materials and facilitate participation in recycling programs, whether they are commingled, dual or multiple streams. Some consensus on labeling was reached among a set of NGO and US companies in the How2Recycle program. This program provides a label for packaging to distinguish how to prepare the material for recycling, how to recycle, the type of material, and the components of the package. Figure 13.37 shows the basic characteristics of this program. In short, the label has four main sections indicating (a) how to prepare the material for recycling, (b) information about how to recycle, (c) the type of recyclable material, and (d) an identification of the section or part of the package.

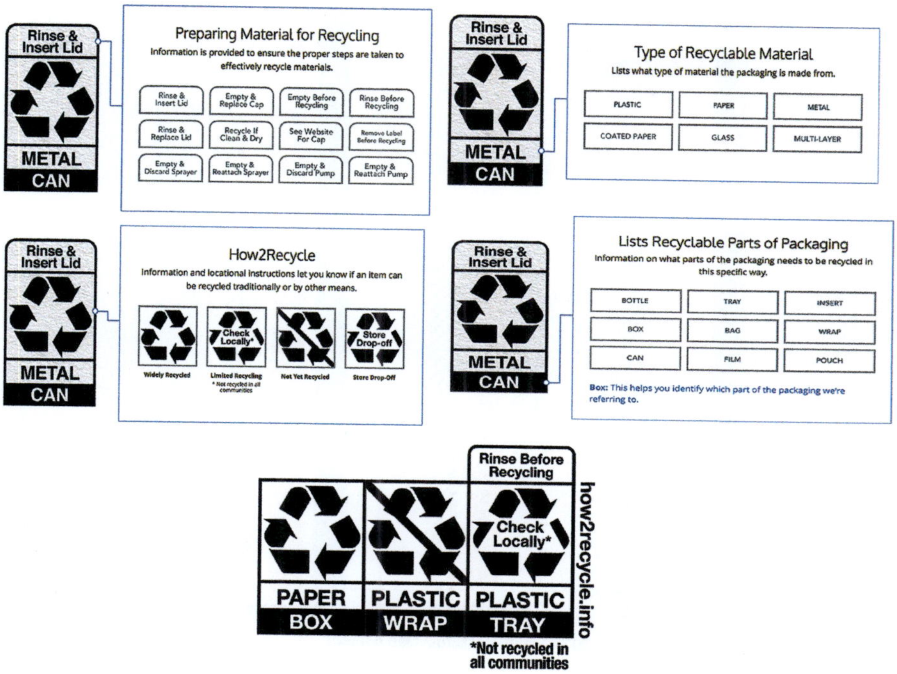

Figure 13.37 How2Recycle label system in the US and Canada; adapted from [77]. Courtesy of How2Recycle. Note: How2Recycle began in 2008 and is a project of the Sustainable Packaging Coalition.

13.10 Case Study 13.1. Environmental Footprint of Recycling Polymeric Resins

A large number of LCA studies have been conducted to determine the environmental effects of recycling containers. Mostly all of the studies demonstrated that using recycled resins ends in a lower environmental footprint for the system. Figure 13.38 shows the EFP of HDPE and PET resins and 100% recycled PET resins. PET resins have lower human health and higher climate change indicators than HDPE. The 100% Recycled PET resin has lower EFP in all the four impact categories. We can conclude that any amount of recycled PET resins introduced in a product will reduce the EFP of the PET product as long as the use of the recycled resin does not require extensive technologies or adaptations.

13.11 Case Study 13.2. End-of-Life Scenario of PLA, PET, and PS Clamshells

Madival et al. [78] compared the environmental footprint of PLA, PET, and PS thermoformed clamshell containers used for the packaging of strawberries. The scope of the study ranged from the extraction of the raw materials for the three polymers and the processes of their resin production, through container formation, followed by their end-of-life disposal. The scope included consideration of global warming, acidification, ozone depletion, aquatic eutrophication,

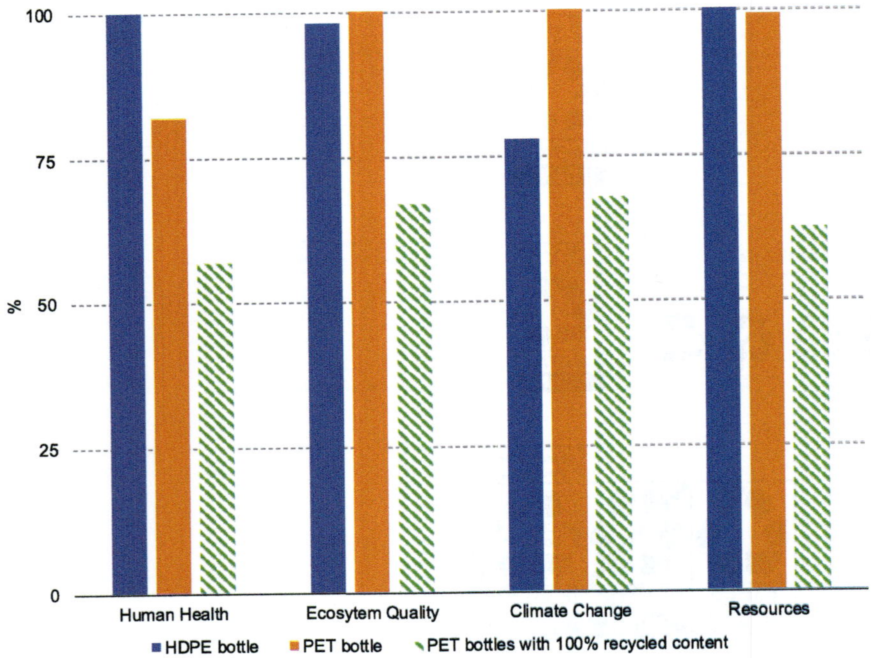

Figure 13.38 Environmental footprint damage assessment indicators – end point – of bottles made of HDPE, PET, and recycled PET resins; original data obtained from Simapro 9.1.1 and reported according to IMPACT 2002 + V2.05.

nonrenewable energy, land occupation, respiratory organics, respiratory inorganics, and aquatic ecotoxicity as impact categories. The functional unit was chosen as 1,000 containers of capacity 0.4536 kg (1 lb) each for the packaging of strawberries. Figure 13.39 shows the life cycle inventory flow chart for PLA, PET, and PS product/package systems. The life cycle flow for the fossil polymers – PET and PS – starts with the extraction of crude oil and the cracking of the extracted oil. For PET, after the extraction of crude oil, resin production included conversion of associated gases into ethylene, naphtha to benzene and ethylene oxide, ethylene oxide to ethylene glycol, and eventually to PET. For PS, after cracking, oil is transformed into ethylene, which is then reacted with benzene to convert to ethyl benzene, followed by styrene and finally polymerized to polystyrene resin. For PLA, the life cycle begins with corn growing and harvesting. The harvested corn is sent to a corn-wet mill where cornstarch is separated and converted to dextrose. The dextrose obtained is converted to lactic acid which, through lactide production, is polymerized to PLA. The resins of the three polymers then go to a converter/container manufacturer. The resin is extruded into sheets. These sheets undergo a thermoforming operation and are converted into containers. Then, they are sent to the strawberry producers, the containers are filled and distributed to the market, where they are bought by consumers. After strawberry consumption the containers reach the end-of- life stage where they end up in a landfill or a recycling center or are incinerated. The energy recovered from the incineration is credited to the energy consumption used for the polymer manufacture.

The authors considered five different end-of-life scenarios in terms of landfill, incineration, and recycling of containers. Composting was not considered as an end-of-life scenario because no emission data was available at that time and no extensive available collection of organics was

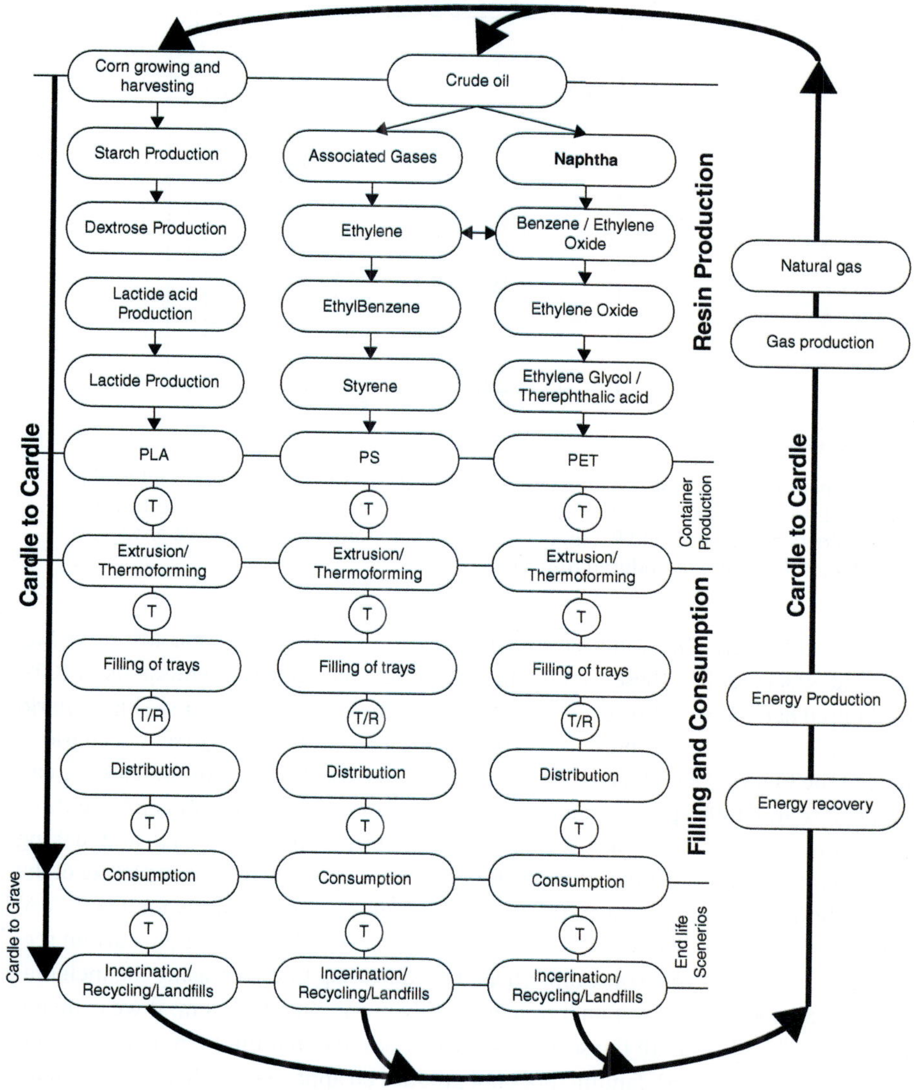

Figure 13.39 Life cycle inventory flow chart for PLA, PET, and PS thermoformed containers from cradle to cradle; reproduced from [78].

available in the country. The first four scenarios were hypothetical; the last one was the scenario of disposal in the US in 2008.

Figure 13.40a and 13.40b show the carbon dioxide equivalent emission and energy consumption for cradle-to-grave analysis of 1,000 containers of PLA, PET, and PS clamshell containers. In the five end-of-life scenarios about carbon dioxide equivalent emission, the values for CO_2-eq emissions for PET were the highest, and the values for PS were lower than PET, showing the role that polymer downgauging and lightweighting play on the reduction of CO_2 emissions. PS container weight was 21% lighter than PET. The reduction of CO_2 emissions can be correlated to the fact that 32.64 kg of PET and 24.96 kg of PS were used for manufacturing 1,000 containers of each. The CO_2 emission

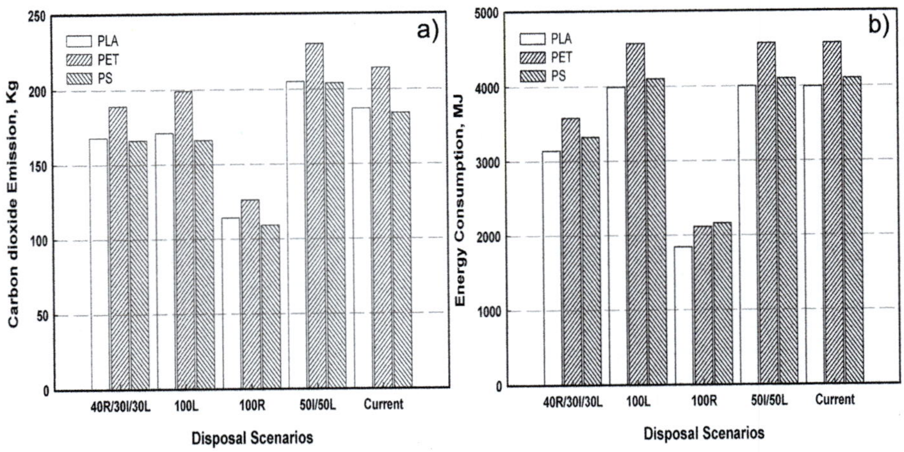

Figure 13.40 Environmental footprint of PLA, PET, and PS clamshell considering different end-of-life scenario (R = recycling; L = landfilling; I = incineration); (a) carbon dioxide emission for cradle-to-grave analysis of 1,000 containers of PLA, PET, and PS; (b) energy consumption for cradle-to-grave analysis of 1,000 containers of PLA, PET, and PS; reproduced from [78].

values for PLA, PET, and PS were almost similar for scenarios I and II, indicating that 40% recycling, 30% incineration, and 30% landfill of these containers have almost the same CO_2 emissions as 100% of landfill. One-hundred percent recycling of containers had the lowest CO_2 emissions. This scenario was introduced to indicate an ideal condition. Scenario IV (50I/50l) and the current scenario V (23.5I/76.5l) showed the highest CO_2 emissions for PLA, PET, and PS. However, in this case we can observe that an increase in landfill percentage reduces the amount of CO_2 emissions for the three polymers due to the sequestration of carbon, providing an interesting point of discussion and making it undeniable that looking at only one indicator such as CO_2 does not provide the entire picture of the EFP. Figure 13.40b shows the energy consumption for the cradle-to-grave analysis for 1,000 PLA, PET, and PS containers. For PLA, PET, and PS, scenarios II (100l), IV (50I/50l), and the current scenario (23.5I/76.5l) have similar energy consumption values, indicating that increasing the incineration percent from 0 to 50 does not recover energy significantly in the overall system. If we compare the energy consumption for PLA for all the scenarios, we can observe that the current scenario for PLA (i.e., 23.5% incineration and 76.5% landfill) does not produce an appreciable advantage when we consider energy consumption with respect to other alternatives such as PS and PET containers. Therefore, to reduce the energy consumption for PLA containers, recycling should be established.

13.12 Study Questions

1. Define recycling as described by the US EPA.
2. What are the main basic reasons for people to recycle?
3. What are the requirements for successful recycling?
4. Discuss the main factors affecting consumer engagement in recycling.
5. Define diversion rates, participation rates, and collection rates.

6 Discuss the main methods of collection of household recyclables.
7 Explain how a deposit system works, such as the bottle deposit system. What are the advantages and disadvantages of a deposit system?
8 Explain the concept of extended producer responsibility (EPR).
9 Describe the process of sortation in an MRF.
10 Explain the needed reprocessing steps for OCC and PET.
11 Explain the influence that end markets have on the availability and use of recycled materials.
12 Discuss the main recycled materials in the US according to the EPA yearly report.
13 What is the main difference between closed- and open-loop recycling?
14 What are the main metals recycled in the US MSW?
15 Describe the main steps in recycling steel and aluminum.
16 What was the average can recycling rate in the US in 2015 and who reports these values?
17 What are the main type of glass containers recycled in the US?
18 Discuss the main factors playing an important role in glass recycling.
19 Is paper, paperboard, or corrugated board recycled the most in the US? Explain why.
20 What are the main uses of recycled paper?
21 Discuss the recycling rates for paper around the world.
22 Discuss the main issues with recycled paper.
23 What are the main plastics recycled in the US?
24 What are the main differences between primary and secondary recycling?
25 What are the main methods to recycle polymer resins?
26 What is the main plastic resin code system? Identify each symbol.
27 What are the four main general approaches to use of recycled plastics in food packaging?
28 Interpret and discuss the labels below created by the How2Recycle program.

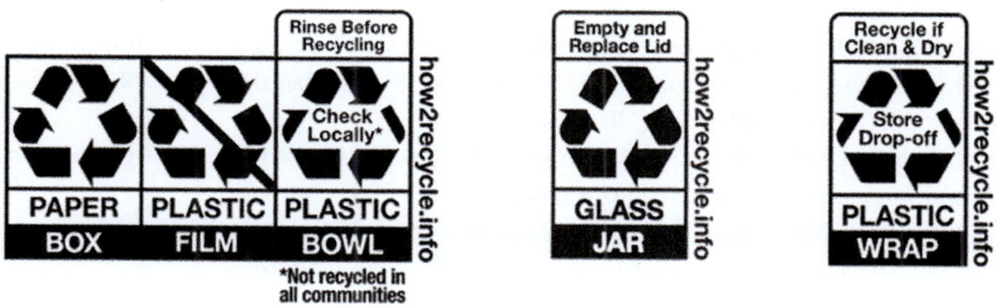

13.13 Additional Resources

1 Virtual tour of the city of Phoenix Materials Recovery Facility
https://www.youtube.com/watch?v=YxnxXSg6dmo

2 Virtual Tour of a Recycling Center – ReCommunity
https://www.youtube.com/watch?v=zgLW9CSvpRw

3 Material Recovery Facility: GreenWaste Recovery Opens 90-tph Mixed Waste MRF
https://www.youtube.com/watch?v=4FpsH_ETT7c

4 MSU MRF Floor Tour
https://www.youtube.com/watch?v=8cv59pmqctk
MSU Virtual Tour facility
https://www.youtube.com/watch?v=k0UKuBejPzw
Merchandise Processing
https://www.youtube.com/watch?v=dv_8O5hoDUM&t=3s
Recycling at MSU
https://www.youtube.com/watch?v=QFyKEncOzWk

5 AMP Robotics Introduction
https://vimeo.com/342840855
https://vimeo.com/334715931

6 David Allaway – Oregon DEQ – Rethinking Recycling.
https://nerc.org/documents/Webinars/RethinkingRecycling/RethinkingRecycling.mp4

7 Evolution of Recycling in Region 5
https://www.youtube.com/watch?v=lKOt6TG1N5E

References

1 de Thomas, D. and Matta, T. (2019). *ASTRX review of material flow at MRFs and reprocessors*, 41. https://astrx.org/resources/astrx-study-of-material-recoverability (accessed November 17, 2021).

2 *Is Switzerland the world champion of recycling?* (2020). https://www.swissinfo.ch/eng/fact-check_is-switzerland-the-world-champion-of-recycling/42382610 (accessed June 19, 2020).

3 Zug, C. (2018). *Information leaflet on waste collection and treatment 2012 Site Regulations for the Waste Collection Center (« Oekihof »)*, 0–2.

4 New York Times. (1986). "Garbage Lottery" spurs recycling in an Illinois City. *New York Times* 12.

5 Europen. (2014). *Extended producer responsibility (EPR) for used packaging*, 2–5. https://www.europen-packaging.eu/wp-content/uploads/2021/04/EUROPEN-factsheet-on-EPR-for-used-packaging.pdf (accessed January 22, 2022).

6 Kaza, S., Yao, L., Bhada-Tata, P., and Van Woerden, F. (2018). *What a waste 2.0: A global snapshot of solid waste management to 2050*. The World Bank.

7 RRS. (2016). *2015-2016 centralized study on availability of recycling*, 30. https://recycle.com/spc-recycling-access-study (accessed January 22, 2022).

8 Bottle Bills in the News. (2020). Delaware. https://www.bottlebill.org/news/locations/delaware.htm (accessed March 9, 2021).

9 Food & Wine. (2020). *Florida proposes 20-cent deposit on bottles and cans*. https://www.foodandwine.com/news/florida-bottle-deposit (accessed March 9, 2021).

10 Waste Dive (n.d.). *What's next for California container recycling after rePlanet's closure?* https://www.wastedive.com/news/whats-next-for-california-container-recycling-after-replanets-closure/560916 (accessed March 9, 2021).

11 Can Manufacturers Institute. (2019). Deposit Laws Washington, DC. https://www.cancentral.com/Deposits (accessed March 9, 2021).

12 The New York Times. (2019). Beverage *companies* embrace recycling, until it costs them . https://www.nytimes.com/2019/07/04/business/plastic-recycling-bottle-bills.html (accessed October 19, 2021).

13 Food Packaging Forum. (2020). *Beverage container associations urge support for deposit programs)*. https://www.foodpackagingforum.org/news/beverage-container-associations-urge-support-for-deposit-programs (accessed October 19, 2021).

14 The Salt. (2019). *Oregon bottle deposit system hits 90 percent redemption rate:* NPR. https://www.npr.org/sections/thesalt/2019/02/04/688656261/oregon-bottle-deposit-system-hits-90-percent-redemption-rate (accessed March 9, 2021).

15 Gitlitz, J. (2013). Bottled up: Beverage container recycling stagnates (2000-2010). *Contain Recycl Inst* 2013: 35. http://www.container-recycling.org/images/stories/PDF/BottledUp-BCR2000-2010.pdf (accessed November 16, 2021).

16 Bottle Bill (2021). Proposed Laws. *Bottle bill resource guide.* https://www.bottlebill.org/index.php/current-and-proposed-laws/proposed-laws (accessed March 9, 2021).

17 Product Stewardship Institute (PSI). (2019). https://www.productstewardship.us (accessed March 9, 2021).

18 U.S. State EPR Laws - Product Stewardship Institute (PSI). (2021). https://www.productstewardship.us/page/State_EPR_Laws_Map(accessed March 9, 2021).

19 Product stewardship and extended producer responsibility (EPR). (2021). https://www.calrecycle.ca.gov/epr (accessed March 9, 2021).

20 Los Angeles Times (2019). *California single-use plastic ban comes down to the wire.* https://www.latimes.com/environment/story/2019-09-04/plastics-single-use-containers-california-recycling-crisis (accessed March 9, 2021).

21 Pittman-Polletta, B. (2010). *Oregon's first-in-the nation paint stewardship law makes it easy to recycle.* oregonlive.com. https://www.oregonlive.com/environment/2010/07/oregons_first-in-the_nation_pa.html (accessed March 9, 2021).

22 Paintcare.org. (2020). *PaintCare Inc. About.* PaintCare Inc. https://www.paintcare.org/about/#/overview (accessed August 13, 2020).

23 Maine Government. (2021). *Maine Legislature.* http://mainelegislature.org/legis/bills/bills_124th/billpdfs/HP115901.pdf (accessed March 9, 2021).

24 Maine's Product Stewardship, Bureau of Remediation and Waste Management, Maine Department of Environmental Protection (2021). https://www.maine.gov/dep/waste/productstewardship/index.html (accessed March 9, 2021).

25 Smalley M. (2021).*Year packed with packaging regulations.* https://www.recyclingtoday.com/article/northeast-recycling-council-webinar-extended-producer-responsibility-packaging-laws (accessed March 9, 2021).

26 Stewardship Ontario. (2021). https://stewardshipontario.ca (accessed March 9, 2021).

27 Stewardship Ontario. (2021). *Fees and Payments: Stewardship Ontario.* https://stewardshipontario.ca/stewards-bluebox/fees-and-payments (accessed March 9, 2021).

28 AMP Robotics. (2020). *AMP Robotics.* https://www.amprobotics.com (accessed March 10, 2021).

29 Porter, R.C. (2002). *The Economics of Waste.* Routledge. Taylor & Francis Group. Washington, DC.

30 ASTRX. (2018). *Navigating the recycling system,* 1. https://astrx.org/wp-content/uploads/2017/01/NavigatingRecyclingSystem_FINAL_May10.pdf (accessed November 30, 2021).

31 Shabecoff, P. (1987). With no room at the Dump, U.S. Faces a Garbage crisis. *New York Times* 1: B8.

32 US Environmental Protection Agency. (2019). *Advancing sustainable materials management: 2016 and 2017 Tables and Figures.* https://doi.org/EPA530F-18-004.

33 US Environmental Protection Agency (2021). *National Framework for Advancing the U.S. Recycling System.* https://www.epa.gov/sites/default/files/2019-11/documents/national_framework.pdf (accessed March 18, 2021).

34 Steel Recycling Institute. (2020). *Packaging.* SRI - Steel Recycling Institute. https://www.steelsustainability.org/packaging (accessed August 21, 2020).

35 Steel Recycling Institute. (2013). *Steel recycling rates steel: North America's most recycled material 2013 steel reinforcement rate 2013:1me.* http://www.recycle-steel.org/~/media/Files/SRI/Releases/SteelRecyclingRatesSheet.pdf?la=en (accessed November 10, 2021).

36 Crawford, G.L. (2000). Steel recycling. In: *McGraw Hill Recycl. Handb*, 2nd e (ed. L. Hf). New York, NY: McGraw Hill.

37 Aluminum, W. (2017). *Global aluminium flow model 2017*. https://international-aluminium.org/statistics/primary-aluminium-production (accessed November 16, 2021).

38 The Aluminum Association. (2021). https://www.aluminum.org (accessed March 10, 2021).

39 Container Recycling Institute - CRI. (2021). https://www.container-recycling.org (accessed March 10, 2021).

40 The Aluminum Can Advantage. (2019). *The Aluminum Association*. https://www.aluminum.org/aluminum-can-advantage (accessed October 21, 2019).

41 Alupro. (2021). *Aluminium drink cans hit 75% recycling rate, an increase of 38% in just eight years*. https://alupro.org.uk/aluminium-drink-cans-hit-75-recycling-rate-an-increase-of-38-in-just-eight-years (accessed March 10, 2021).

42 Latest News. (2021). *The aluminium packaging recycling organisation*. https://alupro.org.uk/news (accessed March 10, 2021).

43 *Remelters IAIO of EAR and. Global Aluminium Recycling: A Cornerstone of Sustainable Development*. (2021). www.world-aluminium.org (accessed October 29, 2020).

44 Statistics - FEVE. (2021). https://feve.org/about-glass/statistics (accessed March 15, 2021).

45 *Paper - Paperboard Recovery*. (2021). https://www.paperrecycles.org/statistics/paper-paperboard-recovery (accessed March 15, 2021).

46 Markets for mixed paper. (2021). Recycling Today. https://www.recyclingtoday.com/article/making-marketable-mixed-paper-bales (accessed March 15, 2021).

47 CEPI (2019). *Key Statistics 2019*. European Pulp & paper Industry. https://www.cepi.org/wp-content/uploads/2020/07/Final-Key-Statistics-2019.pdf (accessed March 15, 2021).

48 Villanueva, A. and Wenzel, H. (2007). Paper waste - Recycling, incineration or landfilling? A review of existing life cycle assessments. *Waste Manag* 27. https://doi.org/10.1016/j.wasman.2007.02.019.

49 Fengler, R. and Gruber, L. (2020). Mineral oil migration from paper-based packaging into food, investigated by means of food simulants and model substances. *Food Addit Contam Part A* 37: 845–857. https://doi.org/10.1080/19440049.2020.1714750.

50 Pirow, R., Blume, A., Hellwig, N., Herzler, M., Huhse, B., Hutzler, C. et al. (2019). Mineral oil in food, cosmetic products, and in products regulated by other legislations. *Crit Rev Toxicol* 49: 742–789. https://doi.org/10.1080/10408444.2019.1694862.

51 Koster, S., Varela, J., Stadler, R.H., Moulin, J., Cruz-Hernandez, C., Hielscher, J. et al. (2020). Mineral oil hydrocarbons in foods: Is the data reliable? *Food Addit Contam Part A* 37: 69–83. https://doi.org/10.1080/19440049.2019.1678770.

52 Sturaro, A., Rella, R., Parvoli, G., Ferrara, D., and Tisato, F. (2006). Contamination of dry foods with trimethyldiphenylmethanes by migration from recycled paper and board packaging. *Food Addit Contam* 23: 431–436. https://doi.org/10.1080/02652030500526052.

53 Zabaleta, I., Blanco-Zubiaguirre, L., Baharli, E.N., Olivares, M., Prieto, A., Zuloaga, O. et al. (2020). Occurrence of per- and polyfluorinated compounds in paper and board packaging materials and migration to food simulants and foodstuffs. *Food Chem* 321: 126746. https://doi.org/10.1016/J.FOODCHEM.2020.126746.

54 Schaider, L.A., Balan, S.A., Blum, A., Andrews, D.Q., Strynar, M.J., Dickinson, M.E. et al. (2017). Fluorinated compounds in U.S. fast food packaging EPA public access author manuscript graphical abstract. *Env Sci Technol Lett* 4: 105–111. https://doi.org/10.1021/acs.estlett.6b00435.

55 Suciu, N.A., Tiberto, F., Vasileiadis, S., Lamastra, L., and Trevisan, M. (2013). Recycled paper–paperboard for food contact materials: Contaminants suspected and migration into foods and food simulant. *Food Chem* 141: 4146–4151. https://doi.org/10.1016/J.FOODCHEM.2013.07.014.

56 Office of Food Additive Safety, Food and Drug Administration. (2020). *Guidance for industry: Preparation of premarket submissions for food contact substances (Chemistry recommendations)*. FDA. https://www.

fda.gov/regulatory-information/search-fda-guidance-documents/guidance-industry-preparation-premarket-submissions-food-contact-substances-chemistry#aii6 (accessed November 13, 2020).
57 Peters, R.J.B., Groeneveld, I., Sanchez, P.L., Gebbink, W., Gersen, A., de Nijs, M. et al. (2019). Review of analytical approaches for the identification of non-intentionally added substances in paper and board food contact materials. *Trends Food Sci Technol* 85: 44–54. https://doi.org/10.1016/J.TIFS.2018.12.010.
58 Mouw, S. and Penner, R. (2013). *Plastic Recycling Collection National Reach Study: 2012 Update.* www.moorerecycling.com (accessed November 13, 2020).
59 US Environmental Protection Agency. (2020). *2018 national post-consumer non-bottle rigid plastic recycling report.* https://archive.epa.gov/epawaste/conserve/smm/wastewise/web/html/buyq_a.html (accessed March 16, 2021).
60 US Environmental Protection Agency. (2020). *2018 national post-consumer plastic bag & film recycling report.* https://archive.epa.gov/epawaste/conserve/smm/wastewise/web/html/buyq_a.html (accessed March 17, 2021).
61 Plastic Film Recycling. (2020). *Find a drop off location.* Plastic Film Recycling. https://www.plasticfilmrecycling.org/recycling-bags-and-wraps/find-drop-off-location (accessed November 13, 2020).
62 EPS Industry Alliance (EPS-IA). (2019). *Take a look at EPS recycling 2019.* www.epsindustry.org. (accessed November 13, 2020).
63 Terracycle. (2020). *Recycle everything with TerraCycle.* https://www.terracycle.com/en-US (accessed November 13, 2020).
64 Britt, P., Byers, J., Chen, E., Coates, G., Coughlin, B., Ellison, C. et al. (2019). *Basic energy sciences roundtable: Chemical upcycling of polymers.*
65 Maurya, A., Bhattacharya, A., and Khare, S.K. (2020). Enzymatic remediation of polyethylene terephthalate (PET)–Based polymers for effective management of plastic wastes: An overview. *Front Bioeng Biotechnol* 8: 1332. https://doi.org/10.3389/fbioe.2020.602325.
66 *ASTM D7611/D7611M – 20 Standard Practice for Coding Plastic Manufactured Articles for Resin Identification 1.* (2022). https://doi.org/10.1520/D7611_D7611M-20.
67 American Chemistry Council. (2020). *Plastic packaging resins resin codes descriptions properties product applications products made with recycled content*.* https://plastics.americanchemistry.com/Plastic-Resin-Codes-PDF (accessed November 13, 2020).
68 RecycleReady Technology. (2019). *Sustainable PE packaging.* Dow Inc. https://www.dow.com/en-us/brand/recycleready.html? (accessed March 17, 2021).
69 Plastic Lumber Trade Association. (2021). http://www.plasticlumber.org (accessed March 17, 2021).
70 The Association of Plastic Recyclers. (2017). *2017 United States National Postconsumer Plastic Bottle Recycling Report.*
71 Food Navigator. (2011). *Envision ramps up production of food grade recycled HDPE* . https://www.foodnavigator.com/Article/2011/01/20/Envision-ramps-up-production-of-food-grade-recycled-HDPE (accessed March 17, 2021).
72 Dairy Reporter. (2008). *UK milk bottle manufacturer to close loop with recycled HDPE.* https://www.dairyreporter.com/Article/2008/10/28/UK-milk-bottle-manufacturer-to-close-loop-with-recycled-HDPE (accessed March 17, 2021).
73 Coca-Cola. (2021). *Coke Bottle made with Plastic from Sea.* Press Release. https://www.coca-colacompany.com/press-releases/a-coke-bottle-made-with-plastic-from-the-sea (accessed March 17, 2021).
74 US Food & Drug Administration. (2021).*Guidance for industry: Use of recycled plastics in food packaging (Chemistry Considerations).* https://www.fda.gov/regulatory-information/search-fda-guidance-documents/guidance-industry-use-recycled-plastics-food-packaging-chemistry-considerations (accessed March 17, 2021).

75 US Food & Drug Administration. (2021). Recycled Plastics in Food Packaging. https://www.fda.gov/food/packaging-food-contact-substances-fcs/recycled-plastics-food-packaging (accessed March 17, 2021).

76 U.S. FDA. (2021). *Submissions on post-consumer recycled (PCR) plastics for food-contact articles.* https://www.cfsanappsexternal.fda.gov/scripts/fdcc/index.cfm?set=RecycledPlastics (accessed October 21, 2021).

77 Sustainable Packaging Coalition. (2020). *About How2Recycle.* How2Recycle. https://how2recycle.info/about (accessed November 20, 2020).

78 Madival, S., Auras, R., Singh, S.P., and Narayan, R. (2009). Assessment of the environmental profile of PLA, PET and PS clamshell containers using LCA methodology. *J Clean Prod* 17: 1183–1194. https://doi.org/10.1016/J.JCLEPRO.2009.03.015.

14

Aerobic and Anaerobic Biodegradation

14.1 Learning Objectives

After learning from this chapter, you are expected to:

- Understand the differences between anaerobic and aerobic biodegradation.
- Understand the requirements for compostable packages.
- Define the tests required for compostability certification according to ASTM and ISO standards.
- Differentiate between the biodegradation of paper and polymers.
- Understand the difference between bio-based, biodegradable, and compostable polymers.

14.2 Introduction

As the amount of single-use packaging reaching the environment and polluting water reservoirs has increased, considerable attention has been given to creating packaging materials for the circular economy. To meet the requirements of a circular economy, one of the main attributes is producing materials from renewable resources and fully recovering them at the end of life so that the carbon cycle is fully restored, as shown in Figure 14.1. When using renewable-carbon feedstocks to manufacture materials instead of fossil-carbon feedstocks, there is an intrinsic zero material carbon footprint value proposition; in other words, the carbon footprint reduction arises from the material itself and not necessarily from the process of converting the feedstock to products (process carbon footprint). Fossil resources could be considered renewable, but it takes more than a million years for biomass to be converted into fossil fuels. Since the rate of consumption is much greater than the rate of replenishment, mass imbalance occurs in the carbon cycle. In contrast, biodegradable materials made from bio-based resources, such as starch-heavy crops (e.g., starches from field/feed corn and others), can be produced and converted into biomass in similar time frames [1].

In the previous chapters, we discussed the option of reusing and recycling packaging materials. However, in several situations, such as contaminated single-use packaging for food and agriculture applications, it is not easy to reuse or recycle the materials since it takes resources to clean the material for reuse or recycling. Packaging materials are susceptible to degradation during and after their useful life. This degradation process can deteriorate the properties of materials, or it can be used to degrade them at the end of their useful life. Hence, increasing attention has been

Life Cycle of Sustainable Packaging: From Design to End of Life, First Edition. Rafael A. Auras and Susan E M. Selke.

Companion Website: www.wiley.com/go/Auras/lifecycleofsustainablepackaging

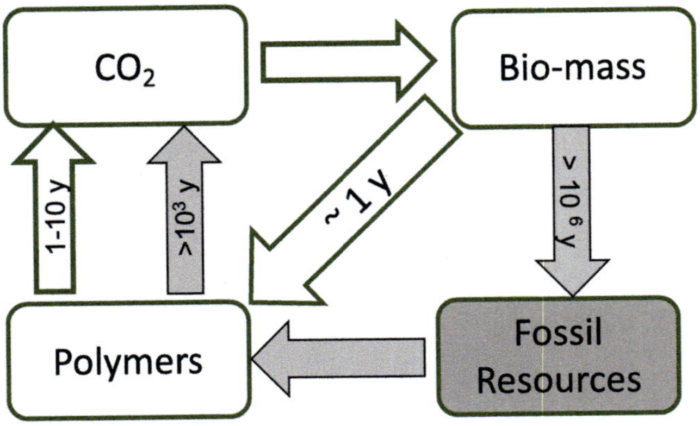

Figure 14.1 Carbon cycle of bio-based and fossil-based materials.

given to broadening the spectrum of recovering materials through degradation processes such as aerobic and anaerobic biodegradation that can handle contamination. Degradation is generally defined as the loss of desired properties (including appearance). This definition applies regardless of the material or system being considered. Figure 14.2 shows the main degradation mechanisms of materials. Degradation can be divided into abiotic and biotic processes. The abiotic degradation process occurs due to mechanical action, heat, radiation, or chemical reaction with, for example, O_2, O_3, atmospheric pollutants, water, acids, and bases. The biotic degradation process occurs due to the reaction of the material with biological organisms. Aerobic biodegradation occurs in the presence of oxygen. In contrast, anaerobic biodegradation happens in an oxygen-absent environment.

In this chapter, we will mostly concentrate on biotic recovery methods such as composting, which is an aerobic process, and anaerobic digestion, an anaerobic process. An important disclaimer must be provided here. We address biodegradation as a controlled disposal route for materials. We are not discussing or encouraging ways to biodegrade packaging materials in open and uncontrolled environments where they may show up as litter. One of the main design flaws of the

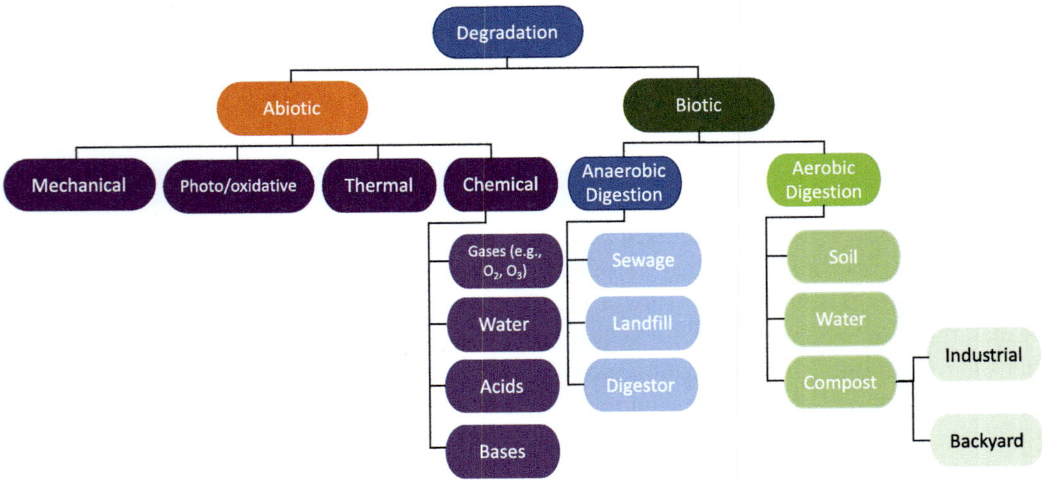

Figure 14.2 Main degradation process of materials.

last decade, it seems, is the desire to design biodegradable materials to biodegrade in the open environment where these materials end up as litter. *Biodegradable litter is litter.* No packaging material should be disposed of in an uncontrolled waste management environment. Designing materials to encourage littering should be discouraged and even prohibited (See Chapter 16). Efforts and actions must be taken to provide proper disposal routes and manage waste scenarios for each material.

14.3 Aerobic Biodegradation

In aerobic biodegradation ($O_2 > 6\%$ according to ASTM D 5338), microorganisms break the materials down using O_2 during the metabolic process. As a result, CO_2 gas and H_2O are produced and released to the atmosphere as shown in Equation 14.1.

$$C_{material} + O_2 \rightarrow CO_2 + H_2O + C_{biomass} + C_{residue} + \text{salts} \tag{14.1}$$

The aerobic degradation processes are associated with the microbial population's capacity to produce the needed enzymes for the degradation of each substrate ($C_{material}$). Carbon is mostly used as an energy source. Figure 14.3 shows a general path for aerobic biodegradation through hydrolyzable and nonhydrolyzable substrate bonds. Besides the carbon source, microorganisms need macronutrients such as nitrogen, phosphorous, and potassium, as well as other trace elements for growth. A small fraction of the carbon is incorporated into microbial cells. The rest is used to

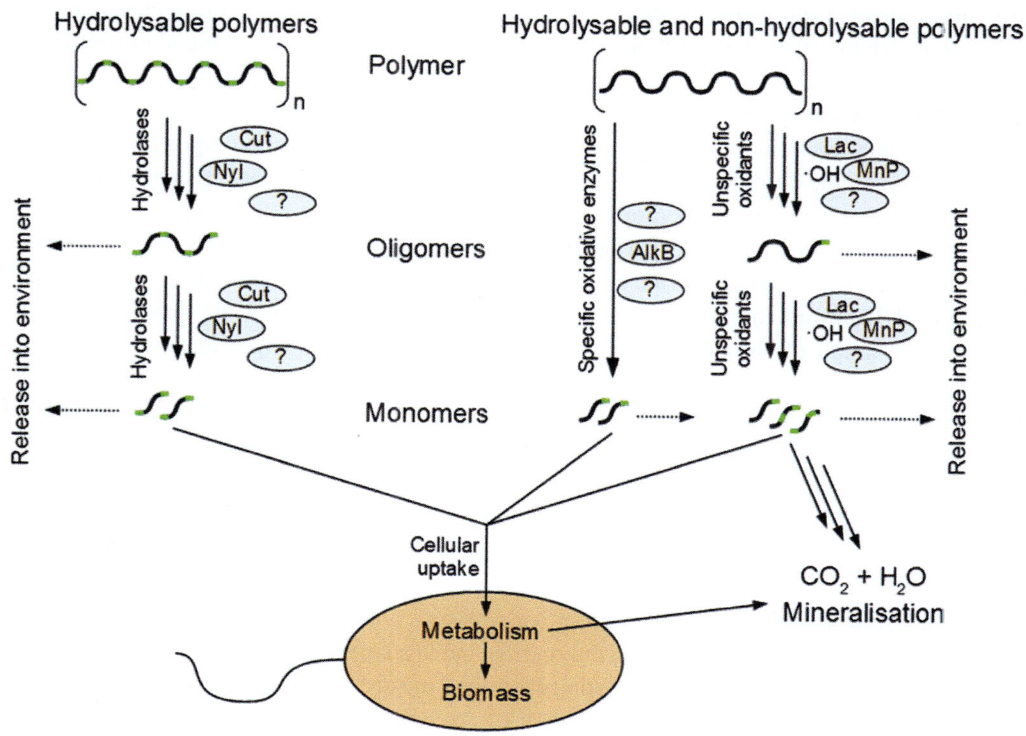

Figure 14.3 Pathways of aerobic microbial degradation of substrates. The decomposition of the substrates may be initiated by the presence of hydrolyzable bonds or oxidation; reproduced from [2] with permission from Elsevier.

provide energy for microbial reproduction and metabolism, and the waste energy is released as heat. Nitrogen is a crucial component for microorganisms since it is a component of proteins, nucleic acid, amino acids, enzymes, and coenzymes, all used during cell growth and functioning. In nitrogen-deficient environments, the degradation process is slow; otherwise, it is released as ammonia gas or other nitrogen compounds. Biodegradation in a compost pile is predominantly aerobic, as will be explained next.

14.3.1 Composting

Composting is a mostly aerobic process and is described as nature's way of recycling. ASTM describes composting as "a process where biodegradable materials, such as manure and leaves, are decomposed and transferred into a humus-like substance called compost, CO_2, water and minerals by microorganisms through a controlled biological process" [3]. According to the new definition by the US Composting Council (USCC), compost "is the product manufactured through the controlled aerobic, biological decomposition of biodegradable materials. The product has undergone mesophilic and thermophilic temperatures, which significantly reduces the viability of pathogens and weed seeds and stabilizes the carbon such that it is beneficial to plant growth. Compost is typically used as a soil amendment, but may also contribute plant nutrients" [4].

Compost can be produced in home/backyard composting or in industrial composting operations. Optimal O_2 of more than 12% is needed for efficient composting operations. If the O_2 is below the optimal such as in a static pile between 5% and 10% O_2, microbes cannot reproduce, and the procedure is prolonged. The optimal C/N ratio has been reported as around 25–40. Moisture content must ideally range between 45% and 65% to avoid drying where the microbes become dormant or excessive moisture that can hinder proper aeration and enhance leaching of nutrients.

14.3.1.1 Home/Backyard Composting

The USCC also defines home composting, also referred to as "backyard" or "composting at home," as "the natural aerobic decomposition of organic wastes or materials usually in small scale composters by 'slow-stack' treatment methods where temperatures are in the psychrophilic (0 – 15°C) to mesophilic (15 – 45°C) ranges" [5]. The compost produced is a result of privately or domestic generated organic waste, which can be applied to private property soils without any commercial transaction [6]. Backyard composting is conducted in bins, piles, or composting tumblers. Figure 14.4 shows an example of a home composting operation. For backyard composting, yard waste and food waste are normally used as ingredients. Using the correct proportions of yard waste is very important to obtain the right C/N ratio, and hence good-quality compost. Turning is usually done manually unless the bins are equipped with rotating drums (tumblers). Backyard composting involves lower temperatures than commercial composting. Since the decomposition is taking place at the mesophilic phase (15–45°C), a longer time is required for the conversion of waste to organic matter to be completed. It may take up to 12 months for complete composting.

14.3.1.2 Industrial Composting

Commercial or industrial composting is mainly carried out at a large scale with heavy machinery for shredding and turning. Commercial composting operations mostly use three types of wastes: manure, yard, and food waste. Manure waste is fecal and urinal excretions of livestock and is usually rich in nitrogen. Therefore, to maintain an adequate C/N ratio, it has to be mixed with other materials that are rich in carbon. Yard waste is vegetative waste from the maintenance of landscaped areas such as lawns and gardens. Yard waste typically includes leaves, grass clippings, tree trunks, and pruning wastes. Food waste is uneaten food (postconsumer waste) or wastes from a food preparation process (preconsumer waste) from domestic, commercial, or institutional sources, such as restaurants or cafeterias. Food waste often contains various types of noncompostable materials such as packaging, which have to be mechanically sorted out before the waste is composted (see Figure 14.5).

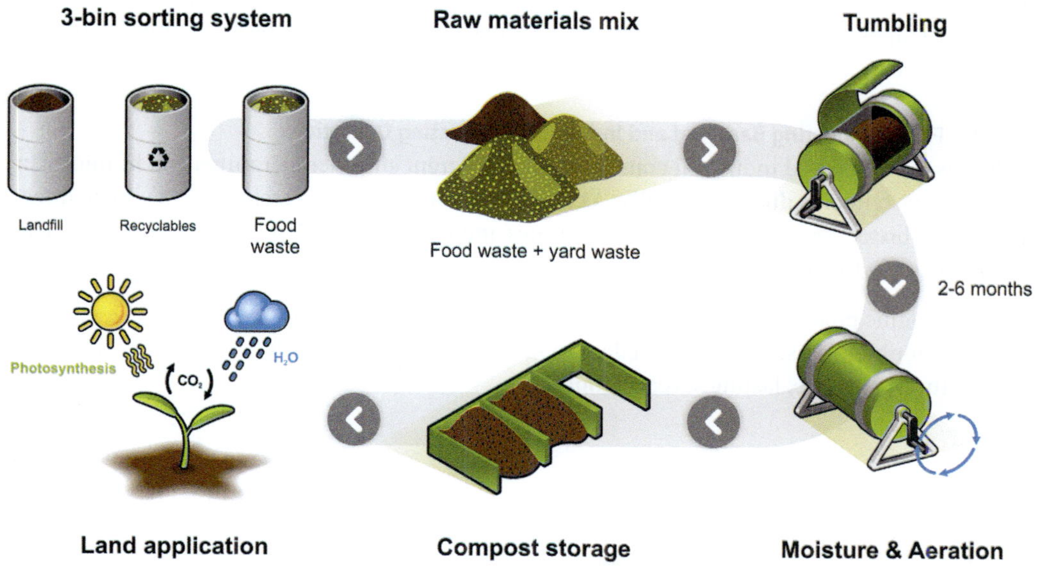

Figure 14.4 Home or backyard composting process; reproduced from [3] with permission from John Wiley & Sons, Inc.

Figure 14.5 Unloading of a food waste container to be composted, containing food waste and packaging materials.

The compost produced in an industrial composting facility is sold as an amended soil [6]. Figure 14.6 depicts an example of an industrial composting operation.

Several composting technologies are used for industrial composting operations as shown in Figure 14.7. In windrow composting, organic waste is formed into rows of long piles and periodically turned manually or mechanically. In aerated static piles, organic waste is mixed in a large pile and designed so that air can pass from the bottom to the top of the pile. Piles are often aerated with a network of pipes or with air blowers that are activated by a timer or sensor. In in-vessel composting, organic waste is fed into a drum, silo, or concrete-lined trench.

14.3.1.3 Factors Affecting Backyard and Industrial Composting Operations

Conditions for home and industrial composting are different since temperatures and humidity in these two environments differ drastically. As described, composting is a technique in which different microbes break down organic carbon substrates and produce CO_2, H_2O, heat, and humus. The industrial composting process includes three main stages:

- The mesophilic phase, lasting for several days
- The thermophilic phase, lasting from days to several months and finally
- The maturation phase, lasting several months

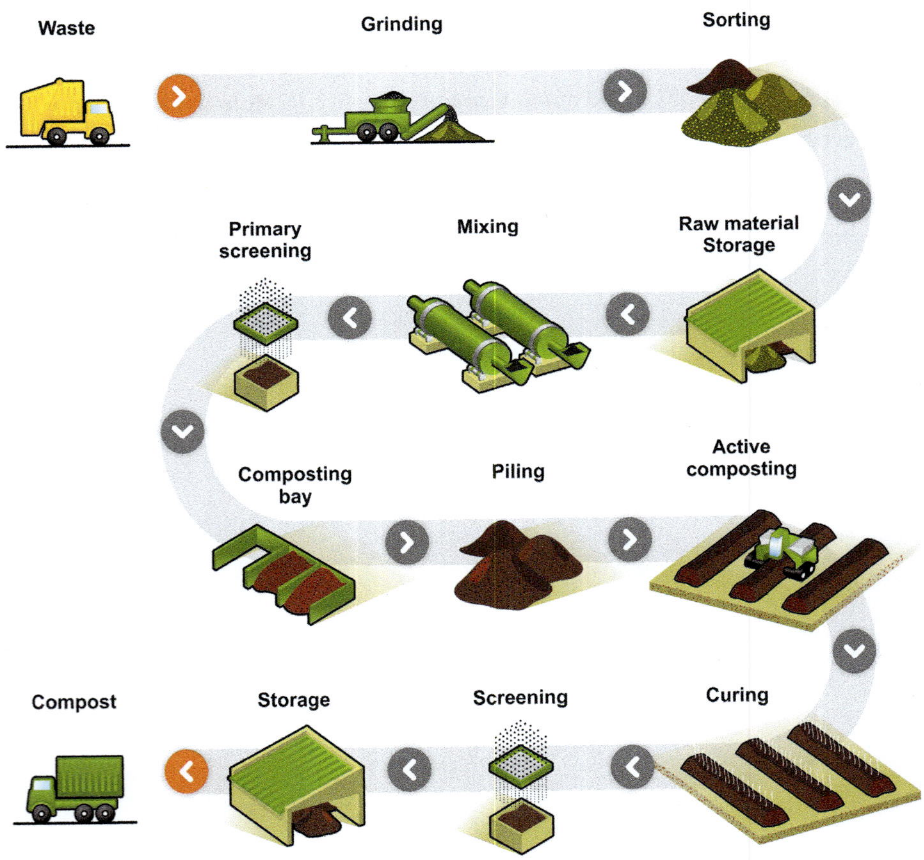

Figure 14.6 Industrial composting process; reproduced from [3] with permission from John Wiley & Sons, Inc.

Figure 14.7 Industrial composting technologies: (a) windrow composting piles, (b) windrow (compost) turner, (c) aerated static piles, (d) in-vessel composting. Windrow compost turner picture by Dave Parker; reproduced under CC BY 2.0.

Distinctive communities of microbes dominate each composting stage (Figure 14.8). The composting process length and temperature depend on the type of organic matter, the aeration process, and agitation. At the beginning of the process, the substrates are at room temperature and generally have low pH. At this early stage, easily biodegradable materials (starch, monosaccharides, lipids) are used by microorganisms. During this mesophilic stage, microbes biologically oxidize available carbon to CO_2 in their cells. Since the process is exothermic, the temperature of the compost rises quickly. Above 40 °C, the thermophilic microbes become active, and the higher

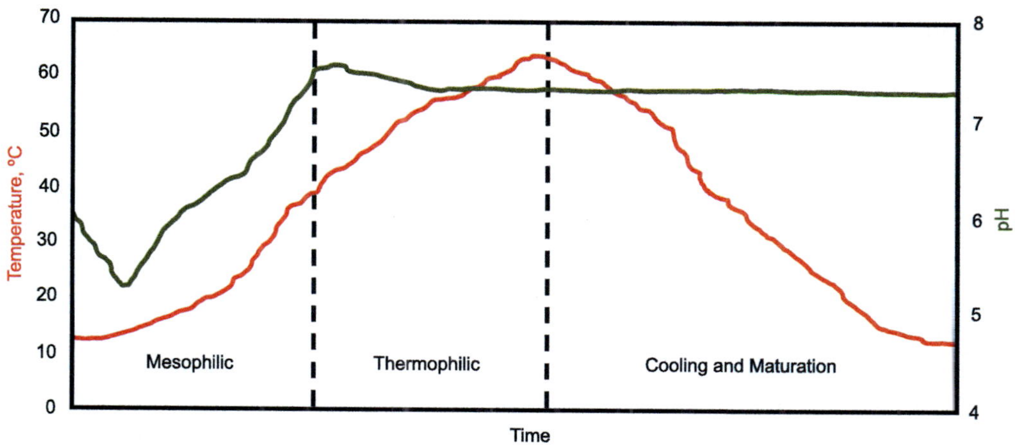

Figure 14.8 Temperature and pH variation during the composting process; adapted from [7].

temperature accelerates the breakdown of proteins, fats, and other carbohydrates such as cellulose and hemicellulose, resulting in the liberation of ammonia and an increase of pH.

After the easily degradable carbon sources are degraded, more resistant compounds such as hemicellulose and lignin are degraded (see Section 14.7) and partly transformed into humus. Humus is the end product of the humification degradation process, producing inert humic substances, which are a great reservoir of organic carbon in soils. Humus has been divided into three main fractions. Humin (nonsoluble in water at any pH), humic acids – HA – (nonsoluble in water under acidic conditions), and fulvic acids – FA – (soluble in water at any pH). Immature compost has a high level of fulvic acids and a low level of humic acids. In more mature compost, the amount of humic acid rises. So, the maturity of compost can be rated by the humification index (C_{HA}/C_{FA}).

When the temperature reaches above 55 °C, human and plant pathogens are destroyed. For safety reasons, several certifying entities require that the temperature get above a specific temperature, such as 55 °C, and remain at that level for a set period, such as 15 days, to ensure that the resulting compost is pathogen free. Temperatures in industrial compost facilities commonly reach around 70 °C. Since the material volume and abundance of microorganisms are lower in backyard composting, it is challenging to maintain these conditions for several days in backyard composting conditions. As composting continues, the compost temperature slowly drops, and mesophilic microbes take over for the final maturation stage of the organic matter. Regardless of whether the composting is done at home, in a community backyard, or an industrial facility, the process must ensure a succession of microbial communities (i.e., mesophiles, thermophiles, and corresponding temperature regimes to operate). These factors provide and guarantee the safety and quality of the composting process and the final product. Most industrial composting systems work using this regime. However, backyard composting is further from this ideal situation due to volume and user operation. Standards to delineate what conditions materials should meet to be certified as compostable in industrial composting operations are available, and will be discussed later [8]. However, the same requirements and standards are not available in the US and worldwide for backyard compostable materials. It is crucial to understand clearly, define, and characterize industrial and backyard composting regarding the compost matrix, time, temperature, moisture, and other operating parameters required for certification.

14.3.2 Agricultural Soils

Several biodegradable materials such as paper and plastic are being used in the agriculture industry. Specifically, the use of plastic in such applications is called plasticulture and defined as "the use of plastic in agriculture." It includes products such as drip irrigation tape, greenhouses, hoop-houses, silage bags, row covers, hay bale wraps, plastic trays and pots, and mulch films. Most of the plastic materials used in agriculture are based on polyolefins, but there also has been an increased focus on the use of biodegradable polymers and paper structures [9, 10]. A plasticulture setting can be considered as a special case of biodegradation in soil. One of the most studied applications in this area has been polymeric mulch films, which have several steps for biodegradation, as shown in Figure 14.9. Mulch film biodegradation first involves an extensive period of photodegradation, when the mulch film starts cross-linking and eroding, followed by an intensive period of biodegradation when the mulch films are subjected to: (i) colonization of the polymer surface by microorganisms, (ii) breakdown of the polymer into low molecular weight compounds due to the action of extracellular enzymes, and (iii) assimilation of these low molecular weight compounds by microorganisms, resulting in the evolution of CO_2 [11, 12]. A detailed discussion of biodegradation in this environment is out of the scope of this chapter, but several peer-reviewed reports can

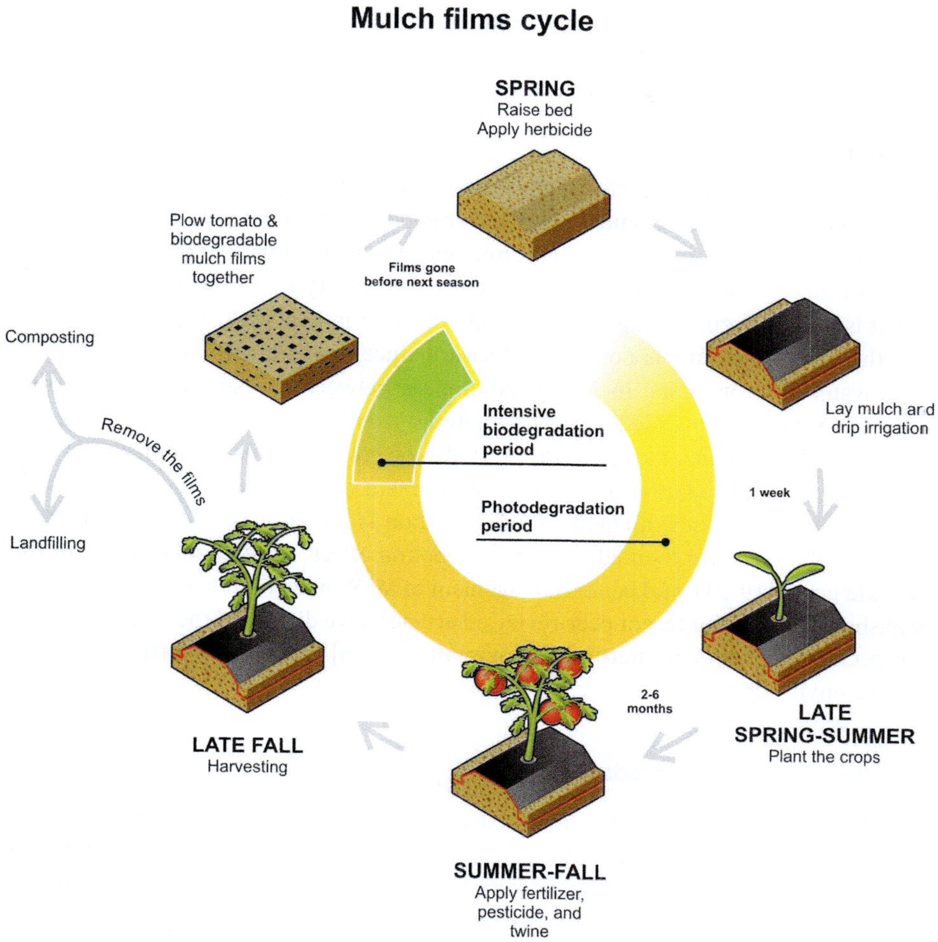

Figure 14.9 Biodegradable films used in the agriculture industry. An example of mulch films used in tomato production in the US Great Lakes region; reproduced from [14] with permission of Elsevier.

be consulted to gain further insight into this field of application for biodegradable polymers and materials in general [10, 11, 13].

14.3.3 Other Mostly Aerobic Degradation Environments

Materials can biodegrade in several other environments, even though these are not initially intended for end disposal. So, we are providing a short discussion of them.

14.3.3.1 Soil Biodegradation

The main parameters to classify a soil are based on granularity and porosity due to the amount of clay, sand, and silt. These components can produce a variety of soils from 100% sand, silt, or clay, to those made up of different proportions of all these components. The formed soil and its environment are a heterogeneous matrix affected by a number of climatic and seasonal factors that affect the formation and activity of the diverse microorganisms it contains. The nature of the soil and the variety of these external conditions create habitats where differing mixed microorganism populations can grow. The main factors playing a significant role in the biodegradation process in soil are soil texture and structure, water content, organic matter, pH, temperature, oxygen, and sunlight [15]. The degradation of materials in soil has been reviewed [15, 16].

14.3.3.2 Aquatic Biodegradation

Aquatic environments such as streams, rivers, lakes, and oceans are not generally the initial main target for biodegradation of consumer and nonconsumer goods made of different materials; however, many consumer goods end up in these environments after being littered in soil and drainage systems. Almost 8 million tons of plastic ends up in our oceans every year in addition to the 150 million tons that is currently circulating overall in aquatic environments [17, 18]. However, biodegradation in these environments is not straightforward since temperatures and oxygen availability in these water environments, including differences between the surface and differing depths, vary widely. When the intended use of the material part, such as fishing nets, lures, and buoys, is in aquatic environments, that purpose must also be considered in design for the water environment. Ultimate biodegradation in a water environment should not be the goal for such consumer and nonconsumer goods. If aquatic biodegradation is intentionally designed for these parts, which can be obtained and then advertised, a "rebound effect" of material disposal in the environment could occur [19–21], and better management of MSW in low-income economies may not be accomplished [22]. Several recent peer-reviewed articles have discussed the biodegradation of materials, especially polymers, in marine environments [23–26]. Chapter 17 further discusses littering in marine environments.

14.3.4 Measuring Aerobic Biodegradation

Several methodologies, both qualitative and quantitative, have been used to describe the biodegradation process. Combining these methodologies provides the best results since the different methods, when used together, help to recognize if there is any disagreement among the achieved results and better understand the biodegradation mechanisms.

The primary methodologies to assess and report biodegradation in aerobic conditions can be divided into five groups and have been documented in several reviews [27]. The oldest and most common method is reducing weight or mass loss of the material under biodegradation. Significant deterioration of mechanical properties has also been reported as a degree of biodegradation. Visualization, mass loss, and deterioration of mechanical properties are approximate methods for the assessment of biodegradation. These methods are more related to the physical degradation of the material and not to the biological processes conducted by a microorganism population.

Supporting quantitative methods such as $\delta^{13}C$, CO_2 evolution, and molecular weight reduction with qualitative methodologies such as SEM, visual observation, and spectroscopy help corroborate the biodegradation of the material under study. As mentioned earlier, during aerobic biodegradation CO_2 gas is released, and therefore, it can be used as an indicator of biodegradation. The aerobic biodegradability of a material can easily be measured as percentage of mineralization, which is the percentage of carbon in the molecules converted into CO_2 by microorganisms, compared to the theoretically available carbon. The equipment used for measuring biodegradation is called a respirometer or respirometric system, which consists of three major components: (1) an air supply; (2) an air-tight closed vessel called the "bioreactor," which contains the compost mixture and/or test or reference materials; and (3) a measuring device, to quantify the amount of O_2 uptake or CO_2 release. Currently, there are three methods of measuring biodegradation, categorized by the methodology used to quantify CO_2 from the exhaust of the bioreactor.

These methods are: (1) cumulative measurement respirometric (CMR) system, (2) direct measurement respirometric (DMR) system, and (3) gravimetric measurement respirometric (GMR) system. In the CMR, the CO_2 from the exhaust is trapped in a basic solution, such as sodium hydroxide (NaOH) or barium hydroxide ($Ba(OH)_2$), and it is titrated at a predetermined time to measure the amount of the trapped CO_2. In the DMR, the concentration of CO_2 is directly measured from the exhaust of the bioreactor using gas chromatography (GC), or an in-line infrared gas analyzer, such as a nondispersive infrared (NDIR) gas analyzer. According to ASTM D 5338, for this method the CO_2 concentration of the exhaust gas of each bioreactor has to be measured at least once every six hours. Finally, in the GMR, the CO_2 from the exhaust is captured in CO_2 absorption columns, and the amount of evolved CO_2 is measured by the weight increase of basic filled columns. Figure 14.10 shows an example of an in-house built equipment (DMR). Additional descriptions of other settings to measure biodegradation can be found elsewhere [3, 27, 28].

14.3.5 Standards and Certifications for Aerobic Biodegradable Materials

Several countries and global regions have standards and regulations to certify bio-based and biodegradable materials, including paper and plastics, before they can be used in their markets and end up in any of the environments described above. We have previously reviewed the main requirements for these certification processes [3].

There are two major standard series involving compostable plastics, ASTM International standards, and ISO and CEN standards (from the International Organization for Standardization and European Committee for Standardization). Both series of standards are equivalent. They include three criteria to identify plastics and paper-based materials as compostable: (1) disintegration, (2) biodegradation, and (3) ecotoxicity. The industrial compostability certification follows the requirements of ASTM D6400 and ASTM D6868. Figure 14.11a shows a basic diagram of the requirements.

Disintegration and biodegradation of the test materials have to be tested based on ASTM D 5338 or ISO 14855 where the test materials are incubated with the compost mixture in bioreactors for a minimum of 45 days or until reaching plateau stage for not more than 180 days at 58 ± 2 °C. First, to pass the disintegration criterion, the fraction of material with size greater than 2 mm should be no more than 10% of the original dry weight after sieving the final compost. Second, to pass the biodegradation requirement, the percentage of mineralization of the test material must be greater

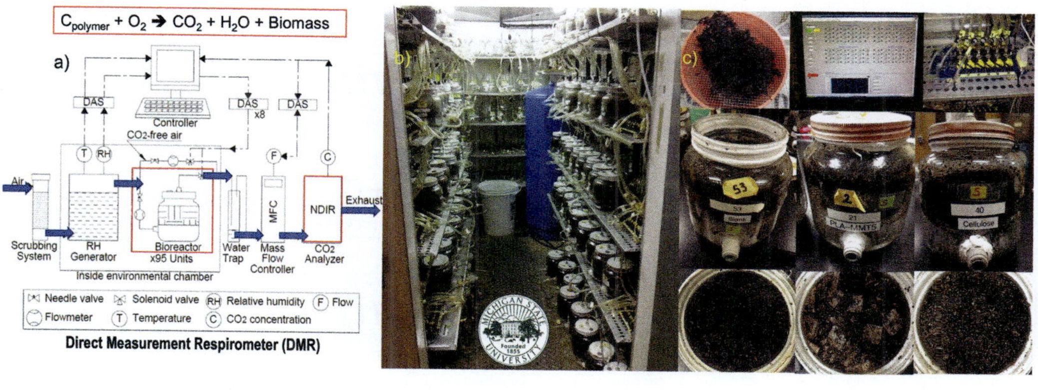

Figure 14.10 In-house built equipment for testing biodegradation at mesophilic and thermophilic composting conditions by analysis of evolved CO_2. (a) Schematic of diagram, (b) actual unit at MSU, (c) bioreactors, and acquisition and controlling system.

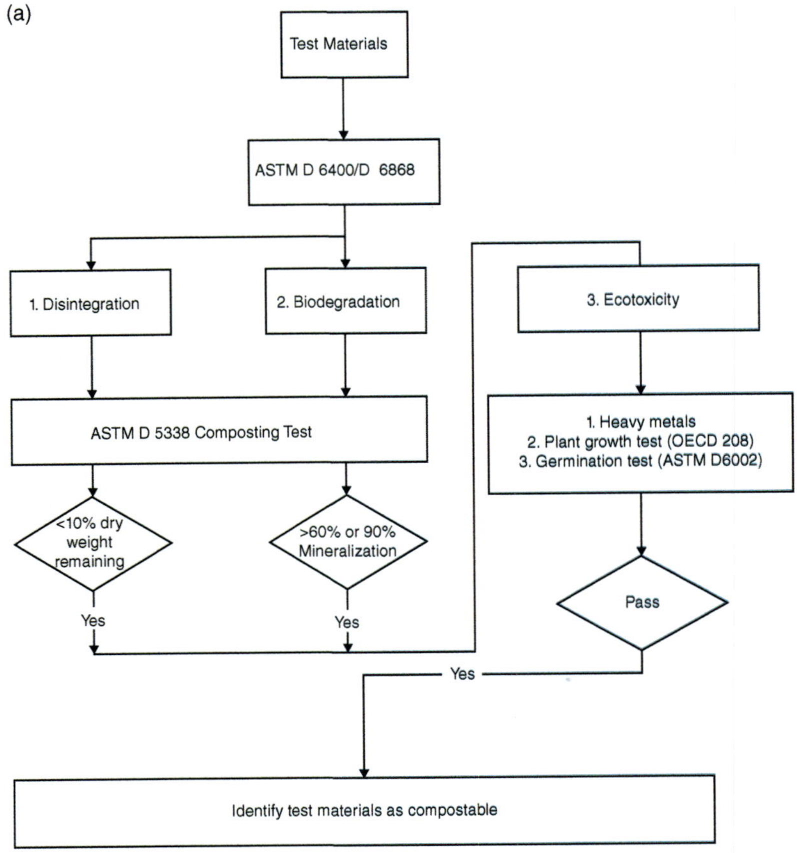

(b)	WORLWIDE	EUROPE	US	AUSTRALIA
Plastics	ISO 17088	EN 14995	ASTM D6400	AS 4736
Packaging	ISO 18606	EN 13432		

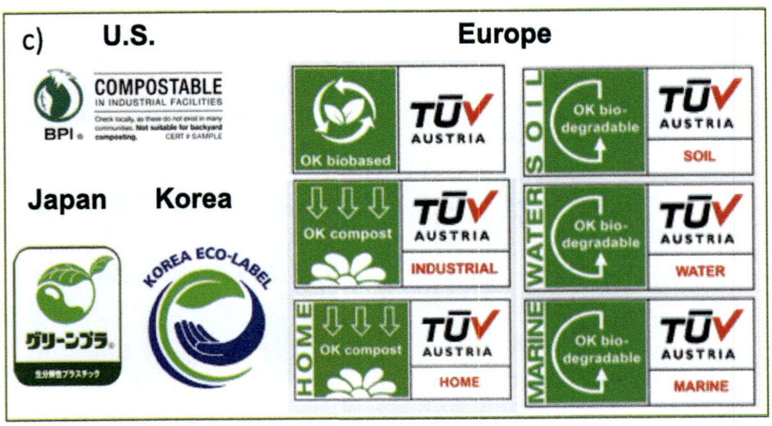

Figure 14.11 Framework review for biodegradable polymer and paper-based materials certification: (a) steps to certify a polymer as compostable according to ASTM D6400; reproduced from [3]; (b) main international and regional standards to certify plastics and packaging as compostable; (c) main US, Japan, Korea, and Europe labels for compostability and biodegradation in several environments. Note: Logos are reproduced from the US: BPI: https://bpiworld.org, JAPAN GREEN PLA: http://www.jbpaweb.net, KOREA: http://www.koreabiopack.org/info/info.php?id=bp11, EUROPE: Vincotte https://www.tuv-at.be/green-marks.

than 60% for a material containing a single polymer (homopolymer or random copolymer) or 90% for a material containing a block/graft copolymer or blend. Basically, it requires that the amount of CO_2 emitted from the samples is at least 90% of the mineralization of cellulose when cellulose reaches 70%. Figure 14.12a shows the evolved CO_2 in gram for a blank compost, cellulose – positive control, and poly(lactic acid) –PLA. Figure 14.12b shows the mineralization values, which are estimated by Equation 14.2.

$$\%\ \text{Mineralization} = \frac{A(CO_2)_T - A(CO_2)_B}{M_{TOT} \times C_{TOT} \times \frac{44}{12}} \times 100 \qquad (14.2)$$

where % Mineralization is the percentage of carbon atoms converted to CO_2, $A(CO_2)_T$ is the average cumulative mass of CO_2 (g) evolved from the sample, $A(CO_2)_B$ the average cumulative mass of CO_2 (g) evolved from the blank, M_{TOT} the mass of test material (g), C_{TOT} the proportion of total organic carbon in the total mass of test material (g/g), 44 the molecular weight of carbon dioxide, and 12 the atomic weight of carbon. The denominator in Equation 14.2 is known as the theoretical amount of carbon dioxide $Th(CO_2)$ that can be produced by the test material, i.e., when 100% of the carbon molecules present in the test material were converted to CO_2.

Figure 14.13 shows that several regions can be differentiated in the biodegradation of polymers such as cellulose and PLA. An initial *lag phase* occurs when the microorganisms are able to start assimilating the available polymers. In the case of cellulose, which is readily biodegradable, microorganisms are able to digest it from the beginning so that the evolved CO_2 increases right from the start. In the case of high molecular weight (M_w) PLA, until the M_w is below ~ 10 kDa, microorganisms are not able to digest PLA so a lag phase of around 20 days is observed in which no CO_2 evolution happens (Figure 14.13 b). Since the mineralization is the subtraction of the value with the blank (Equation 14.2), negative mineralization is recorded since the blank produced more CO_2 than the sample. After the PLA M_w is below ~ 10 kDa, microorganisms are able to start using PLA oligomers as a food source and a fast mineralization phase is observed, known as the biodegradation phase. Finally, after all the available carbon is consumed, a plateau phase is obtained.

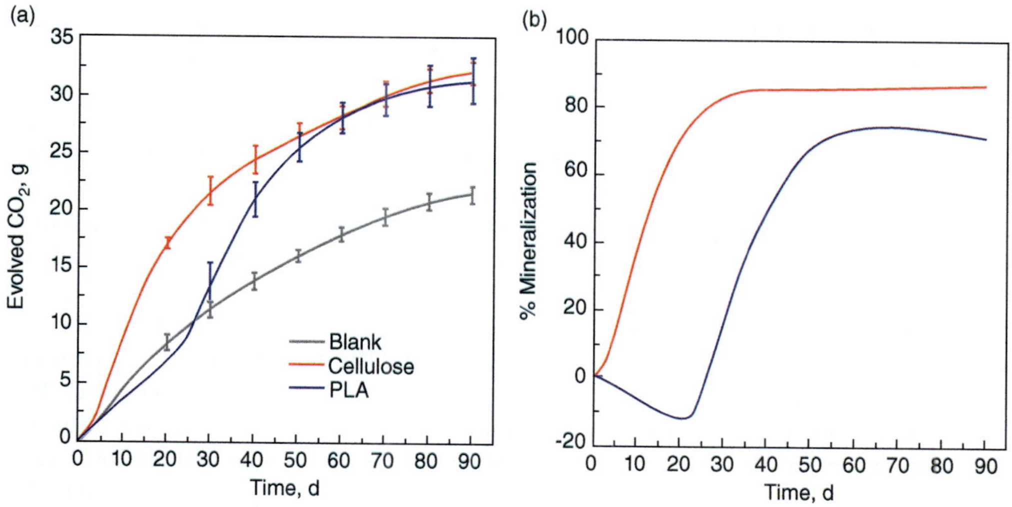

Figure 14.12 (a) Evolved CO_2 of blank compost, cellulose, and poly(lactic acid) – PLA – films; (b) mineralization of cellulose and PLA films; reproduced from [29] with permission from Elsevier.

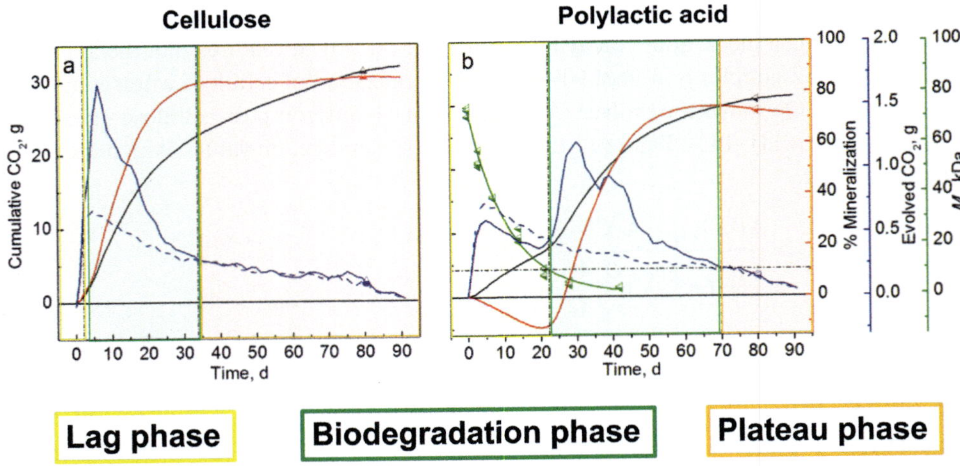

Figure 14.13 Biodegradation regions of cellulose and poly(lactic acid) – PLA – films; adapted from [29].

To pass the ecotoxicity requirement, the concentration of heavy metals in the plastic materials must be less than 50% of those in Table 3 of 40 CFR Part 503.13 for the US, in Table I of the Trade Memorandum T-4-93, or less than that in Annex A.1 in EN 13432, and the germination rate and plant biomass of the sample compost must not be less than 90% of those of the control compost. The heavy metals whose concentrations are limited by the standards include arsenic (As), cadmium (Cd), copper (Cu), fluorine (F), lead (Pb), mercury (Hg), nickel (Ni), selenium (Se), and zinc (Zn). Additional determination of chromium (Cr) and molybdenum (Mo) concentrations in the polymer are required to pass the EN 13432 standard. Furthermore, Canada limits the amount of cobalt (Co). Overall, the maximum limits of heavy metal content in the packaging material in the European standard (CEN), DIN V 54900 (from the German Institute for Standardization), and GreenPla® (from Japan Bioplastics Association) are much stricter than those in the ASTM International and the Canadian standards. Table 14.1 shows the metal limits in the different standards.

Plant germination and growth rate is tested according to OECD 208 and the testing scheme is illustrated in Figure 14.14. The germination rate and plant biomass of the sample compost must not be less than 90% of those of the control compost. Two plant species, a *Dicotyledonae* such as *Cucumis sativus* and a *monocotyledonous* such as *Avena sativa*, should be used.

After certified labs test the sample, the collected information together with the sample and the sample composition are submitted to the certifying institution. In the US, the Biodegradable Product Institute (BPI) certifies samples and products for bio-based content and compostability in industrial composting facilities [16]. In Europe, the certification is mostly run by the TÜV AUSTRIA Group [30]. Most certifications in other regions are based on the structure provided by the ASTM and ISO Standards framework.

A list of current US-certified products accepted for industrial compostability can be found at the BPI homepage [31]. A list from TÜV AUSTRIA can also be found [30]. The BPI does not certify home composting, nor soil, water, and marine biodegradability at the time of this chapter writing. In Europe, TÜV provides certification for industrial and home composting, and soil, water, and marine biodegradation.

Agreement among regions about the requirements for industrial composting and bio-based content certification are consistent and very similar; however, no such agreement exists for backyard, soil, water, and marine biodegradation. Standards are evolving, and this is an area of much discussion among the members of standards organizations such as ASTM and ISO. Companies with international presence and products must be aware of this situation if they are to certify their

Table 14.1 Metal limits required for compostable samples; adapted from [3] and updated.

Element	Limit US	Limit Canada	DIN V 54900-1	EN 13432	Green Pla
	(ppm)	(ppm)	(ppm)	(ppm)	(ppm)
Arsenic	20.5	37.5	NS	5	3.5
Cadmium	19.5	10	0.3	0.5	0.5
Calcium	NS	NS	NS	NS	NS
Chromium	NS	350	30	50	50
Cobalt	NS	75	NS	NS	NS
Copper	750	378	23	50	37.5
Iron	NS	NS	NS	NS	NS
Lead	150	252	30	50	50
Mercury	8.5	2.5	0.3	0.5	0.5
Molybdenum	NS	10	NS	1	1
Nickel	210	90	15	25	25
Selenium	50	7	NS	0.75	0.75
Zinc	1400	934	100	150	150
Flourine	NS	NS	NS	100	100

Note: NS: Not Specified;
Metal limits for US are 50% of those proscribed by Table 40 CFR Part 503.13. Metal limits for Canada are 50% of those proscribed in Table 1 of the Trade Memorandum T-4-93 "Safety standards for fertilizers and supplements." Effective October 26, 2020. Data provided for a maximum acceptable concentration of a metal-based on annual application rate (mg metal/kg product) of 4,400 kg/ha per year. The Trade Memorandum also requests the testing of Thallium (Tl) < 2.5 and Vanadium (V) < 328 for special products, but not all.

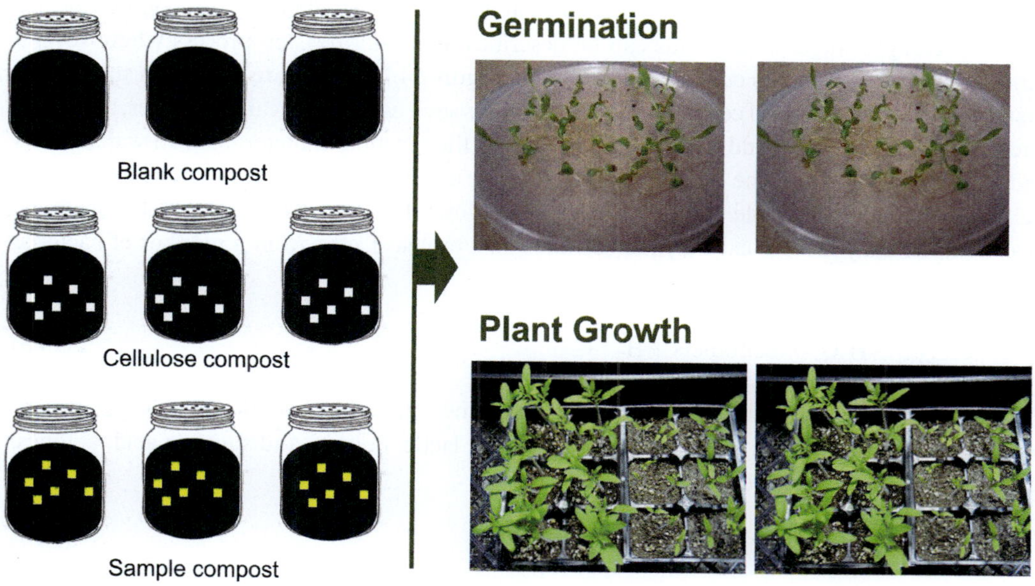

Figure 14.14 Germination and plant growth test according to OECD 208.

products in several markets for biodegradation with any of these end-of-life scenarios. The readers are directed to several peer-reviewed articles to gain additional insight on how to get material certified for bio-based content and biodegradation in different environments [29, 32].

14.3.6 Bio-based Carbon Content

To determine the amount of bio-based carbon in a sample, determinations of the ratio of $^{14}C/^{12}C$ are conducted. The basic concept of how the bio-based content in a sample can be measured is by using carbon-14 (^{14}C) radiation to separate modern carbon from fossil carbon. The fossil carbon is the carbon that contains essentially no radiocarbon or ^{14}C, since the age of fossil carbon (millions of years) is much greater than the half-life of C of 5730 years. Practically, to obtain this measurement a material is combusted in an oxygen-rich environment to form CO_2 gas, which later is trapped in a CO_2 trapping reagent. The trapping reagent then is further analyzed using liquid scintillation counters, which measure the sum of photons generated from the decay of C. The counts are compared to the C reference SRM 990c (oxalic acid), which is a Standard Reference Material supplied by the National Institute of Standards and Technology, Gaithersburg, MD. If the sample has the same count as that of the reference or 100% ^{14}C, it indicates that the sample contains only modern carbon or 100% bio-based carbon. On the other hand, 0% ^{14}C carbon signifies that the polymer contains no ^{14}C carbon indicating that the carbon atoms in the polymer matrices are entirely from fossil or petroleum-based carbon. The percentage of ^{14}C has to be corrected for the nuclear bomb testing effects from the 1950s that have significantly increased the ^{14}C in the atmosphere by multiplying the percentage of carbon by 0.93. In the US, the bio-based content certification is based on ASTM D6866, and very few labs perform the test.

14.4 Anaerobic Biodegradation

Anaerobic digestion is a complicated biological process in which bacteria break down organic matter in several steps. Anaerobic digestion has become one of the most common methods of sludge stabilization. Several complex feedstocks, such as agricultural wastes from animal manure and food processing wastes, have been treated by anaerobic biodegradation. The anaerobic degradation of complex organic materials can be described as a multiple-step process where several series and parallel reactions occur, as illustrated in Figure 14.15. There are four main stages. The first stage is the hydrolysis of complex organic materials such as proteins, carbohydrates, and lipids due to the action of water and enzymes. Hydrolysis is the reaction of materials with water. Acids and bases can accelerate the reaction. Enzymes can also accelerate or drive the process [33]. Cellulose, starch, and simple sugars can be broken down by water and enzymes. In anaerobic digestion, the enzymes are exoenzymes (cellulosome, protease, etc.) from a number of bacteria, protozoa, and fungi as indicated by:

$$C_{\text{material}} + H_2O \rightarrow \text{monomers} + H_2 \tag{14.3}$$

The second stage is acidogenesis where soluble monomers are converted into small organic compounds such as short chain acids – propionic, formic, lactic, butyric, and succinic acid, ketones, and alcohols – methanol and ethanol.

$$C_6H_{12}O_6 + 2H_2 \rightarrow 2CH_3CH_2COOH + 2H_2O \tag{14.4}$$

$$C_6H_{12}O_6 \rightarrow 2CH_3CH_2OH + 2CO_2 \tag{14.5}$$

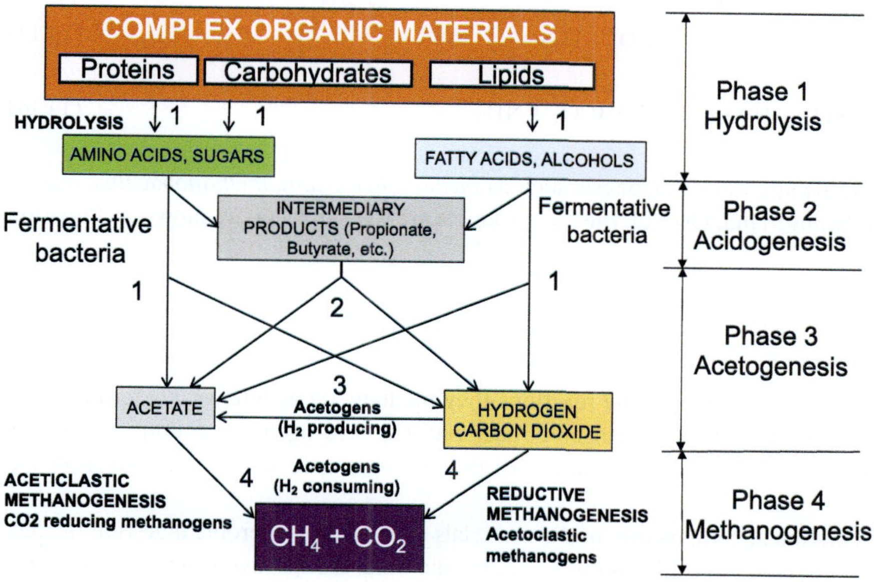

Figure 14.15 Reaction scheme for anaerobic digestion of complex organic materials; adapted from [34].

The third stage is acetogenesis where acetogenic bacteria – microorganisms that generate acetate (CH_3COO^-) – create acetic acid, CO_2, and H_2. Several bacteria contribute to this process such as *Syntrophobacter wolinii* (propionate decomposer), *Syntrophomonos wolfei* (butyrate decomposer), *Clostridium spp., peptococcus anaerobes, lactobacillus,* and *actinomyces* all are acid formers.

$$CH_3CH_2CO_6 + 3H_2O \rightarrow CH_3COO^- + H^+ + HCO_3^- + 3H_2 \tag{14.6}$$

$$C_6H_{12}O_6 + 2H_2O \rightarrow 2CH_3COOH + 2CO_2 + 4H_2 \tag{14.7}$$

$$CH_3CH_2OH + 2H_2O \rightarrow CH_3COO^- + 2H_2 + H^+ \tag{14.8}$$

$$2HCO_3^- + 4H_2 + H^+ \rightarrow CH_3COO^- + 4H_2O \tag{14.9}$$

During the last stage, methanogenesis, several reactions occur using the intermediate products of Equation 14.3 to Equation 14.9, and the main product is methane. Conversion of acetate to CH_4 occurs through the methanogenesis process, and reduction of several compounds is involved in producing CH_4 [34].

$$2CH_3CH_2OH + CO_2 \rightarrow 2CH_3COOH + CH_4 \tag{14.10}$$

$$CH_3COOH \rightarrow 2CH_4 + 2CO_2 \tag{14.11}$$

$$CO_2 + 4H_2 \rightarrow CH_4 + 2H_2O \tag{14.12}$$

$$CH_3OO^- + SO_4{}^{2-} + H^+ \rightarrow 2HCO_3 + H_2S \quad (14.13)$$

$$CH_3OO^- + NO^- + H_2O + H^+ \rightarrow 2HCO_3 + NH_4{}^+ \quad (14.14)$$

Several bacteria contribute to methagonesis such as *methanobacterium*, *methanobacillus*, *methanococcus*, and *methanosarcina*. Overall, the end products are CH_4 and CO_2 as shown in Equation 14.15.

$$C_{material} + H_2O \rightarrow CH_4 + CO_2 + C_{residue} + H_2S + NH_4{}^+ + H_2O + salts \quad (14.15)$$

The major groupings of bacteria and the reactions they mediate are as follows: (1) fermentative bacteria; (2) hydrogen-producing acetogenic bacteria; (3) hydrogen-consuming acetogenic bacteria; (4) carbon dioxide-reducing methanogens; and (5) aceticlastic methanogens, as shown in Figure 14.15.

The nature and chemical composition of the materials used in the anaerobic digestion process dictate the operative degradation subprocesses and the microbial groups involved in the anaerobic conversion of these substrates. Table 14.2 shows the chemical composition of various materials commonly used as substrates for anaerobic digestion. A significant component of activated sludge is crude protein. Meat-packing waste and primary sludge have a relatively high lipids content. Lignin and hemicellulose are substantial components of animal waste and municipal refuse.

Since anaerobic digestion produces CH_4 and CO_2 as biogas, they can be used to produce energy. Generally, large anaerobic digestor reactors are used for this purpose where manure, wastewater biosolids, food waste, and other organics are introduced to produce biogas and generate energy. Figure 14.16a shows the introduction of multiple organic materials in one digester, a practice called codigestion. Figure 14.16b shows the MSU anaerobic digestor unit which deals with manure from the MSU stock animals and food wastes from the student cafeterias. The MSU digester utilizes roughly 17,000 tons of organic waste from MSU and the greater Lansing area per year as feedstock to produce biogas used to generate over 2.8 million kWh of electricity per year. The digestor is a codigestor. The primary feedstocks include dairy manure from the MSU Dairy Teaching and Research Center, food waste from several campus dining halls, fruit and vegetable waste from the

Table 14.2 Composition of various substrates commonly found in anaerobic digestion process, expressed as % of dry matter; adapted from [34].

Component	Primary Municipal Sludge	Activated Sludge*	Municipal Refuse	Meat-packing Waste	Cattle Manure	Chicken Manure
Volatile solids	75.0	79.0	82.1	92.0	72.0	76.0
Lipids	10.3	5.8	6.36	54.6	3.5	1.5
Cellulose	32.2	9.7[b]	37.0	-	17.0	28.3
Hemicellulose	2.5	-	16.5	-	19.0	11.9
Lignin	13.6	-	-	-	6.8	9.2
Crude protein	19.0	53.7	6.07	28.7	19.0	28.8
Volatile acids	6.4[a]	-	-	-	-	-
Ash	25.0	21.0	18.4	8.0	28.0	24.0

[a]Expressed as total free fatty acids;.[b]measured as total carbohydrates;
*biological sludge derived from laboratory units fed with totally soluble feed.

Figure 14.16 Anaerobic digestor, (a) main substrates introduced in anaerobic digestor; reproduced from the US EPA [35]; (b) MSU anaerobic digestor to deal with cow manure and food waste from the student cafeteria.

Meijer Distribution Center in Lansing, and fats, oil, and grease from local restaurants. The mix of solids and liquid remaining after digestion (digestate) is pumped to a solid–liquid separator. Separated solids are composted at the MSU composting facility.

Additional anaerobic biodegradation environments are created in sewage, landfills, and some treatment reservoirs. Landfills will be discussed in Chapter 16.

14.4.1 Standards and Certifications for Anaerobic Biodegradable Materials

Several standards are used for measuring the anaerobic biodegradation of organic materials as shown in Table 14.3. The standards developed to understand the anaerobic biodegradation of plastics are mostly to understand their end of life, mainly in a landfill. However, several laboratory methods are currently used to measure the anaerobic biodegradation of many substrates. Most of the ASTM test methods are intended to simulate biologically active landfills or accelerated bioreactors where moisture and temperature are controlled, and biogas recovery is promoted. Extensive reports have been provided on the evaluation of plastic in these environments.

14.5 Main Factors Affecting Aerobic and Anaerobic Biodegradation

Biodegradation can be defined as "those processes which result from an attack on the material by living organisms, e.g., bacteria, fungi, insects, and rodents. The term is, however, usually restricted to the degradation caused by microorganisms."

Biodegradation is, fundamentally, a chemical process involving *enzymes* produced by microorganisms. A characteristic of enzyme reactions is that they tend to be highly specific – tailored to particular chemical bonds and particular configurations of structures surrounding the target bonds. Enzyme reactions are favored by conditions that accelerate microorganism growth: high relative humidity, moderately increased temperature, and darkness.

Biodegradability of organic materials depends on the availability of both specific sites for reaction, the enzyme in question, and environmental conditions, primarily moisture, and temperature. Moisture is essential in two regards. Most fundamentally, microorganisms require adequate levels of water to be able to grow and reproduce. Additionally, many biodegradation reactions involve

Table 14.3 Standards for biodegradation of plastic under anaerobic conditions.

Test Standards	Title	Obtained Information
ASTM D5511	Standard Test Method for Determining Anaerobic Biodegradation of Plastic Materials under High-Solids Anaerobic-Digestion Conditions	Test duration, % biodegradation as a function of time, % CH_4, and % CO_2 in evolved gas [36].
ASTM D5526	Standard Test Method for Determining Anaerobic Biodegradation of Plastic Materials under Accelerated Landfill Conditions	Test duration, % biodegradation as a function of time, % CH_4, and % CO_2 in evolved gas [37].
ASTM D7475	Standard Test Method for Determining the Aerobic Degradation and Anaerobic Biodegradation of Plastic Materials under Accelerated Bioreactor Landfill Conditions	Temperature range of the test as a function of time, test duration, % biodegradation as a function of time, % CH_4, and % CO_2 in headspace, changes in molecular weight, weight, tensile, and other properties of the samples [38].

Note: ASTM D5210 – Standard Test Method for Determining the Anaerobic Biodegradation of Plastic Materials in the Presence of Municipal Sewage Sludge was withdrawn in 2016.

hydrolysis. In some cases, prior hydrolysis produces molecular fragments small enough to be susceptible to microbiological attack.

For the most part, current synthetic polymers are resistant to biodegradation, while natural polymers are relatively susceptible. One fundamental reason is that although microorganisms are remarkably adaptable, in most cases for synthetic polymers, no polymer-specific enzymes have had time to evolve, owing to the relatively short period these molecules have existed and the relatively low exposure of microorganisms to these polymers in the absence of more easily consumed food sources. In other words, there have been few drivers for microorganisms to evolve in ways that permit them to degrade these particular types of chemical structures. In synthetic polymers, biodegradability is enhanced by low molecular weight, the presence of certain end groups, high polarity, and high hydrophilicity ("water-lovingness").

The biodegradation process and inherent biodegradability of materials are influenced by the type of material and environmental factors. The material factors can be divided into volume and surface aspects. The parameters associated with volume aspects can be categorized with material structure, such as morphology, crystallinity, constitutional unit, flexibility, cross-linking rate, molecular weight, tacticity, density, shape, and polarity; and the surface aspects are mostly related to hydrophobicity, roughness, and polarity.

The material is initially, and sometimes simultaneously, affected by one or several degradation mechanisms such as mechanical, thermal, photo, chemical, and biodegradation. These mechanisms act alone or produce a synergistic effect that accelerates the material degradation. All these mechanisms are affected by the external abiotic and biotic environmental factors. The main abiotic aspects are temperature, moisture, pH, radiation, substrate concentrations, abrasion, and oxygen. The main biotic aspects correspond to the density and diversity of the microbial population and the type of enzymes. The interrelationship between the environment and material is reflected in the rate and efficiency of the whole degradation process. Figure 14.17 shows a pictorial view of the main external environment and internal material factors affecting biodegradation.

The intensity or level of exposure to each of the external factors, plus the time that the material is exposed to them, are linked to the rate of biodegradation. So, environments can exert different

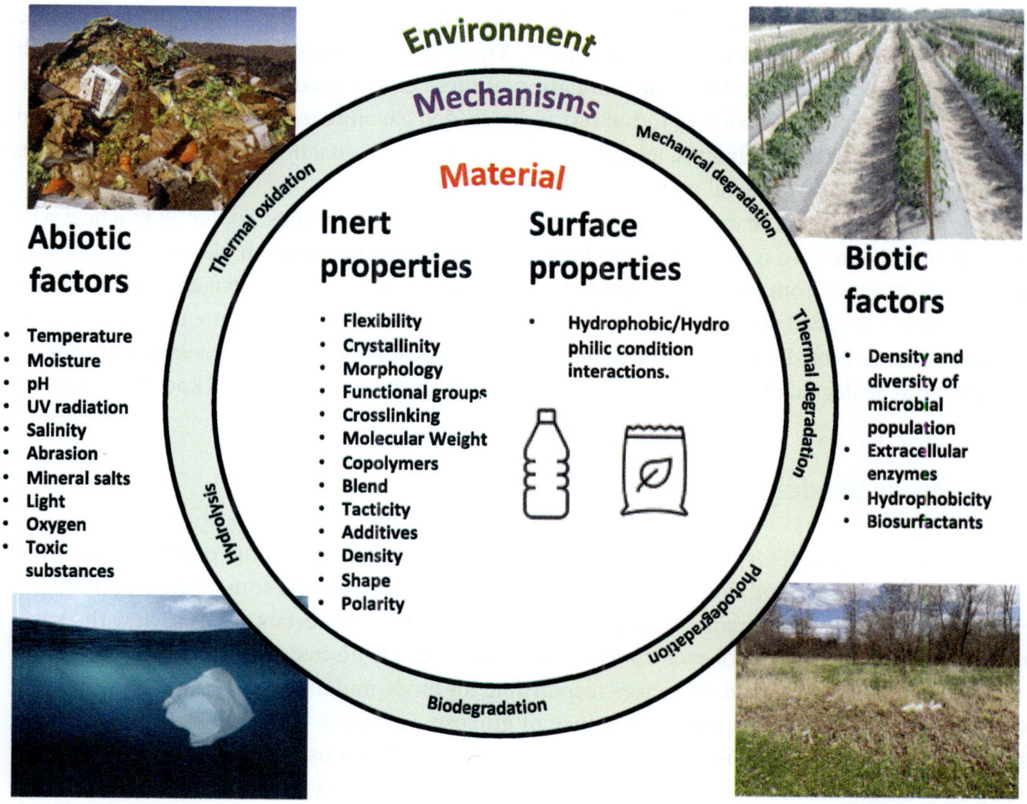

Figure 14.17 Main external environmental and internal factors affecting the biodegradation of polymers.

levels of these factors and affect the rate of the degradation process in very different ways. For example, soil, compost, and marine environments are different and impact material biodegradation differently. Substrates can go from just being reduced to small pieces with a potential to form micro- or nanostructures at the end of their life, to being cleaved and assimilable as molecules for biological steps for microorganisms, or they can linger in the environment concentrating different amounts of pollutants [32, 39, 40]. A full discussion of these factors and their roles is outside the scope of this chapter, but they are essential to understand when developing biodegradable polymers. Many review papers have discussed these factors, which the reader is encouraged to consult to gain additional insights [10, 15, 41–53].

14.6 Biodegradation of Packaging Materials

Metals, including ferrous and nonferrous materials used in packaging, are not truly biodegradable. They can be oxidized (rusted) but are not fully biodegradable by microorganisms. Special metal alloys mainly containing magnesium (Mg) designed for disintegration and often called biodegradable metals are suggested as revolutionary biomaterials for medical therapies, including bone grafting therapies. Mg and its alloys have been one of the most attractive biodegradable metals due to their nontoxicity and properties matching with the human body. As single compounds, Fe, Al, and Mg are essential nutrients for the biodegradation steps of microorganisms. However, at high concentrations, they may be toxic. Heavy metals such as lead, mercury, cadmium, and arsenic are considered pollutants, as discussed in Chapter 5. These heavy metals are generally found with organic pollutants in soil due to the accumulation of these contaminants due to emissions from, for example, the rapid expansion of industrial areas, leaded gasoline in the past, coal combustion residues, spillage of petrochemicals, and atmospheric deposition, so a need for remediation of the environment due to heavy metal contamination is common.

Glass is an inert material that stays intact in the environment for centuries. So, for all practical purposes and this chapter, we consider glass to break but not decompose. Broken glass has been found dating back to 2000 BC from glassmaking operations in Egypt. Glass is one of the longest-lasting human-made materials. Glass will never biodegrade but will slowly erode to particles that are not recognizable. Extensive work has been done to understand the effect of assimilating sodium and potassium from glass, but no practical biodegradation of glass has been reported, to the best of the authors' knowledge [54].

14.7 Paper Biodegradation

Paper is produced from fibers from several crops, but primarily for packaging, paper is made from wood cells that contain cellulose, hemicellulose, and lignin, as well as some minor extractives. Before discussing the biodegradation of paper and lignocellulose materials, let us review the basic structure and biodegradation of each compound since as already discussed, the chemical structure of the substrate is one of the main factors responsible for biodegradation.

Cellulose resembles the chemical structure of glucose. Cellulose is the most abundant macromolecule on earth. Cellulose is a linear homopolymer of glucose joined by β-1-4 linkages. Cellulose is a high molecular weight compound with a molecular weight larger than 100,000 Da. Cellulose provides great strength to paper. Each glucose molecule is "upside down" in relation to the neighboring glucose molecule so that the repeating unit is cellobiose (Figure 14.18). In contrast to other carbohydrates such as starch, the structure of cellulose in plants and in paper is able to provide a great deal of strength. The fibrils of cellulose have crystalline and amorphous regions. Depending on their origin and treatment, the crystallinity of cellulose can vary from fully amorphous to fully crystalline. Higher crystallinity makes cellulose resistant to chemical attack. During the wood pulping process, the cellulose is liberated from the lignin – the glue holding the fibers together.

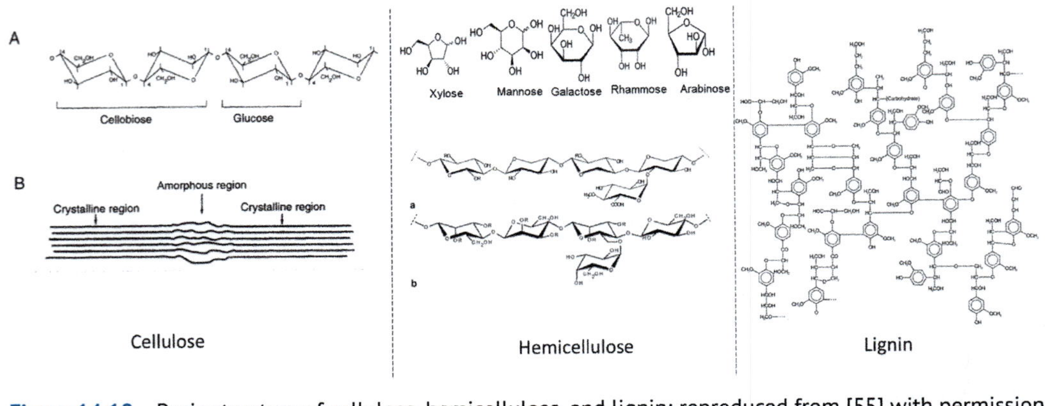

Figure 14.18 Basic structure of cellulose, hemicellulose, and lignin; reproduced from [55] with permission from Springer Nature.

In the secondary wall of plant cells, cellulose forms several sheets organized as parallel microfibrils. These microfibrils are embedded in a matrix of hemicellulose and lignin, as shown in Figure 14.19. Pure cellulose is available in several forms, such as cotton, filter paper, and Avicel. Amorphous forms of cellulose are used for biodegradation studies due to their fast rate of hydrolysis. Cellulolytic microorganisms can be found in all biota where cellulosic waste is present. Mixed populations of microorganisms are present, containing both cellulolytic and noncellulolytic species, which tend to interact synergistically. So, these interactions lead to the complete biodegradation of cellulose, which is ultimately converted to CO_2 and H_2O in an aerobic environment and CO_2, CH_4, and H_2O in an anaerobic environment, as previously explained.

Most of the cellulose biodegradation occurs by cellulolytic microorganisms belonging to eubacteria and fungi. Also, some anaerobic protozoa and slime molds are able to degrade cellulose.

Aerobic biodegradation of cellulose mostly occurs by cellulolytic bacteria; several species identified as the genera *Cellulomona*, *Pseudomonas*, *Thermomonospora*, and *Microbispora* have been shown to biodegrade cellulose [56].

In anaerobic environments, anaerobic cellulolytic microorganisms degrade cellulose. First, glucose derived from cellulose is fermented by cellulolytic and associated saccharolytic microorganisms, producing hydrogen, carbon dioxide, and various organic compounds including acids and alcohols (phase 1 – hydrolysis – Figure 14.15). In the second phase, these compounds are converted into acetate and carbon dioxide – acidogenesis. This step is thermodynamically unfavorable unless the partial pressure of hydrogen is maintained at deficient levels. Excess reducing equivalents must therefore be transferred to appropriate acceptors such as sulfate or carbon dioxide. Sulfate reducers reduce sulfate into hydrogen sulfide. CO_2 is converted to acetate by acetogens – phase 3 and into methane in methanogenesis – phase 4. Anaerobic biodegradation of cellulose occurs in manure, compost, and sludges. Hydrolysis of cellulose by animals in their gastrointestinal tract plays an essential role in the nutrition of animals. Few microorganisms produce a complete set of enzymes to degrade cellulose efficiently, but a larger group of microorganisms have a subset of cellulase enzymes [56].

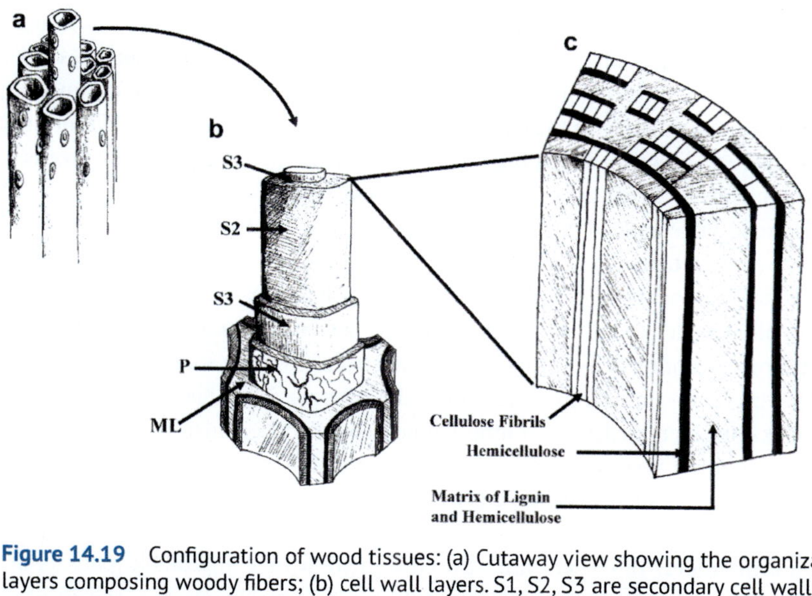

Figure 14.19 Configuration of wood tissues: (a) Cutaway view showing the organization of the cell wall layers composing woody fibers; (b) cell wall layers. S1, S2, S3 are secondary cell wall layers, P is the primary wall, ML is the middle lamella; (c) suggested relationship of lignin and hemicellulose to the cellulose microfibrils in the secondary walls; reproduced from [56] with permission from Springer Nature.

Hemicellulose is a complex carbohydrate polymer with lower molecular weight than cellulose, and makes up to 30% of the total dry weight of wood. Hemicellulose consists of D-xylose, D-mannose, D-glucose, D-galactose, L-arabinose, 4-O-methyl-glucuronic, D-galacturonic, and D-glucuronic acids. The sugars are linked by β-1,4 and sometime β-1,3-glycosidic bonds (Figure 14.18). In hardwood, the main component of cellulose is glucuronoxylan, and glucomannan is predominant in softwood (Figure 14.18). The main difference between cellulose and hemicellulose is that hemicellulose has a branched structure with short lateral sugar chains, which are easily hydrolyzed.

Hemicellulose biodegradation produces mostly monomeric sugars and acetic acid. Xylan is the main carbohydrate found in hemicellulose; its degradation requires the action of several hydrolytic enzymes. Two main enzymes, endo-1,4-β-xynalase and xylan 1,4,β-xylosidase, work together to degrade hemicellulose. Endo-1,4-β-xynalase produces oligosaccharides from the cleavage of xylan and xylan 1,4,β-xylosidase works on degrading the produced oligosaccharides. Several endoxylanases have been isolated able to degrade hemicellulose [55].

Lignin together with cellulose is the most abundant polymer in nature, and is present in the cellular cell wall, providing structural support, resistance against microbial attack, and impermeability. Lignin is an amorphous heteropolymer, nonsoluble in water, optically inactive, consisting of phenylpropane units linked by different bonds. Lignin is produced by the synthesis and dehydrogenization of several alcohols such as coniferyl, coumaroyl, and sinapyl alcohols. Coniferyl alcohols are the main component for softwood lignin, while coumaroyl and sinapyl are the main hardwood lignin components. The final structure of the polymerization produces a heterogenous polymer joined by C–C and aryl–ether linkages (Figure 14.18).

Due to the complexity of the heterogenous lignin polymer, its high molecular weight, and insolubility in water, it is difficult to degrade. Several extracellular, oxidative, and nonspecific enzymes, which release highly unstable products further undergoing oxidative reactions, are involved in the initial depolymerization of lignin. White-rot fungi – *Phanerochaete chrysosporium*, the most extensive – are the most cited microorganisms able to degrade lignin from wood. The main enzymes associated with ligninolysis are peroxidases (lignin peroxidases and manganese-dependent peroxidases) and laccases (blue-copper phenoloxidases).

Figure 14.20 shows a general overview of the degradation of lignocellulose materials in aerobic and anaerobic processes. Most of the biodegradation of lignin has been studied in aerobic processes [55, 56]. Cellulose and hemicellulose can be depolymerized through aerobic and anaerobic pathways. The degradation processes are associated with the capacity of producing the enzymes needed for the degradation of each substrate. The presence of coatings can interfere with the biodegradation process and the ability of microorganisms to attack lignocellulose materials.

A large list of certified compostable molded fiber containers can be found at the BPI homepage [31]. Most of these containers are used for take-out food and have short shelf-life requirements. To protect the containers from water damage, a coating is generally introduced, polymer-based materials that are hydrophobic and will not allow water to diffuse through. The primary marketing tool of these containers is to appeal to consumers wanting to use containers derived from natural resources, in this case, fiber-based containers. However, some initial push-back from composter operators has occurred. As the amount of compostable packaging increased, the resulting compost could not be sold to organic farmers. One of the primary users of compost is organic farmers, but the National Organic Program considers compostable packages and containers such as plates, cups, cutlery, and plastic bags as synthetic materials that cannot be used in compost for organic production, so this limitation hinders the acceptance of compostable packages. Furthermore, producers are also concerned with compostable packaging designed to hold wet or greasy food that contains chemicals known as "forever chemicals" – specifically PFAS, which can transfer into finished compost and contaminate soil and waterways [57].

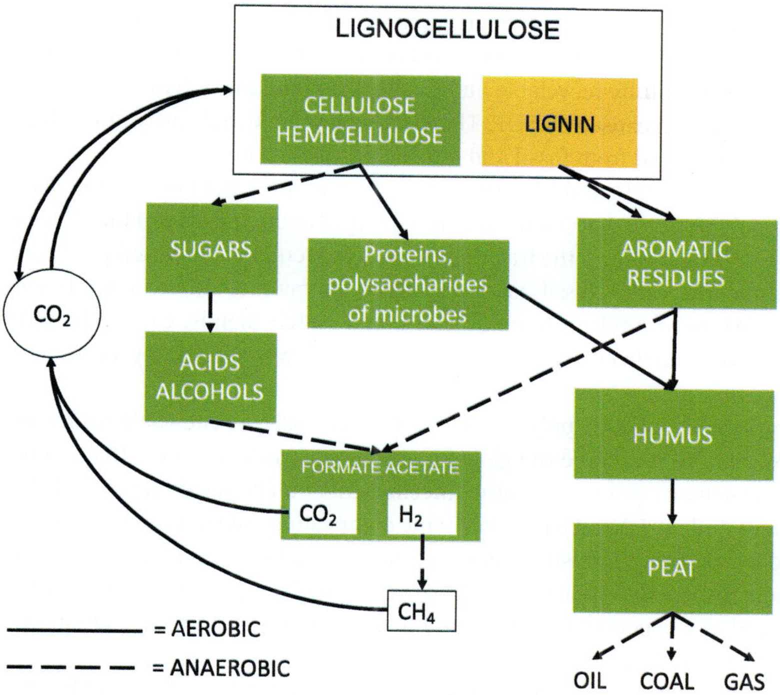

Figure 14.20 Carbon cycle of lignocellulose; adapted from [7].

Additionally, the companies recognized that a small amount of conventional plastic could be commingled, and it is impossible to sort or screen out. However, these issues have been considered, and the BPI, for example, is not certifying products containing high levels of fluorine and PFAS.

14.8 Polymer Biodegradation

The precise nature of plastic – providing easy processability, durability, low cost, and abundance – favors its use in almost all durable and nondurable goods, containers, and packaging, and by all industrial sectors such as medical, packaging, automotive, agriculture, and electrical components. On the other hand, the ability of plastic to persist even in severe environments has led to white pollution (i.e., leakage and accumulation of plastics in the environment) and consumer concerns. Plastic pollution is perceived by many as a global crisis leading to ecological imbalance. A paradigm shift in consumer perspectives and awareness due to the growing environmental concerns has pushed industries to design polymers that leave a minimal carbon footprint, especially to replace flexible and rigid single-use plastics (SUP).

Increasing amounts of SUP end up in landfills as a percentage of MSW, in the environment, in drainage systems all around the world, and ultimately leaking into the oceans. SUP has been one of the main offenders blamed for white pollution, and it has increased to dramatic proportions [58]. Most fossil-carbon plastics were not conceived while thinking about their end of life, so they do not break down to monomers and they do not biodegrade. Thus, there is an imperative need to design and produce sustainable plastics to create a long-term circular economy. Ideally, these plastics would be obtained from benign and renewable resources; have the required mechanical, thermal and barrier performance properties; can be adequately recovered after their useful life through biological, chemical, and mechanical recycling; and do not harm the environment if they end up as litter. The development of fully recyclable and easily biodegradable polymers has great potential to mitigate plastic pollution caused by SUP. These fully recyclable and easily biodegradable polymers for SUP can be produced from fossil and renewable resources.

Polyolefins (i.e., LDPE, LLDPE, HDPE, and PP) currently represent more than 60% of the plastics used [59]. These fossil polymers are derived from different types of chemicals produced from the cracking of petroleum or nowadays from the fracking of gas. Production of bio-based polymers is several orders of magnitude lower than fossil-based polymers; the main commercial bio-based polymers are poly(lactic acid), poly(ethylene terephthalate), and starch blends. Currently, bio-based polymers are mostly derived from starch-derived sugars from starch-heavy crops not intended for human consumption.

Among all these polymers, biodegradable polymers derived from fossil or renewable resources are seen as a promising solution to overcome the global plastic waste problem since they can be recovered by all waste management options, including mechanical and chemical recycling if the collected volume is sufficient, industrial and backyard/home composting, and energy recovery. If these polymers are created from a benign constitutional unit or building block from the onset, they can be deployed as a socially responsible choice and nondamaging to the environment. So, biodegradable polymers are viewed as a dependable present and future option to achieve sustainability and the long-term circular economy of plastics.

Biodegradable polymers can be produced from fossil and natural resources, as shown in Figure 14.21. However, the main benefits are obtained when biodegradable polymers are produced from renewable resources since they are able to replenish the carbon cycle (i.e., the times needed to create them and to convert them to biomass are roughly equivalent; Figure 14.1) [1].

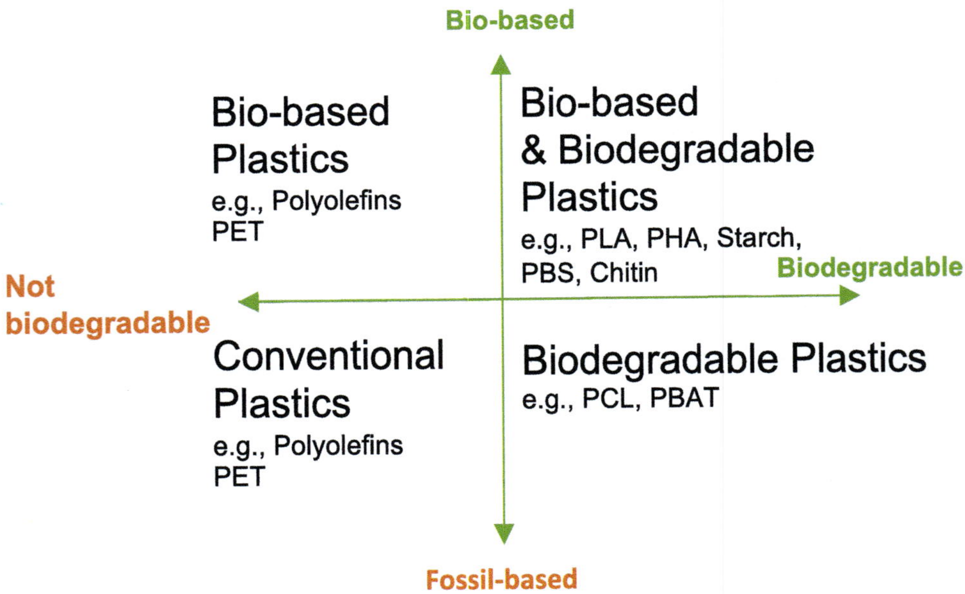

Figure 14.21 Classification of biopolymers according to their sources and biodegradation.

A few definitions are important to recall so that we can avoid any confusion.

- Bio-based plastic: A plastic produced from renewable resources.
- Degradable plastic: A plastic designed to undergo a significant change in its chemical structure under specified environmental conditions, resulting in a loss of some properties.
- Biodegradable plastic: A degradable plastic in which the degradation results from the action of naturally occurring microorganisms such as bacteria, fungi, and algae.
- Compostable plastic: A plastic that undergoes degradation by biological processes during composting to yield carbon dioxide, water, inorganic compounds, and biomass at a rate consistent with other known compostable materials and leaves no visually distinguishable or toxic residues.

Figure 14.22 shows the different sources where bio-based polymers can be extracted directly from biomass such as starch, and cellulose-based polymers, produced directly by microbes such as PHAs or bacterial cellulose, and derived from bio-derived chemical blocks such as PLA and bio-based PE.

Several companies have been intensively working for the last two decades on producing biodegradable polymers. Table 14.4 lists some examples of commercially available biodegradable polymers made from fossil and bio-based resources. Global companies, such as BASF SE (Germany), NatureWorks LLC (USA), Eastman Chemical LLC (USA), Biome Technologies PLC (UK), Mitsubishi Chemical Corporation (Japan), and Plantic Technologies Ltd. (Australia), are developing new biodegradable plastics derived from fossil- and bio-based resources.

As new targets to reduce or to avoid plastic reaching the environment are globally implemented, fully recyclable bio-based and biodegradable polymers could significantly help to achieve these goals. Among these polymers, thermoplastic starch (TPS), poly(lactic acid) – PLA, bio-poly(butylene succinate) – Bio-PBS, and polyhydroxyalkanoates (PHAs) produced preferably from renewable resources can make a difference in reducing the amount of plastic reaching the MSW stream and environment.

Narancic et al. evaluated the biodegradation of several plastics in aerobic and anaerobic managed and unmanaged environments [86]. They reported that PLA, PCL, PBS, PHO (polyhydroxyoctanoate), PHB, and TPS all are compostable in industrial composting settings (Table 14.5). PCL, PHB, and TPS were the only ones fully compostable in in-home composting. However, several blends of these polymers with PLA and PHO were able to biodegrade in home composting. Only PHB and

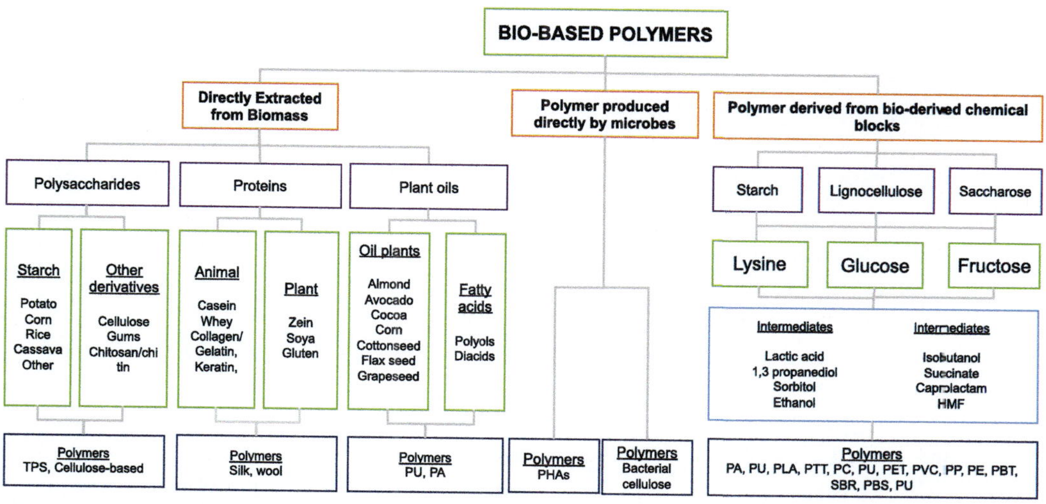

Figure 14.22 Sources to produce bio-based polymers.

Table 14.4 Commercially available biodegradable polymers derived from fossil-based and bio-based resources.

	Polymer	Company	Trade Name	Ref.
Fossil-based polymers	PBAT: Poly(butylene adipate-co-terephthalate)	BASF	Ecoflex®	[60]
	PTMAT: Poly(tetramethylene adipate-co-terephthalate)	Eastman (Novamont)	Eastar Bio™	[61]
		Jinhui ZhaoLong	Ecoworld	[62]
		Dupont™	Biomax®	[63]
	PVOH: Poly(vinyl alcohol)	Kuraray Poval, Selvol, Gohsenol, Mowiol, Exceval, Polyviol, Sinopac, Elvanol	(multiple names)	[64]
	PCL: Polycaprolactone	BASF	Capromer™	[65, 66, 67, 68, 69]
		eSun	eSUNXXXXC	
		Perstorp	Capa™	
		Daicel	Ɛ-caprolactone	
		Sumitomo	Caprolactam	
	PBS: Poly(butylene succinate)	Cortec	Eco Film™	
		Mitsubishi Chemical	GS PLA®	[70]
		BASF-Corbion	Succinity®	[71]
Bio-based polymers	PHAs: Poly(hydroxyalkanoate) family	Yield10Science	Myrel™	[72]
		Mango, Redwood, CA, US	Mango Materials	[73]
		TianAn Biologic Materials Co. Ltd, China	TianAn Biopolymer ENMAT™	[74]
		Kaneka	PHBH™	[75]
	PHB	Danimer Scientific	Nodax™	[76]
	PLA: Poly(lactic acid) and PLA blends	Natureworks LLC	Ingeo™	[77]
		Total Corbion	Luminy® Puralact®	[78]
		Pyramid Technologies		[79]
		Synbra	Synterra®	[79]
	Starch polymers	BioLogiQ	NuPlastiQ GP®	[80]
		Novamont™	Mater-BI®	[81]
	Bio-polybutylene succinate (Bio-PBS)	Mitsubishi Chemical	BioPBS™	[70, 82, 83]
	Cellulose-based polymers	Eastman	Treva™	[84]
	PDO: 1,3-Propanediol – ingredient copolymer	Dupont™	Susterra® Zemea®	[85]
Blends of bio-based and fossil polymers	Starch + PCL	Novamont™	Mater-Bi®	[70]
	PLA + PBAT	BASF	Ecovio®	
	PBS blend	Cortec	Eco Works®	

Disclaimer: Some of the reported companies are listed as producers of the polymers, but some production plants are not always working and/or are currently sitting idle.

Table 14.5 Biodegradation of bioplastics in managed and unmanaged environments tested according to international biodegradation standards; adapted from [86].

	Managed Environments			Unmanaged Environments			
Polymer or Blend	Industrial Composting	Home Composting	Anaerobic Digestion	Marine	Fresh Water	Anaerobic Aquatic Digestion	Soil
PLA	Green	Orange	Green	Orange	Orange	Orange	Orange
PCL	Green	Green	Green	Orange	Orange	Orange	Green
PBS	Green	Orange	Orange	Orange	Orange	Orange	Orange
PHO	Green	Orange	Orange	Orange	Orange	Orange	Orange
PHB	Green	Green	Green	Green	Green	Green	Green
TPS	Green	Green	Green	Green	Green	Green	Green
PLA-PCL (80/20)	Green	Green	Green	Orange	Orange	Orange	Orange
PLA-PBS(80/20)	Green	Orange	Orange	Orange	Orange	Orange	Orange
PLA-PHB (80/20)	Green	Green	Orange	Orange	Orange	Orange	Orange
PLA-PHO (85/15)	Green	Green	Orange	Orange	Orange	Green	Green
PHB-PHO (60/40)	Green	Green	Green	Orange	Orange	Orange	Green
PHB-PCL (60/40)	Green	Green	Green	Orange	Orange	Orange	Orange
PHB-PBS (50/50)	Green	Green	Orange	Orange	Orange	Orange	Orange
PCL-PHO (85/15)	Green	Green	Green	Orange	Orange	Orange	Orange
PCL-TPS (70/30)	Green	Green	Green	Orange	Orange	Orange	Green

PLA: Poly(lactic acid); PCL: polycaprolactone; PBS: poly(butylene succinate); PHO: polyhydroxyoctanoate; TPS: thermoplastic starch; PHB: polyhydroxybutyrate.

Green color indicates that the polymer passes a respective standard.
Orange color indicates that the polymer or blend fail a respective standard.

TPS can biodegrade in unmanaged environments. Evaluation of biodegradable polymers is a very active area of research.

14.9 Case Study 14.1. Biodegradation of Poly(butylene adipate-*co*-terephthalate) – PBAT – Films in Yard, Food, and Manure Compost

PBAT is a polymer derived from fossil resources as indicated in Table 14.4. PBAT is an aliphatic aromatic copolyester certified as compostable by the Biodegradable Products Institute (BPI) according to the ASTM D6400 specification [3]. PBAT is a linear random copolyester consisting of two types of dimer. The rigid section BT is an ester repeat unit consisting of 1,4 butanediol and

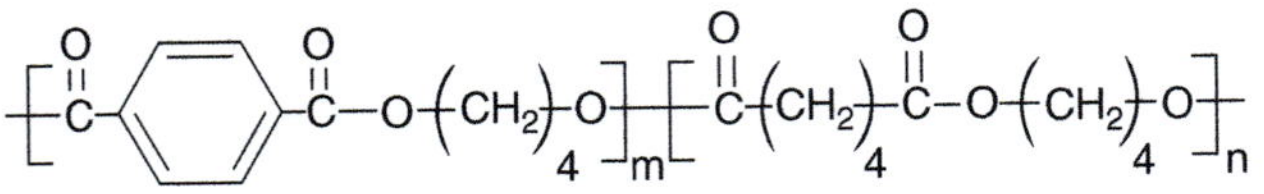

Figure 14.23 Constitutional unit of poly(butylene adipate-co-terephthalate) – PBAT.

terephthalic acid monomers, while the flexible section BA consists of 1,4 butanediol and adipic acid monomers (Figure 14.23.)

According to ASTM D6400, to be compostable, a material must possess three essential qualities: the ability to disintegrate in a composting environment quickly, inherent biodegradability, which is tested using the ASTM D5338 method, and absence of ecotoxicity or any adverse effects of the final compost after decomposition. Biodegradability of PBAT in composting conditions is affected by both abiotic and biotic factors of the environment, such as temperature, moisture, pH, bio-surfactant and enzymes, and by polymer characteristics, such as chain flexibility, crystallinity, regularity and heterogeneity, functional groups, and molecular weight. Kijchavengkul et al. demonstrated that cross-linking affected the biodegradation of PBAT films tested according to ASTM D5338 [87]. Figure 14.24 shows that PBAT biodegradation is largely reduced when the cross-linking degree is higher than 30%. So, to pass the requirements of ASTM D5338 to reach 60% at 45 days when cellulose already reached 100%, PBAT should have below 30% cross-linking rate.

Among biotic factors, extracellular enzymes produced by different microorganisms may have active sites with different specificity, such as complementary shape, charge, and hydrophilicity/hydrophobicity, and hence have more capability to biodegrade certain polymers. For example, *Aspergillus niger* and *Aspergillus flavus* fungi, present in soil and crops, respectively, produce enzymes that more easily digest aliphatic polyesters derived from 6 to 12 carbon di-acid monomers than those produced from other monomers. This enzymatic specificity may contribute to different degradation rates of biodegradable polyesters in differing compost environments. So, we can expect that the biodegradation of PBAT in different compost environments differs. Figure 14.25 shows the biodegradation of PBAT in manure, food, and yard compost. Mineralization of PBAT in manure compost is faster than in food and yard compost, indicating that the microbial composition population plays a very important role in the biodegradation of PBAT [88]. This can be extrapolated to other polymers.

14.10 Case Study 14.2. Anaerobic Degradation of PLA Films

The anaerobic biodegradation of PLA was tested in accelerated landfill conditions (ALC test) according to ASTM D5526 at 21°C for 6 months, and in high accelerated landfill conditions (HSAD) test according to ISO1595 by Kolstad et al. [89]. The ALC test matrix and inoculum was prepared by adding waste (vegetable, fruit, and garden) and composting. The inoculum sludge for the ALC test was from an anaerobic sewage water treatment plant. The HSAD test matrix and inoculum was obtained from a lab digester with maize silage as the main substrates. Two sample PLA films, fully amorphous and crystalline, were tested and cellulose was used as a control. Figure 14.26 a and b show the biodegradation of the PLA samples in the ALC and HSAD test. Figure 14.26a shows that the amorphous and semicrystalline PLA did not biodegrade after 400 days at 41°C in the ALC environment. In the case of the HSAD (Figure 14.26b), the amorphous PLA showed some biodegradation. However, there was only minor biodegradation of the semicrystalline PLA. This

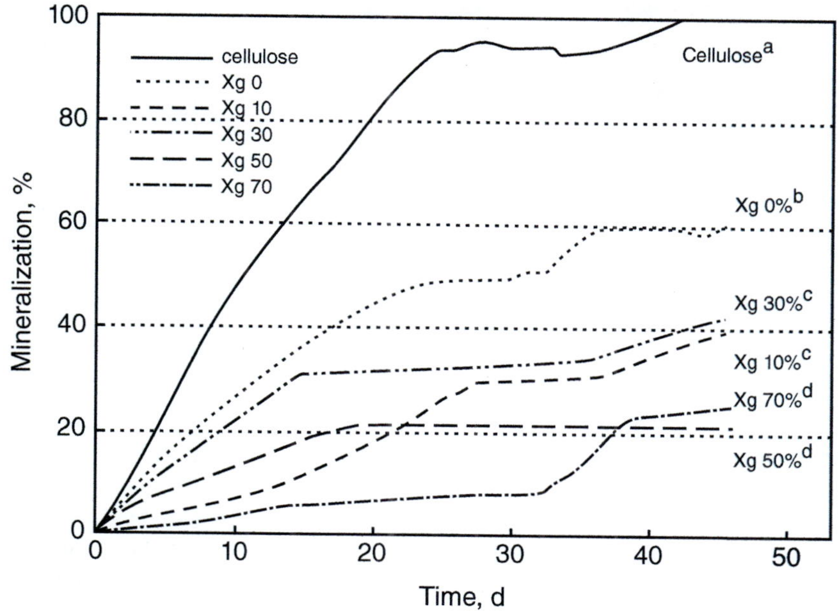

Figure 14.24 Percentage of mineralization of cross-linked PBAT films compared to noncross-linked PBAT films and cellulose after an incubation period of 45 days in manure compost mixture; reproduced from [87] with permission from Elsevier.

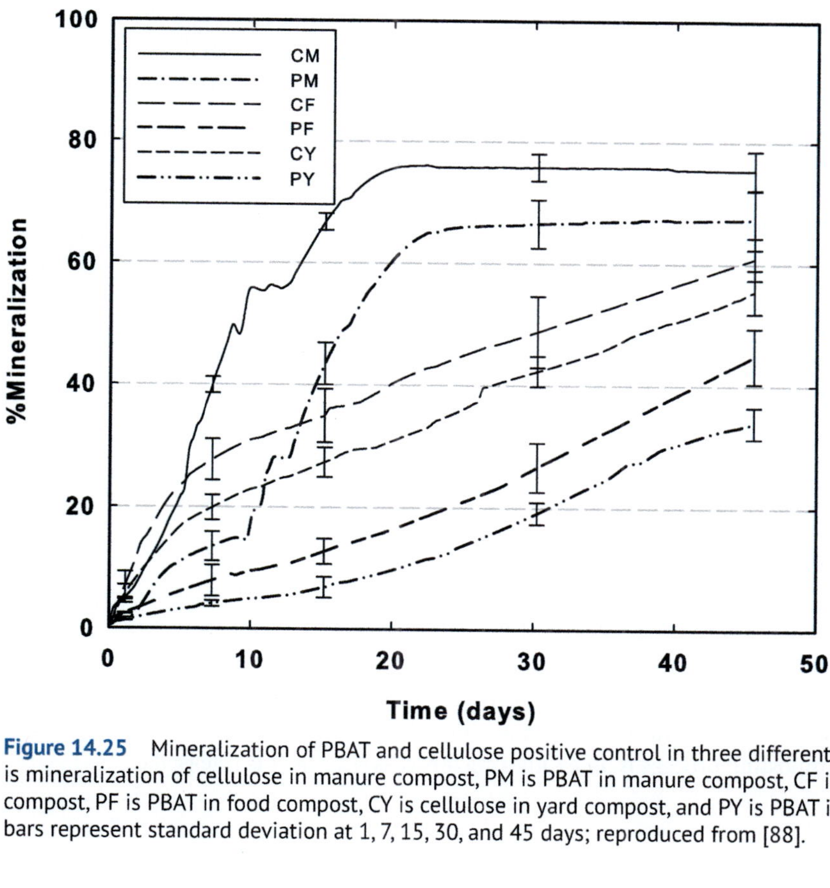

Figure 14.25 Mineralization of PBAT and cellulose positive control in three different composts, where CM is mineralization of cellulose in manure compost, PM is PBAT in manure compost, CF is cellulose in food compost, PF is PBAT in food compost, CY is cellulose in yard compost, and PY is PBAT in yard compost. Error bars represent standard deviation at 1, 7, 15, 30, and 45 days; reproduced from [88].

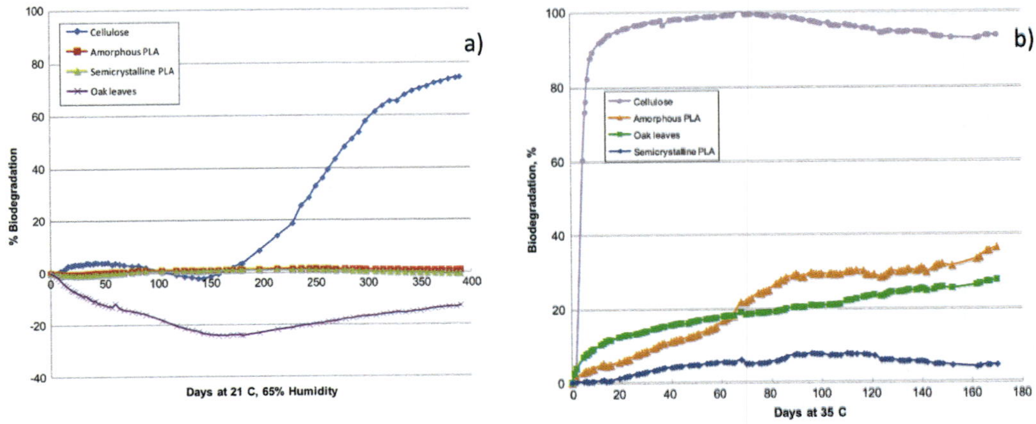

Figure 14.26 Biodegradation in anaerobic conditions: (a) calculated biodegradation for ALC test at 65% humidity, (b) calculated biodegradation for HSAD test; reproduced from [89] with permission from Elsevier.

information is useful to indicate that if PLA is disposed in a landfill, biodegradation should not be expected, and it would require a chemical hydrolysis step prior to any biodegradation. So, PLA samples may persist for several decades in a landfill before any considerable evolution of gas is detected.

14.11 Study Questions

1 Explain the carbon cycle of bio-based and fossil-based materials.

2 A bio-based and a fossil-based polymer have the same carbon cycle.
- **A** True
- **B** False

3 Degradation mechanisms can be divided into abiotic and biotic.
- **A** True
- **B** False

4 Define the main degradation mechanism.

5 During aerobic biodegradation, CO_2, H_2O, $C_{biomass}$, $C_{residues}$, and salts are produced.
- **A** True
- **B** False

6 Nitrogen is an essential component during aerobic biodegradation.
- **A** True
- **B** False

7 Describe the chemical reaction occurring during aerobic biodegradation.

8 Define composting according to the ASTM standard.

9 What are the main factors affecting a composting operation?

10 Provide a short description of home or backyard composting.

11 Provide a short description of industrial composting.

12 What are the main methods of industrial composting?

13 The industrial composting process is an exothermic process (i.e., heat is released).
- **A** True
- **B** False

14 What are the main phases of a composting operation?
15 Humic substances are a great reservoir of carbon.
 A True
 B False
16 Several certifying entities require that the temperature of compost exceeds a specific temperature, such as 55 °C, and remains at that level for a set period, such as 15 day, to ensure that the resulting compost is pathogen free.
 A True
 B False
17 What are the main environments in which biodegradation can occur?
18 During the biodegradation of mulch films, what are the main factors to consider?
19 Which biodegradable materials are used in agriculture applications?
 A Paper
 B Plastic
 C Coated paper
 D All of the above
20 Photodegradation is a significant process for agriculture films.
 A True
 B False
21 What is a respirometer?
22 What are the main standards to certify compostability in the US?
23 Describe the main requirements for certifying a biodegradable material as compostable.
24 Describe the main phases of biodegradation.
25 Cellulose is used as a positive control for biodegradation studies according to ASTM D5338.
 A True
 B False
26 The biodegradation process of cellulose follows three main regions: lag, biodegradation, and plateau phases.
 A True
 B False
27 Which are the main organizations providing compostability certification?
28 Heavy metals such as Ar, Cd, Pb, and Hg are not regulated for samples certified as compostable.
 A True
 B False
29 Germination of one plant species must be tested for compostable certification.
 A True
 B False
30 Describe the anaerobic degradation process.
31 What are the main standards for testing anaerobic degradation in the US?
32 Describe the main factors affecting aerobic and anaerobic biodegradation.
33 Are ferrous and nonferrous metals used in packaging biodegradable?
34 Is glass used in packaging biodegradable?
35 Provide examples of biodegradable lignocellulose materials.
36 Provide examples of biodegradable polymers.
37 Define bio-based plastic.
38 Define degradable plastic.

39 Define biodegradable plastic.
40 Define compostable plastic.
41 List three companies producing compostable plastics.
42 Is PLA compostable in home-composting?

14.12 Additional Resources

1 Composting at Home
https://www.epa.gov/recycle/composting-home

2 BPI
https://bpiworld.org

3 Anaerobic digestion (AD)
https://www.epa.gov/anaerobic-digestion

4 Sustainable Management of Food – Types of Composting and Understanding the Process
https://www.epa.gov/sustainable-management-food/types-composting-and-understanding-process

5 Byocycle
https://www.biocycle.net

6 Great compost in five easy steps
https://www.canr.msu.edu/news/great_compost_in_five_easy_steps

7 Cornell Composting
http://compost.css.cornell.edu

8 US Composting Council
https://www.compostingcouncil.org

9 CCET guideline series on intermediate municipal solid waste treatment technologies – Composting
https://www.ccet.jp/sites/default/files/2020-10/Composting%20guideline_web_201014.pdf
Video: https://www.youtube.com/watch?v=kuPFklbNxQs

10 Anaerobic Digestion Research and Eduation Center (ADREC)
https://www.egr.msu.edu/bae/adrec

References

1 Narayan, R. (2011). Carbon footprint of bioplastics using biocarbon content analysis and life-cycle assessment. *MRS Bull* 36: 716–721. DOI: 10.1557/mrs.2011.210.

2 Schlosser, D. (2021). *Cultivation of Filamentous Fungi for attack on synthetic polymers via biological fenton chemistry*, 648, 1st e. Elsevier Inc. DOI: 10.1016/bs.mie.2020.12.006.

3 Kijchavengkul, T. and Auras, R. (2008). Perspective compostability of polymers. *Polym Int* 57: 793–804. DOI: 10.1002/pi.2420.

4 USCC (2019). New compost definition – Results from USCC work with AAPFCO. US Composting Council. https://www.compostingcouncil.org/news/467612/New-Compost-Definition–Results-From-USCC-Work-with-AAPFCO.htm (accessed November 16, 2021).

5 Mitaftsi, O. and Smith, S.R. (2006). Quantifying household waste diversion from landfill disposal by home composting and kerbside collection.

6 Australian Standard. (2010). AS 5810 - Biodegradable plastics — Biodegradable plastics suitable for home composting, 1–19.

7 Tuomela, M., Vikman, M., Hatakka, A., and Itavaara, M. (2000). Biodegradation of lignin in a compost environment: A review. *Bioresour Technol* 72: 169–183. DOI: 10.1016/S0960-8524(99)00104-2.

8 Kijchavengkul, T., Auras, R., Rubino, M., Ngouajio, M., and Thomas Fernandez, R. (2006). Development of an automatic laboratory-scale respirometric system to measure polymer biodegradability. *Polym Test* 25: 1006–1016. DOI: 10.1016/j.polymertesting.2006.06.008.

9 Qi, Y., Yang, X., Pelaez, A.M., Huerta Lwanga, E., Beriot, N., Gertsen, H. et al. (2018). Macro- and micro- plastics in soil-plant system: Effects of plastic mulch film residues on wheat (Triticum aestivum) growth. *Sci Total Environ* 645: 1048–1056. DOI: 10.1016/J.SCITOTENV.2018.07.229.

10 Sander, M. (2019). Biodegradation of polymeric mulch films in agricultural soils: Concepts, knowledge gaps, and future research directions. *Environ Sci Technol* 53: 2304–2315. DOI: 10.1021/acs.est.8b05208.

11 Zumstein, M.T., Schintlmeister, A., Nelson, T.F., Baumgartner, R., Woebken, D., Wagner, M. et al. (2018). Biodegradation of synthetic polymers in soils: Tracking carbon into CO_2 and microbial biomass. *Sci Adv* 4: eaas9024. DOI: 10.1126/sciadv.aas9024.

12 Kyrikou, I. and Briassoulis, D. (2007). Biodegradation of agricultural plastic films: A critical review. *J Polym Environ* 15: 125–150. DOI: 10.1007/s10924-007-0053-8.

13 Kijchavengkul, T., Selke, S., Ngouajio, M., Fernandez, R.T.T., Rubino, M., Auras, R. et al. (2011). Formulation selection of aliphatic aromatic biodegradable polyester film exposed to UV/solar radiation. *Polym Degrad Stab* 96: 1919–1926. DOI: 10.1016/j.polymdegradstab.2011.07.001.

14 Kijchavengkul, T., Auras, R., Rubino, M., Ngouajio, M., and Fernandez, R.T. (2008). Assessment of aliphatic–aromatic copolyester biodegradable mulch films. Part I: Field study. *Chemosphere* 71: 942–953. DOI: 10.1016/j.chemosphere.2007.10.074.

15 Emadian, S.M., Onay, T.T., and Demirel, B. (2017). Biodegradation of bioplastics in natural environments. *Waste Manag* 59: 526–536. DOI: 10.1016/j.wasman.2016.10.006.

16 Selke, S., Auras, R., Nguyen, T.A., Castro Aguirre, E., Cheruvathur, R., and Liu, Y. (2015). Additives to make plastic biodegradable don't cut it. Conversat. https://theconversation.com/additives-to-make-plastic-biodegradable-dont-cut-it-39212 (accessed October 21, 2020).

17 Beckly, J. (2018). Marine Plastics. Smithsonian Ocean. https://ocean.si.edu/conservation/pollution/marine-plastics (accessed October 21, 2020).

18 McKinsey & Company Center for Business and Environment, Ocean Conservancy (2015). Stemming the tide: Land-based strategies for a plastic-free ocean.

19 Economic and Social Research Council (2020). The rebound effect - Economic and Social Research Council. https://esrc.ukri.org/about-us/50-years-of-esrc/50-achievements/the-rebound-effect (accessed October 21, 2020).

20 Sorrell, S. (2020). Mapping rebound effects from sustainable behaviours. Key Concepts and Literature Review. https://www.sustainablelifestyles.ac.uk/sites/default/files/projectdocs/slrg_working_paper_01-10.pdf (accessed October 21, 2020).

21 Sustainable Lifestyles Research Group (2020). Mapping rebound effects from sustainable behaviours. www.sustainablelifestyles.ac.uk. https://www.sustainablelifestyles.ac.uk/sites/default/files/projectdocs/slrg_working_paper_01-10.pdf (accessed October 21, 2020).

22 Kaza, S., Yao, L., Bhada-Tata, P., and Van Woerden, F. (2018). What a Waste 2.0: A Global Snapshot of Solid Waste Management to 2050. The World Bank. DOI: 10.1596/978-1-4648-1329-0.

23 Amaral-Zettler, L.A., Zettler, E.R., and Mincer, T.J. (2020). Ecology of the plastisphere. *Nat Rev Microbiol* 18: 139–151. DOI: 10.1038/s41579-019-0308-0.

24 Min, K., Cuiffi, J.D., and Mathers, R.T. (2020). Ranking environmental degradation trends of plastic marine debris based on physical properties and molecular structure. *Nat Commun* 11. DOI: 10.1038/s41467-020-14538-z.
25 Ostle, C., Thompson, R.C., Broughton, D., Gregory, L., Wootton, M., and Johns, D.G. (2019). The rise in ocean plastics evidenced from a 60-year time series. *Nat Commun* 10: 8–13. DOI: 10.1038/s41467-019-09506-1.
26 Jâms, I.B., Windsor, F.M., Poudevigne-Durance, T., Ormerod, S.J., and Durance, I. (2020). Estimating the size distribution of plastics ingested by animals. *Nat Commun* 11: 1594. DOI: 10.1038/s41467-020-15406-6.
27 Ruggero, F., Gori, R., and Lubello, C. (2019). Methodologies to assess biodegradation of bioplastics during aerobic composting and anaerobic digestion: A review. *Waste Manag Res* 37: 959–975. DOI: 10.1177/0734242X19854127.
28 Kale, G., Auras, R., Singh, S.P., and Narayan, R. (2007). Biodegradability of polylactide bottles in real and simulated composting conditions. *Polym Test* 26: 1049–1061. DOI: 10.1016/j.polymertesting.2007.07.006.
29 Castro-Aguirre, E., Auras, R., Selke, S., Rubino, M., and Marsh, T. (2017). Insights on the aerobic biodegradation of polymers by analysis of evolved carbon dioxide in simulated composting conditions. *Polym Degrad Stab* 137: 251–271. DOI: 10.1016/j.polymdegradstab.2017.01.017.
30 TÜV AUSTRIA (2021). Certified Products. https://www.tuv-at.be/green-marks/certified-products (accessed July 29, 2021).
31 BPI (2021). BPI. https://products.bpiworld.org/?s=1&search=&type=1 (accessed April 12, 2021).
32 Ramesh Kumar, S., Shaiju, P., O'Connor, K.E., and RB, P. (2020). Bio-based and biodegradable polymers - State-of-the-art, challenges and emerging trends. *Curr Opin Green Sustain Chem* 21: 75–81. DOI: 10.1016/J.COGSC.2019.12.005.
33 PSU (2020). Anaerobic Digestion. EGEE 439: Alternative Fuels from Biomass Sources. https://www.e-education.psu.edu/egee439/node/727 (accessed April 7, 2021).
34 Pavlostathis, S.G. and Giraldo-Gomez, E. (1991). Critical reviews in environmental science and technology kinetics of anaerobic treatment: A critical review kinetics of anaerobic treatment: A critical review. *Crit Rev Environ Control* 21: 411–490. DOI: 10.1080/10643389109388424.
35 US EPA O (2021). How does anaerobic digestion work? https://www.epa.gov/agstar/how-does-anaerobic-digestion-work (accessed April 7, 2021).
36 ASTM International (2012). Standard test method for determining anaerobic biodegradation of plastic materials under high-solids anaerobic-digestion conditions, ASTM D5511-12. DOI: 10.1520/D5511-12.2.
37 ASTM International (2012). Standard test method for determining anaerobic biodegradation of plastic materials under accelerated landfill conditions, ASTM D5526-12. DOI: 10.1520/D5526-12.2.
38 ASTM International (2011). Standard test method for determining the aerobic degradation and anaerobic biodegradation of plastic materials under accelerated bioreactor landfill conditions, ASTM D7475-11. DOI: 10.1520/D7475-11.2.
39 Tsang, Y.F., Kumar, V., Samadar, P., Yang, Y., Lee, J., Ok, Y.S. et al. (2019). Production of bioplastic through food waste valorization. *Environ Int* 127: 625–644. DOI: 10.1016/J.ENVINT.2019.03.076.
40 Kumar, V., Kumar, S., and Singh, D. (2020). Microbial polyhydroxyalkanoates from extreme niches: Bioprospection status, opportunities and challenges. *Int J Biol Macromol* 147: 1255–1267. DOI: 10.1016/J.IJBIOMAC.2019.09.253.
41 Iwata, T. (2015). Biodegradable and bio-based polymers: Future prospects of eco-friendly plastics. *Angew Chemie Int Ed* 54: 3210–3215. DOI: 10.1002/anie.201410770.

42 Lambert, S. and Wagner, M. (2017). Environmental performance of bio-based and biodegradable plastics: The road ahead. *Chem Soc Rev* 46: 6855–6871. DOI: 10.1039/c7cs00149e.

43 Karamanlioglu, M., Preziosi, R., and Robson, G.D. (2017). Abiotic and biotic environmental degradation of the bioplastic polymer poly(lactic acid): A review. *Polym Degrad Stab* 137: 122–130. DOI: 10.1016/j.polymdegradstab.2017.01.009.

44 Qi, X., Ren, Y., and Wang, X. (2017). New advances in the biodegradation of Poly(lactic) acid. *Int Biodeterior Biodegrad* 117: 215–223. DOI: 10.1016/j.ibiod.2017.01.010.

45 Pathak, V.M. (2017). Navneet. Review on the current status of polymer degradation: A microbial approach. *Bioresour Bioprocess* 4. DOI: 10.1186/s40643-017-0145-9.

46 Bläsing, M. and Amelung, W. (2018). Plastics in soil: Analytical methods and possible sources. *Sci Total Environ* 612: 422–435. DOI: 10.1016/j.scitotenv.2017.08.086.

47 Ng, E.L., Huerta Lwanga, E., Eldridge, S.M., Johnston, P., Hu, H.W., Geissen, V. et al. (2018). An overview of microplastic and nanoplastic pollution in agroecosystems. *Sci Total Environ* 627: 1377–1388. DOI: 10.1016/j.scitotenv.2018.01.341.

48 Jaiswal, S., Sharma, B., and Shukla, P. (2020). Integrated approaches in microbial degradation of plastics. *Environ Technol Innov* 17: 100567. DOI: 10.1016/j.eti.2019.100567.

49 Narancic, T. and O'Connor, K.E. (2019). Plastic waste as a global challenge: Are biodegradable plastics the answer to the plastic waste problem? *Microbiol (United Kingdom)* 165: 129–137. DOI: 10.1099/mic.0.000749.

50 Ahmed, T., Shahid, M., Azeem, F., Rasul, I., Shah, A.A., Noman, M. et al. (2018). Biodegradation of plastics: Current scenario and future prospects for environmental safety. *Environ Sci Pollut Res* 25: 7287–7298. DOI: 10.1007/s11356-018-1234-9.

51 Brodhagen, M., Peyron, M., Miles, C., and Inglis, D.A. (2015). Biodegradable plastic agricultural mulches and key features of microbial degradation. *Appl Microbiol Biotechnol* 99: 1039–1056. DOI: 10.1007/s00253-014-6267-5.

52 Haider, T.P., Völker, C., Kramm, J., Landfester, K., and Wurm, F.R. (2019). Plastics of the future? The impact of biodegradable polymers on the environment and on society. *Angew Chemie - Int Ed* 58: 50–62. DOI: 10.1002/anie.201805766.

53 Hong, M. and Chen, E.Y.X. (2019). Future directions for sustainable polymers. *Trends Chem* 1: 148–151. DOI: 10.1016/j.trechm.2019.03.004.

54 Xu, L., Li, K., Liang, Y., and Zhao, L. (2018). *Screening and evaluation of the biodegradation of glass fiber by Bacillus Mucilaginosus*, 122–127. Atlantis Press. DOI: 10.2991/bbbs-18.2018.27.

55 Pérez, J., Muñoz-Dorado, J., de la Rubia, T., and Martínez, J. (2002). Biodegradation and biological treatments of cellulose, hemicellulose and lignin: An overview. *Int Microbiol* 5: 53–63. DOI: 10.1007/s10123-002-0062-3.

56 Béguin, P. and Aubert, J.-P. (1994). The biological degradation of cellulose. *FEMS Microbiol Rev* 13: 25–58. DOI: 10.1111/j.1574-6976.1994.tb00033.x.

57 Eater (2020). The environmental impact of compostable take-out containers. https://www.eater.com/2020/1/15/21065446/compostable-take-out-containers (accessed April 15, 2021).

58 Ellen MacArthur Foundation (2020). The new plastics economy global commitment, 75. https://www.ellenmacarthurfoundation.org/assets/downloads/Global-Commitment-2019-Progress-Report.pdf (accessed November 25, 2019).

59 Jambeck, J., Geyer, R., Wilcox, C., Siegler, T.R., Perryman, M., Andrady, A. et al. (2015). Plastic waste inputs from land into the ocean. *Science (80-)* 347: 768–771.

60 BASF (2020). Ecoflex®. https://plastics-rubber.basf.com/global/en/performance_polymers/products/ecoflex.html (accessed October 15, 2020).

61 Eastar Copolyesters (2020). Eastman Chemical Company. https://www.eastman.com/Brands/Eastar/Pages/Overview.aspx (accessed October 15, 2021).

62 JinHui ZhaoLong Cop (2021). Biodegradable Polymer Introduction-Product Introduction-Jin Hui Zhao Long High Tech Co., Ltd. http://www.jinhuizhaolong.com/en/jingcai/3.html (accessed November 15, 2021).

63 Biomax® TPS 2501 Biodegradable Polymers DuPont Packaging & Industrial Polymers (2013). http://www.b2bpolymers.com/TDS/DuPont_Biomax_TPS_2501.pdf (accessed October 15, 2019).

64 Kuraray, P.K. (2021). About Kuraray Polyvinyl Alcohol. https://www.kuraray-poval.com/products/kuraray-poval/kuraray-poval-k (accessed November 15, 2021).

65 BASF (2021). Poly-Caprolactone, BASF. ChemPoint. https://www.google.com/url?sa=t&rct=j&q=&esrc=s&source=web&cd=&ved=2ahUKEwiWltPFyKD0AhVHHc0KHecOCQgQFnoECAcQAQ&url=https%3A%2F%2Fchemicals.basf.com%2Fglobal%2Fintermediates%2Fbrochues-for-special-applications%2FBASF_Capromer_Polycaprolactones.pdf&usg=AOvV (accessed November 15, 2021).

66 eSUN (2021). Polycaprolactone|esun. https://www.brightcn.net/en/product/item-71.html (accessed November 15, 2021).

67 Perstorp (2021). Price increases globally on all Capa™ Caprolactone grades from October 12th, 2017. https://www.pcimag.com/ext/resources/PCI/Home/Files/PDFs/Virtual_Supplier_Brochures/Perstorp.pdf (accessed November 15, 2021).

68 Daicel (2021). Daicel Corporation. Organic Chemicals. https://www.daicel.com/yuuki/en/product/index.php?act=list_view (accessed November 15, 2021).

69 Sumitomo (2021). Caprolactam | Petrochemicals & Plastics. Sumitomo Chemical. https://www.sumitomo-chem.co.jp/english/products/detail/en_a02005.html (accessed November 15, 2021).

70 Mitsubishi Chemical (2021). Biodegradable Polymer "BioPBS™" | Products | Mitsubishi Chemical Corporation. https://www.m-chemical.co.jp/en/products/departments/mcc/sustainable/product/1201025_7964.html (accessed November 15, 2021).

71 Hummelsberger, M. (2021). Value of Biobased in a Period of Low Oil Prices. https://www.bio.org/sites/default/files/legacy/bioorg/docs/MarkusHummelsberger.pdf (accessed November 15, 2021).

72 Yield10Bioscience (2021). Corporate Overview - Yield10 Bio Agricultural Company. https://www.yield10bio.com (accessed November 15, 2021).

73 Mango Materials (2021). Technology - Mango Materials. Mango Materials. https://www.mangomaterials.com (accessed November 15, 2021).

74 TianAn Biopolymer (2021). TianAn Biopolymer: Nature's Eco-Friendly Solution. http://www.tianan-enmat.com (accessed November 19, 2021).

75 Kaneka (2021). Biodegradable polymer | Material Solutions Unit | Business. Kaneka Corporation. https://www.kaneka.co.jp/en/business/material/nbd_001.html (accessed November 15, 2021).

76 Danimer Scientific (2021). The Story of Nodax™ PHA. Danimer Scientific. https://danimerscientific.com/building-tomorrows-sustainability (accessed November 15, 2021).

77 NatureWorks (2021). NatureWorks | Beauty and Household. https://www.natureworksllc.com/Ingeo-in-Use/Beauty-and-Household (accessed November 15, 2021).

78 Total Corbion (2021). Packaging disposables | Total Corbion. https://www.total-corbion.com (accessed November 15, 2021).

79 News, B. 2020 Polylactic acid production, price and market. https://bioplasticsnews.com/polylactic-acid-or-polylactide-pla (accessed November 15, 2021).

80 BioLogiQ 2021. NuPlastiQ | BioLogiQ. https://www.biologiq.com (accessed November 15, 2021).

81 Novamont (2021). Mater-Bi - biodegradable and compostable bioplastics. Novamont. https://www.novamont.com/eng/mater-bi (accessed November 15, 2021).

82 PTT MCC Biochem (n.d.). Reinventing Living Green. www.pttmcc.com (accessed May 2, 2020).

83 PTT MCC Biochem (2021). Introducing BioPBS. https://www.pttmcc.com/what-is-biopbs (accessed November 15, 2021).

84 Eastman (2021). Engineering Bioplastics | Treva Thermoplastics | Eastman. https://www.eastman.com/Brands/Treva/Pages/Overview.aspx (accessed November 15, 2021).

85 DupontTM (2021). Propanediol | Susterra® renewably sourced propanediol. https://duponttateandlyle.com/susterra (accessed November 15, 2021).

86 Narancic, T., Verstichel, S., Chaganti, S.R., Morales-Gamez, L., Kenny, S.T., De Wilde, B. et al. (2018). Biodegradable plastic blends create new possibilities for end-of-life management of plastics but they are not a panacea for plastic pollution. DOI: 10.1021/acs.est.8b02963.

87 Kijchavengkul, T., Auras, R., Rubino, M., Ngouajio, M., and Fernandez, R.T. (2008). Assessment of aliphatic–aromatic copolyester biodegradable mulch films. Part II: Laboratory simulated conditions. *Chemosphere* 71: 1607–1616. DOI: 10.1016/j.chemosphere.2008.01.037.

88 Kijchavengkul, T., Auras, R., Rubino, M., Selke, S., Ngouajio, M., and Fernandez, R.T.T. (2010). Biodegradation and hydrolysis rate of aliphatic aromatic polyester. *Polym Degrad Stab* 95: 2641–2647. DOI: 10.1016/j.polymdegradstab.2010.07.018.

89 Kolstad, J.J., Vink, E.T.H., De Wilde, B., and Debeer, L. (2012). Assessment of anaerobic degradation of Ingeo^{TM} polylactides under accelerated landfill conditions. *Polym Degrad Stab* 97: 1131–1141. DOI: 10.1016/j.polymdegradstab.2012.04.003.

15

Incineration of Municipal Solid Waste with Energy Recovery

15.1 Learning Objectives

After learning from this chapter, you are expected to:

- Understand the difference between incineration of municipal solid waste (MSW) and incineration of MSW with energy recovery.
- Describe the main advantages and disadvantages of incineration of MSW.
- Understand the approach to incineration in different countries.
- Describe the different types of combustion units.
- Describe the main sections of an MSW combustion facility and waste-to-energy (WtE) facilities.
- Describe the main differences between the bottom, fly, and mixed ashes.
- Understand the treatments and uses of ash.
- Describe the incineration of the main packaging materials.
- Recognize the environmental footprint of energy generation using WtE facilities.

15.2 Introduction

Incineration with energy recovery is considered as one of the key components of the waste management hierarchy for MSW in the US and around the world. The US EPA ranks energy recovery from MSW below source reduction (reduction covered in Chapter 11 and reuse covered in Chapter 12) and recycling (covered in Chapters 13 and 14). It defines *energy conversion from waste* as "the conversion of non-recyclable waste materials into useable heat, electricity, or fuel through a variety of processes, including combustion, gasification, pyrolization, anaerobic digestion, and landfill gas (LFG) recovery" [1]. Waste minimization, the term employed in the Resource Conservation and Recovery Act (RCRA) statute, is described to include both source reduction and certain types of environmentally sound recycling. The EPA's highest priority is to achieve waste minimization through source reduction. Recycling activities involve the generation and active management of wastes, so they do not constitute waste minimization. The treatment for the purposes of destruction or disposal is not part of recycling or of waste minimization.

However, current trends indicate a shift from single solutions for waste management, such as landfills. Appropriate components of waste management may include waste minimization,

Life Cycle of Sustainable Packaging: From Design to End of Life, First Edition. Rafael A. Auras and Susan E. M. Selke.

Companion Website: www.wiley.com/go/Auras/lifecycleofsustainablepackaging

recycling, materials recovery, composting, biogas production, refuse-derived fuel (RDF), waste-to-energy incineration, residual landfills, etc. These approaches favor the integration of incineration within a range of complementary approaches. In this process, there is a tendency to replace mass burn incineration with more specific and efficient techniques such as RDF incineration, gasification, or pyrolysis, as later discussed.

In the US, the first incinerator was built in 1885 in Governors Island in New York City. By the mid-twentieth century, hundreds of incinerators were in operation; however, a good understanding and control of the impacts of the water and air emissions from these units did not occur until around the 1960s. When the Clean Air Act (CAA) came into effect in 1970, incinerator facilities were required to comply with new air emission requirements (see Chapter 7). Consequently, many incineration facilities closed since they were no longer economical to operate with the newly required air pollution control systems.

Combustion of MSW still grew until 1990, when the US was combusting more than 15% of all MSW, with most incinerators recovering energy and having pollution control systems. In 1990, due to mercury and dioxin emissions concerns in the US, the EPA enacted the Maximum Achievable Control Technology (MACT) regulations in the CAA Amendments. The CAA air pollution control methodology required the US EPA to determine the emissions reductions of various sources. "To determine the MACT floor, EPA ranks the existing sources in each source category using emission reduction data gathered from questionnaires and other sources. The agency averages the best 12 percent of emission reductions from the ranking list to arrive at the MACT floor. These standards require emitters to use the best control technologies already demonstrated by industry sources" [1]. There is an exception for small industries.

"Section 112 (c) of the CAA requires the US EPA to publish a list of industry group (major source and area source) categories and subcategories that employ, manufacture, or emit hazardous air pollutants (HAP)" [2]. So, the 1990 Amendments to the CAA reflected a different approach for regulating air toxic compounds. At that time, the EPA was charged to identify categories of primary sources that emit any of the 189 toxic air pollutants listed in the CAA. In July 1992, the EPA published 174 source categories including 166 major source categories and 8 area source categories (any source not considered a major source). MSW incineration units were classified under section 129 and require a permit regardless of size. A major source emits "more than 10 tons per year of a single air toxic or 25 tons per year of any combination of HAP" [2]. As a result, most of the existing incineration facilities had to be retrofitted with air pollution control systems or shut down their operations.

Although currently MSW combustion (i.e., the confined and controlled burning of MSW) is not as prevalent in the US as in 1990, several countries such as Japan, Sweden, Denmark, Norway, and Switzerland where populations are denser, and land is limited, consider waste-to-energy (WtE) one of the main strategies to avoid MSW ending in landfills. Furthermore, several of these countries, such as Sweden and Norway, have laws banning any MSW from being sent to landfills. Thus, the amount of MSW combusted and energy recovered from MSW in those countries is larger than in the US (Figure 15.1). Since the US has a large extension of land, space limitations have not been the main hindering factor in adopting WtE (except in certain regions such as the northeast). Landfilling, as explained in Chapter 16, in the US is often considered a more practical option due to the low economic cost of building an MSW landfill versus an MSW combustion facility, and low tipping fees – the amount paid per ton of trash. In 2020, the average tipping fee for landfills was \$53.72/ton with ranges between \$30.36/ton in Mississippi and \$72.03/ton in the Pacific region, which includes Alaska, Arizona, California, Hawaii, Idaho, Oregon, and Washington. Several reports have stated that tipping fees for incineration are around 25% higher. However, landfill tipping fees in regions with WtE facilities are higher [3–5].

In 2018, about 12% of the generated MSW in the US was burned in waste-to-energy plants. The US currently has 75 WtE plants as shown in Figure 15.2. The majority of the WtE plants (69 out of 75) are located in states that classify energy produced from the WtE plants as renewable energy. The majority (59 out of 75) generate only electricity, 15 produce both heat and power, and 3 plants export steam directly to local users. The gross electric capacity of the WtE plants is "2,547 megawatts (MW) and equivalent combined heat and power (CHP) capacity is 2,747 MW" [7]. A typical WtE plant generates about 550 kilowatt hours (kWh) of energy per ton of waste. Of the WtE plants, 41 are privately owned and 35 are public. Operation is conducted by 65 private companies and 11 public entities.

The mean cost of energy production from MSW has been reported as 18 ¢/kWh [7]. The precise retail rates of purchased energy vary by state, time of day, and demand. "Rates typically range from 2¢/kWh–7¢/kWh for the largest industrial customers to 8¢/kWh – 11¢/kWh or more for retail consumers" [7]. Moreover, long-term power purchase contract prices have dropped even lower at the wholesale market level due to the large deployment of natural gas, wind, and solar sources. Contractual wind prices have been reported as "2¢/kWh and solar at approximately 3¢/kWh" [7]. At these price points WtE plants cannot compete. Furthermore, when there is a need for peak capacity that obtains lucrative prices, the increasing number of flexibly operating natural gas turbine plants are able to meet this demand. "Incinerators require long lead times for start-up, making them unable to react to and take advantage of sudden increases in electricity demand" [7]. Furthermore, most of the WtE facilities available in the US need improvements to keep operating

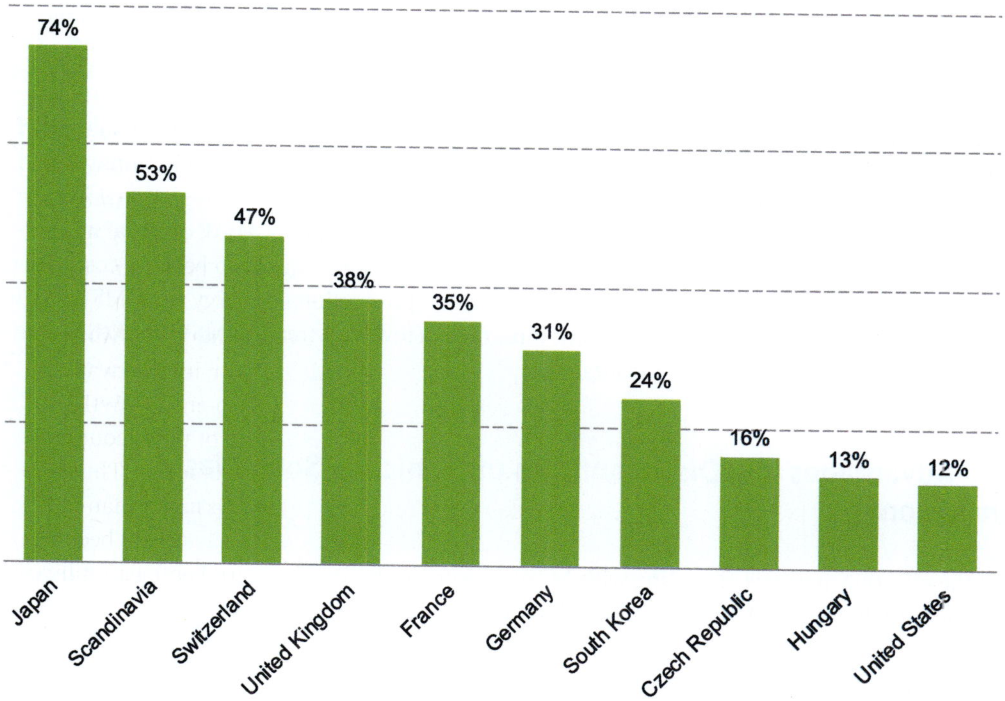

Figure 15.1 Percentage of total municipal solid waste (MSW) that is burned with energy recovery in designated countries; adapted from [6]. Note: Scandinavia includes Denmark, Norway, and Sweden. Data year for South Korea is 2016; for Japan is 2017; for all other countries is 2018.

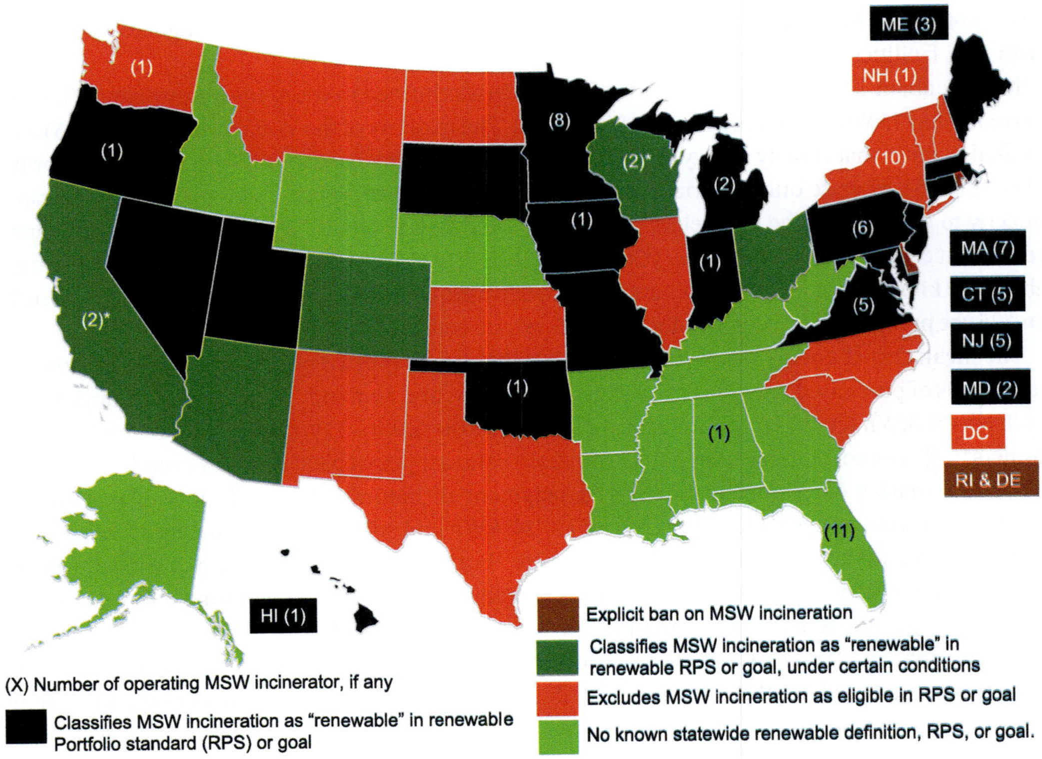

Figure 15.2 Municipal solid waste (MSW) incineration in state "renewable" energy definitions; adapted from [7]. (*) Policy grandfathers existing, but not new incinerators. Note: States excluding MSW incineration from RPS may still count other forms of incineration (e.g., biomass) and/or subsidize MSW incineration through other state or local policies.

and in general energy generation from WtE plants is more expensive compared to other sources of energy such as solar, onshore and offshore wind, natural gas, and coal due to the additional cost of pollution control systems and low heating value of the MSW, as discussed in Section 15.7. An integrated approach is needed if WtE is to become a solution for reducing MSW disposal in landfills. The technology for production of electricity in MSW WtE plants needs to become cost competitive with other power options in the market. Currently, production of energy from MSW WtE plants is only cost competitive in specific niche situations. However, treating MSW in WtE plants has several advantages as well as disadvantages.

15.3 Advantages and Disadvantages of Municipal Solid Waste Incineration

One of the effective methods of dealing with MSW to decrease their volume and potential harm is incineration. Incineration of MSW has several advantages and disadvantages.
Advantages:

- Volume and weight are reduced to a fraction of their original amount.
- Immediate reduction in amount of disposed waste as compared to landfill.
- Waste can be incinerated on site.
- Air discharges can be controlled for minimal impact with proper air pollution control systems.

- Ashes are non-putrescible – inert material.
- Technology exists to destroy even the most hazardous materials.
- Disposal of ash requires only a small area.
- Heat and energy recovery are possible, reducing cost.

Disadvantages:

- Capital cost of installing a WtE plant is high.
- Trained workers are required.
- Not all materials reaching a WtE plant are incinerable (e.g., construction and demolition wastes).
- Supplemental fuel is required to start and sometimes to sustain the incineration process.

15.4 Types of Waste Combustion Units

Based on the type and volume of waste combusted, combustion units can be classified as:

- Large, Small, and Very Small Municipal Solid Waste Incinerators: Large MSW combustors are incinerators capable of burning greater than 250 tons of MSW per day and can burn household, commercial, and/or institutional waste. Section 129 of the CAA provides regulations under section 111 of the CAA limiting emissions of nine air pollutants "(cadmium, carbon monoxide, hydrogen chloride, lead, mercury, nitrogen oxides, particulate matter, polychlorinated dibenzodioxins and polychlorinated dibenzofurans, and sulfur dioxide) from four categories of solid waste incineration units: municipal solid waste; hospital, medical and infectious solid waste; commercial and industrial solid waste; and other solid waste" [8]. Further detail about these units will be discussed in the Section 15.5.
- Hospital/Medical/Infectious Waste Incinerators: "Hospital/medical/infectious waste incinerators (called medical waste incinerators) are incinerators used by hospitals, health care facilities, and commercial waste disposal companies to burn hospital, medical, and infectious wastes" [9]. The CAA section 129 provides emission guidelines (EG) and federal implementation plans (FIP) for hospital/medical/infectious waste incineration units (HMIWI). The rules also set restrictions for the nine pollutants listed above described in the CAA section 129 [9].
- Commercial and/or Industrial Incinerators: EPA has regulations under sections 111 and 129 of the CAA guidelines for commercial and/or industrial solid waste incineration units (CISWI) [8]. New source performance standards (NSPS) and EG to reduce air pollution from CISWI were established in June 2016. "Those standards and guidelines applied to incinerators used by commercial and industrial facilities to burn nonhazardous solid waste. The NSPS and EG were designed to substantially reduce emissions of a number of harmful air pollutants, such as lead, cadmium, mercury, and dioxins/furans, which are known or suspected to cause adverse health and environmental effects" [8].
- Institutional Waste Incinerators: Institutional waste incinerators are used by institutions, such as schools, churches, and prisons, to burn the waste generated by these facilities. They belong to the category of Combustors, along with the Very Small Municipal Waste Combustors, known as Other Solid Waste Incineration (OSWI) units. The EPA enforced the NSPS and EG for new and existing OSWI units. These rules for OSWI units fulfill the requirements of sections 111 and 129 of the CAA. The final rules stated that the incineration of nonhazardous solid wastes should protect public health by reducing emissions of air pollutants [10].
- Hazardous Waste and Manufacturing Waste Incinerators: Incineration is a common technique employed to destroy hazardous waste. "These incinerators are different than those

described in the proceeding paragraphs due to kinds of waste that they burn. Manufacturing Waste Incinerators (MWI) can burn non-household waste from a given manufacturing process. This type of combustor is regulated under the RCRA or the hazardous waste regulations" [11]. Subpart O of 40 CFR Parts 264/265 sets standards for hazardous waste incinerators that primarily regulate emissions that result from the combustion process. Specifically, it restricts the emissions of organics, HCl, and particulate matter (PM) and fugitive emissions.

- Boilers and Industrial Furnaces: Another class of combustion units that can burn wastes regulated under RCRA is boilers and industrial furnaces (BIFs). "BIFs are typically used to burn hazardous waste for their significant energy and material recovery potential, with waste treatment being a secondary benefit" [12]. These units can be found in industrial settings, such as paper plants burning black liquor, a by-product from the kraft pulping process. The waste needs to have an energy value, i.e., BTUs, before it can be burned in these units. The recovered energy is reused in the industrial process. They can combust solid or liquid hazardous waste. Solid waste incinerators tend to be classified according to the type of waste that they are designed to burn (i.e., refuse or industrial waste). However, more than one waste type can often be burned in a unit. "Boilers typically combust waste for energy recovery, while industrial furnaces burn waste for both energy or material recovery. A boiler is defined as an enclosed device that uses controlled flame combustion to recover and export energy in the form of steam, heated fluid, or heated gases" [12].

  Incinerators designed for destruction of solid waste as opposed to incinerators that burn a designated material such as coal or methane are the most difficult to design and operate due to the nature of the waste. Incinerators are used mainly for waste destruction. "An industrial furnace is a unit that is an integral part of a manufacturing process and uses thermal treatment to recover materials and energy. The following units are considered industrial furnaces: a) cement, lime, phosphate, and aggregate kilns, b) coke ovens, c) blast furnaces, d) smelting, melting, and refining furnaces, e) titanium dioxide chloride process oxidation reactors, f) methane reforming furnaces, g) halogen acid furnaces, h) pulping liquor recovery furnaces, i) combustion device used in the recovery of sulfur values from spent sulfuric acid" [12].

  40 CFR Parts 260, 261, 264, 265, 266, 270, and 271 control emissions of toxic organic compounds, toxic metals, HCl, ClO_2, and particulate matter from BIFs. Furthermore, this rule subjects owners and operators of BIFs to the standards applicable to hazardous waste treatment, storage, and disposal [12].

- Hazardous Waste Incinerators: "Incinerators are used to burn hazardous waste primarily for waste destruction/treatment purposes; however, some energy or material recovery can occur. When performed properly, incineration destroys the toxic organic constituents in hazardous waste and reduces the volume of the waste. Since metals will not combust, incineration is not an effective method for treating metal-bearing hazardous wastes. Several types of hazardous waste incinerators are: a) rotary kilns, b) fluidized bed units, c) liquid injection units, d) fixed hearth units. The RCRA regulations governing incinerators can be found at 40 CFR Part 264 Subpart O-Incinerators and Part 265 Subpart O-Incinerators" [13]. A detailed discussion of these incinerators is outside the scope of this book.

15.5 Municipal Solid Waste Combustion Plants

Most MSW combustion plants in the US use waste to fuel a mass burn facility, usually with minimal sorting of wastes before it enters the combustion chamber. Several types of technology are used in MSW incinerator facilities, and nearly all incorporate energy recovery. These are typically continuous incinerators including stoker-type incinerators, fluidized bed-type incinerators, and gasification melting furnaces. Stoker-type incinerators (stoker refers to a "set of grates") are the most common. In this type of MSW incinerator facility, the MSW is burned in a single combustion chamber in an excess of air (needed due to the inconsistency of the solid waste) (see Figure 15.3). Once the waste is delivered to the receiving area of the MSW combustion plant, it may be mixed, or selected items may be sorted out. It is common for large items that were not separated at the source to be removed, and there may be magnetic separation for ferrous metals. Sometimes there is additional sorting of high-value materials from selected loads, such as removal of corrugated from collection originated in retail stores. Negative airflow may be used to carry dust and odor from the receiving area into the combustion chamber, to eliminate its spread outside the facility. After the mixed waste material enters the combustion chamber, the solids are burned on an inclined moving grate that shakes or moves to agitate the waste and mix it with injected air, resulting in mostly complete combustion. The combustion chamber has three main stages: dry, combustion, and burn-out zones. Even waste with high moisture can be combusted in this type of chamber. The design and operating conditions of these three zones must be correctly adjusted according to the amount and type of waste. After the MSW is combusted, generated NO_X exits with the burn gases and is neutralized by injection of ammonia or urea. Formation of dioxins occurs primarily on the heat exchange surfaces of the boiler, at around 300–400°C. There, if ash and products produced due to incomplete combustion of carbonaceous nature (for example soot) are deposited on the tubes, and with the presence of copper chloride in the fumes, dioxins can be formed. So, dioxins and furans are destroyed by exposing the flue gases to temperatures of 850°C (1,562 °F) for at least 2 seconds under turbulent air flow, which is designed to remove more than 99% of produced dioxins and furans. To monitor the combustion conditions, O_2, CO_2, and CO are continuously monitored. According to Japanese guidelines, to prevent dioxin formation, CO should be lower than 30 ppm (35.7 mg/m^{3}N) and O_2

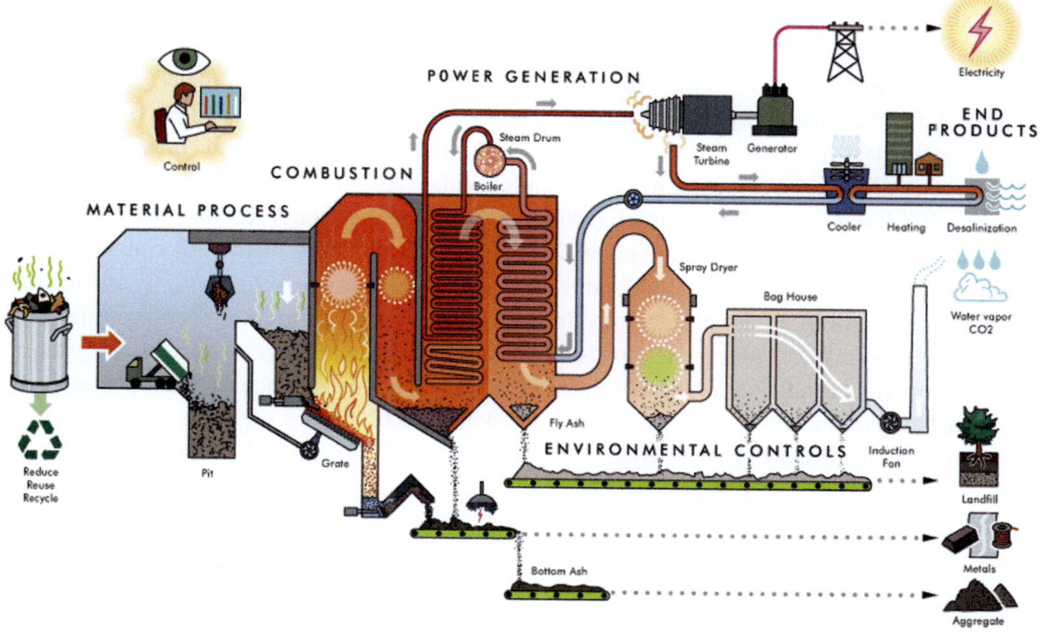

Figure 15.3 Municipal solid waste incineration facility; reproduced from [15]. Courtesy of Deltaway Energy International.

below 12% – average concentration in 4h [14]. Next the flue gases move forward to the boiler to generate superheated steam to power the steam turbine generator. The cooled steam is cycled back into water through a condenser, or it can be diverted as a heat source for buildings. The flue gases coming out of the boiler are diverted to a spray dryer. These gases contain fly ash, which can contain mercury and heavy metals. To capture them, activated carbon is injected into the hot gases to absorb and remove heavy metals, such as mercury and cadmium. Additionally, the acidic gases are neutralized by injecting lime or NaOH, producing gypsum and removing HCl. The bottom ash, the unburned remains of combustion, can be passed by magnets and eddy current separators to remove both ferrous and nonferrous metals for further recycling. "Fly ash is captured throughout the process. The finest airborne particulates are removed in a filter baghouse, where an induction fan draws air through fabric bags toward the stack or chimney" [15]. Captured fly ash, along with bottom ash, is often sent to landfills allowed to accept ash in the US.

Bottom ash coming from the stoker grate sifting, the end of the grate, and boiler ashes are captured and combined (Figure 15.3). Bottom ash and fly ash may be treated, and their disposal is regulated globally. In the US, the EPA considers ash from WtE as hazardous waste and it is regulated accordingly as disposal of combustion units. The EPA estimates that around 80% of WtE facilities combine the bottom and fly ash for disposal.

The amount of ash generated ranges from 15 to 25% wt. and from 5 to 15% v/v. of the MSW processed. Fly ash "mostly contains fine particles removed from the flue gas and includes residues from other air pollution control devices, such as scrubbers. Fly ash is 10-20 % by weight of the total ash" [16]. The rest is bottom ash. The main chemical components of bottom ash are silica, calcium, iron oxide, and aluminum oxide. Bottom ash has a moisture content between 22% and 62% by dry weight. The chemical composition of the ash can change depending on the type of MSW feedstock and how the combustion process is managed. In the US, the majority of the bottom ash is sent to landfills. Ash residues from the combustion of MSW have densities ranging from 65 to 75 lb/ft^3 (1,041–1,201 kg/m^3) and normally are around 20% moisture [16], which is comparable to Portland cement, 94 lb/ft^3 (1,505 kg/m^3).

Bottom ash can be used for several purposes such as landfill and aggregates for road construction, as asphaltic mixtures, Portland cement, building blocks, or other purposes. In the US, landfilling has been the most common method to dispose ash. Sometime ash is commingled with regular waste or it may be disposed in selected landfill cells for landfills allowed to take them (see Chapter 16 for landfill classification). However, codisposal of ash with MSW has raised concern since the acids generated by the decomposition of MSW could escalate concentrations of soluble toxic metals in the landfill leachate, needing additional stringent containment, better leachate treatment, and increasing groundwater monitoring systems. However, the alkalinity of the ash can neutralize acids and reduce acid leaching. Table 15.1 shows several approaches to dealing with fly ash and the air pollution control (APC) waste generated in MSW incineration facilities in several countries.

One of the objectives of WtE incineration is the recovery of energy from waste by generating steam. The most important conditions for the boiler are the steam generation conditions (i.e., temperature and pressure). Until around 1990, steam generation was done at 300°C and 3 MPa or less. Modern WtE plants generate steam at 400°C and 4 MPa, which increases power generation by more than 20% [14].

15.6 Refuse Derived Fuel

Burning of MSW is typically a low-efficiency process. MSW WtE plants can burn MSW directly. However, not all of the waste is combustible and water content can be high, leading to low heating

Table 15.1 Waste management of fly ash (FA) and air pollution control (APC) from municipal solid waste incineration facilities in several countries; adapted from [17].

US	The APC residues and bottom ash are mixed at most MSW incineration (MSWI) plants and disposed as "combined ash." The most frequent disposal option are landfills that receive only MSW from incineration plants, known as monofils.
Canada	The APC residues are disposed in a hazardous waste landfill after treatment.
Sweden	The APC residues are disposed in secure landfills after their treatment.
Denmark	The APC residues and fly ash are classified as a special hazardous waste and are currently exported or stored temporarily in big bags.
Germany	The APC residues are mainly disposed of in underground disposal sites, such as abandoned salt mines.
Netherlands	Flue-gas cleaning wastes are disposed temporarily in large, sealed bags at a controlled landfill until better options are available. The utilization of APC residues is presently not considered.
France	After industrial solidification and stabilization processes based on the properties of hydraulic binders, the waste is stored in confined cavities in a specific landfill (French class I and II).
Italy	Various technologies have been proposed, but the most widely adopted comprises solidification with a variety of hydraulic binders (such as cement and/or lime, blast furnace slag, etc.).
Portugal	The APC residues are treated with hydraulic binders (solidification/stabilization method) and landfilled in specific sites (monofills).
Japan	The MSW incineration fly ash and APC residues are considered as hazardous, and before intermediate landfill treatments must be performed, such as melting, solidification with cement, stabilization using chemical agents or extraction with acid or other solvents. Melting slag may be used in road construction, and materials after solidification and stabilization with cement are landfilled.

value and a large amount of ash. An alternative approach is to use separation processes to produce solid particles, known as refuse derived fuel (RDF), that have higher heating value and are easier to handle (see Figure 15.4).

The larger the moisture content of the waste, the more fuel is required to destroy the waste. So, wastes with large water content (e.g., 95%) are poor candidates for incineration. MSW in the US is composed of ~52% combustible matter, 26% moisture, and 22% ash and inert materials. Biomass materials account for about 64% of the weight of the combustible MSW and for about 51% of the electricity generated. The rest of the combustible MSW is nonbiomass combustible materials, largely plastics.

15.7 Energy Recovery from Burning MSW

While burning MSW eliminates large amounts of refuse; relatively little energy is recovered. When performed correctly, incineration annihilates the toxic organic constituents in hazardous waste and decreases the volume of the waste. Accurate data on the quantity and quality of MSW is needed. Several types of waste are incinerated, and they have been previously classified from 0 to 5 as trash, rubbish, refuse, garbage, animal solids and organic wastes, gaseous, liquid or semiliquid wastes, and semisolid and solid wastes, respectively [19]. However, nowadays waste tends to be classified according

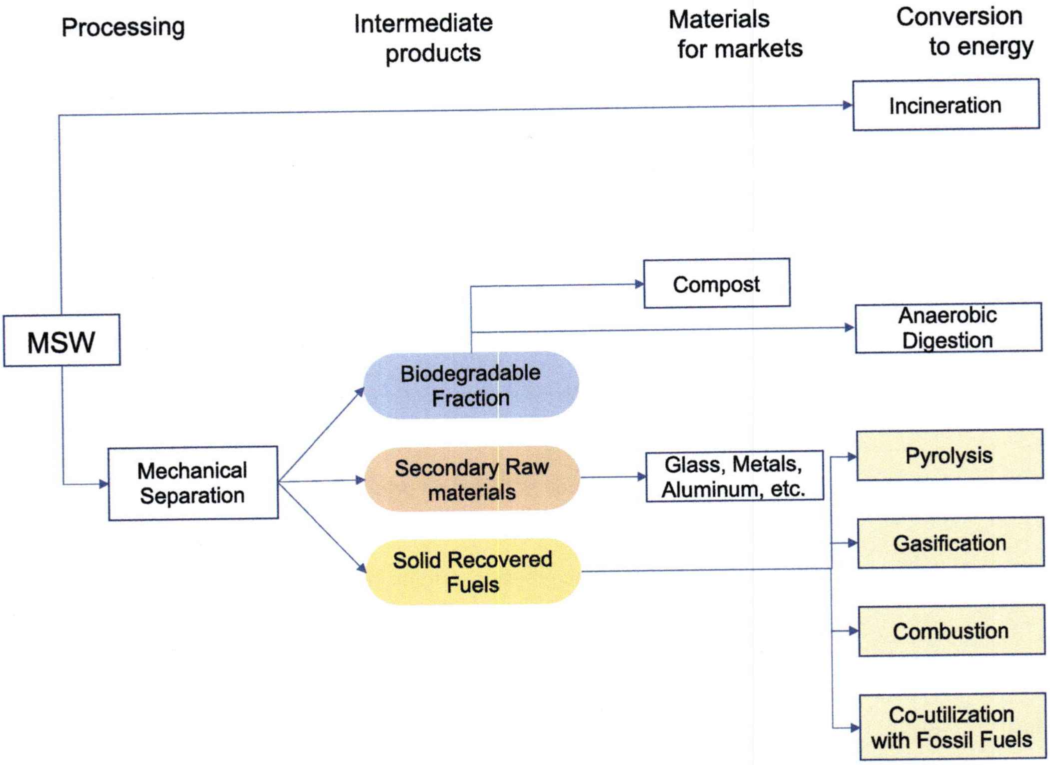

Figure 15.4 Pathway for municipal solid waste treatment for recovery and recycling processes; adapted from [18].

to its energy contribution. Data on waste composition and the lower calorific value (LCV) – a parameter determined by subtracting the heat of vaporization of the water from the higher heating value of the waste – are crucial to optimize energy generation and destroy waste. Several studies reported that the LCV must be around 7,000 kJ/kg and never below 6,000 kJ/kg for WtE incineration with energy recovery to be feasible. Figure 15.5 shows a triangle diagram indicating the region where WtE incineration and incineration of waste can be acceptable as well as the MSW composition in different low-, middle-, and high-income countries. Incineration technologies can be applied to waste with a moisture content less than 75% and volatile content about or more than 20%, corresponding to an LCV of around 3,352 kJ/kg or more. If energy recovery is part of the process, moisture content must be below 65% and volatile content about or more than 30%, corresponding to an LCV of 6,285 kJ/kg or more. If waste has high moisture content due to a large proportion of food and kitchen waste, it is recommended that collected waste should be source separated to increase the LCV value.

Typically, MSW has an average heating value of 8–12 MJ/kg, as compared "with 19 MJ/kg for dry wood, 15 MJ/kg for lignite or 22 MJ/kg for coal for power plants. Mixed plastics have an average heat value of 33 MJ/kg. Wet compostable material is in the range of 4 to 6 MJ/kg, while for comparison, natural gas has a value of about 50 MJ/kg" [19]. Table 15.2 shows a detailed list of calorific value for several materials including fuels, such as heating oil, coal, paper, and MSW. Since metals will not combust, incineration is not an effective method for treating metal-bearing hazardous wastes.

Overall, in 2018 in the US, from the 292 million tons generated of MSW, 34.56 million tons were combusted with energy recovery (Figure 15.6). Glass represents around 5%; metals 8.5%; plastics ~16%; paper and paperboard 12%; rubber, leather, and textiles 16.5%; food ~22%; and others 4.2%.

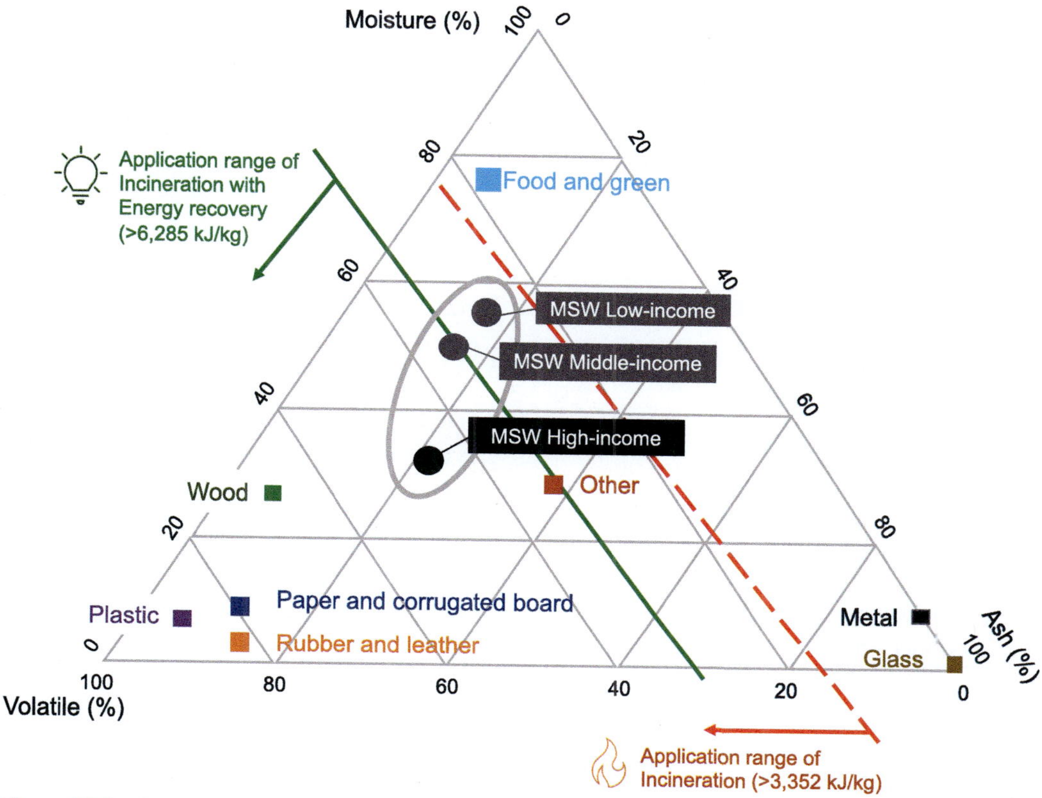

Figure 15.5 Composition of application range for feasible incineration and incineration with energy recovery for municipal solid waste generated in different country groups; adapted from [14, 20].

The total heating value of the solid waste can be estimated by adding the typical heating value of MSW components as shown in Table 15.3. Trained operators can mix incoming solid waste to maximize the total energy contribution. A value of 13.96 MJ/kg of MSW as estimated in Table 15.3 is very close to the previously reported values of 10.5 MJ/kg for MSW in Table 15.2.

Table 15.2 Typical moisture and energy of municipal solid waste components; adapted from [19, 21].

Component	Typical Moisture, %	Typical Energy, MJ/kg
Food wastes	70	4.6
Paper	6	16.7
Corrugated board	5	16.3
Mixed plastics	2	32.6
Polyethylene (PE)	2	45
Polypropylene (PP)	2	43
Polystyrene (PS)	3	40
Polyamide	8	30
Polycarbonate (PC)	8	30
Poly(ethylene terephthalate) – PET	2.5	22

(Continued)

Table 15.2 (Continued)

Component	Typical Moisture, %	Typical Energy, MJ/kg
Polyvinyl chloride (PVC)	N/A	18
Polytetrafluoroethylene	N/A	4
Polyvinyl alcohol (PVOH)	8	25
Polylactide	2.5	18
Textiles	10	17.5
Rubber	2	23.3
Leather	10	17.5
Garden trimmings	60	6.5
Wood	20	18.7
Coal	2	30
Heating oil	N/A	43
Glass	2	0.14
Tin cans	2	0.7
Municipal solid waste	20	10.5
Dirt, ashes, brick, etc.	8	7.0

N/A: Not available.

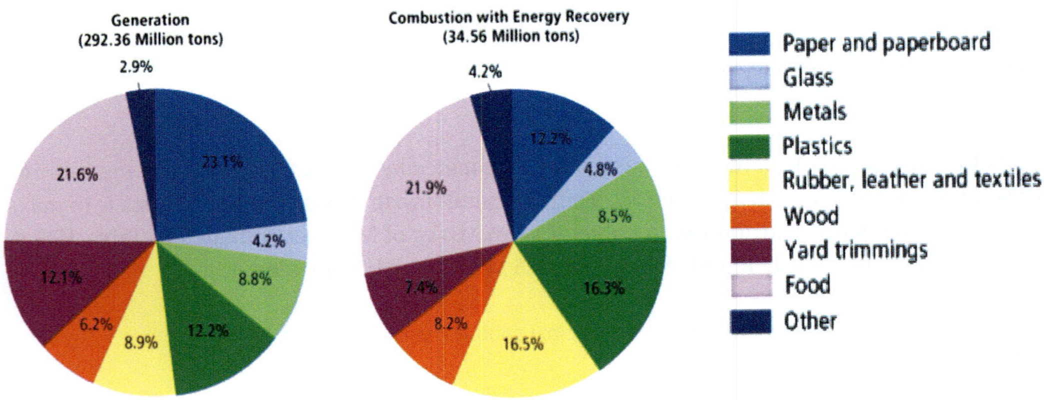

Figure 15.6 Materials generated and combusted with energy recovery in the municipal solid waste in the US, 2018; reproduced from [22].

Table 15.3 Estimation of total energy contribution (MJ/kg) in the US municipal solid waste in 2018.

Component	Solid Waste, %	Inherent Energy, MJ/kg	Total Energy Contribution, MJ/kg
Paper and paperboard	12.2	16.5	2.01
Glass	4.8	0.1	0.01
Metals	8.5	0.7	0.06

(Continued)

Table 15.3 (Continued)

Component	Solid Waste, %	Inherent Energy, MJ/kg	Total Energy Contribution, MJ/kg
Plastics	16.3	32.6	5.31
Rubber, leather, and textiles	16.5	19.4	3.21
Wood	8.2	18.7	1.53
Yard trimmings	7.4	6.5	0.48
Food wastes	21.9	4.6	1.01
Other	4.9	7.0	0.34
Total	100		13.96

Note: Values for rubber, leather, and textiles were considered an average of the three categories.

15.8 Incineration of Metals

As explained, incineration of MSW creates a large concentration of ferrous and nonferrous metals in the bottom and fly ash. The melting and vaporization temperature of metals and nonmetal materials can be directly correlated with where the metals end up (see Table 15.4). Most WtE incineration processes operate at temperatures around 850°C and above, so thermal destruction of most organic compounds occurs, which happens between 590°C and 650°C (1,100°F and 1,200°F). Waste is retained around 2 seconds or longer with sufficient turbulence in the combustion zone. Most hazardous waste incinerators operate at 980–1,200°C (1,800–2,200°F) to ensure nearly complete destruction of the organics in the waste. Mercury and zinc vaporization temperatures are 357°C and 906°C, respectively. Thus, these metals are easily carried out of the incinerator in the flue gases, making it extremely important to have efficient air pollution control systems (see Figure 15.3).

Table 15.5 shows the concentration of metals in combined ash, fly ash, and fly ash/scrubber ash (combination between ashes coming from fly ash and scrubbing ash). Combined bottom ash has mostly mineral metals found in common earth. We can observe that the amount of lead, cadmium, and mercury concentrations are higher in the fly ash as well as chloride and sulfate due to large fraction of soluble salts in the waste. The potential toxic metals coming from packaging in fly ash are derived mostly from pigments, fillers, and inks used in paper and plastic products.

Table 15.6 shows the amount of metal incinerated MSW in the US since 1960 coming from packaging materials. We can observe that total steel packaging represents around 1.35% of the total amount combusted for containers and packaging and that total aluminum packaging represents around 3.5% of the total amount combusted for containers and packaging, which is expected as steel and aluminum do not provide significant energy. So, these materials are often separated before MSW is combusted. The largest amount of ferrous metals (iron and steel) incinerated in the MSW in 2018 came from appliances, furniture, tires, and miscellaneous durable goods (2,550 thousand tons of metals of the 25,600 thousand tons of metals generated in 2018).

15.9 Incineration of Glass

As shown in Figure 15.6 and Table 15.3, glass has no significant heating value; it is already oxidized. The amount of glass incinerated with energy recovery is mostly due to glass from containers

Table 15.4 Melting and evaporation temperatures of ferrous and nonferrous metals.

	Melting Temperature		Boiling Point, Vaporization Temperature	
	°C	°F	°C	°F
Aluminum	659	1,218	2,327	4,221
Brass	930	1,706	1,100	2,012
Cobalt	1,480	2,696	2,900	5,252
Copper	1,083	1,981	2,595	4,703
Gold	1,063	1,945	2,600	4,712
Lead	328	622	1,750	3,182
Manganese	1,260	2,300	2,150	3,902
Magnesium	650	1,202	1,110	2,030
Mercury	−38.8	−37.8	357	675
Molybdenum	2,620	4,748	5,560	10,040
Nickel	1,455	2,651	2,730	4,946
Silver	962	1,764	1,950	3,542
Titanium	1,700	3,092	3,260	5,900
Tungsten	3,370	6,098	5,930	10,706
Zinc	420	788	906	1,663

Table 15.5 Total metals (ppm) in combined bottom ash, fly ash, and fly ash/scrubber residues; adapted from [16, 23].

Metal, ppm	Bottom Ash [23]	Combined Bottom Ash [16]	Fly Ash [16]	Fly Ash/ Scrubber [16]
Al	14,000 –79,000	17,857	27,500	12,714
Ca	8,600–170,000	33,642	64,857	176,428
Fe	3,100–150,000	20,428	10,857	3,314
K	660–16,000	3,071	36,928	8,450
Mg	240–260			
Mn	7.7–3,200	538	649	365
Na	2,200–42,000	3,828	26,928	1,292
P	440–10,500			
Si	4,300–308,000			
		Other Minor Metal Components		
Ag			31–95	
As	0.12–190		31–95	
Ba	69–5,700		90–1,800	
Cd	0.3–70	35	250–450	160

(Continued)

Table 15.5 (Continued)

Metal, ppm	Bottom Ash [23]	Combined Bottom Ash [16]	Fly Ash [16]	Fly Ash/ Scrubber [16]
Co	10–40		29–69	
Cu	190–25,000		860–1,400	
Cr	20–3,400		140–530	
Hg	N/A	N/A	3	73
Mo	2.5–280			
Ni	7.0–4,300		95–240	
Pb	74–14,000	3,142	7,400–19,000	3,257
Se	0.05–10		6.1–3.1	
Sn	2.0–470			
Sr			80–250	
Tl	0.008–0.23			
V	16–120			
Zn	10–20,000	4,107	53,500	9,143
Other Components Able to Complex with Metals				
Chloride		928	65,428	164,285
Sulfate		7	33	8

N/A: Not available.

and packaging that goes into mass burn facilities. Table 15.7 shows that glass represented around 15% of the MSW combusted in 2018. Of this amount, 54% corresponded to beer and soft drink containers, 18% to wine and liquor bottles, and 28% to other bottles and jars. There is no value obtained from incinerating glass. Since bottom ash has a high concentration of Si, several uses of bottom ash have been developed for building materials such as road materials and cement fillers [24].

15.10 Incineration of Paper, Paperboard, and Corrugated Board

Although paper, paperboard, and corrugated board can be recycled, they are also ideal materials for incineration with energy recovery due to their calorific value (Tables 15.3 and 15.8). In the US MSW, paper, paperboard, and corrugated board represent 21% of the total combusted containers and packaging. Corrugated boxes are highly recycled in the US, so not many go to WtE. Overall, paper represents around 23% of the MSW in 2018, but only 12.2% of the incinerated material. It is interesting to note that paper is one of the main contributors of Cd, Cr, Hg, and Pb from printed articles [25].

15.11 Incineration of Plastics

Plastic represents 16.3% of the MSW incinerated (Figure 15.6 and Table 15.9). Plastics from packaging represent around 33% of the total amount of containers and packaging incinerated. Among these, plastics from bags, sacks, and wraps and other general packaging categories represent more than 65%.

Table 15.6 Metals combusted with energy recovery in the municipal solid waste, 1960 to 2018, with detail on containers and packaging.

Products	Thousands of Tons						
	1960	1970	1980	1990	2000	2010	2018
Steel Packaging							
Beer and soft drink cans	0	10	10	20	<0.05	<0.05	<0.05
Cans	0	10	50	330	210	130	90
Other steel packaging	0	<0.05	<0.05	20	20	20	20
Total steel packaging	0	20	60	370	230	150	110
Aluminum Packaging							
Beer and soft drink cans	0	<0.05	10	100	130	120	130
Other cans	0	<0.05	<0.05	<0.05	10	10	20
Foil and closures	0	<0.05	10	50	70	80	100
Total aluminum packaging	0	<0.05	20	150	210	210	250
Total containers and packaging	0	150	880	8,110	9,110	6,870	7,420
Total MSW combusted with energy recovery – weight	0	450	2,760	29,760	33,730	29,310	34,550

Table 15.7 Glass combusted with energy recovery in the municipal solid waste, 1960 to 2018, with detail on containers and packaging.

Products	Thousands of Tons						
	1960	1970	1980	1990	2000	2010	2018
Glass packaging							
Beer and soft drink bottles*	0	20	120	640	810	590	550
Wine and liquor bottles	0	10	50	310	290	210	210
Other bottles and jars	0	20	100	620	490	310	550
Total glass packaging	0	50	270	1,570	1,590	1,110	1,310
Total containers and packaging	0	150	880	8,110	9,110	6,870	7,420
Total MSW combusted with energy recovery – weight	0	450	2,760	29,760	33,730	29,310	34,550

* Includes carbonated drinks and non-carbonated water, teas, flavored drinks, and ready-to-drink alcoholic coolers and cocktails.

Burning plastics to recover their energy may seem beneficial due to the high internal energy of plastics and potential claimed environmental benefits when incinerated in high efficiency and high electricity-to-heat ratio WtE plants. Plastics such as polyethylene have a large energy value, e.g., PE = 45 MJ/kg, which is similar to natural gas since the composition is carbon and hydrogen. According to Covanta, a leading WtE operator in the US operating two-thirds of WtE facilities, "these facilities reduce CO_2 emissions by about a ton for every ton of waste that is not landfilled. Because

Table 15.8 Paper, paperboard, and corrugated boxes combusted with energy recovery in municipal solid waste, 1960 to 2018, with detail on containers and packaging.

Products	Thousands of Tons						
	1960	1970	1980	1990	2000	2010	2018
Paper and paperboard packaging							
Corrugated boxes	0	40	210	2,120	1,920	770	230
Subtotal other paper and paperboard packaging	0	30	170	1,380	1,750	1,150	1,340
Total paper and board packaging	0	70	380	3,500	3,670	1,920	1,570
Total containers and packaging	0	150	880	8,110	9,110	6,870	7,420
Total MSW combusted with energy recovery – weight	0	450	2,760	29,760	33,730	29,310	34,550

Table 15.9 Plastics combusted with energy recovery in the municipals solid waste, 1960 to 2018, with detail on containers and packaging.

Products	Thousands of Tons						
	1960	1970	1980	1990	2000	2010	2018
PET bottles and jars			<0.05	50	260	330	440
HDPE natural bottles			<0.05	90	90	100	100
Other containers	0	<0.05	20	240	300	270	330
Bags and sacks							
Wraps							
Subtotal bags, sacks, and wraps			30	410	780	620	740
Other plastics packaging	0	<0.05	20	340	530	770	850
Total plastics packaging			70	1,130	1,960	2,090	2,460
Total containers and packaging	0	150	880	8,110	9,110	6,870	7,420
Total MSW combusted with energy recovery – weight	0	450	2,760	29,760	33,730	29,310	34,550

fossil fuels are not being burned for energy, methane is not being generated in a landfill, and metals are being collected for recycling" [26]. However, critics warn that even the most advanced WtE facilities are not completely clean and always pollute, unlike other forms of energy such as wind and solar, which have lower emissions and pollution. The main concerns for WtE are the incomplete burning and the emissions from uncontrolled incinerators that can release dioxin and heavy metals to the environment. However, new WtE plants use scrubbers, precipitators, and several types of filters, as discussed in Section 15.5, to catch toxic compounds and prevent release of toxins to the environment.

So, a highly concentrated waste with plastics can improve the calorific value in MSW; however, existing WtE facilities in the US cannot accommodate all the plastics that are available and piling up, since the plants are designed to generate a specific amount of energy per day. Plastics have a higher energy content than most trash. If a WtE facility processes more plastics, it will take in less waste overall to maintain target electricity production. This can be a problem since WtE facilities generate most of their revenues from tipping fees by getting rid of waste. Covanta "generated 70% of its income in 2017 from such fees, and only 19% came from the electricity it sold. Another 5% came from the metals it recovered. WtE plants exist to dispose of waste, so sanitation is their primary objective, and materials and energy recovery are secondary benefits" [26]. So, to incinerate more plastics, the US would need to build more WtE plants to process more trash overall. To increase the amount of plastic processed, the US would need to embrace WtE technology. Currently, "only about 13% of US garbage is burned for energy. In the European Union, the figure is double. Germany incinerates about a third of its waste for power; Norway and Sweden, more than half" [26]. Consumers and NGOs are becoming more concerned about the effect of WtE facilities on the overall circular economy and on emissions contributing to climate change.

15.12 Case Study 15.1. Burning of Poly(vinyl chloride) – PVC

PVC is a widely used polymer around the world for the packaging, construction, automotive, electronics, and the pharmaceutical industry. After its use, PVC enters the waste stream and due to the presence of chlorine in its main structure, burning can result in production of dioxins and chlorinated furans as well as carbon monoxide. Additionally, the heavy metals in additives used to process PVC can be released. Even though PVC can provide calorific value to the MSW (see Table 15.2), many organizations have complained about burning of PVC. Dioxins and furans are toxic man-made products that can cause diseases such as cancer and birth defects. In WtE facilities, by products of burning PVC have been identified; however, it has also been shown that increasing PVC from 0.7% to 5% in the MSW stream did not significantly increase the concentration of dioxins and furan isomers in well-run facilities [27]. Proponents of continued burning of PVC have stated that dioxins are also produced due to the presence of plants and that the technology to scrub dioxins from air emissions has advanced to deal with these emissions [27]. Finally, it should be clear at this point that burning PVC in a WtE facility is not the same as burning PVC in an open environment where the toxic compounds are released to the environment.

15.13 Case Study 15.2. Comparison of Emissions from Waste-to-Energy Facilities with Those from Fossil Fuels and Their Greenhouse Gas Emissions

Production of energy emits chemical compounds and pollutants. Table 15.10 provides a comparison between the emissions from WtE facilities and from burning residual oil, and bituminous and lignite coal. It can be observed that based on equivalent electric power generation, WtE facilities generally have much lower emissions than those of the fossil fuels listed. However, fracking has made gas extensively available and cheap. So, evaluations of energy generation looking at indicators such as energy indices have been reported by several authors with some reporting that WtE is more sustainable overall [28].

Pfadt-Trilling et al. (2021) reported the GHG emissions from electricity generation in the New York region and indicated a large variability in GHG emissions from different sources. They reported a large reduction of GHG emissions by using a coproduct scenario in WtE. In this case,

Table 15.10 Comparison of emissions from WtE facilities with those from fossil fuels in lb/1,000 megawatt-hour; adapted from [25].

	Residual Oil	Bituminous Coal, Pulverized	Lignite Coal, Pulverized	WtE (Mass Burn/ Refuse Derive Fuel)
Arsenic	0.22	0.46	0.91	<0.033
Beryllium	0.06	0.03	0.06	<0.017
Cadmium	0.18	0.10	0.11	0.063
Chromium	0.24	4.56	570	<0.19
Copper	3.19	2.28	3.42	0.43
Mercury	0.04	0.23	0.23	0.17
Nickel	1,436	3.42	3.42	0.84
Lead	0.34	0.87	0.11	0.44
Selenium	NR	0.29	0.29	<0.022
Vanadium	3.4	4.0	4.0	0.025
Zinc	0.47	8.0	8.0	1.23
Particulate	1,030	440	440	125

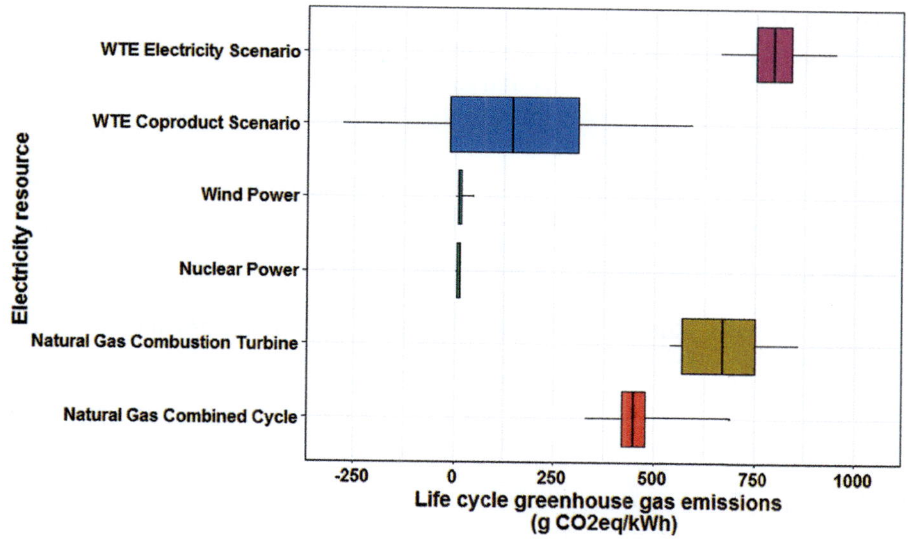

Figure 15.7 Comparison of the life cycle climate change impact interquartile range results of WtE compared with various other methods for electrical generation utilized in the New York region; reproduced from [29] with permission from American Chemical Society.

coproduct refers to incorporating the benefits of waste reduction and metals recycling happening in the WtE (see Figure 15.7). Their study suggests that there may be climate change benefits from WtE as a waste management practice that generates electricity by crediting the system for GHG emissions avoided from landfilling. However, as a method of electricity generation alone, without considering MSW as an inevitable nuisance that must be managed, the GHG emissions of WtE-sourced electricity are at least one order of magnitude larger than any other renewable source of electricity generation.

15.14 Study Questions

1. Define energy conversion from waste as described by the US EPA.
2. Is energy recovery from waste considered a waste minimization strategy by the EPA?
3. RDFs stand for:
   - A Radio-device-frequency
   - B Radio-dot-face
   - C Refused-do-fuels
   - D Refused-derived-fuels
   - E None of the above
4. When was the first US incinerator built, and when were emissions produced by waste-to-energy plants first regulated?
5. How does the EPA control the emissions from incinerators?
6. Do MSW incineration facilities need a permit to operate?
7. List some countries in which waste-to-energy is more prevalent than in the US.
8. Is energy generated in waste-to-energy plants considered renewable in the US?
9. Is energy produced from WtE plants cost competitive?
10. What are the main advantages and disadvantages of municipal solid waste incineration?
11. What are the main types of combustion units?
12. Explain how a municipal solid waste plant works.
13. What is the main method to dispose fly and bottom ash?
14. Describe the impact of moisture and volatile content on burning waste.
15. Polyethylene has similar calorific value to natural gas.
    - A True
    - B False
16. Estimate the total energy contribution of the municipal solid waste shown in the table below. What is the main difference with the results shown in Table 15.3?

Component	Solid Waste, %	Inherent Energy, MJ/kg	Total Energy Contribution, MJ/kg
Paper and paperboard	20	16.5	
Glass	1	0.1	
Plastics	30	32.6	
Ruber, leather, and textiles	28	19.4	
Wood	1	18.7	
Yard trimmings	5.1	6.5	
Food wastes	10	4.6	
Other	4.9	7.0	
Total	100		

17. What type of air pollution control systems are required to avoid the emission of heavy metals? Which type of ash contains larger amounts of heavy metals?
18. Incinerating glass provides a lot of caloric value.
    - A True
    - B False

19 What types of emissions are the main concern in MSW incineration of printed articles?
20 What are the main compounds emitted from burning PVC?
21 Under what conditions are the greenhouse gas emissions from generation of energy from WtE plants similar to those from standard energy production from other sources such as wind, nuclear, and natural gas?

15.15 Additional Resources

1 EPA – Energy Recovery from the Combustion of Municipal Solid Waste (MSW) https://www.epa.gov/smm/energy-recovery-combustion-municipal-solid-waste-msw
2 CCET guideline series on intermediate municipal solid waste treatment technologies. Waste-to-Energy Incineration
Report: https://www.iges.or.jp/en/pub/ccet-guideline-wtei/en
Video: https://www.youtube.com/watch?v=8Easjj70Ghk
3 DeltaWay – Waste and Biomass Power Plant Design and Operation https://deltawayenergy.com/2018/08/waste-to-energy-how-it-works/#tip10
4 How waste-to-energy plants work? https://www.eia.gov/energyexplained/biomass/waste-to-energy-in-depth.php
5 How waste-to-energy plant works? https://youtu.be/lmtOuAed5nM
6 Waste-to-energy plant operation. https://youtu.be/1HpxZ9ilSzQ
7 How waste-to-energy can change our future? https://youtu.be/UYvprn_RRs8

References

1 US Environmental Protection Agency. (2020). Sustainable materials management: Non-hazardous materials and waste management hierarchy. https://www.epa.gov/smm/sustainable-materials-management-non-hazardous-materials-and-waste-management-hierarchy (accessed May 13, 2021).
2 US Environmental ProtectionAgency. (2020). Energy Recovery from the Combustion of Municipal Solid Waste (MSW) | Sustainable Materials Management | US EPA. https://www.epa.gov/smm/energy-recovery-combustion-municipal-solid-waste-msw (accessed April 29, 2021).
3 Analysis of MSW Landfill Tipping Fees-April. (2019). The environmental research & education foundation (EREF) maintains a database of municipal solid waste 2019. https://erefdn.org/product/analysis-msw-landfill-tipping-fees-2 (accessed May 13, 2021).
4 Energy Justice Network. (2021). Trash incineration is the most expensive way to manage waste. http://www.energyjustice.net/incineration/expensive-waste (accessed May 13, 2021).
5 The High Cost of Waste Incineration. (2021). https://zerowasteworld.org/wp-content/uploads/The-High-Cost-of-Waste-Incineration-March-30.pdf (accessed May 13, 2021).
6 US Energy Information Administration. (2021). Waste-to-energy (MSW). https://www.eia.gov/energyexplained/biomass/waste-to-energy.php (accessed April 29, 2021).
7 U.S. Department of Energy. (2019). Waste-to-energy from municipal solid wastes. https://www.energy.gov/sites/prod/files/2019/08/f66/BETO–Waste-to-Energy-Report-August–2019.pdf (accessed April 29, 2021).
8 US Environmental Protection Agency. (2020). Commercial and industrial solid waste incineration units (CISWI): New source performance standards (NSPS) and Emission guidelines (EG) for existing sources. https://www.epa.gov/stationary-sources-air-pollution/commercial-and-industrial-solid-waste-incineration-units-ciswi-new (accessed April 30, 2021).

9 US Environmental Protection Agency. (2020). Hospital, medical, and infectious waste incinerators (HMIWI): New source performance standards (NSPS), emission guidelines, and federal plan requirements regulations. https://www.epa.gov/stationary-sources-air-pollution/hospital-medical-and-infectious-waste-incinerators-hmiwi-new-source (accessed April 30, 2021).
10 US Environmental Protection Agency. (2020). Other solid waste incinerators (OSWI): New source performance standards (NSPS) and Emission guidelines (EG) for existing sources. https://www.epa.gov/stationary-sources-air-pollution/other-solid-waste-incinerators-oswi-new-source-performance (accessed April 30, 2021).
11 US Environmental Protection Agency. (1999). RCRA, superfund & EPCRA hotline training module introduction to: Hazardous waste incinerators. Solid Waste Emerg Response. https://www.epa.gov/hwpermitting/resource-conservation-and-recovery-act-rcra-training-module-about-hazardous-waste (accessed November 15, 2020).
12 U.S. Environmental Proctection Agency. (2005). Hazardous waste identification (40 CFR Parts 261). Solid Waste Emerg Response; EPA530-K-0.
13 Electronic Code of Federal Regulations. (2021). 40 CFR Part 264. E-CFR. https://www.ecfr.gov/current/title-40/chapter-I/subchapter-I/part-264 (accessed April 30, 2021).
14 Institute for Global Environmental Challenges. (2020). CCET guideline series on intermediate municipal solid waste treatment technologies. https://www.iges.or.jp/en/pub/ccet-guideline-wtei/en (accessed May 10, 2021).
15 Deltaway. (2020). Waste-to-energy: How it works. https://deltawayenergy.com/2018/08/waste-to-energy-how-it-works/#tip10 (accessed April 30, 2021).
16 Hasselriis, F. (2002). Waste-to-energy combustion. Part 13 B. Ash management and disposal. In: *Handb. Solid Waste Manag*, 2nd e (ed. G. Tchobanoglous and F. Kreith). New York, NY: McGraw-Hill.
17 Quina, M.J., Bordado, J.C., and Quinta-Ferreira, R.M. (2008). Treatment and use of air pollution control residues from MSW incineration: An overview. *Waste Manag* 28: 2097–2121. https://doi.org/10.1016/j.wasman.2007.08.030.
18 IEA Bioenergy. (2003). A position paper prepared by IEA bioenergy municipal solid waste and its role in sustainability. https://www.ieabioenergy.com/blog/publications/position-paper-municipal-solid-waste-and-its-role-in-sustainability-2 (accessed April 29, 2021).
19 Brunner, C.H. (2002). Waste-to-energy: Part 13 A incineration technologies. In: *Handb. Solid Waste Manag*, 2nd e (ed. G. Tchobanoglous and F. Kreith). New York, NY: McGraw Hill.
20 Kawai, K., Huong, L.T.M., Yamada, M., and Osako, M. (2016). Proximate composition of household waste and applicability of waste management technologies by source separation in Hanoi, Vietnam. *J Mater Cycles Waste Manag* 18: 517–526. https://doi.org/10.1007/s10163-014-0348-5.
21 Castro-Aguirre, E., Iñiguez-Franco, F., Samsudin, H., Fang, X., and Auras, R. (2016). Poly(lactic acid)—Mass production, processing, industrial applications, and end of life. *Adv Drug Deliv Rev* 107: 333–366. https://doi.org/10.1016/j.addr.2016.03.010.
22 US Environmental Protection Agency. (2020). Advancing sustainable materials management: 2018 Tables and Figures. https://www.epa.gov/sites/default/files/2021-01/documents/2018_tables_and_figures_dec_2020_fnl_508.pdf (accessed April 29, 2021).
23 Šyc, M., Simon, F.G., Hykš, J., Braga, R., Biganzoli, L., Costa, G. et al. (2020). Metal recovery from incineration bottom ash: State-of-the-art and recent developments. *J Hazard Mater* 393. https://doi.org/10.1016/j.jhazmat.2020.122433.
24 Joseph, A.M., Snellings, R., Van den Heede, P., Matthys, S., and De Belie, N. (2018). The use of municipal solidwaste incineration ash in various building materials: A Belgian point of view. *Materials (Basel)* 11. https://doi.org/10.3390/ma11010141.

25 Hasselriis, F. (2002). Waste-to-energy combustion. Part 13 C. Emission control. In: *Handb. Solid Waste Manag*, 2nd e (ed. G. Tchobanoglous and F. Kreith), 13.121. New York, NY: McGraw-Hill.

26 Alex Tullo. (2018). Should plastics be a source of energy? *C&EN Glob Enterp* 96: 34–39. https://doi.org/10.1021/cen-09638-cover.

27 PVC: To burn or not to burn? (2020). Waste Management World. https://waste-management-world.com/artikel/pvc-to-burn-or-not-to-burn (accessed May 12, 2021).

28 Yazdani, S., Salimipour, E., and Moghaddam, M.S. (2020). A comparison between a natural gas power plant and a municipal solid waste incineration power plant based on an emergy analysis. *J Clean Prod* 274: 123158. https://doi.org/10.1016/j.jclepro.2020.123158.

29 Pfadt-Trilling, A.R., Volk, T.A., and Fortier, M.O.P. (2021). Climate change impacts of electricity generated at a waste-to-energy facility. *Environ Sci Technol* 55: 1436–1445. https://doi.org/10.1021/acs.est.0c03477.

16

Landfill

16.1 Learning Objectives

After learning from this chapter, you are expected to:

- Understand the main purpose of disposing of municipal solid waste (MSW) through a landfill in the US and around the world.
- Describe the role of landfills in US MSW management.
- Understand the basic economics of landfill waste and tipping fees.
- Define the main terms related to landfills.
- Describe the main advantages and disadvantages of disposing MSW in landfills.
- Differentiate between solid and hazardous wastes.
- Describe the main landfill types.
- Describe how the location of a landfill is decided, and how a landfill is built, operated, and closed.
- Explain the main parts of a modern landfill in the US.
- Describe the main emissions from a landfill.
- Explain the generation of energy from landfill gases.
- Explain the landfilling of metal, glass, paper, and plastics.

16.2 Introduction

Disposing of municipal solid waste (MSW) in a landfill is the last strategy of the US MSW integrated waste management system. A municipal solid waste landfill (MSWLF) can be described as an area of land or an excavation designated for the purpose of receiving and containing household waste and other types of nonhazardous waste, such as commercial solid waste, nonhazardous sludge, bottom, and mixed ashes. Many substances such as paper, appliances, packaging including plastic containers, food scraps, yard trimmings, and additional wastes from home, commerce, and industry are accepted in MSWLF. Several treatments can be conducted to reduce the volume and toxicity of wastes prior to their disposal in MSWLF. These treatments can include physical – shredding, chemical – incineration, and biological – anaerobic digestion.

Modern landfills are designed facilities to ensure compliance with state and federal regulations. In the past, the term sanitary landfill was used. Landfills accepting MSW are mostly regulated by state, tribal, and local governments. However, the US EPA establishes national standards that

Life Cycle of Sustainable Packaging: From Design to End of Life, First Edition. Rafael A. Auras and Susan E. M. Selke.

Companion Website: www.wiley.com/go/Auras/lifecycleofsustainablepackaging

every landfill is required to meet to be operational. The federal landfill rules through the Resource Conservation and Recovery Act (RCRA) eliminated the open dumps of the past and provide the modern regulations for landfills [1]. Landfills must be designed, operated, and closed under strict regulatory requirements. Methane gas, produced due to decomposition of organic waste, must be collected in all new MSWLFs. After a landfill is covered, the lot can be used for several purposes such as green spaces, recreation sites, parks, ski slopes, or golf courses.

Currently, in the US, we have 550 operational landfills, 481 candidate landfills, and 596 closed landfills with gas collection; totaling 1,596 landfills [2], as shown in Figure 16.1. The Landfill Methane Outreach Program (LMOP) tracks the landfill gas (LFG) energy projects as well as the MSWLFs in the US. The intention of the LMOP is to track landfill projects in various stages such as planning, under construction, operational, and closed. Landfills are the most common form of waste disposal in the US. Different states within the US differ appreciably in the proportion of MSW that is landfilled. The total number of landfills in the US steadily decreased between 1988 and 2000 and is now down to 1,031 open and planned landfills.

Figure 16.1 also shows a map of the average landfill tipping fees in the US in 2020 in US$/ton. We can observe that Alaska, Washington, Hawaii, Rhode Island, and Vermont have the highest average tipping fees in the nation (113 US$/ton). The Northeast and Pacific had notably higher tipping fees in 2020 than the rest of the US (average: Northeast = 68.9 US$/ton; Pacific = 72 US$/ton; Mountains/Plains and Midwest = 47.8 US$/ton; Southeast = 46.3 US$/ton; and South Central 39.7 US$/ton). The average regional tip fees increased in the Northeast and the Southeast by 3.25% and 2.2%, respectively, from 2019 to 2020 [3].

There are differences between states in the amount of available landfill space. Even though the overall number of landfills has decreased, the average size of the new opened landfills is much larger than the average size of the landfills that are closing. So, landfill capacity did not drop substantially through the years in the US. The US has two of the largest landfills in the world, Apex Regional (Las Vegas, NV) with 2,200 acres and Puente Hills, Los Angeles, CA with 630 acres. The second largest landfill in the world is in Bordo Poniente, Mexico City, Mexico with 927 acres. An indirect measure of the decline in concern over landfill space availability can be found in the most recent EPA report, which includes data on national average landfill tipping fees, adjusted to 2020, and shown in Figure 16.2. We can see that tipping fees rose dramatically between 1986 and 1995, dropped between 1995 and about 2004, and have risen somewhat since then, but only to about the same rate as in 1995. The five-year trends in average tip fees suggest a long-term increase in the national average MSW tip fees despite slight drops in 2018 and 2020.

Small landfills were reported to accept an average of 26,480 tons/year, while medium landfills and large landfills accepted 166,516 tons/year and 809,455 tons/year, respectively, in 2018 [3]. Construction and demolition debris (C&D) tipping fees were on average the same as MSW. C&D represents twice the amount of waste generation in the US as MSW (see Chapter 10 for more information about C&D).

Figure 16.3 represents the main disposal methods of waste around the world as previously discussed in Chapter 10. Although open dumps and landfills represent the largest methods to deal with MSW around the world, their contribution in different regions varies. North America (excluding Mexico) does not have open dumps, but mostly landfills. Latin America and the Caribbean still disposed 26.8% in open dumps in 2016. Overall, open dumps still represented the largest category of management of MSW around the world with 33%, unspecified landfill accounted for 25.2% and sanitary landfill with gas collection represented only 7.7% in 2016 [4].

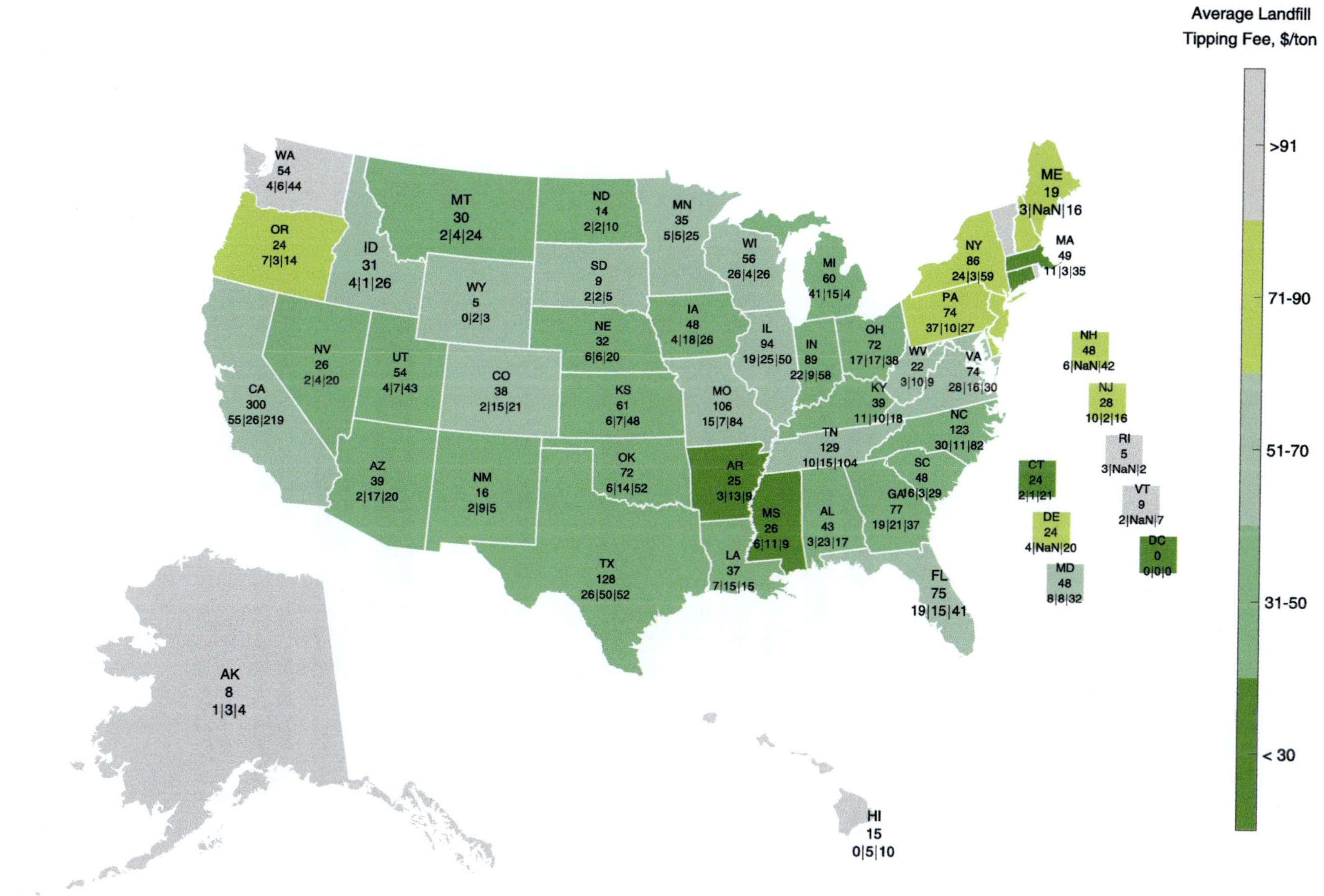

Figure 16.1 Number of total – top, bottom (operational | candidate | closed) landfills in each state in the US as of March 2021, and tipping fees indicated by color as December 2020. Information obtained from the US EPA. NaN: The Landfill Methane Outreach Program does not have information on candidate landfills in this state [2]. Landfill tipping fee data obtained from [3].

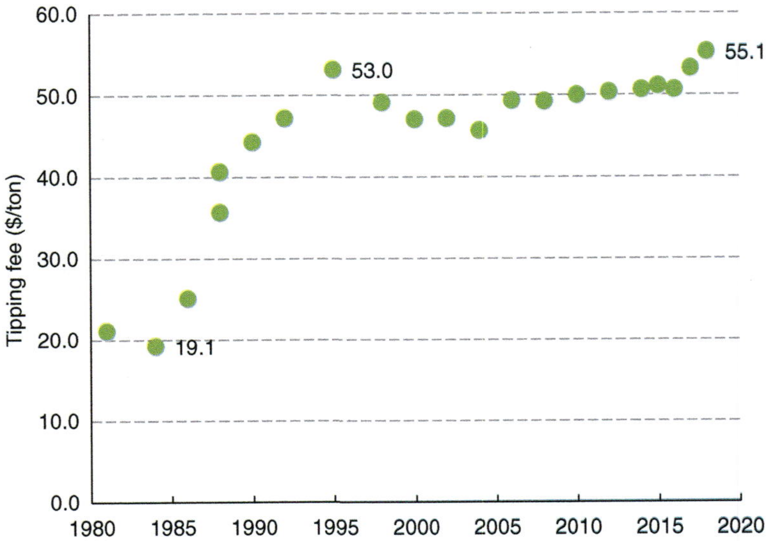

Figure 16.2 US average landfill tipping fees since 1980.

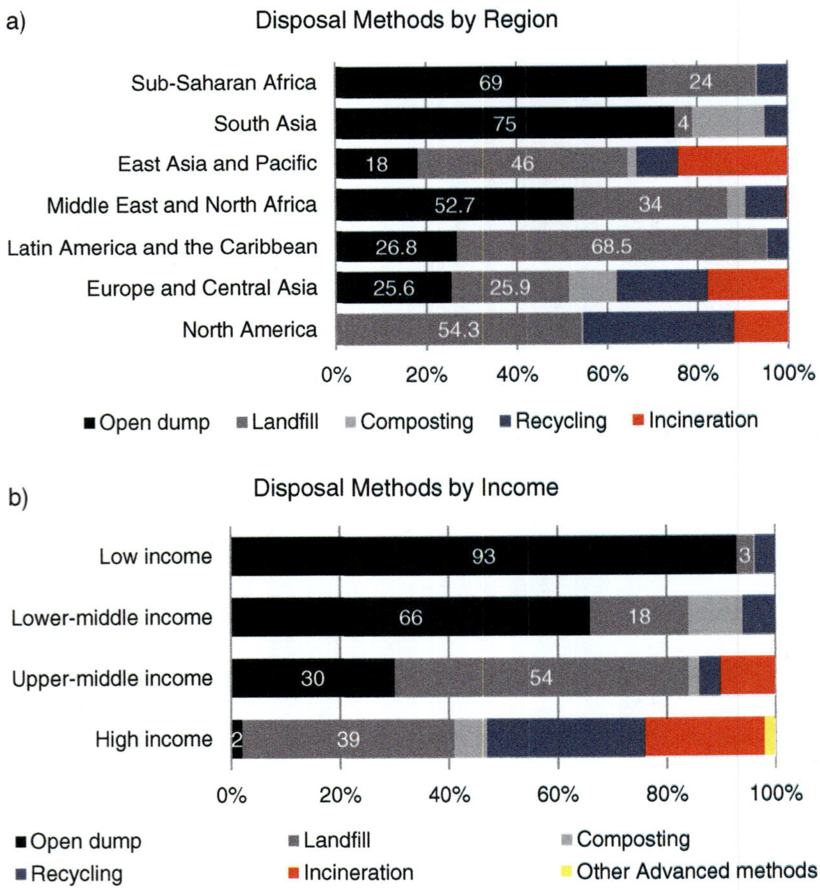

Figure 16.3 Disposal methods around the world in 2016; (a) by region, (b) by income; data obtained from [4].

Dumping of waste in open spaces is still the prevalent method to deal with waste in lower-income countries. Open dumps can represent as high as 69% and 75% of the waste management system in sub-Saharan Africa and South Asia, respectively. The risks and costs of dumpsites are being increasingly recognized, so governments are pursuing other sustainable waste disposal methods. See Chapter 10 for a discussion of MSW disposal methods in different economies.

In this chapter, we will concentrate on discussing landfills and their regulations as these relate to the US framework and its requirements. However, the majority of the information about their operation and requirements can be translated to different regions.

16.3 Definition of Terms

Before we continue discussing landfills and disposal of MSW in landfills, several terms specific to RCRA Title 40: Protection of Environment Part 258 are relevant and must be defined to facilitate the understanding of the following sections.

Landfills: areas for disposal of various types of waste such as MSW and incinerator ash.

Monofills: areas for disposal of selected specific materials.

Sanitary landfills: all modern landfills.

Secure landfills: areas for the disposal of hazardous wastes.

Dumps: places where waste is dumped on or into the ground in no organized manner.

Active life: period of operation beginning with the initial acceptance of solid waste and ending at completion of closure activities.

Active portion: portion of the landfill that is currently receiving waste.

Commercial solid waste: solid waste generated by stores, offices, restaurants, warehouses, and other nonmanufacturing activities, excluding residential and industrial wastes.

Household waste: "solid waste including garbage, trash, and sanitary waste in septic tanks derived from households such as single and multiple residences, hotels, motels, bunkhouses, ranger stations, crew quarters, campgrounds, picnic grounds, and day-use recreation areas" [1].

Industrial solid waste: "solid waste generated by manufacturing or industrial processes that is not a hazardous waste regulated under RCRA subtitle C" [1].

Cell: volume of material placed in a landfill during one operating period.

Daily cover: 6–12 in. (0.15–0.30 m) of native soil or alternative material between layers of MSW to avoid blowing of materials, reduce odors, and control entry of water.

Lift: complete layer of cells over the active area. Landfills are composed of a series of lifts.

Bench (or terrace): area of the landfill with height that may exceed 50–75 ft. They are constructed to preserve the slope stability of the landfill and to place drainage channels.

Final lift: final cover of a landfill with a cover layer.

Landfill liners: materials to line the bottom (and often sides) of a landfill.

Landfill cover: final cover of the surface of a landfill.

Lateral expansion: horizontal expansion of the waste boundaries.

Leachate: liquid that collects at the bottom of the landfill. Mostly forms when rainwater runs through wastes placed in the landfill. When the filtered liquid interacts or contacts the buried wastes, it leaches and can extract chemicals from the waste.

Runoff: any liquid such as rainwater and leachate running over the landfill surface.

Landfill gas: mixture of gases (mostly CH_4; CO_2, O_2, organics) that forms in a landfill.

Environmental monitoring: collection and analysis of water and air samples from the landfill.

Landfill closure: steps to close a landfill after operation ceases.

Postclosure care: long-term care of the landfill.

16.4 Advantages and Disadvantages of Disposing Municipal Solid Waste in Landfills

One of the effective methods of dealing with MSW is landfilling waste. Landfilling of MSW has several advantages and disadvantages. Among them:
Advantages:

- Landfills are a relatively inexpensive way to dispose waste.
- Waste disposal volume can increase with minimal addition of people/equipment.
- Gas produced from the decomposition of organic wastes can be collected and used for producing energy.
- They require little infrastructure to operate.
- After closure they can be used for recreation or other purposes.
- They are often readily available in the US around big cities, which generate the largest amount of MSW.
- They can create local employment.

Disadvantages

- They result in a large area loaded with waste.
- Completed landfill areas can collapse, gases escaping, requiring maintenance.
- Methane ignites easily when present.
- If methane is not collected for energy generation or heat, it is a potent greenhouse gas (see Chapter 8).
- They increase the risk of contaminated soil and water if not properly designed and managed (see Chapters 5 and 6).
- They can affect surrounding wildlife and human health. Animals might dig in the trash to find food.
- Smells can easily escape landfills, affecting surrounding areas.
- They generate leachate that must be treated.
- They can delay the adoption of more sophisticated methods to reduce waste creation.
- They are expensive to initially build, and proper long-term management of the area is not always assured.

16.5 Classification of Landfills

"Sanitary landfills" (better known simply as "landfills") were a major improvement over the open dumps that they replaced in the US. Current modern landfills are operated more like advanced chemical reactors, and they are regulated under:

a) RCRA Subtitle D (nonhazardous solid waste),
b) RCRA Subtitle C (hazardous solid waste), or
c) Toxic Substances Control Act (TSCA).

16.5.1 Landfills Regulated under RCRA – Subtitle D

Subtitle D regulates nonhazardous solid waste from households, industries, commercial facilities, or some small-quantity generators. The regulation focuses on state and local governments as the main entities to control the planning, regulation, and implementation of management systems for landfill disposal of nonhazardous solid waste, such as household garbage and nonhazardous industrial solid waste. Subtitle D bans open dumping of waste and sets minimum federal criteria for the operation of MSWLFs, including planning, location, operation, financial assurance, corrective action, and closure requirements. The landfills regulated under Subtitle D are:

16.5.1.1 Municipal Solid Waste Landfill

MSWLFs are specifically designed to receive household waste, as well as other types of nonhazardous wastes. Some materials are banned from disposal in MSWLFs such as household items like cleaners/chemicals, paints, batteries, pesticides and motor oil. Title 40 of CFR part 258 addresses seven major regulatory categories of MSWLFs, which are: "general scope, location restrictions, operating criteria and practices, design criteria, groundwater monitoring and corrective action, closure and post-closure care requirements, and financial assurance criteria" [5].

Bioreactor landfill (BF): BFs are a kind of MSWLF operating to swiftly convert and degrade organic waste. A BF is a MSWLF in which liquids from the leachate and other sources such as stormwater, wastewater, and wastewater treatment plant sludges are added to help bacteria break down the waste. By adding liquid and air, the biodegradation process is enhanced by purposeful control of the oxygen level and moisture content to facilitate microbiological processes. This bioreactor concept differs from the traditional "dry tomb" MSWLF approach and from regular landfills where leachate is mostly recirculated. Three main types of BFs are known:

Aerobic: in this BF, the leachate is removed from the bottom layer of the landfill through the collection tubes, and then it is piped to storage tanks, and then in a controlled manner recirculated into the landfill. Meanwhile, air is injected into the disposed waste in the MSWLF using vertical or horizontal wells to promote aerobic biodegradation.

Anaerobic: in this BF, moisture is added to the disposed waste in the MSWLF in the form of recirculated leachate as well as other sources to obtain optimal moisture levels. Biodegradation occurs in the absence of oxygen (anaerobically) producing LFG used for energy projects.

Hybrid: in this BF, a mixture between sequential aerobic and anaerobic degradation process occurs. The disposed waste in the top part of the MSWLF is rapidly decomposed and gas is collected from the lower sections. In this hybrid process, methanogenesis is accelerated (see Chapter 14 for a description of methanogenesis).

16.5.1.2 Industrial Waste Landfill (IWLF)

These landfills are intended to collect commercial and institutional nonhazardous waste (i.e., industrial waste), a significant share of solid waste. If the IWLFs are collecting hazardous waste, they are regulated by Subtitle C for hazardous waste landfills (HWLFs).

C&D debris landfill: This kind of IWLF is exclusively intended for C&D materials consisting of debris generated during construction, remodeling, and demolition of buildings, roads and bridges. C&D amounts to 600 M tons per year, and only 22% is reused to manufacture products [5]. C&D includes steel, wood products, drywall and plaster (mostly gypsum), brick and clay tile, asphalt shingles, concrete, and asphalt concrete. Other types of C&D materials can include metals, plastics, glass, salvaged building components such as plumbing fixtures, doors and windows, trees, stumps, earth and rock from clearing sites [5]. C&D landfills do not receive hazardous waste or industrial solid waste, unless those landfills meet certain standards and are permitted to receive such wastes.

Coal combustion residual (CCR) landfills: CCR, commonly known as coal ash including fly and bottom ashes, boiler slag, and flue gas desulfurization material, are created from burning coal in power plants. Since coal ash contains contaminants such as mercury, cadmium, and arsenic, which can pollute surface and groundwater and the air, their disposal must be properly managed. CCR is still one of the largest types of industrial waste generated in the US but is declining. They were initially regulated under Subtitle D of RCRA in 2008. The EPA established updated requirements for the disposal of CCR in landfills that became effective in 2015 [6]. The new rule established nationally applicable minimum criteria for the safe disposal of CCR in landfills.

16.5.2 Landfills Regulated under RCRA – Subtitle C

Subtitle C of RCRA regulates landfills that store hazardous waste. The main goal of the Subtitle C regulations is to assure that hazardous wastes are treated in a way that protects human health and the environment. Subtitle C regulates the production, transfer, treatment, storage, and disposal of hazardous wastes. Landfills included under Subtitle C are the following:

16.5.2.1 Hazardous Waste Landfills

These types of LFs are for disposal of hazardous wastes and MSW cannot be disposed in them. Hazardous wastes (as described in Section 16.5) are classified as: (a) *Characteristic hazardous wastes* (CHWs). CHWs are designated hazardous because they exhibit dangerous properties such as corrosivity, ignitability, reactivity, or toxicity. CHWs cannot be delisted since they have dangerous properties; (b) *Listed hazardous wastes* (LHWs). LHWs are designated hazardous because the processes that generate them have typically produced wastes with dangerous properties such as those mentioned for CHWs. Due to historic knowledge, LHWs are assumed to possess dangerous properties.

16.5.3 Landfills Regulated under the *Toxic Substances Control Act*

Landfills for specific substances known to be toxic or problematic in handling or for the environment are regulated under TSCA.

Polychlorinated biphenyl (PCB) landfills: PCBs are a group of man-made organic chemicals consisting of carbon, hydrogen, and chlorine atoms. PCBs have no known taste or smell and range in consistency from an oil to a waxy solid. PCBs were manufactured from 1929 to 1979 (see Case Study 6.1 about water pollution in the Pine River in Chapter 6). They have a range of toxicity. PCBs were used in many industrial and commercial applications, such as fluids for electrical, heat transfer and hydraulic equipment, plasticizers in paints, plastics and rubber products, pigments, and dyes. PCBs are regulated by the TSCA. Facilities can be permitted by EPA to dispose of PCB waste. EPA maintains a searchable list of disposal facilities approved to dispose PCBs under the EPA Regional Office [7].

If there are not state-approved programs for landfill disposal, the federal requirements must be met by waste facilities. The different types of landfill can be summarized as indicated in Table 16.1.

Table 16.1 Unofficial classification of landfill types, but commonly used by landfill operators.

Class	Type	Characteristics
I	Hazardous waste	Hazardous wastes such as PCB.
II	Designated waste	Nonhazardous wastes releasing compounds in concentrations in excess of applicable water quality, such as combustion ashes and asbestos.
III	Municipal solid waste	Household and commercial waste.
IV	Green waste	Controlled disposal of organic materials.

While this classification is not an official EPA scheme, it is normally used by landfill operators. Class IV is an emerging type of landfill in some states where a designated area is allocated for the decomposition of green waste. This type of disposal is controversial. Furthermore, some landfill operators with MSWLF gas production have been lobbying to introduce more organics to landfill to increase the production of LFG (see Case Study 16.1).

16.6 Location, Building, Operation, Closure, and Financial Assurance of Landfills

The disposal, construction, and requirements of surface impoundments, waste piles, land treatment units, as well as landfills are codified in RCRA (40 CFR part 264/265), Subparts K through N [8]. Before defining landfills, we need to introduce simpler structures, surface impoundments, and waste piles, which are used in temporary settings.

Surface impoundments are similar to landfills, and they are either a normal topographic depression, man-made excavation, or diked area molded mainly of material such as soil. The units are for different uses than landfill since surface impoundments are usually used for transitory storage or treatment, whereas landfills are areas selected for final waste disposal. While the design and operating standards are very similar, the closure and postclosure standards greatly differ.

Waste piles are noncontainerized piles of solid, nonflowing hazardous waste, which are temporarily located in a specific location. So, waste piles are not regulated for closure since they are finally disposed through the waste management system.

Landfills are more complex operations than surface impoundments, but they are subject to similar minimum technological requirements (MTRs) according to the RCRA. The structure of landfills has evolved over time, as regulators and landfill operators have became more knowledgeable about the hazards presented by landfills, procedures have been developed for ways to reduce those hazards.

Location of landfills is strictly regulated. They must be located away from airports to reduce likelihood of bird/aircraft collisions, as birds like to congregate around landfills to scavenge waste food or rodents. If landfills are close to an airport due to space limitations, they must have a bird management program. MSWLFs and their lateral expansions should not restrict the flow of water in 100-year flood plains or reduce the water storage capacity of a floodplain. The construction and operation of MSWLFs should not violate any applicable state water quality standards or jeopardize wetlands and their habitats. MSWLFs must protect ecological resources by controlling for erosion, stability of the landfill, and the displacement of native wetlands. MSWLFs should not impact wildlife and aquatic resources. Additionally, MSWLFs should not be located in seismic impact zones or fault zones (i.e., fracture or a zone of fracture in any material) or in unstable areas, such as areas prone to sinkholes.

Building of landfills: Modern landfill design incorporates double liners; usually one is a synthetic layer and one a clay layer, but sometimes both are synthetic, usually high-density polyethylene; mostly when there is not a clay layer underneath. The double liner system is intended to prevent the migration of hazardous constituents into the soil. Additionally, there should be 3 ft. of compacted soil material on top of the liner before laying any waste. The double liner must be equipped with a leachate collection and removal system (LCRS), also used as a leakage detection system. The LCRS and the leak detection system drainage layers must be constructed with a bottom slope of at least 1%, be made of materials chemically resistant to the wastes placed in the unit, and be designed to remove the liquids at a particular minimum rate. The LCRS itself must be designed to collect liquids in a slump and then pump them out to a station. The LCRS is situated above the top liners to allow the LCRS to collect the greatest amount of leachate. Figure 16.4 shows an example of the LCRS for a landfill with a clay layer at the bottom.

Operation of landfills: MSWLFs are highly regulated. Owners of LFs must have a program for detecting and preventing the disposal of regulated hazardous waste. Random inspection of incoming loads must be conducted, and the records of the inspections recorded. Daily solid waste must be covered with at least 6 in. of soil (or alternative cover) at the end of each operating day or at regular intervals to regulate for disease vectors such as rodents, flies, mosquitoes, or other animals, fires, odors, blowing litter, and scavenging. The owner of the landfill must have a disease vector control program so that vectors capable of transmitting disease to humans are controlled. Since the explosion of gases may occur from landfills due to the decomposition of organic waste, the operator must have a system to prevent and remediate explosions. This can be done by guaranteeing that the concentration of methane gas does not exceed the lower explosive limit for methane. Air emissions from landfills are regulated by section 110 of the Clean Air Act as discussed in Chapter 7, and the contamination of surface water is regulated by the Clean Water Act as well as by Section 402 of the NPDES requirements. The landfills must provide for the collection, removal, and treatment of leachate. Landfills must also provide control over the methane that is generated in the landfill once it becomes anaerobic (without oxygen). Modern landfills build systems for leachate and gas collection into the landfill. In older landfills, these systems may need to be added afterward. It is much more economical to design them into the landfill than to retrofit.

Landfills are also required to provide periodic testing of the groundwater around the landfill, as an early detection system for pollutants, through a groundwater monitoring and corrective action system. Samples must be collected to analyze any processes, physical, chemical, or biological, affecting the production of contaminants, their fate, and their transport. This collection is done by having a system of wells, as shown in Figure 16.4, installed at appropriate locations and depths so that groundwater samples from the surrounding uppermost aquifer can be collected. The operator of the MSWLF must show that the groundwater is not affected due to the presence, operation, and

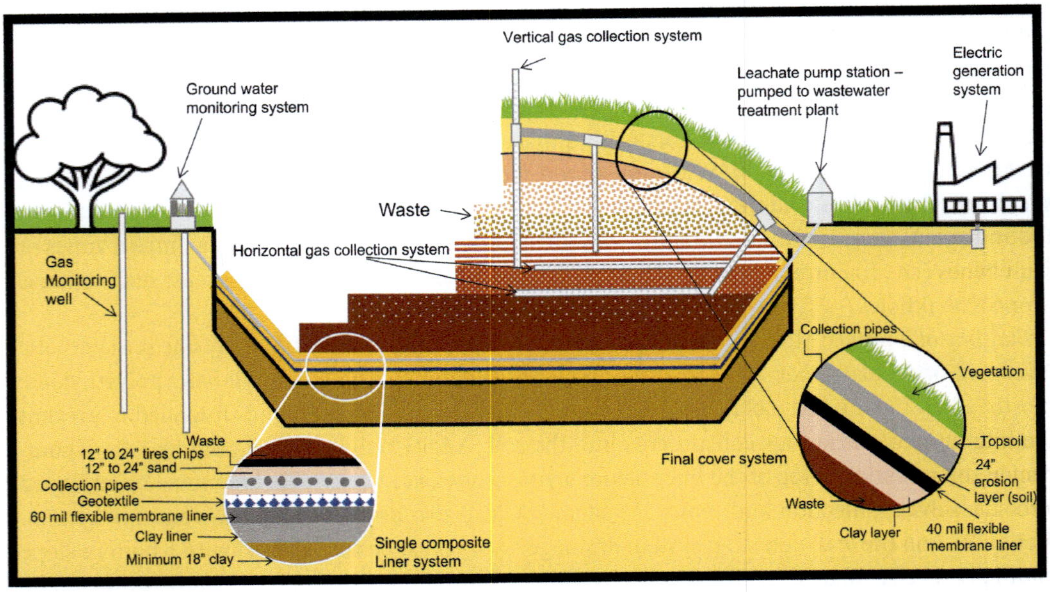

Figure 16.4 Landfill cross and main sections, including the design of a single and a double composite liner system and the final cover system.

later closure of the landfill. Groundwater samples must pass the requirements of the state. If contamination is detected, an assessment monitoring and remediation program is required as described in RCRA Title 40 Subpart E 258 [1].

Closure of landfills: A final cover system is added at the end of the life of landfills to minimize infiltration and erosion. The final cover system must be designed and constructed with set water permeability specifications, and it must contain an infiltration layer that contains a minimum thickness of earthen material. The landfill cap includes a barrier layer to help keep water out, a layer of soil (usually on top of a drainage layer), and then vegetation. Vegetation is important to stabilize the soil cover and prevent erosion, as well as to improve the esthetics of the area. There are limits on the kind of vegetation that is considered appropriate for a landfill cover so that the roots do not perforate the cover liners. Generally, anything with deep roots that might penetrate the cap is considered unacceptable. Postclosure care of the landfill area is required for usually 30 years.

Financial assurance criteria: By law, landfill operators must provide for the "post-closure" care of the landfill. The length of time varies by state law but is usually between 20 and 50 years. During this time period, the owner of the landfill must maintain the land, and continue the leachate collection and methane collection or flaring, as well as monitoring the groundwater near the landfill. The main goal of the financial assurance requirement is to prove that the funds needed to cover the costs of closure, postclosure care, and corrective action for known releases will be available whenever they are required. In general, this assurance is established through a trust fund. The trust operations are legalized and scrutinized by a federal or state agency.

In the past, it was not uncommon to build factories, or even residences, on top of closed landfills. Now, with modern landfill design and a greater understanding of what does and does not happen inside landfills, this is usually not acceptable due to dangers of structure collapse and emissions. It is desirable to get some kind of use out of the land, but buildings should be avoided. Golf courses or ski hills are less problematic uses. Even here, however, it is important to realize that there will be changes in the land over time. In particular, there will be subsidence (sinking), usually in an uneven manner, as decay processes proceed.

In some areas in the US, mostly in Florida but also to some extent in New York and elsewhere, landfills have been "mined." When a landfill is mined, it is dug up, the material it contains sorted out, and fresh MSW placed in the landfill. The major goal is usually to recover the valuable space occupied by the old, full landfill. When the waste is dug up, much of it has generally decayed and become a humus-like material, which can be used as daily cover for the fresh garbage. Usually there is some recovery of materials from the landfill, especially ferrous metals, which can readily be separated using an electromagnet and aluminum. In some cases, materials such as paper and plastic are sent to incineration. Landfill mining is still quite rare in the US. In most cases, it is not cost-effective.

16.7 Emissions from Landfills

Landfills are concentrated areas where solid wastes are deposited, including organics and liquids from households. Organics are prone to decomposition by bacteria and production of LFG. Liquids percolate through the landfill and are collected as leachate on the bottom of the landfill. The production of these emissions is time-bound, and they evolve depending on the conditions in the landfill. Landfills are regulated as point source pollutant facilities. The main emissions to consider are air and leachate emissions. Air emissions are controlled by the CAA, and liquid effluent emissions and limitations to effects on surface and groundwater are regulated by the CWA.

16.7.1 Air Emissions

Most LFGs are created by *bacterial decomposition*, *volatilization*, and *chemical reactions*. *Bacterial decomposition* occurs when organic waste is broken down by bacteria naturally existing in the waste and in the soil used to cover the landfill. Organic wastes include food, garden waste, street sweepings, textiles, and wood and paper products. Bacteria decompose organic waste in five phases, and the composition of the gas changes during each phase. A description of biodegradation under anaerobic conditions is provided in Chapter 14.

Several states prohibit the deposit of garden waste in landfills since they are readily biodegradable and produce a large amount of LFGs. The more biodegradable compounds that are disposed of in the landfill, the more LFGs are produced. In the case that other volatile chemicals are disposed of in landfill, the more likely nonmethane organic compounds (NMOCs) and other gases will be produced either through volatilization or chemical reactions.

Volatilization is the process whereby certain wastes, particularly organic compounds, change from a liquid or a solid into a vapor. NMOCs in LFG may be the result of the volatilization of certain chemicals disposed of in the landfill. *Chemical reactions* occur when certain chemicals present in waste contact each other and undergo chemical changes. For example, if chlorine bleach and ammonia come into contact with each other within the landfill, a harmful gas is produced.

Figure 16.5 shows a visualization of the aerobic and anaerobic gas production stages in a landfill. Gas composition changes depending on the stage of the landfill. During phase I, the decomposition is mostly aerobic, so bacteria that are active in the presence of oxygen decompose organic matter producing CO_2, humus, and water. Nitrogen content is high at the beginning but drops rapidly. Phase I – initial adjustment – can last from days to a month depending on the amount of oxygen present when the waste was disposed of. Oxygen concentration varies depending on compaction levels.

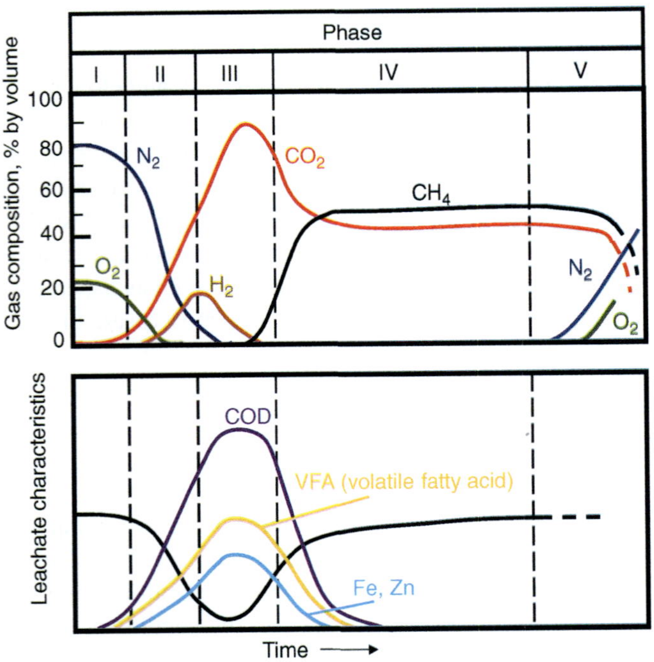

Figure 16.5 Evolution of landfill gases (top) and composition of leachate (bottom) as a function of time (phase I – initial adjustment, II – transition, III – acid phase, IV – methane fermentation, and V – maturation phase); adapted from [11].

During phase II – transition phase, oxygen is used up and anaerobic decomposition starts. Bacteria able to grow under anaerobic conditions convert compounds into acetic, lactic, and formic acids, and alcohols such as methanol and ethanol. So, the pH of the landfill drops. After the acids mix with the moisture in the environment, some nutrients dissolve producing N_2 and phosphorus, which are food for bacteria, producing as byproducts CO_2 and H_2. During phase III – acid phase, the bacteria consume the organic acids produced in phase II, forming acetates and organic acids. In this phase, the landfill becomes more neutral and methane-producing bacteria become established. Methane- and acid-producing bacteria have a cooperative or mutually beneficial relationship. The acid-producing bacteria produce several compounds needed by the methanogenic bacteria. The methanogenic bacteria ingest the CO_2 and acetate, which could be toxic to the acid-producing bacteria in excess, as discussed in Chapter 14. During phase IV – methane fermentation phase – a steady state of production of methane is reached (and also CO_2). Typically, this steady phase runs for about 20 years. However, gas can continue to be emitted for 50 or more years after the waste is deposited in the landfill. Gas creation may last longer if larger amounts of organic waste were present in the original waste [9] or if decomposition is slow, such as in cold or dry climates. Finally, during phase V – maturation phase – the amount of CH_4 and CO_2 declines and N_2 and O_2 return to normal concentrations. A recent review has summarized and detailed the types of microbiological reactions present in a landfill and the production of LFGs [10].

The main compounds of LFGs are carbon dioxide, methane, nitrogen, and hydrogen sulfide, as shown in Table 16.2. Several factors affect the production of gases and the period spent in each phase. The main factors are the waste composition, the age of the refuse, the presence of oxygen, the moisture content, and the temperature.

Table 16.2 Typical landfill gas components.

Component	Percentage by Volume	Release Kinetics Comments as Described in Figure 16.5	Characteristics
Methane	45–60	During phase III to V.	It is a naturally occurring gas, colorless, and odorless. Landfills are the single largest source of US man-made methane emissions. At high concentrations, it can explode.
Carbon dioxide	40–60	During phase I to V, starting as soon as the biodegradation starts.	It is naturally found at small concentrations in the atmosphere around 0.044%, and it is colorless, slightly acidic, and odorless.
Nitrogen	2–5	Starting high in phase I and reducing to the end of phase IV, may increase in phase V.	Nitrogen comprises approximately 79% of the atmosphere. It is odorless, tasteless, and colorless.
Oxygen	0.1–1	Reduce from 21% in phase I to almost 0% in phase II. It may increase in phase V.	O_2 comprises around 21% of the atmosphere. It is also odorless, tasteless, and colorless.
Ammonia	0.1–1	As soon as the decomposition starts in phase I.	Ammonia is a colorless gas with a pungent odor.

(Continued)

Table 16.2 (Continued)

Component	Percentage by Volume	Release Kinetics Comments as Described in Figure 16.5	Characteristics
Nonmethane organic compounds	0.01–0.6	As soon as the decomposition starts in phase I.	NMOCs are organic compounds that contain carbons. Methane is not considered an NMOC. Common NMOCs in landfills are acrylonitrile, benzene, dichloromethane, trichloroethylene, carbonyl sulfide, ethyl benzene, hexane, methyl ethyl ketone, tetrachlorobenzene, toluene, vinyl chloride, and xylenes.
Sulfides	0–1	As soon as the decomposition starts in phase I.	Sulfides such as hydrogen sulfide, dimethyl sulfide, and mercaptan are gases with rotten-egg smell. Sulfides can cause unpleasant odors at even low concentration.
Hydrogen	0–0.2	Starts increasing in phase I and reaches a maximum at the beginning of phase II.	Hydrogen is an odorless, colorless gas.
Carbon monoxide	0–0.2	Throughout the process.	CO is an odorless, colorless gas.

Waste composition: The more organic waste that is added to a landfill, the more LFG is produced by the bacteria during decomposition. The presence of particular types of organic waste containing nutrients, such as sodium, potassium, calcium, and magnesium, helps bacteria prosper so that LFG production increases. In contrast, some wastes contain compounds such as a high salt concentration that harm methane-producing bacteria, causing less gas to be produced.

Age of refuse: Waste buried in the last 10 years produces more LFG through bacterial decomposition, volatilization, and chemical reactions than waste buried more than 10 years ago. Peak LFG production typically occurs from 5 to 7 years after the waste is buried. Most LFGs has been produced by 20 years from initially dumping the waste. However, small quantities of LFG might continue to be emitted from a landfill for 50 or more years. Since landfills are closed by section, different portions of the landfill might be in different phases of the decomposition process at the same time, depending on when the waste was originally placed in each area.

Presence of oxygen. Methane is not produced in a landfill if oxygen is present. The higher the O_2 concentration, the longer the aerobic phase I lasts, as shown in Figure 16.5. When waste is initially deposited in the landfill, waste undergoes aerobic degradation, consuming the oxygen from the surrounding environment until it is fully depleted, as described in Chapter 14. Afterward, typically within less than 1 year, methane starts being generated through anaerobic degradation. Highly compacted waste contains less O_2, so methane production (phase III) starts sooner. Any O_2 remaining in the landfill will slow methane production, hindering economic gain for the landfill operator, which is highly correlated to the LFG collection system, so there is an incentive to reach phase III sooner.

Moisture content. The presence of moisture in a landfill increases LFG production since it boosts bacterial decomposition (Figure 16.6). Moisture might also promote chemical reactions that produce LFG. Moisture content around 40% or higher will maximize the gas production in a capped

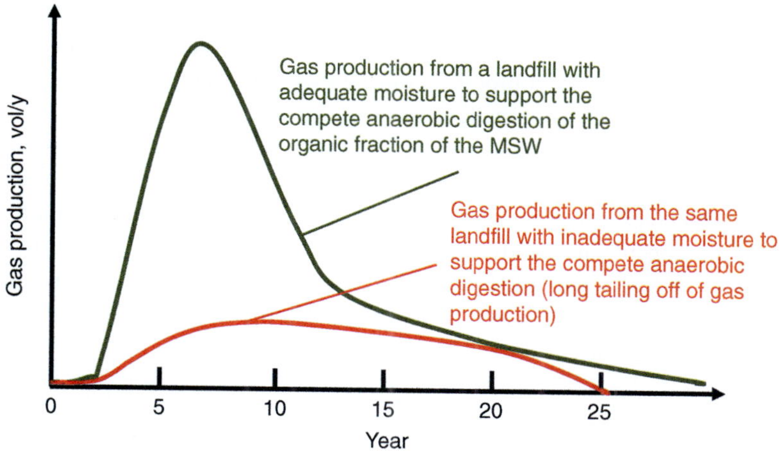

Figure 16.6 Effect of reduced moisture content on the production of landfill gas; adapted from [11].

landfill. Therefore, some landfill leachate is pumped into the landfill to accelerate the biodegradation process. However, too much moisture can also decrease biodegradation. Waste compaction, in this scenario, slows gas production since it increases the density of the landfill content, decreasing the rate at which water can permeate through the waste. So, there is an optimal condition for LFG operation.

Temperature. The biodegradation process is temperature dependent (see Chapter 14). As the landfill temperature increases, bacterial activity surges, resulting in amplified LFG production. Increased temperature may also intensify rates of volatilization and chemical reactions. However, too high temperatures can have a deleterious effect and reduce biodegradation. Bacterial activity drops off dramatically below 10°C (50°F). Deep landfills are less subject to temperature variation. When reaching steady state, temperature is typically stable between 25°C and 45°C (77°F and 113°F). However, temperatures as high as 70°C (122°F) have been reported. As temperature rises, the NMOCs rise. A rule of thumb is NMOCs double with every 10°C (18°F). NMOCs emitted from landfills must comply with the New Source Performance Standards (NSPS) [12].

16.7.2 Leachate

Liquid effluents from landfills are called leachate and are defined as "the liquid formed when rainwater filters through wastes placed in a landfill. When this liquid comes in contact with buried wastes, it leaches, or draws out, chemicals or constituents from those wastes" [1]. Leachate must be collected, controlled, and monitored through the groundwater monitoring requirement systems. A typical composition of landfill leachate is shown in Table 16.3, and Figure 16.5 bottom shows the evolution of most of these compounds during the life of a landfill. There is a symbiotic relation between the gas composition and the leachate characteristics. During phase I – initial adjustment – the pH is neutral, and the biological oxygen demand (BOD) and the chemical oxygen demand (COD) are low. As decomposition proceeds and the landfill moves from aerobic to anaerobic, the COD and BOD increase, and the leachate becomes more acidic; this phase is known as the acid phase – phase III. In this process, there is hydrolysis of higher-molecular mass compounds such as polymers, proteins, and lipids into compounds suitable as sources of energy. This process is followed by acidogenesis – see Chapter 14 – where compounds such as acetic acid and small concentrations of fulvic acid, and other more complex organic acids are produced (Table 16.3). CO_2 is the main gas generated in phase III. The microorganisms involved in this stage are normally known as acidogens or acid formers. Due to acid produced during this stage, the pH will drop to around 5, and the COD, BOD, and conductivity of the leachate will increase. Due to the low pH, a number of heavy metals will be dissolved, and their concentration will increase (Figure 16.5 and Tables 16.3 and 16.4). Many important nutrients will be removed through the leachate. So, if leachate is not circulated, they will be lost from the system. During phase IV – methane formation phase – bacteria known as methanogenic are involved and CH_4 and CO_2 are the predominantly produced gases. The COD, BOD, and VFA (volatile fatty acids) are reduced from the leachate, as well as the heavy metals. During the final maturation phase (phase V), the leachate may have a high concentration of humic and fulvic acids.

Table 16.3 Typical leachate characteristics; adapted from [13].

	Age of Landfill		
Parameter	**Young**	**Medium**	**Old**
pH	5.7–8.0	6.4–8.0	6.6–8.3
BOD, gL^{-1}	7.5–17	0.37–1.1	0.07–0.26
COD, gL^{-1}	10–48	1.2–22	0.67–1.9
NH_3-N, gL^{-1}	0.04–1.0	0.03–3.0	0.01–0.9
Organics, %	0.9–2.5 VFA	5–30 VFA Humic and fulvic acid	50–100 VFA Humic and fulvic acid

Table 16.4 Typical leachate inorganic characteristics; adapted from [13].

Parameter	Range, mgL^{-1}
Hardness	400–2,000
TDS	0–42,300
Calcium	100–1,000
Chloride	20–2,500
Iron	0.2–5,500
Lead	0–5
Magnesium	16.5–15,600
Manganese	0.06–1,400
Potassium	3–3,800
Sodium	0–7,700
Zinc	0–1,350
Mercury	0.0046–0.00715
Arsenic	0.10–0.44

16.8 Energy Recovery from Landfills

LFG is a major source of energy. Currently, no new landfills are being authorized without energy recovery in the US. The main goal of the LFG operation project is to capture methane from the decomposition of organic materials in landfills. The EPA has a national project to monitor and to make sure that LFG is collected and transformed into heat or energy. So, modern MSWLFs are designed as bioreactors that can quickly transform and degrade organic wastes. The increase in waste degradation and stabilization is accomplished through the recirculation of leachate and, in some cases, air to initially accelerate microbial processes.

LFG is harvested using vertical and horizontal pipes and wells (Figure 16.4). LFG is pulled from the landfill using vacuum, and it goes through several steps to prepare for use as electricity or direct use. The gas used for electricity is prepared (right mix of O_2 and C to optimize combustion) and fed to an electric turbine generator, and the direct-use gas is fed through pipes to power combustion equipment such as boilers. Some of the advantages of using LFG are that it is reliable and consistent, it can be used in place of fossil fuels, and it has around 500–600 Btu/ft^3 (22.4 MJ/m^3). Landfills can typically produce power for about 20 years after they are closed. However, some landfills have been reported to produce gas for longer periods of time.

Michigan has 41 such projects in operation and 15 in consideration, ranking the state second in the country, behind only California, which has 55 in operation and 26 pending (Figure 16.1). In fact, Michigan was among the early states to adopt this technology, and one of the local landfill operators (Granger, Lansing, MI) had one of the first few such sites in the US. In 2020, the US had 561 LFG sites operational, 11 under construction, and 54 planned, with the majority of them for energy generation. Location of the projects in the US is approved and reported by the EPA as shown in Figure 16.7.

According to the landfill methane outreach program (LMOP), the program reduced methane emissions from hundreds of US landfills and avoided emissions totaling approximately 37.4 million metric tons of CO_2 equivalent (MMTCO$_2$-eq) in 2019, as shown in Figure 16.8.

16.9 Landfilling of Municipal Solid Waste

In the US, landfill has represented the main route for dealing with MSW. Table 16.5 shows that in 2018, 50% of all MSW was landfilled. However, this represents a large reduction since 1960 when 94% of the MSW was landfilled. Containers and packaging reduced from 30% being landfilled in 1960 to 21% in 2018. Although the products landfilled under the category of containers and packaging decreased by 9%, the products under the category of durable goods increased by 14%. Recycling of products in the category of durable goods increased in the 2000s but has been constant since 2010. The recycling of products in the category of containers and packaging increased dramatically from 1980 to 2010 but has also been almost constant since 2010 (see Chapter 13).

Table 16.6 shows a more detailed observation of the types of material landfilled in the US. The amount of paper and paperboard reaching landfill in 1960 was around 30%. Nowadays, it is about 12% as a large amount of paper and paperboard is being recycled (see Chapter 13). Plastics were not an essential component of MSW until 1970. In 2018, they represented 18% of the materials disposed of in landfill.

Interestingly, when looking at organic waste (i.e., food waste and yard trimmings) food waste represented 15% in 1960, and in 2018 represented 24%. On the other hand, yard trimmings were a

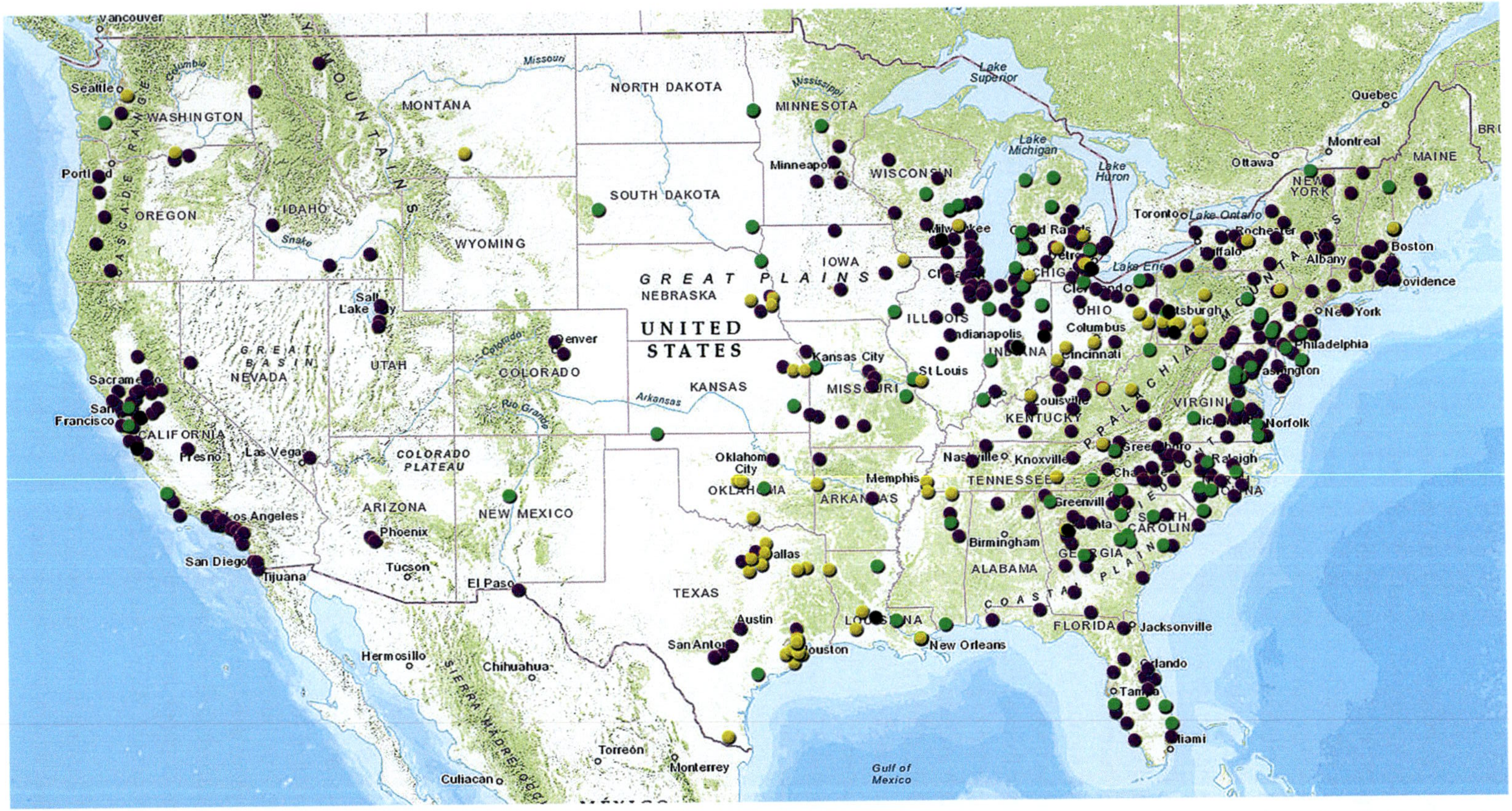

Figure 16.7 Landfill methane outreach program (LMOP); reproduced from [14]. Note: Pink dot indicates electricity generation, green dot indicates direct use for heat, yellow dot indicates Renewable Natural Gas (RNG) - pipeline injection, and black dot indicates Renewable Natural Gas (RNG) - local use.

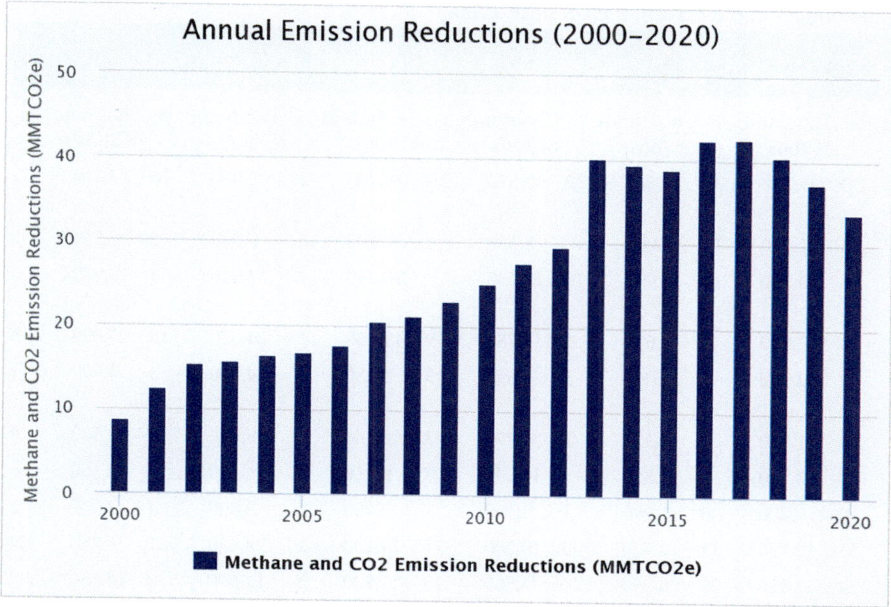

Figure 16.8 Methane and CO_2 emission reductions in MMTCO_2-eq from landfill in the US since 2000; reproduced from [14].

Table 16.5 Amount of MSW* by category landfilled in the US since 1960.

Products	1960		1970		1980		1990		2000		2010		2018	
	Thousands of Tons	%	Thousands of Tons	%	Thousands of Tons	%	Thousands of Tons	%	Thousands of Tons	%	Thousands of Tons	%	Thousands of Tons	%
Durable goods	9,570	12	13,660	12	20,000	15	21,870	15	26,030	19	32,890	24	37,410	26
Nondurable goods	14,940	18	21,240	19	29,170	22	35,990	25	37,450	27	28,030	21	29,160	20
Containers and packaging	24,500	30	40,060	36	43,300	32	39,640	27	37,860	27	31,920	23	30,530	21
Total other wastes	33,500	41	37,630	33	41,890	31	47,770	33	38,920	28	43,470	32	49,080	34
Total MSW landfilled – weight	82,510	94	112,590	93	134,360	89	145,270	70	140,260	58	136,310	54	146,180	50
Total MSW generated	88,120		121,060		151,640		208,270		243,450		251,050		292,360	

*landfilling after recycling, composting, other food management pathways and combustion with energy recovery.

Table 16.6 Materials landfilled* in the MSW in the US since 1960.

Materials	1960		1970		1980		1990		2000		2010		2018	
	Thousands of Tons	%	Thousands of Tons	%	Thousands of Tons	%	Thousands of Tons	%	Thousands of Tons	%	Thousands of Tons	%	Thousands of Tons	%
Paper and paperboard	24,910	30	37,390	33	42,560	32	43,570	30	40,450	29	22,000	16	17,220	12
Glass	6,620	8	12,520	11	14,080	10	8,660	6	8,100	6	7,030	5	7,550	5
Total metals	10,770	13	13,290	12	13,990	10	10,530	7	10,290	7	12,220	9	13,930	10
Plastics	390	0	2,900	3	6,670	5	13,780	9	19,950	14	24,370	18	27,030	18
Rubber and leather	1,510	2	2,710	2	4,000	3	4,590	3	3,880	3	4,400	3	4,990	3
Textiles	1,710	2	1,970	2	2,320	2	4,270	3	6,280	4	8,900	7	11,300	8
Wood	3,030	4	3,710	3	6,860	5	10,000	7	9,910	7	11,120	8	12,150	8
Other **	70	0	470	0	1,990	1	2,100	1	2,480	2	2,800	2	2,930	2
Food	12,200	15	12,750	11	12,740	9	19,800	14	24,200	17	28,620	21	35,280	24
Yard trimmings	20,000	24	23,110	21	26,950	20	25,560	18	11,900	8	11,690	9	10,530	7
Mis-cellaneous inorganic wastes	1,300	2	1,770	2	2,200	2	2,410	2	2,820	2	3,160	2	3,270	2
Total MSW landfilled – weight	82,510	94	112,590	93	134,360	89	145,270	70	140,260	58	136,310	54	146,180	50
Total MSW generated	88,120		121,060		151,640		208,270		243,450		251,050		292,360	

*landfilling after recycling, composting, other food management pathways and combustion with energy recovery; **Includes electrolytes in batteries and fluff pulp, feces and urine in disposable paper.

large category until 1990 (18%); however, after regulations were introduced in several states banning yard trimming disposal in landfills, it declined to 7% of the landfilled MSW in 2018. Most of the yard trimmings in 2018 were recovered through composting.

16.10 Landfilling of Metals

Since metals can be endlessly recycled, the recovery of metals from MSW has increased. The amount of ferrous metal in MSW has increased since 1980 by around 50% (Figure 16.9); however, the amount of ferrous metal reaching landfills has been relatively constant since 1960 at around 10,000 thousand tons due to increased recycling.

Because of the high concentration of metals in most MSW landfills, there is still debate about whether discharging toxic metals contained in the leachate will lead to an environmental risk. The main concern about high amounts of heavy metals in landfills is their dissolution during phase III – acid phase and their concentration increase in the leachate (Figure 16.5). Iron is a ferrous metal, and it is the most predominant metal in landfills. Iron concentration changes widely from milligrams to tens of grams per liter of leachate. Iron as Fe(II) has toxic effects on fish and animals and can pollute

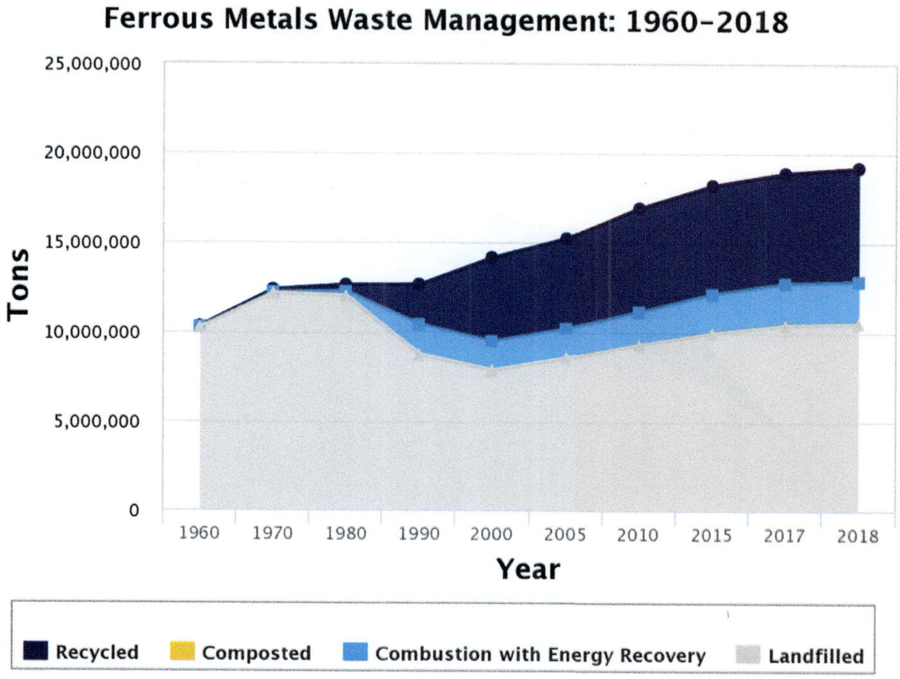

Figure 16.9 Ferrous metals waste management since 1960; reproduced from [16].

water streams. Extensive research has been conducted to determine the amount of Fe(II) or Fe(III) available in leachate depending on the type of landfill (i.e., recirculated landfill (RL) or conventional landfill (CL)), and it has been shown that there is great variation of iron content depending on the MSW composition and the period of decomposition [15]. So, if the landfill is not managed properly, iron can permeate the bottom liners and contaminate groundwater.

The amount of aluminum present in landfills in 2018 was more than seven times the amount in landfills in 1960 (Figure 16.10). Aluminum can enter the MSWLF from regular MSW and from minor contribution from incineration facilities. The main other sources of aluminum increase in the MSW can be mostly attributed to a large amount of the total amount of aluminum present in the containers and packaging category and recycling has not kept up with increased generation. Aluminum is mostly an inert material; however, in the landfill environment, which is a complex active chemical reactor (see Figure 16.5), several reactions can occur. Aluminum can be dormant in a landfill for several years; however, if the pH increases to more than 8.5, aluminum can experience several exothermic chemical reactions slowing microbial activities or ignite and pyrolize the surrounding MSW although rare. These underground reactions can consume MSW and result in MSW settlement, damaging the interior of the landfill. High temperatures may compromise the integrity of the landfill liner system and other components such as the liner or the tubes used for gas collection. The related reactions of aluminum can also release potentially toxic and/or flammable gases and noxious odors [17].

Generally, if the landfill only accepts MSW, the pH is lower and there is no chance of chemical reactions for aluminum [17]. Iron and aluminum reactions with water are exothermic. The reaction of iron with water produces iron rust, which is not a soluble product.

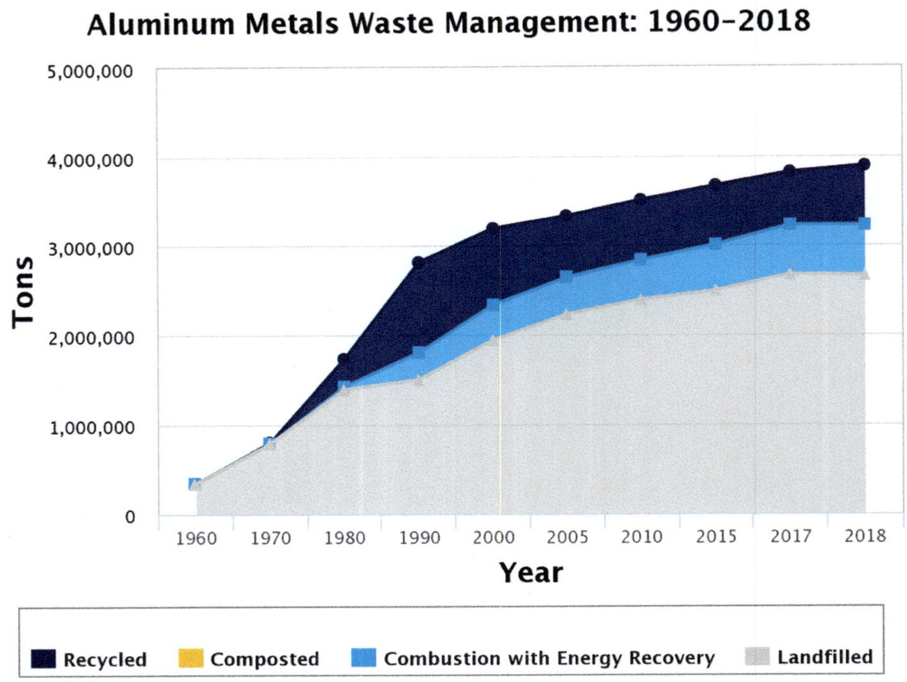

Figure 16.10 Amount of aluminum in MSW in the US since 1960.

Other nonferrous metals such as lead, copper, and zinc originate mostly in durable products such as appliances and electronics. Other than aluminum, lead–acid batteries from passenger cars, trucks, and motorcycle are the most prevalent nonferrous metal in MSW. Figure 16.11 shows that the amount of generated nonferrous metals present in MSW has increased since 1960. The largest source of this increase is lead–acid batteries. The good news is that the majority of these batteries are now recycled. When we change a battery, we pay a core charge – a deposit to properly dispose the batteries, so most lead–acid batteries are now recycled in the US (see Chapter 13). "In all 50 states, retailers can charge a battery core charge and in over 30 states, it is required by law" [18].

16.11 Landfilling of Glass

Since glass is an inert material, its decomposition is almost nil. The surface of some glass bottles will absorb moisture over time, causing devitrification – a process that causes the outer layer of glass bottles to crystallize and flake off. When the cost of glass recycling is prohibitive due to long transportation distances, low prices, or wrong color (too much green glass, which is hard to recycle in the US), it is sent to the landfill and can be used as aggregate and supporting material between layers of MSW.

As shown in Table 16.6, the percentage of glass disposed in landfill increased from 1960 to 1980 from 8% to 10%, but since then it has decreased and maintained around 5%. The amount of glass disposed in landfills, recycled, or sent to incineration has not changed much since 1990 as shown in Figure 16.12. The US disposed around 13 million metric tons of glass annually, and more than half is disposed in landfill.

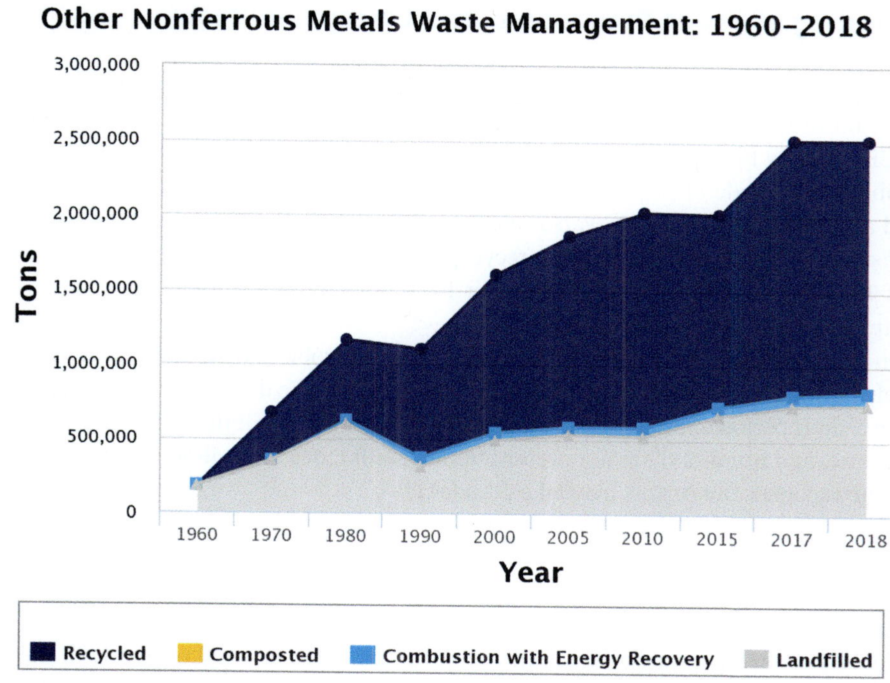

Figure 16.11 Nonferrous metals waste management in the US MSW since 1960.

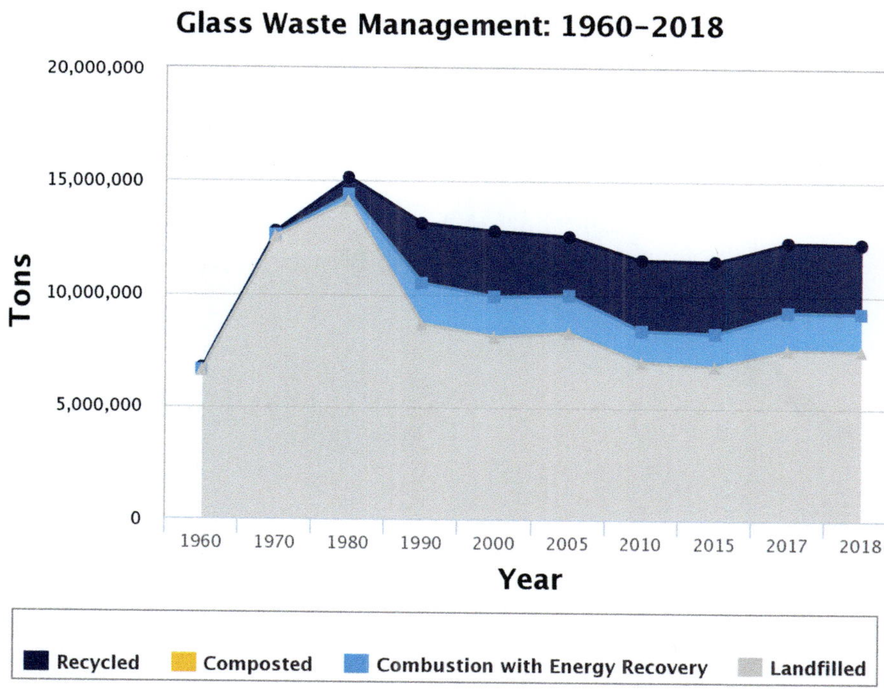

Figure 16.12 Amount of glass in the MSW in the US since 1960.

16.12 Landfilling of Paper, Paperboard, and Corrugated Board

As postconsumer paper, paperboard, and corrugated board (PP&CB) has become a prevalent material, the amount of PP&CB reaching landfill increased until about 1980, but since then the amount of PP&CB reaching the landfill has been reduced due to the increase of recycling (Figure 16.13). Nondurable goods made of paper, excluding newspapers, had a recycling rate of 43.1%. Newspapers had a recycling rate of 64.8%. Paper and paperboard from the containers and packaging category, excluding corrugated boxes, had a recycling rate of 20.8 %. Corrugated boxes had a recycling rate of 96.5% in 2018, as discussed in Chapter 13. A large effort has been conducted by commodity groups to keep PP&CB out of landfills.

Decomposition of cellulose-based materials in aerobic and anaerobic environments was discussed extensively in Chapter 14. In short, if the landfill is fully active, the decomposition of paper should happen in a short period of time. However, it has been reported that if the right conditions (i.e., pH, temperature, and humidity) are not present in the landfill, decomposition of day-to-day products such as newspapers and carrots may take decades [19].

16.13 Landfilling of Plastics

Plastic represents 18% of the material landfilled in the US in 2018. The amount of plastic in the landfill has increased in the MSW since 1970. The containers and packaging category represents the largest growing category for plastics, representing almost 28% of the plastic landfilled. Although plastics such as HDPE and PET are increasingly being recycled, most of the single-use plastic such

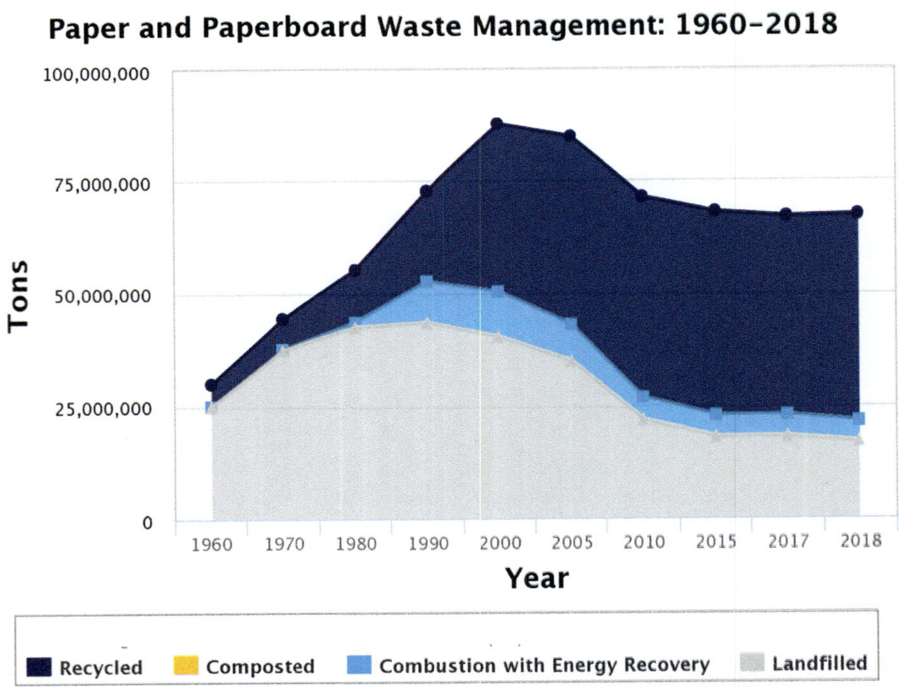

Figure 16.13 Amount of paper, paperboard, and corrugated board in landfill in the US since 1960.

as bags, sacks, and other containers ends in the landfill, resulting in a very low recycling rate for plastic (see Figure 16.14).

Some initial research conducting mining of landfills has shown that PE and PP are the most common polymers excavated from landfill waste. This is not a surprise since PE and PP are the most used polymers. The authors have indicated a potential to mine landfill to recover these polymers [20].

Common fossil-based plastics such as polyolefins (LDPE, HDPE, PP), PVC, PS, polyurethane (PU), and PET have shown small degradation rates (<15%) in simulated and real landfill conditions [21], which can be mostly attributed to the degradation of oxidative products. Addition of special degradation-promoting additives has not been shown to produce degradation of PE and PET in simulated landfill – see Case Study 16.2 [22]. So, most currently used plastics ending in landfills will be captured for centuries or millennia before they break down sufficiently to be assimilated by microorganisms (see Chapter 14).

However, if common polymers start to break down or fragment, there are concerns that have been raised by scientists and consumers about the presence of microplastics in the environment and in landfill leachate. Silva et al. reported concentrations of microplastics between 0 and 291 particles L^{-1} in landfill leachate [23], which could end up in the water supply.

Most studies of the anaerobic digestion of polymers in accelerated landfill conditions reported in the literature examine biodegradable polyesters, such as PLA, PHB, and PCL, which have ester groups that can be consumed by microorganisms [21]. Kolstad et al. concluded that amorphous PLA generated a small amount of gas when tested under anaerobic conditions at 35°C, but that semicrystalline PLA did not produce a significant generation of gas [24].

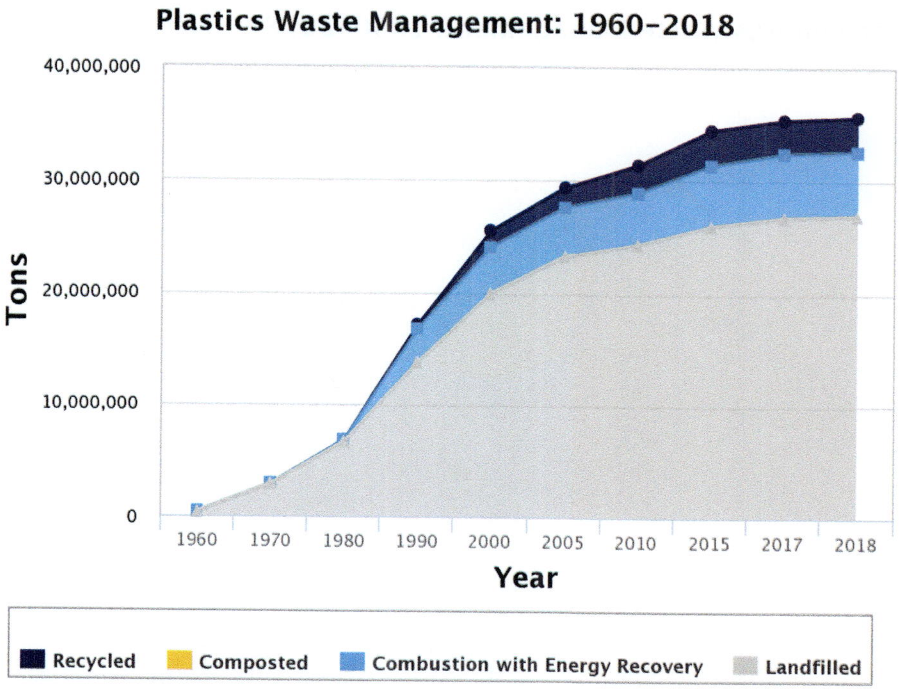

Figure 16.14 Plastic in MSW in the US since 1960.

16.14 Case Study 16.1. Landfilling of Yard Trimmings

LFG production is directly related to the amount of organic waste in the landfill. As discussed, LFG can be used to produce energy, which is classified and sold as renewable energy in many states. So, there is an incentive for LF owners to increase the amount of organics disposed of in MSWLF to produce more LFG. One of the easy ways to introduce organic waste in landfill is through the collection of yard trimmings, which include grass, leaves, and tree and brush trimmings from residential, institutional, and commercial sources.

Figure 16.15 shows that the amount of yard trimming disposed in landfills was reduced tremendously from 1990 to 2000 in the US, when several states passed legislation banning the disposal of yard trimmings in landfills. Michigan banned the disposal of yard trimmings or clippings in MSWLF in 1995. However, landfill operators, which generally own integrated waste management companies operating organic and household waste collection as well as MRFs, complained that it was difficult to compost during the winter months in Michigan, from December to March. So, they lobbied for a bill to introduce yard clippings in landfills during this period [25]. However, the major composters in the state complained that the December 1 deadline would ultimately reduce the volume of leaves collected by composters, which is vital since leaves are quickly decomposed and essential for healthy compost. Additionally, many composters and several organizations were concerned that this bill would create a loophole in the MI legislation of sending yard waste to landfill. Ultimately, the bill did not pass. However, the underlying questions still linger in many states with pressure from some stakeholders to introduce more organic wastes in landfills to generate LFG and increase revenue. A considerable consequence of introducing more organic waste in landfills is that the main gas produced from landfill decomposition is methane, a very potent

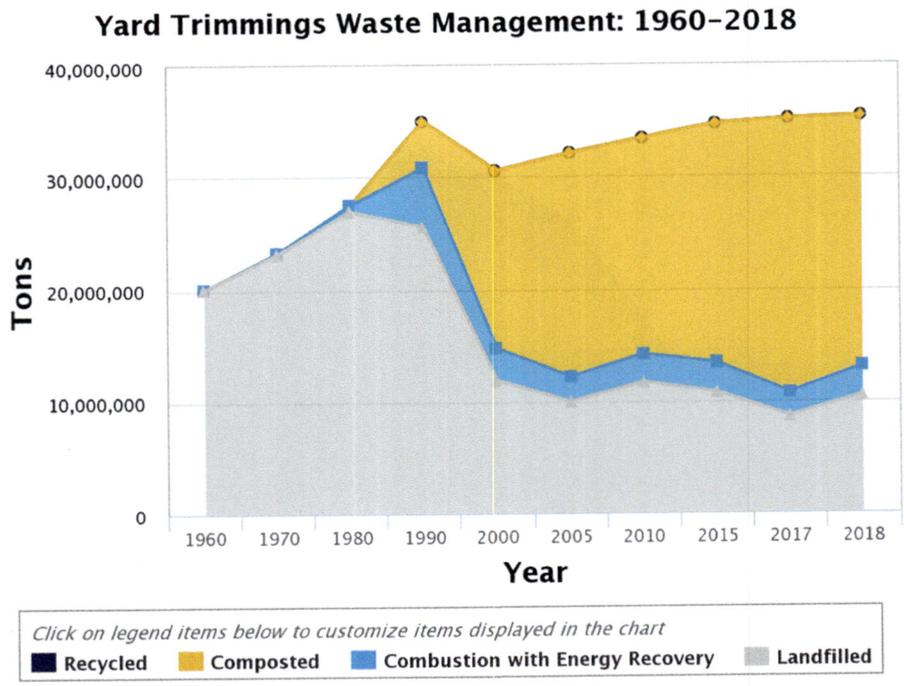

Figure 16.15 Amount of yard trimmings in MSW in the US since 1960.

greenhouse gas (see Chapter 8), which may not be completely collected in current landfills. Rigorous research about the true impact of introducing more organic waste in MSWLF with gas collection is still lacking.

16.15 Case Study 16.2. Evaluation of Biodegradation of Polyethylene and Poly(ethylene terephthalate) in Simulated Landfill Environments

Evaluating the degradation of polymers under anaerobic digestion conditions is important for several end-of-life scenarios, such as marine disposal, anaerobic digestion units designed for energy recovery, and landfills. Most of these scenarios can be operated at low mesophilic conditions (~35°C) to high thermophilic conditions (~55°C and more). Most degradable materials degrading through anaerobic digestion go through four main stages as described in Chapter 14: (1) hydrolysis of complex organic materials such as proteins, carbohydrates, and lipids into amino acids, carbohydrates, fatty acids, and alcohols; (2) fermentation of amino acids and carbohydrates producing short-chain fatty acids, succinate, amino valerate, and H_2 and resulting in ethanol, acetate, H_2, and CO_2; (3) anaerobic oxidation of long-chain fatty acids and alcohols with end products of acetate and propionate; and (4) anaerobic oxidation of short-chain fatty acids such as propionate and butyrate to acetate and H_2 called methanogenesis, producing the end products of CH_4 and CO_2.

Extensive work has been conducted to understand the anaerobic biodegradation of PE and PET in landfills and simulated landfill conditions since these are the main polymers entering landfills and are highly used for single-use packaging applications. Additionally, a number of companies have promoted the addition of additives able to degrade polyolefins and polyester. Therefore, we conducted a simulated anaerobic biodegradation study to determine the biodegradation of PE films and PET sheets by tracking the evolution of total gas produced by these films and sheets under anaerobic digestion environments at 35 and 50°C, which represented a modified simulated biologically active landfill, for 464 days. PE films made of LDPE and LLDPE resins blended at a ratio of 70:30 by weight were produced. One and five percent by weight of the biodegradation-promoting additive masterbatches were added to this blend for extrusion. Additives for PE films were obtained from three different companies as explained elsewhere [22]. PE films with a thickness of around 22.9 ± 5.1 μm were produced by cast extrusion. PET resin was mixed with 1 and 5 wt.% of different biodegradation-promoting additive masterbatches from two companies as also detailed elsewhere [22]. The overall thickness of the PET sheet was 234 ± 15 μm, also produced by cast extrusion.

Figure 16.16 shows that the total accumulated gas for PE and PET films at 35°C and 50°C were not significantly different than the blank samples at days 252 and 464 (at $P = 0.05$). Furthermore, optical microscopy confirmed that negligible amounts of biofilm were formed on the surfaces of the plastic samples retrieved. After 464 days we decided to spike selected bioreactors – PE, PE E5, PE S5, PE W5, PET E5, PET W5, and the blank (M1) – with cornstarch. The increase in gas production indicated that the microorganisms present in the bioreactors did become active when a food source was provided. The additional total accumulated gas for the PE films and PET sheets in the spiked bioreactors at 35 and 50°C were not significantly different from the blank samples at day 514 (at $P = 0.05$).

As we can observe in Figure 16.16, no additional hydrolysis and/or anaerobic oxidation of the PE films and PET sheets was observed to provide additional intermediate products to increase the final methanogenesis of these reactors. Therefore, we concluded that this long-term gas evolution test suggested that PE films and PET sheets without and with the biodegradation-promoting additives used in this study disposed in an active anaerobic environment with temperature up to

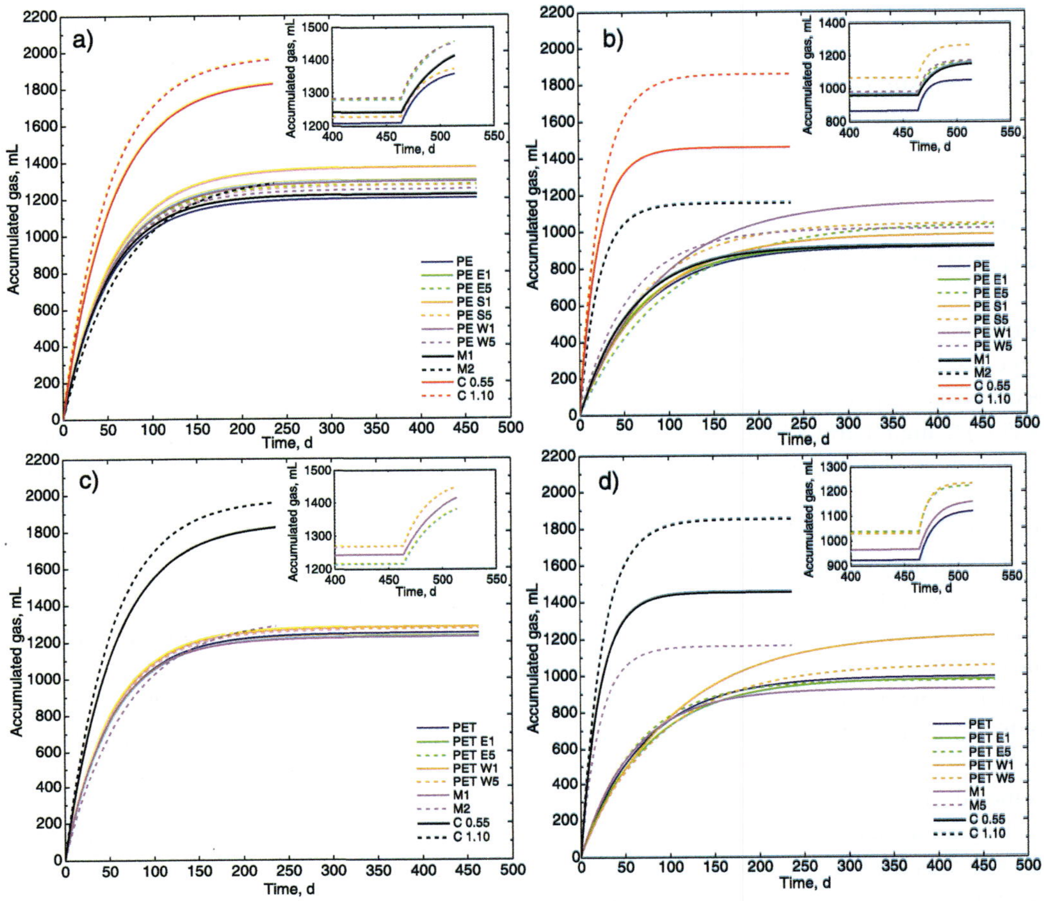

Figure 16.16 Anaerobic biodegradation of PE films and PET sheets. Total accumulated gas in mL vs time in d for PE films (a @ 35°C and b @ 50°C) and PET sheets (c @ 35°C and d @ 50°C) exposed to liquid anaerobic digestion environments created from fresh dairy manure obtained from the MSU dairy farm, inoculum from the MSU anaerobic digestor and water added to create a 5% (w/v) total solids mixture. Total accumulated gas for PE films and PET sheets were not significantly different than the blank (M1 and M2) samples (at $P = 0.05$). The total accumulated gas for cellulose 1.10 g and 0.55 g was significantly higher than the blanks (M1 and M2) and PE films and PET sheets at both temperatures. Inserts show selective spiking of bioreactors at 464 days with 0.55 g of cornstarch; reproduced from [22] with permission of the American Chemical Society.

50°C would not appreciably degrade at least for 500 days, and no evidence was found to suggest that additional exposure for a moderate period of time would change the result.

16.16 Study Questions

1 Define a municipal solid waste landfill.
2 How many MSWLFs are operational in the US?
3 Methane gas collection is optional for all new MSWLFs approved in the US.
   A True
   B False

4 Landfill tipping fees are the same across the US.
  A True
  B False
5 The main disposal method for low and lower-middle income economies in the world in 2016 was:
  A Recycling
  B Composting
  C Incineration
  D Landfill
  E Open dump
  F Other advanced methods
6 Define landfills, dumps, and cells.
7 List the advantages and disadvantages of disposing MSW in a landfill.
8 Explain the process to identify hazardous waste.
9 Explain how landfills are regulated.
10 Provide examples of landfills regulated under RCRA Subtitle D.
11 What are C&D debris landfills?
12 What are the differences between surface impoundments, waste piles, and landfills?
13 Draw the main parts of a landfill.
14 What are the main emissions from a landfill?
15 What are the main resources recovered from current landfills?
16 Describe the main product categories landfilled in the US.
17 Plastics are the largest materials landfilled in the MSW in 2018.
  A True
  B False
18 What is main concern about deposit of heavy metals in landfills?
19 What are the main sources of aluminum in the landfill?
20 Aluminum reactions are dangerous at pH ≤ 8.
  A True
  B False
21 Describe the presence of paper and paperboard and plastic waste in the MSW.
22 Common plastic such as polyethylene and polyethylene terephthalate are readily biodegradable in MSWLFs.
  A True
  B False
23 Explain the pros and cons of introducing yard trimmings in MSWLFs.

16.17 Additional Resources

1 EPA – Electronic Code of Federal Regulations – Title 40: PART 258—CRITERIA FOR MUNICIPAL SOLID WASTE LANDFILLS
https://www.ecfr.gov/current/title-40/chapter-I/subchapter-I/part-258

2 Landfills – EPA
https://www.epa.gov/landfills

3 EPA – Landfill Methane Outreach Program (LMOP)
https://www.epa.gov/lmop

4 How a Landfill works?
https://www.youtube.com/watch?v=RmLhY7f8WxQ
5 What happens at the Landfill?
https://www.eia.gov/energyexplained/biomass/waste-to-energy-in-depth.php
6 What is a Landfill? Republic Services
https://www.youtube.com/watch?v=821RZegnHdM

References

1 US Environmental Protection Agency. (2021). Resource conservation and recovery act (RCRA) training modules. https://www.epa.gov/rcra/resource-conservation-and-recovery-act-rcra-training-modules (accessed May 15, 2021).
2 US Environmental Protection Agency. (2021). Project and landfill data by state. https://www.epa.gov/lmop/project-and-landfill-data-state (accessed May 17, 2021).
3 The environmental research & education foundation (EREF). (2019). Analysis of MSW landfill tipping fees-april. The environmental research & education foundation (EREF) maintains a database of Municipal Solid Waste 2019. https://erefdn.org/product/analysis-msw-landfill-tipping-fees-2 (accessed May 13, 2021).
4 Kaza, S., Yao, L., Bhada-Tata, P., and Van Woerden, F. (2018). What a waste 2.0: A global snapshot of solid waste management to 2050. The World Bank. https://doi.org/10.1596/978-1-4648-1329-0 (accessed May 13, 2021).
5 Electronic Code of Federal Regulations. (2021). 40 CFR Part 264. E-CFR. https://www.ecfr.gov/current/title-40/chapter-I/subchapter-I/part-264 (accessed April 30, 2021).
6 Coal Ash (Coal Combustion Residuals, or CCR) | US EPA. (2019). https://www.epa.gov/coalash (accessed May 18, 2021).
7 Disposal and Storage of Polychlorinated Biphenyl (PCB) Waste | Polychlorinated Biphenyls (PCBs) | US EPA. (2021). https://www.epa.gov/pcbs/disposal-and-storage-polychlorinated-biphenyl-pcb-waste (accessed May 18, 2021).
8 US Environmental Protection Agency. (2005). Introduction to land disposal units (40 CFR Parts 264/265, Subparts K, L, M, N), 16. https://www.epa.gov/sites/production/files/2015-07/documents/ldu05.pdf (accessed May 18, 2021).
9 Agency for Toxic Substances and Disease Registry. (2008). Landfill gas basics. Landfill Gas Prim. - An Overv. Environ. Heal. Prof., Washington.
10 Meyer-Dombard, D.R., Bogner, J.E., and Malas, J. (2020). A review of landfill microbiology and ecology: A call for modernization with 'next generation' technology. *Front Microbiol* 11: 1127. https://doi.org/10.3389/fmicb.2020.01127.
11 O'leary, P.R. and Tchobanoglous, G. (2002). *Landfilling. Handb. Solid Waste Manag.*, 2nd e, 14.1–14.93. New York, NY: McGraw-Hill.
12 US Environmental Protection Agency. (2021). Municipal solid waste landfills: New source performance standards (NSPS), Emission guidelines (EG) and compliance times. https://www.epa.gov/stationary-sources-air-pollution/municipal-solid-waste-landfills-new-source-performance-standards (accessed May 26, 2021).
13 Greene, M. and Reichert, A. (2016). Landfill Leachate Overview. 18th Annu. New Engl. Pretreat. Coord. Work.
14 US Environmental Protection Agency. (2021). Landfill methane outreach program (LMOP). US EPA. https://www.epa.gov/lmop/lmop-national-map (accessed May 21, 2021).

15 Li, W.B., Yao, J., Xia, F.F., Feng, H.J., Feng, H., Jiang, C.J. et al. (2011). Leaching behavior of iron from simulated landfills with different operational modes. *Bioresour Technol* 102: 7422–7428. https://doi.org/10.1016/j.biortech.2011.05.024.

16 US Environmental Protection Agency. (2020). Advancing sustainable materials management: 2018 tables and figures assessing trends in materials generation and management in the United States. U.S. EPA.

17 Calder, G.V. and Stark, T.D. (2010). Aluminum reactions and problems in municipal solid waste landfills. *Pract Period Hazardous, Toxic, Radioact Waste Manag* 14: 258–265. https://doi.org/10.1061/(ASCE)HZ.1944-8376.0000045.

18 State Recycling Laws. (2021). Battery Council International. https://batterycouncil.org/page/State_Recycling_Laws (accessed May 25, 2021).

19 Rathje, W. and Murphy, C. (2001). *Rubbish!: The Archaeology of Garbage*, 1st e. Phoenix: University of Arizona Press.

20 Canopoli, L., Coulon, F., and Wagland, S.T. (2020). Degradation of excavated polyethylene and polypropylene waste from landfill. *Sci Total Environ* 698: 134125. https://doi.org/10.1016/j.scitotenv.2019.134125.

21 Quecholac-Piña, X., MDC, H.-B., MDC, M.-S., Espinosa-Valdemar, R.M., and Vázquez-Morillas, A. (2020). Degradation of plastics under anaerobic conditions: A short review. *Polymers (Basel)* 12: 1–18. https://doi.org/10.3390/polym12010109.

22 Selke, S., Auras, R., Nguyen, T.A., Castro Aguirre, E., Cheruvathur, R., and Liu, Y. (2015). Evaluation of biodegradation-promoting additives for plastics. *Environ Sci Technol* 49. https://doi.org/10.1021/es504258u.

23 Silva, A.L.P., Prata, J.C., Duarte, A.C., Soares, A.M.V.M., Barceló, D., and Rocha-Santos, T. (2021). Microplastics in landfill leachates: The need for reconnaissance studies and remediation technologies. *Case Stud Chem Environ Eng* 3: 100072. https://doi.org/10.1016/j.cscee.2020.100072.

24 Kolstad, J.J., Vink, E.T.H., De Wilde, B., and Debeer, L. (2012). Assessment of anaerobic degradation of Ingeo^{TM} polylactides under accelerated landfill conditions. *Polym Degrad Stab* 97: 1131–1141. https://doi.org/10.1016/j.polymdegradstab.2012.04.003.

25 De Vuyst, L. (2000). Allow Yard clipping in landfills from December 1 to March 15. https://www.legislature.mi.gov/documents/1999-2000/billanalysis/House/htm/1999-HLA-5641-A.htm (accessed May 25, 2021).

17

Litter and Marine Pollution

17.1 Learning Objectives

After learning from this chapter, you are expected to:

- Understand pollution of the land related to litter.
- Understand pollution of water in rivers related to packaging litter.
- Describe ocean pollution related to packaging litter.
- Recognize the impacts of litter on wildlife.
- Understand the trade-offs between biodegradability and litter.
- Understand potential impacts of microplastics.

17.2 Introduction

We have discussed the role of packaging in generation of packaging waste, as well as its role in prevention of product waste, and the need to evaluate environmental footprints of the whole system to determine which products and processes have lower environmental impacts. Our discussion of packaging wastes has mostly assumed that these wastes enter designated waste management systems. However, this is not always the case as we have discussed in Chapter 10. In this chapter, we will discuss the impacts of packaging wastes that are not properly disposed – that are released as litter.

There are two main pathways for litter. It can occur on land, or in water systems of various kinds. Even when litter occurs on land, the littered objects may not remain on or in the land; they can be carried by wind or water into bodies of water such as rivers and streams, which can then further carry the litter to larger bodies of water such as lakes and oceans. (Or, of course, wind can also carry litter directly to lakes or oceans.)

Litter can pose direct hazards to wildlife, either terrestrial or aquatic, through entrapment or ingestion. It can also impact wildlife more indirectly, by acting as a concentrator for toxic chemicals. We will discuss all of these in this chapter.

Litter is fundamentally a problem of human behavior – the failure to handle wastes appropriately. However, the characteristics of packages and their materials can have a profound influence on both their impacts on the environment, including living organisms, and their persistence in the environments where they end up.

Life Cycle of Sustainable Packaging: From Design to End of Life, First Edition. Rafael A. Auras and Susan E. M. Selke.

Companion Website: www.wiley.com/go/Auras/lifecycleofsustainablepackaging

17.3 Litter in the US and around the World

Littering is a problem all over the world and has presumably existed since before recorded history. When cities were first established, the "normal" route for waste disposal was to throw discards out in the street – something we would now regard as littering. Nevertheless, information about the quantity of litter is hard to come by. Much of what we know comes from cleanup studies in particular locations.

In the US, the leading antilitter organization, and the leading source of national information on littering, is Keep America Beautiful [1]. They have been doing periodic litter surveys since 1969. Their most recent study, in 2020, picked up and quantified litter at a variety of sites, including 240 roadways, 189 waterways, 181 nonroadway land sites, and 126 "behavioral" sites. The full study had not yet been released at the time of this writing but reported highlights include a 54% decrease in roadside litter in the last decade, which still left 50 billion pieces of litter on the ground in the US. Greater detail is available from the previous study, conducted in 2009. That study, which estimated over 51 billion pieces of roadside litter in the US, reported that 91% of the litter was less than 4 in. in size. As shown in Figure 17.1, the largest single category of litter was related to cigarettes. Much of that was cigarette butts. Packaging accounted for more than half of the larger, more visible litter, as shown in Figure 17.2.

One US state that has put out recent information about littering is Texas, which has a long-standing "Don't Mess With Texas" campaign to reduce littering. They report that about 362 million pieces of litter reach Texas roadways each year, 71% of which is 2 in. in size or smaller. They also report that half of the litter comes from cars and trucks. Interestingly, for "visible litter" 35% is "vehicle debris" and 29% is, specifically, tire debris. The age group most responsible for litter is those between 16 and 34 [2].

Cleanup Australia reported its 2020 litter cleanup results from 7,778 sites across the country. There, too, cigarette butts were the most common single item found, 16.2% of the total. Packaging of various types accounted for just under 50% of the total [3].

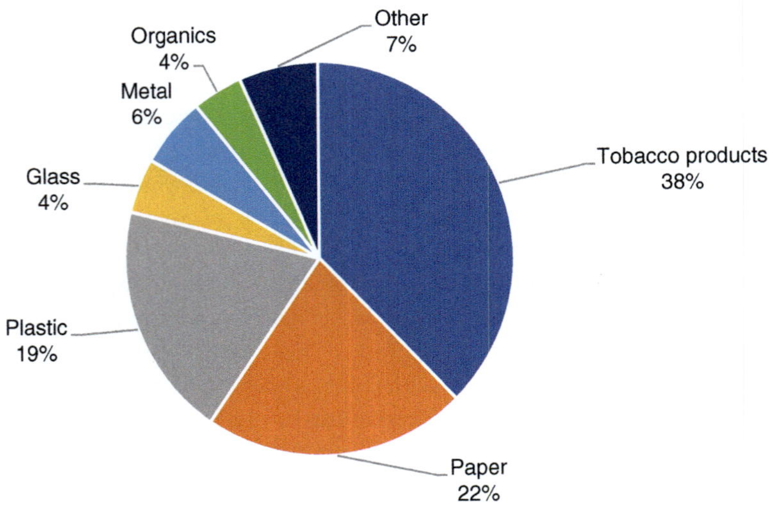

Figure 17.1 Roadside litter in America in 2009; data from Keep America Beautiful [1].

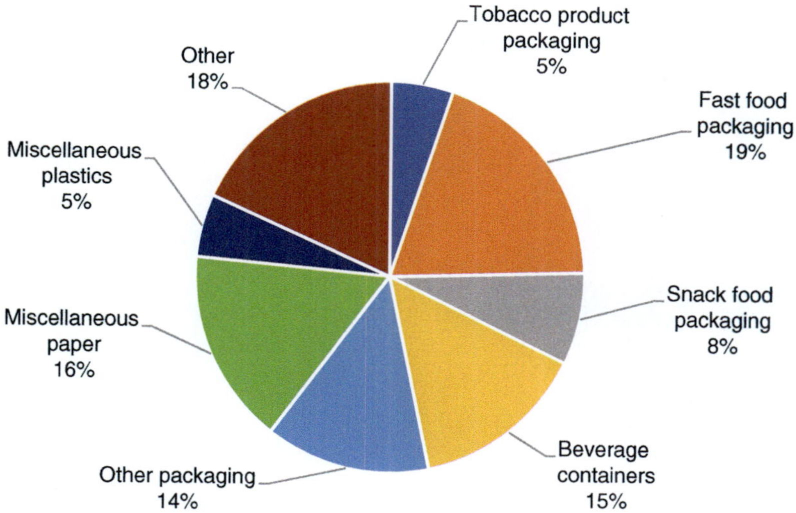

Figure 17.2 US roadside litter 4 in. in size or larger, 2009; data from Keep America Beautiful [1].

Keep Britain Tidy reports that more than 2 million pieces of litter occur every day. They cite "fast-food packaging, sweet wrappers, drinks cans, bottles and cigarette butts" as common littered items [4].

The problem of litter is exacerbated when people do not have easy access (or any access) to well-managed waste management systems, which unfortunately is the case in many parts of the world in low and lower-middle economies (see Chapter 10 for detailed statistics). Industrial or upper-middle- and high-income economies tend to have well-developed systems for waste management, but low- and lower-middle income countries may not. Even when garbage is collected, if the disposal sites themselves are not well managed, items may be released through the action of wind, weather, or animals, becoming litter. In the US, landfill operators are required to apply daily cover and take other actions to control the inadvertent release of items intended to be safely buried to avoid the generation of litter. Even then, such actions are not always 100% successful in avoiding litter. In countries that rely on open dumps rather than sanitary landfills, litter originating in disposal sites is a much larger problem. This will be discussed further in Section 17.4.

Litter is unsightly and annoying; it is also costly. Keep America Beautiful, in its 2009 report, estimated that litter cleanup costs more than $11.5 billion per year to businesses, government, and others, as shown in Figure 17.3 [5]. Keep Britain Tidy estimates the cost of street cleaning related to litter as over £1 billion per year [4].

The direct costs of litter are not its only economic impacts. Keep America Beautiful also discusses indirect costs, such as effects on decreasing property values, reducing the attractiveness of communities for new businesses, decreasing tourism, etc. [1]. Keep Britain Tidy states that litter contributes to further crime and makes people feel less safe. They also say that littered food items attract vermin [4].

The prevalence of beverage containers in the survey of larger litter should be noted (see Figure 17.2). The visibility of beverage containers littered along roadsides, in parks, on beaches, and elsewhere was the driving force behind enactment of beverage container deposit legislation

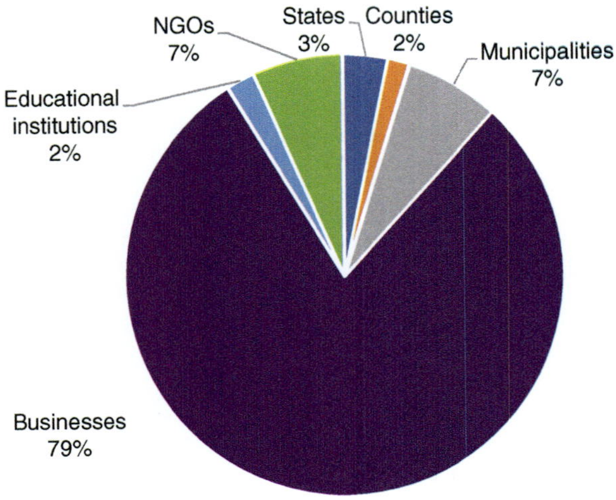

Figure 17.3 Cost of litter cleanup in the US in 2009, totaling $11.5 billion; data from Keep America Beautiful [5].

beginning in the 1970s. The initial set of these laws applied specifically to those containers that were most prevalent in litter: carbonated beverage containers (soft drinks and beer). The deposit provided a monetary incentive for people to not toss away (litter) their empty containers, but rather to retain them to return for a refund of the deposit they had paid. Further, they provided an incentive for others to collect littered containers, to obtain their monetary value. The effect in states that enacted deposits was immediate: littering of containers covered by the deposit dropped massively. Some had hoped that there would be a carryover effect on other littering, that if people were not throwing out their beverage containers, they might also refrain from other types of litter. That evidently did not occur, but the overall effect on reduction of visible litter was still substantial. See Chapter 13 Section 13.3.2.4 for further details on bottle bills.

Other regulations target littering behavior directly. Most states in the US and many countries around the world have laws that prohibit littering and provide penalties for those caught doing so. For example, in Texas, you can be fined $500 for a single littering event – and if you litter something that weighs more than 5 pounds, the fine can increase to $2,000 [2]. In Australia, the fine for littering is A$200, unless it is something such as broken glass, lit cigarette, syringe, etc., that "creates a public risk" in which case it increases to A$500 [6]. The fine for littering in Hong Kong is HK$1,500 [7]. Of course, fines only work if the person doing the littering is caught and charged.

17.4 Marine Litter

Litter is a particular problem when it reaches bodies of water. The oceans used to be regarded as an appropriate disposal site for wastes of various types. They were seen as having limitless capacity for accepting wastes, which would either sink to the bottom or be diluted, and not cause any significant environmental impacts. For example, much of the waste from New York City used to be trucked to garbage barges, hauled out to sea, and dumped. Our understanding of the adverse impacts of such actions has – fortunately – grown considerably since those days.

There are now international agreements in place that prohibit disposal of wastes of most types into the ocean. The International Convention for the Prevention of Pollution from Ships (MARPOL) was adopted in 1973 and has been revised several times through the adoption of annexes. The intent is to prevent both deliberate and accidental ocean pollution from ships and other marine vessels. Most relevant to the issue of litter is Annex V, which focuses on garbage from ships and prohibits the disposal from ships of any form of plastics [8].

Litter that enters the oceans or other bodies of water can have several fates. It can float in the water for long periods of time, carried along by tides and currents, and sometimes accumulating in particular locations. It can sink below the surface, where it may settle out or be carried along to new locations. Even if it settles to the bottom, it may be dislodged by various events such as floods or earthquakes. It can wash up on shorelines, where it may remain, or it may be carried farther onto the land or back into the ocean by wind or water. It can sink to the deep ocean bottom, where it may eventually be covered by sediment. Along the way, it can interact in various ways with other components in the water, or with wildlife, as will be discussed in Sections 17.5 and 17.6. Generally, much of the litter entering small bodies of water will eventually be carried to the ocean.

17.4.1 Shoreline and Beach Litter

Pictures of litter on beaches and shorelines around the world are a potent reminder of how widespread litter problems are in the environment. Such litter directly impacts recreational activities and tourism. In the 2020 International Coastal Cleanup report on littered items collected from beaches as well as from underwater by participating divers, cigarette butts came in as the number two item found. The other nine of the top ten items were all packaging-related, as shown in Figure 17.4 [9].

The relative prevalence of these items varied considerably at different locations around the world, as illustrated in Figure 17.5.

17.4.2 Oceans and Gyres

Particular concern has been expressed about plastics accumulating in "plastic islands" in the ocean. One of the most well-known is the "Great Pacific Garbage Patch" in the northern Pacific

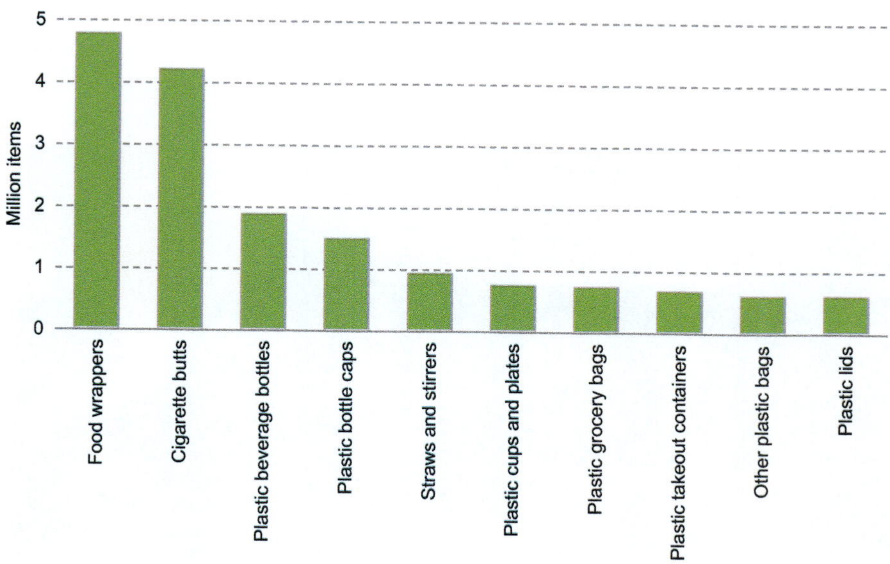

Figure 17.4 Top ten items collected in International Coastal Cleanup 2020; data from Ocean Conservancy [9].

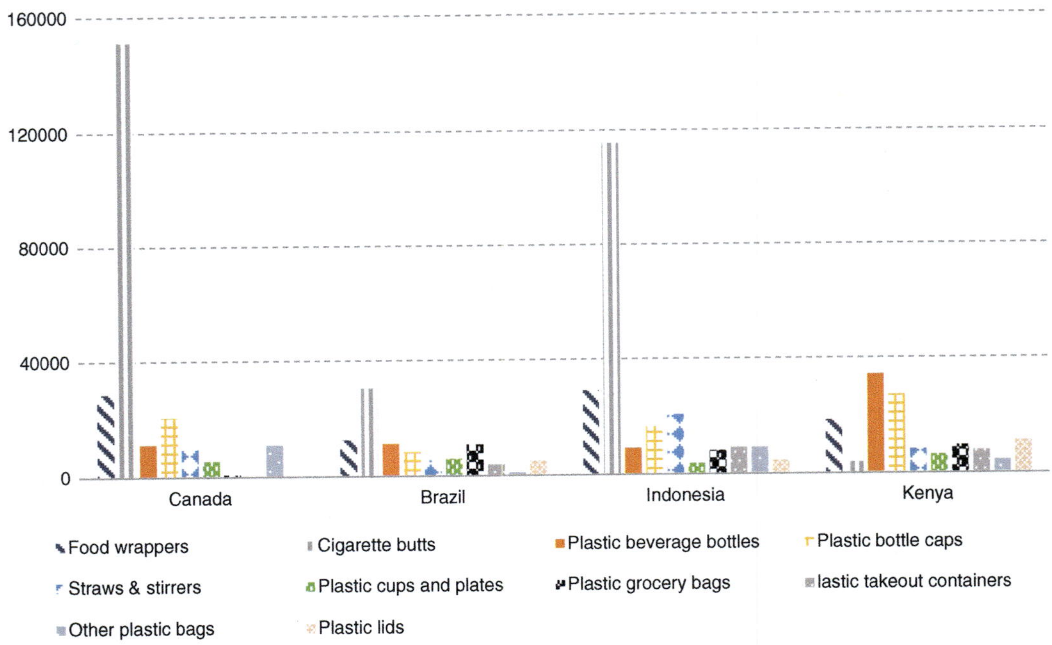

Figure 17.5 Relative proportions of top 10 items collected in International Coastal Cleanup 2020 at selected locations. Data from Ocean Conservancy [9].

Ocean. Ocean debris tends to accumulate in gyres, which are large systems of circulating ocean currents. Debris that gets into a gyre tends to stay in that location, prevented from escape by those circulating currents. Gyres are caused by a combination of winds, the rotation of the earth, and the effect of landmasses. There are five major gyres in earth's oceans [10]. The common perception of the "Great Pacific Garbage Patch" as an "island" of plastic garbage is not correct (Figure 17.6). As stated by the National Oceanic and Atmospheric Administration (NOAA),

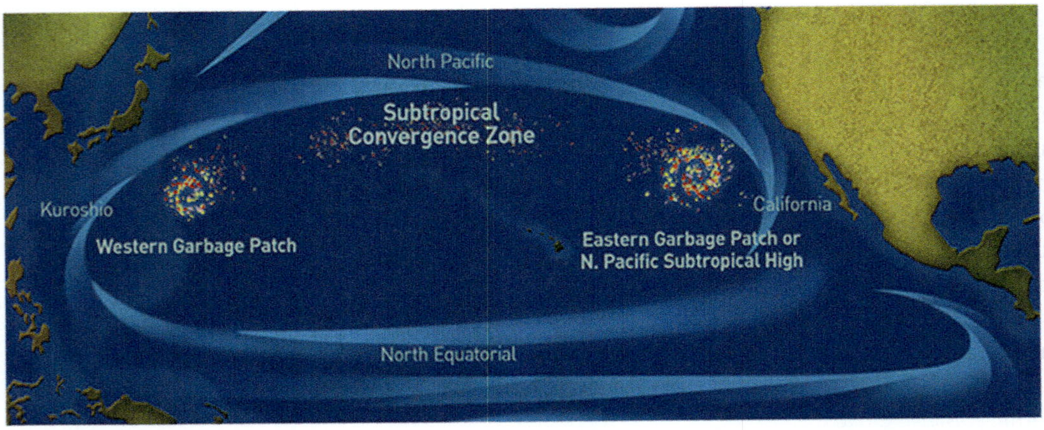

Figure 17.6 Great Pacific garbage patch; reproduced from [11].

"The name 'Pacific Garbage Patch' has led many to believe that this area is a large and continuous patch of easily visible marine debris items such as bottles and other litter—akin to a literal island of trash that should be visible with satellite or aerial photographs. This is not the case. While higher concentrations of litter items can be found in this area, much of the debris is actually small pieces of floating plastic that are not immediately evident to the naked eye.

Ocean debris is continuously mixed by wind and wave action and widely dispersed both over huge surface areas and throughout the top portion of the water column. It is possible to sail through 'garbage patch' areas in the Pacific and see very little or no debris on the water's surface. It is also difficult to estimate the size of these 'patches,' because the borders and content constantly change with ocean currents and winds. Regardless of the exact size, mass, and location of the 'garbage patch,' manmade debris does not belong in our oceans and waterways and must be addressed." [11]

Similar accumulations of floating debris, much of it plastics, have been found in other places in the ocean. Because most plastic does not biodegrade, floating plastic tends to break up into smaller and smaller pieces as it is affected by sunlight and mechanical stress, but it does not "go away." Plastic that is lighter than water, such as polyolefins, tends to float, while heavier plastic tends to sink. However, things that accumulate on the plastic, such as algae growth, can influence the buoyancy of the plastic. Therefore, some plastic and other packaging debris sinks to the bottom, some floats on the surface, and some floats at intermediate depths.

17.4.3 Litter in Other Bodies of Water

Pollution of bodies of water by litter is not limited to the oceans. Heavier litter can accumulate at the bottom of rivers and streams. Floating litter tends to follow the water flow downstream to larger bodies of water, though of course it may wash up on shore. Even litter that has sunk to the bottom may be dislodged and carried downstream as a result of floods or other disruptions.

Litter that enters rivers and streams is likely to end up either in lakes or in the ocean. In large lakes, the effects are similar to litter in the oceans; in smaller lakes, there may be more similarities to rivers and streams.

Much of the focus on litter in non-ocean water has been directed at reducing the amount of litter entering these systems, thereby also impacting the amount of litter reaching the oceans.

17.4.4 Cleanup and Prevention

In recent years, there have been efforts to clean up some of the plastic accumulations in the oceans and other bodies of water, but the task is monumental. Most efforts concentrate on floating plastic, but as mentioned in Section 17.4.1, some is removed by divers in ocean cleanup events such as those sponsored by the Ocean Conservancy. In 2013, The Ocean Cleanup, a nonprofit organization, was founded to directly target removal of plastics from the ocean. It has a goal of cleaning up 50% of ocean plastics every 5 years [12]. 4 Ocean is a business that sells products with a commitment to "pull one pound of trash from the ocean, rivers, and coastlines" for every product they sell [13].

Other organizations focus more on preventing ocean pollution rather than cleanup, or on understanding the impacts of plastic in the oceans. A number of environmental organizations, such as the Nature Conservancy, Environmental Defense Fund, and Natural Resources Defense Council, include ocean pollution within a broader focus on environmental issues. Other organizations such as the Lonely Whale Foundation, Surfrider Foundation, and Oceana focus on ocean health more broadly, while including ocean pollution. The 5 Gyres Institute, founded in

2009, conducts research on plastic pollution in the oceans and "continuously engages diverse stakeholders in understanding the science to drive impact as well as conduct community outreach and citizen science to implement data-driven solutions" [14].

Because much of the problem of floating debris in the oceans relates to plastics packaging, there are also efforts to decrease plastic waste, either through increased recycling and reuse of plastics, or by reduction in use or even bans of single-use plastics. A leading voice in implementation of a "circular economy" for plastics is the Ellen MacArthur Foundation, which advocates for a "New Plastics Economy" and has issued a series of reports on this topic [15].

An alliance of plastics associations around the world, the Global Plastics Associations for Solutions on Marine Litter, was established in 2011. It reported in 2020 that 395 projects on reducing marine litter had been planned, were underway, or had been completed, varying from beach cleanups to expanded waste management capacity, and including both research and education campaigns [16].

17.4.5 Sources of Ocean Plastics

To effectively prevent continued entry of plastics into ocean environments, it is important to understand where it comes from. As discussed, much of the plastic that enters the ocean originates on the land. A common estimate is that 80% of the plastic found in the ocean originates on land, with the remaining 20% coming from water-based activities such as fishing and ships [17, 18].

Inadequacy of waste disposal systems makes it more likely that waste will enter bodies of water and eventually flow into the ocean. Therefore, it is not surprising that much of the waste plastic that enters the oceans comes from parts of the world that have broader problems with waste management. In fact, the World Economic Forum reported in 2018 that 90% of the plastic washed into the ocean by rivers comes from just 10 river systems, 8 in Asia and 2 in Africa [17, 19]. These river systems were all characterized by high populations and "less than ideal" waste management systems.

An Ocean Conservancy study in 2015 estimated that 55–60% of plastic waste "leaking" to the oceans came from five countries: China, Indonesia, The Philippines, Thailand, and Vietnam. They estimated that 75% of leakage from land to the ocean came from uncollected waste, while 25% was from gaps within the collection system. They conducted in-depth studies in China and the Philippines to further analyze leakage sources and found five major types of physical locations that were major contributors. These were low-waste-density rural areas without collection services; medium-waste-density urban areas lacking proper waste management infrastructure, often due to rapid urbanization; high-waste-density urban areas with either overstretched waste management services or with high costs to citizens that discourages their use; illegal dumping by trash haulers; and dump sites on waterways, often open dumps with little or no controls [20]. Additional discussion about the input of plastic to the ocean is presented in the Case Study 17.1.

17.5 Litter and Wildlife

Much of the concern about litter relates to its impact on wildlife, both terrestrial and aquatic. While all types of litter can have impacts, plastic litter is especially problematic. The US EPA discusses five categories of impacts from mismanaged plastic trash in aquatic systems: habitat impacts, chemical impacts, biological impacts, problems associated with ingestion of plastic particles, and human impacts [21].

More generally, direct effects of litter on wildlife can be grouped into two main categories: ingestion and entrapment. The physical form of littered objects can result in entrapment of wildlife that comes into contact with it. While the biggest problem in the ocean comes from discarded or lost fishing nets, which can continue to "ghost fish" for years; plastic packaging also poses risks. Pictures of birds and turtles with their necks caught in plastic ring connectors for beverage cans contributed to the requirement that such connectors sold in the US be photodegradable, so that their physical form would not persist in the environment.

A report by World Wildlife Fund in 2018 stated that over 90% of damage to marine wildlife by waste is due to plastics, with about 700 marine species threatened. A study specifically on the Mediterranean found 35% of the victims of entrapment in plastics were birds and 27% were fish. Invertebrates accounted for 20% and marine mammals for 13% [22].

Wildlife can mistake trash for food. It is well established that marine life such as birds and turtles often ingest plastic, and once ingested, it often accumulates in the digestive tract and may result in starvation. For example, sea turtles can mistake floating plastic bags for jellyfish, one of their favorite foods. Ingestion can also occur indirectly, such as when sea birds ingest fish that have previously ingested plastic particles. Plastic has been found in the stomachs of a variety of species, and it is estimated that 90% of the world's seabirds have plastic in their stomachs [22].

Problems associated with ingestion of plastic are not limited to the direct effects of the plastic. It is well established that plastic, especially in bodies of water, can absorb toxic chemicals, acting as a concentrator. Wildlife that ingests this plastic then also ingests the chemicals that the plastic has absorbed. This is particularly a concern for microplastics, as will be discussed in Section 17.6.

17.6 Microplastics

When plastics are littered, because they generally do not biodegrade, they tend to persist in the environment for a very long time – decades or centuries. However, that does not necessarily mean that they stay intact. All plastics are subject to some degree to photodegradation; the action of light energy results in physical and chemical changes in the plastic that make it brittle. Mechanical stress without light exposure has similar effects (See Chapter 14 section 14.2). The net result is that large plastic objects tend to break down into smaller plastic pieces. With enough time and exposure, the plastic object is no longer recognizable, but the plastic molecules remain largely intact. When the plastic pieces are small enough, they become microplastics – plastic pieces that are very tiny and may be visible only with a microscope.

NOAA defines microplastics as plastic pieces less than 5 mm long [23]. While most microplastics originate from larger plastic pieces that have been broken down, until recently plastic microbeads were commonly used in health and beauty products. Studies done on microplastics in the Great Lakes found high abundance of these particles and documented ingestion by fish and birds, which can mistake these beads for fish eggs, a common food [24]. In the US, the Microbead-Free Waters Act of 2015 banned plastic microbeads in cosmetics and personal care products [23].

Another recently identified major source of microplastics in water is from laundry of synthetic fabrics. A 2021 article *New York Times* report stated that 35% of microplastic pollution in the ocean comes from textiles, amounting to 2.2 million tons of microfibers each year [25].

The effects of microplastics in the environment, including impacts on human health, are not yet well understood but are the focus of a considerable amount of ongoing research. NOAA is one of the leaders in this area, through their Marine Debris Program [23]. The United Nations put out a report on microplastics in 2017 that discussed the current state of knowledge and identified research needs [26].

The problem of microplastics is exacerbated because normal water treatment systems often are not able to remove these tiny particles. Recently, the presence of microplastics has been documented in drinking water, both bottled water and treated tap water [27]. Further discussion is provided in Case Study 17.2.

17.7 Biodegradability and Litter

Because of the connection between nonbiodegradability and environmental persistence, a number of organizations have advocated for replacing single-use plastics, especially in packaging and food service, with biodegradable alternatives. However, this can raise significant issues and have a rebound effect [28].

First, as we have discussed in this book, to fully evaluate the environmental impacts of a product or process, we must look at the entire life cycle. A recent report by the UK Parliament's Environment, Food and Rural Affairs Committee warned that replacing plastic with other materials in food and drink packaging could actually make pollution worse, not better, stating concerns that backlash against plastics is leading to substitution with other materials "without proper consideration of wider environmental consequences, such as higher carbon emissions" [29].

Another concern is unintended impacts on human behavior. If consumers believe that items that they throw away will safely disappear, will that result in increased littering? Littering behavior is inherently difficult to study, since people tend not to litter when others are watching! A 2015 United Nations report cited two studies that indicated that labeling products as "biodegradable" made them more likely to be littered [30].

Another concern is that biodegradation behavior is strongly affected by the exposure environment, as discussed in Chapter 14. For example, objects that are compostable in industrial composting operations may not biodegrade rapidly in an ocean environment. The UN report mentioned previously also discussed the differences in biodegradation behavior of "biodegradable" plastics both between land and water and between differing aquatic environments [30]. See also Chapter 14.

17.8 Case Study 17.1. Emission of Plastic from Rivers to the World's Oceans

As we have discussed in the previous sections, plastics in the marine environment have become a nuisance for coastal areas and a major problem for marine life and increasingly for human health. However, we seem to be impressed by many pictures of animal entangled by plastic without fully understanding the location and the reason of the pollution. Part of solving a problem is understanding the magnitude and the geographical boundary of the locations, so meaningful approaches can be used to fix these issues and the right stakeholders engaged for optimal outcomes.

It is well known that the amount of plastic entering the ocean from rivers is correlated to land-based sources, and they are considered the dominant input of plastic into oceans.

Observed plastic concentrations entering the oceans from rivers differ by several orders of magnitude around the world, depending on factors such as population density, levels of urbanization and industrialization, economy of the coastal area, rainfall rates, waste management systems, and the presence of artificial barriers.

Figure 17.7 shows a global estimate of the mismanaged plastic waste (MPW) reported by Lebreton et al. [31]. The authors estimate that between 1.15 and 2.41 million tons of plastic larger than 0.3 mm and smaller than 0.5 m currently flow from global river systems into the ocean every year. This represents around 12% to 21% of the total amount of plastic produced ending in the ocean every year (about 11 million metric tons in 2016) [32]. However, this amount is estimated to increase to 29

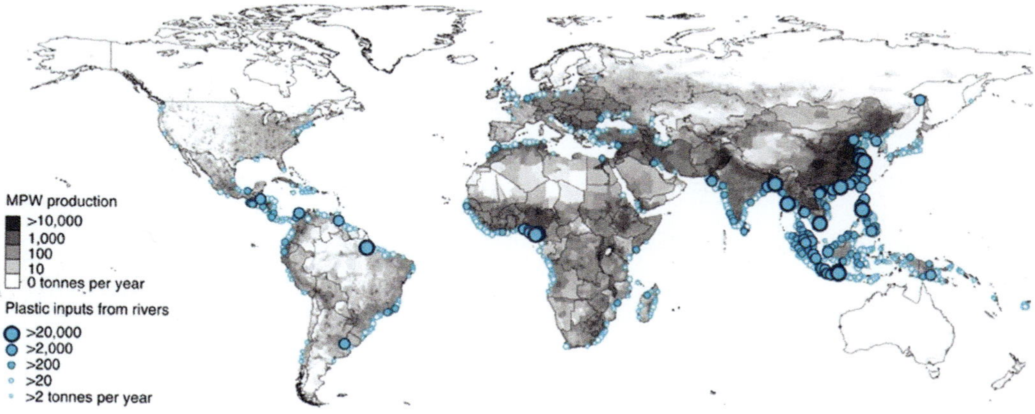

Figure 17.7 Mass of river plastic flowing into oceans in tons per year; reproduced from [31] under CC BY 4.0.

million metric tons by 2040 if business continues as usual. The authors indicated that their estimate was conservative since it does not include plastic outside the sampled size range.

The top five polluting rivers were located in Asia (Yangtze, Xi, Huangpu – China; Ganges – India and Bangladesh; Cross – Nigeria and Cameroon). The top 20 rivers, mostly also located in Asia with the exception of the Amazon and Magdalena, accounted for more than two-thirds (67%) of the global annual input. However, they cover only 2.2% of the continental surface area and represent only 21% of the global population. Moreover, the 122 most contaminating rivers (4% of total landmass surface area and 36% of global population) added more than 90% of the plastic inputs with 1 located in Europe, 8 in Africa, 8 in South and Central America, and 103 rivers in Asia.

The estimated plastic releases from the primary Asian rivers account for 86% of the total global input to the oceans. The rest of the world shares 14% of mass input, with Africa responsible for 7.8%, South America responsible for 4%, Central and North America for 0.95%, Europe for 0.28%, and Australia and the Pacific Region 0.02%. The authors also predicted the effect of seasonal variations, estimating that more than 74% of the total input happens between May and October, with a peak in August, which is regulated mainly by the East Asia monsoon with significant inputs from China.

Although it may seem natural to think that with Central and North America, Europe, Australia, and the Pacific Region being responsible for only 1.25% of the plastic input to the ocean, there is no problem to solve there; however, we should still keep in mind that just 20 polymer producers accounted for more than half of all single-use plastic generated globally in 2019, as recently reported [33]. Eleven of the top 20 polymer producers are based in Asia – five in China, four in Europe, three in North America, one in Latin America, and one in the Middle East. Most of the producers own production plants in many countries and all engage in global polymer trading. Thus, reducing plastic pollution is a global issue, and it should be addressed as such.

17.9 Case Study 17.2. Presence of Microplastics in Drinking Water and Food

An increasing body of research has shown microplastics present in our drinking water and food supply. A recent report by the World Health Organization [27] and a systematic review of 50 research articles by Koelmans et al. [34] reported detection of microplastics in fresh water, drinking water, and wastewater. The reported concluded that in fresh water, the presence of microplastics

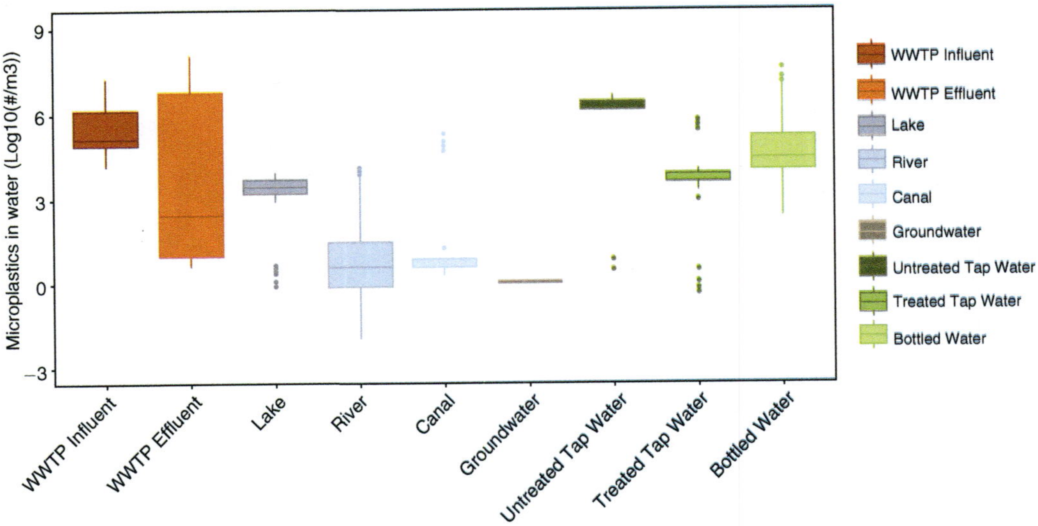

Figure 17.8 Box and whisker plot showing median and variation in microplastic number concentrations in individual samples taken from different water types; reproduced from [34] under CC BY 3.0 IGO license.

was consistent with their production volume and societal uses. The most predominant shapes in drinking water were microfragments and fiber shapes, and the polymers most commonly detected were PET and PP.

In both freshwater and drinking water studies, the smallest detected microplastics were limited by the mesh size used in the sampling study. The highest value detected was on the order of 10^8 #particles/m^3 (Figure 17.8). The average value detected in treated tap water was 10^4 #particles/m^3, mostly due to a small mesh [34]. Effluents from wastewater treatment plants showed values as high as 10^8 #particles/m^3. The authors concluded that microplastics are commonly present in freshwater and drinking water, and the concentrations spanned ten orders of magnitude from 1×10^{-2} to 10^8 #particles/m^3.

Further assessment was conducted regarding the particle size. Particles as small as 1 μm were detected, but this limit was mostly imposed by the detecting method, Figure 17.9. Of the microplastic particles detected, the largest reported frequency was in the following order: PE ≈ PP > PS > PVC > PET, which approximately matches global plastic use and demand. It is important to note that PS, PVC, and PET have density larger than 1 g/cm^3, so they should sink in water.

A similar review of the analytical methods to determine microplastics in food was conducted by Kwon et al. [35]. They reported a large number of studies identifying commercial polymers such as PE, PP, PS, and PET present in table salt (rock, lake, and sea) and seafood such as shrimp, dried fish, crab, clams, oysters, and mussels (Figure 17.10). Concentrations in salts were in the range of 1–19,800 particles/kg depending on the product and identification method [35]. Concentrations in fish were reported from 0.2 to 24 particles/fish. Some values were also expressed per gram of fish. Identification techniques also play a significant role. In the case of processed foods such as beer, honey, milk, sugar, and teabags, concentrations were between 3 and 254 particles/L or between 3 and 249 particles/kg. Values as high 379 particles/kg of sugar were reported. They reported that microplastic particles reported in food were mostly classified as fragments, fibers, pellets, or films. Particle size was not yet fully assessed.

Finally, a large amount of research is being conducted to be able to identify and to measure nano-plastic particles in liquid and solid matrices. However, quantification of nano-plastic particles in the environment and their overall exposure risk and health impact still escapes our scientific understanding.

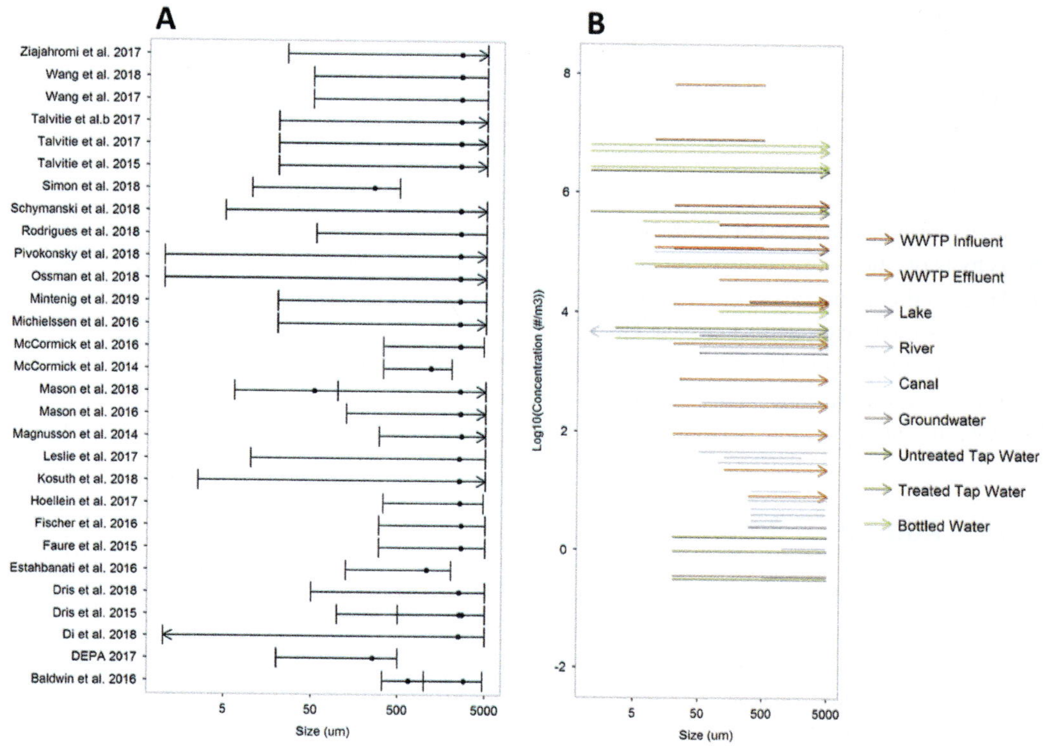

Figure 17.9 Size ranges used (A) and number concentrations per size range reported (B) in studies on microplastics in drinking, surface and wastewaters as referenced in Figure 17.8. "Arrows indicate that no upper or lower size limit was specified, in which case values of 5 mm or 1 μm were assigned, respectively." Panel A: Size ranges per study are ordered alphabetically per author name as reported in [34]. Data points represent the average of the size range. Panel B: reported concentrations as a function of size range. Colors of arrows (Panel B) correspond to colors of the box and whiskers in Figure 17.8. Figures reproduced from [34] under CC BY 3.0 IGO license.

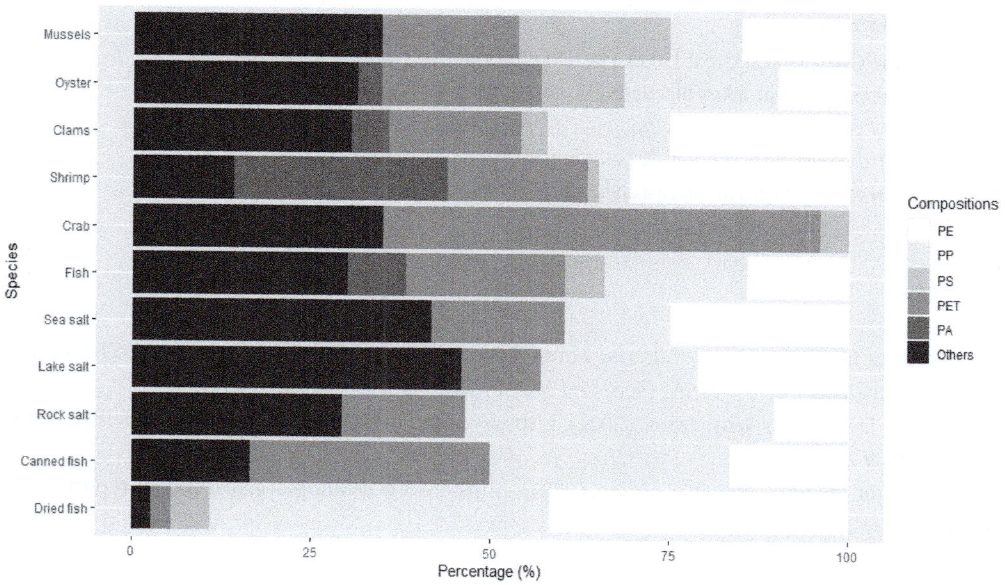

Figure 17.10 Types of plastic materials identified in seafood and salt; reproduced from [35] under CC BY 4.0 license.

17.10 Study Questions

1 What is the main cause of litter?
2 What are the most common types of littered items?
3 What are the costs of litter?
4 What is the connection between beverage container deposit laws and litter?
5 Is disposal of plastic wastes from ships allowed?
6 Why is plastic in the oceans more of a problem than other types of materials?
7 Discuss the main impacts of litter on wildlife.
8 What is the Great Pacific Garbage Patch?
9 What is the relationship between land-based litter and ocean litter?
10 What is a microplastic, and where do microplastics come from?
11 Why are biodegradable plastics not a solution to plastic litter?

17.11 Additional Resources

1 National Geographic, Great Pacific Garbage Patch, https://www.nationalgeographic.org/encyclopedia/great-pacific-garbage-patch
2 The Great Pacific Garbage Patch is Not What You Think It Is, https://www.youtube.com/watch?v=6HBtl4sHTqU
3 Alliance for the Great Lakes, 5 Ways Plastic Pollution is Different in the Great Lakes, https://greatlakes.org/2018/06/5-ways-plastic-pollution-is-different-in-the-great-lakes/?gclid=EAIaIQobChMI_PjprM3V8AIV5TizAB1U1wnEEAAYASAAEgIhU_D_BwE
4 Charles, D., Kimman, L. & Saran, N., *The Plastic Waste Makers Index*, Minderoo Foundation, 2021, https://www.minderoo.org/plastic-waste-makers-index/
5 Gamaralalage, D. & Kato, M., *Strategies to reduce marine plastic pollution from land-based sources in low and middle-income countries*, United Nations Environment Programme, 2019, https://www.iges.or.jp/en/pub/strategies-reduce-marine-plastic-pollution-land-based-sources-low-and-middle-income-countries
6 Alliance for the Great Lakes, Great Lakes Plastic Pollution, https://greatlakes.org/great-lakes-plastic-pollution-fighting-for-plastic-free-water/
7 Pew Charitable Trusts and SystemIQ, *Breaking the Plastic Wave, A comprehensive assessment of pathways towards stopping ocean plastic pollution*, 2020, https://www.pewtrusts.org/-/media/assets/2020/07/breakingtheplasticwave_report.pdf

References

1 Keep America Beautiful. (2021). *Littering Facts, Resources & Statistics From Keep America Beautiful.* https://kab.org/goals/end-littering (accessed May 27, 2021).
2 Litter Facts - Don't Mess With Texas. (2021). https://www.dontmesswithtexas.org/about/litter-facts (accessed May 27, 2021).
3 National major types of rubbish reported. (2021). https://www.cleanup.org.au/rubbish-report (accessed May 27, 2021).

4 Home | Keep Britain Tidy. (2021). https://www.keepbritaintidy.org/home (accessed May 27, 2021).
5 Key Findings: The Costs of Americans Littering. (2010). www.kab.org (accessed May 27, 2021).
6 *Litter Laws | Keep Australia Beautiful WA 2021*. https://www.kabc.wa.gov.au/resources/for-local-government/litter-laws (accessed May 27, 2021).
7 Health & hygiene | Hong Kong Tourism Board. (2021). https://www.discoverhongkong.com/ca/plan/traveller-info/health-hygiene.html (accessed May 27, 2021).
8 International Convention for the Prevention of Pollution from Ships (MARPOL). (2021). https://www.imo.org/en/About/Conventions/Pages/International-Convention-for-the-Prevention-of-Pollution-from-Ships-(MARPOL).aspx (accessed May 27, 2021).
9 Ocean Conservancy. *Together, we are team ocean*. Washington, D.C.: 2020.
10 US National Oceanic and Atmospheric Administration. (2021). *What is a gyre?* https://oceanservice.noaa.gov/facts/gyre.html (accessed May 27, 2021).
11 US National Oceanic and Atmospheric Administration. (2021). *What is the Great Pacific Garbage Patch?* https://oceanservice.noaa.gov/facts/garbagepatch.html (accessed May 27, 2021).
12 The Ocean Cleanup. (2021). https://theoceancleanup.com (accessed May 27, 2021).
13 About Us - 4ocean (2021). https://www.4ocean.com/pages/about (accessed May 27, 2021).
14 *About 5 Gyres* — 5Gyres.org. (2021). https://www.5gyres.org/about-us (accessed May 27, 2021).
15 Ellen MacArthur Foundation. *The new plastics economy global commitment 2020:75*. https://www.ellenmacarthurfoundation.org/assets/downloads/Global-Commitment-2019-Progress-Report.pdf (accessed November 25, 2019).
16 Global Plastics Associations for Solutions on Marine Littter. *Marine litter Solution: The Declaration of the Global Plastics Associations for Solutions on Marine Litter 5th PROGRESS REPORT*. 2020.
17 Schmidt, C., Krauth, T., and Wagner, S. (2017). Export of plastic Debris by rivers into the sea. *Environ Sci Technol* 51: 12246–12253. https://doi.org/10.1021/acs.est.7b02368.
18 Where Does Marine Litter Come From? (2021). https://www.marinelitterthefacts.com/sources (accessed May 27, 2021).
19 *90% of plastic polluting our oceans comes from just 10 rivers | World Economic Forum* n.d. https://www.weforum.org/agenda/2018/06/90-of-plastic-polluting-our-oceans-comes-from-just-10-rivers (accessed May 27, 2021).
20 McKinsey & Company Center for Business and Environment, Ocean Conservancy. *Stemming the tide: Land-based strategies for a plastic-free ocean* 2015:44. https://oceanconservancy.org/wp-content/uploads/2017/04/full-report-stemming-the.pdf (accessed November 22, 2019).
21 US Environmental Protection Agency. (2021). *Impacts of Mismanaged Trash | Trash Free Waters*. https://www.epa.gov/trash-free-waters/impacts-mismanaged-trash (accessed May 27, 2021).
22 World Wild Life Funds. *Out of The Plastic Trap. Saving the Mediterranean from Plastic Pollution*. Washinghton, DC: 2018.
23 US National Oceanic and Atmospheric Administration. (2021). *What are microplastics?* https://oceanservice.noaa.gov/facts/microplastics.html (accessed May 27, 2021).
24 *Microbeads in the Great Lakes - Tip of the Mitt Watershed Council 2021*. https://www.watershedcouncil.org/microbeads.html (accessed May 27, 2021).
25 Your Laundry Sheds Harmful Microfibers. *Here's What You Can Do About It. | Wirecutter 2021*. https://www.nytimes.com/wirecutter/blog/reduce-laundry-microfiber-pollution (accessed May 27, 2021).
26 Kershaw, P.J. *Marine Plastic Debris Global lessons and research to inspire action*. 2016.
27 World Health Organization. *Microplastics in drinking water*. Geneva, Switzerland: 2019.
28 Grubb, M.J. (1990). Communication Energy efficiency and economic fallacies. *Energy Policy* 18: 783–785. https://doi.org/10.1016/0301-4215(90)90031-X.

29 Plastic food and drink packaging - Environment, Food and Rural Affairs Committee - House of Commons. (2021). https://publications.parliament.uk/pa/cm201719/cmselect/cmenvfru/2080/208002.htm (accessed May 27, 2021).

30 Kershaw, P.J. (2020). *Biodegradable Plastics & Marine Litter : Misconceptions, Concerns and Impacts on Marine Environments*. Nairobi, Kenya: United Nations Environment Programme.

31 Lebreton, L.C.M., van der Zwet, J., Damsteeg, J.-W., Slat, B., Andrady, A., and Reisser, J. (2017). River plastic emissions to the world's oceans. *Nat Commun* 8: 15611. https://doi.org/10.1038/ncomms15611.

32 Trusts TPC, SystemIQ. Breaking the Plastic Wave. *A Comprehensive Assessment of Pathways Towards Stopping Ocean Plastic Pollution*. Washington DC: 2020.

33 Dominic, C., Kimman, L., Saran, N., and Minderoo F. The Plastic Waste Makers Index, *Minderoo Foundation* 2021:86.

34 Koelmans, A.A., Mohamed Nor, N.H., Hermsen, E., Kooi, M., Mintenig, S.M., and De France, J. (2019). Microplastics in freshwaters and drinking water: Critical review and assessment of data quality. *Water Res* 155: 410–422. https://doi.org/10.1016/j.watres.2019.02.054.

35 Kwon, J.H., Kim, J.W., Pham, T.D., Tarafdar, A., Hong, S., Chun, S.H. et al. (2020). Microplastics in food: A review on analytical methods and challenges. *Int J Environ Res Public Health* 17: 1–23. https://doi.org/10.3390/ijerph17186710.

18

Keeping in Perspective

18.1 Learning Objectives

After learning from this chapter, you are expected to:

- Determine the environmental footprint of the primary, secondary, and tertiary packaging systems and their trade-offs.
- Understand the environmental footprint of the product/packaging systems and their trade-offs.
- Discuss the role of packaging in waste creation.
- Understand the impact of transportation, consumer behavior, and end of life on the EFP of packaging systems.
- Integrate the concepts learned in previous chapters to develop packaging that contributes to sustainable development.

18.2 Introduction

Throughout the previous chapters, we examined the role of packaging in helping to achieve sustainable development, the use of design thinking in developing packaging systems, the environmental footprint (EFP) of packaging systems, and their end-of-life (EoL) scenarios. By now, you should recognize that packaging serves a variety of purposes and that primary, secondary, and tertiary packages create intrinsic value by the services they provide. However, as stated by Williams et al., the value of packaging systems is created during its delivered service and use, as framed by the service-dominant logic theory [1]. So, a package may not have an intrinsic value or purpose without fulfilling its function and mission of containing, protecting, and communicating, which is an important concept for any packaging professional to remember.

Modern product supply chains are very complex networks that could cover several continents and span many years from production to consumption (e.g., tomato sauce, almonds, olive oil, wine, coffee, salt, computers, screws, and nails). Hence, understanding and evaluating the EFP contribution of each portion of the network to the total EFP of the product/packaging systems is complex but necessary. As discussed in Chapter 8, the CO_2-eq emissions released due to conducting an activity or making a product do not differ based on where they come from, such as the package, product, process, transportation, and/or disposal. So, to limit or reduce greenhouse

Life Cycle of Sustainable Packaging: From Design to End of Life, First Edition. Rafael A. Auras and Susan E. M. Selke.

Companion Website: www.wiley.com/go/Auras/lifecycleofsustainablepackaging

gas (GHG) emissions, our overall goal must be to limit or better to reduce GHG emitted to the environment as well as other pollution. We must use a comprehensive and systematic approach to understand the real climate impact and overall sustainability of a service or activity. In this chapter, we will expand on the contribution of the product/packaging systems, their transportation, and their disposal considering their overall EFPs to keep in perspective overall impacts for decision-making. Mainly, we will focus on assessing the primary, secondary, and tertiary packaging and package/product EFP ratios. We will explore the impact of packaging on waste creation, the impact of transportation on the overall EFP of the product/package, the impact of consumer behavior, and the impact of EoL scenarios on the EFP. Since the EFP of food products has been commonly reported, we will use food examples. However, the considerations apply to most categories of packaged products.

18.3 Environmental Footprint of Primary, Secondary, and Tertiary Packaging Systems

In Chapter 9, we discussed the EFP of a glass, can, and plastic bottle to deliver a drink to the consumer. Table 18.1 shows the overall EFP of 100 glass and PET bottles, and aluminum cans. We can easily observe that the glass bottles have a higher EFP, followed by the plastic-PET bottles and the aluminum cans.

If we observed the contribution of the primary, secondary, and tertiary packaging systems for each container, we see that the primary package provides the largest contribution for each packaging system, followed by the secondary package, transportation, and lastly the tertiary package (see Figure 18.1). So, it would make a lot of sense to reduce the EFP of the primary package if feasible, as hypothetically discussed in Section 18.8. Furthermore, we can also see that the contribution of each phase of the packaging systems contributes differently for each impact indicator and container type, see, for example, the variation between ecosystems, species.yr for glass, aluminum, and plastic.

Table 18.1 Environmental footprint of 100 glass bottles, aluminum cans, and plastic-PET bottles to distribute drinks, as introduced in Chapter 9, Case Study 9.1.

Indicator	Units	Glass Bottle	Aluminum Can	Plastic-PET Bottle
Climate change	kg CO_2-eq	33	14	18
Cumulative energy demand	MJ	593	189	370
Ecosystems	species.yr	5.8E-07	2.6E-07	2.8E-07
Human health	DALY	7.2E-05	2.8E-05	3.2E-05
Water	m^3	0.011	0.004	0.002

Note: Values included the environmental footprint of the primary, secondary, and tertiary packaging systems and transportation.

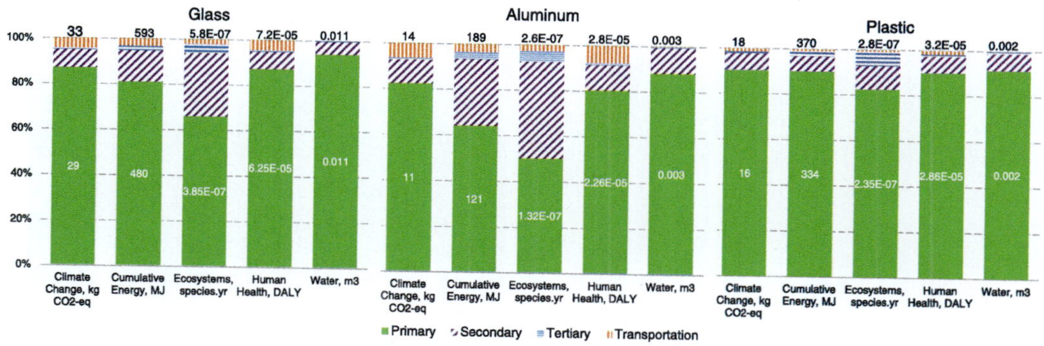

Figure 18.1 Environmental footprint contribution of the primary, secondary, and tertiary packaging systems for 100 glass bottles, aluminum cans, and plastic-PET-bottles.

18.4 Environmental Footprint of the Product/Package System

Let us get back to the supply chain of tea discussed in Chapter 9. The Twinings® Earl Grey tea leaves that we purchase in Michigan are mostly grown in South America. Then, they are shipped to the United Kingdom, mixed with several other components in the UK such as citrus flavors (bergamot) and other teas. After that, the tea is packed in a primary, secondary, and tertiary container and shipped to the US where we can purchase it in a store. So, it is straightforward to visualize that when we consume a cup of tea, we are linked to the EFPs of the entire supply chain of the production, consumption, and disposal of the tea. With this concept in mind and understanding that reduction of GHG emissions should be targeted to any of the most impactful parts of the system, we can explore the ratio between the EFP of the package/product systems and consider the impact of decision-making on limiting or reducing GHG.

Figure 18.2 shows a general contribution of the different stages of the food system in the US. Packaging overall represents around 7% of the overall energy used in US food supply. However, the common perception is that packaging is a larger component, which may be the case for very special products such as maybe a plastic water bottle.

Figure 18.3 shows the GHG distribution of the different stages of the supply chain for several food products, such as fresh produce (spinach, ready-to-eat lettuce, mushroom, tomatoes, potatoes, orange juice), dairy and eggs (milk, cheese, eggs), and meat (ground turkey, pork, and beef). The GHG contribution attributed to the production and processing increases from spinach to beef. Spinach and lettuce production contributes relatively little to the entire GHG emission of the system, with the other stages having larger contributions. Other products show a larger contribution from the production stages and from edible food waste. In the case of orange juice, the transportation stage shows a large impact, mostly due to the modeling conditions assuming that they were produced in a single US location and distributed to the entire continental US.

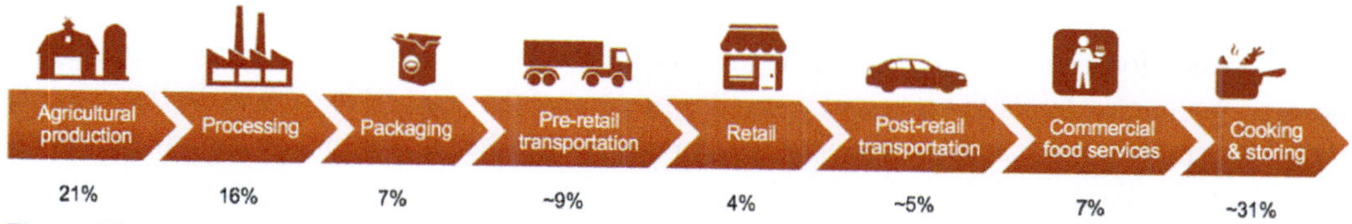

Figure 18.2 Distribution of energy used by the food system in the US; adapted from [2].

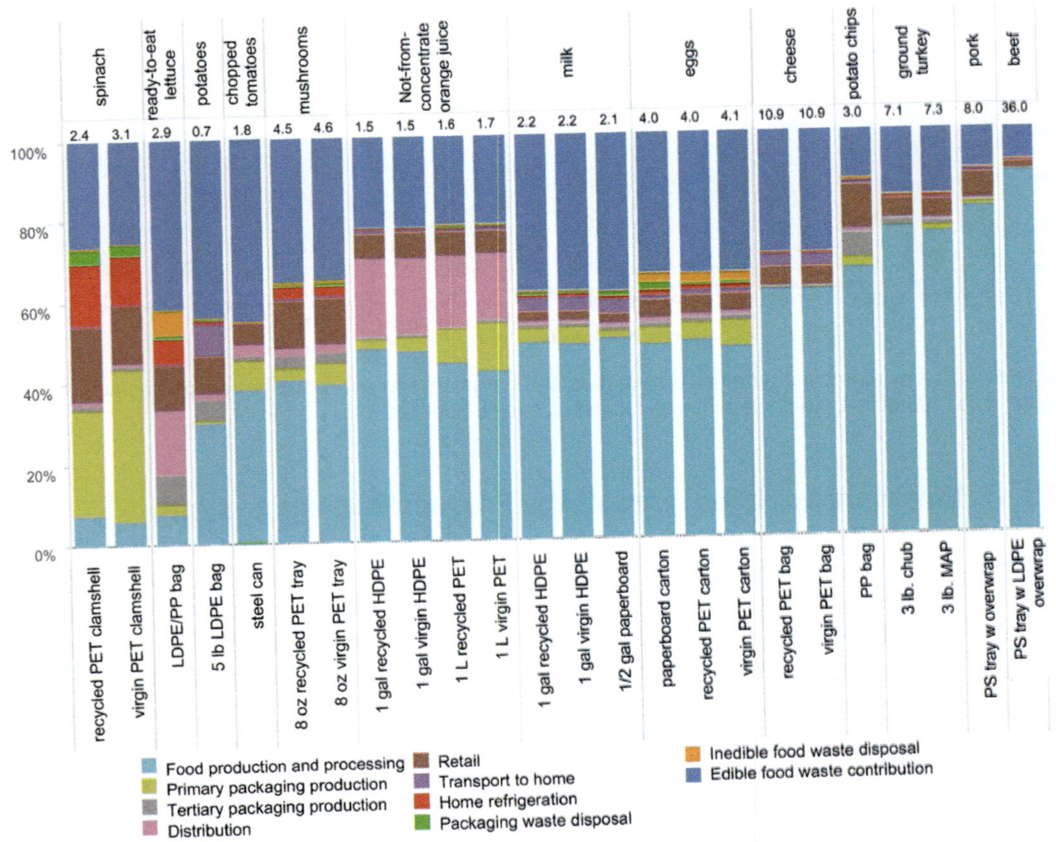

Figure 18.3 Greenhouse gas emissions for different stages of products packaged in different containers; reproduced from [5] with permission from John Wiley & Sons.

Consequently, from an environmental standpoint, not all foods are created equal. Fresh fruits and vegetables account for the lowest GHG impact while products from ruminants have the highest environmental impact [3, 4]. Understanding the environmental impact of food products, such as meat and dairy, could provide a large opportunity for EFP reduction in the food supply chain.

We should not overlook in Figure 18.3 the total contribution of GHG among products (e.g., spinach packaged in a recycled PET clamshell 2.4 kg CO_2-eq and beef packaged in a PS tray with LDPE overwrap 36 kg CO_2-eq) to fully understand how to model consumption. A nutritious diet may contain and require different volume requirements for carbohydrates, proteins, and fibers derived from different products.

Many LCA studies have demonstrated that the EFP of the packaging system is in many cases only a minor portion of the entire environmental profile of the total system. GHG contribution considerations have been calculated for several products, such as beer [6], wine [7], coffee [8], tomatoes [9], orange juice [10], and pork [11]. Figure 18.4 shows the EFP contribution to the total GHG emission for selected products. As the intensity of production and/or transport requirements increase, packaging tends to represent a smaller portion of the total EFP, such as in the case of pork. Therefore, looking at these systems, we can rationalize that depending on the products the more "bang for the buck," regarding limiting and reducing EFP will be to reduce the EFP of the largest contributing stage of the supply chain. For beer, wine, and tomatoes, a reduction of the EFP

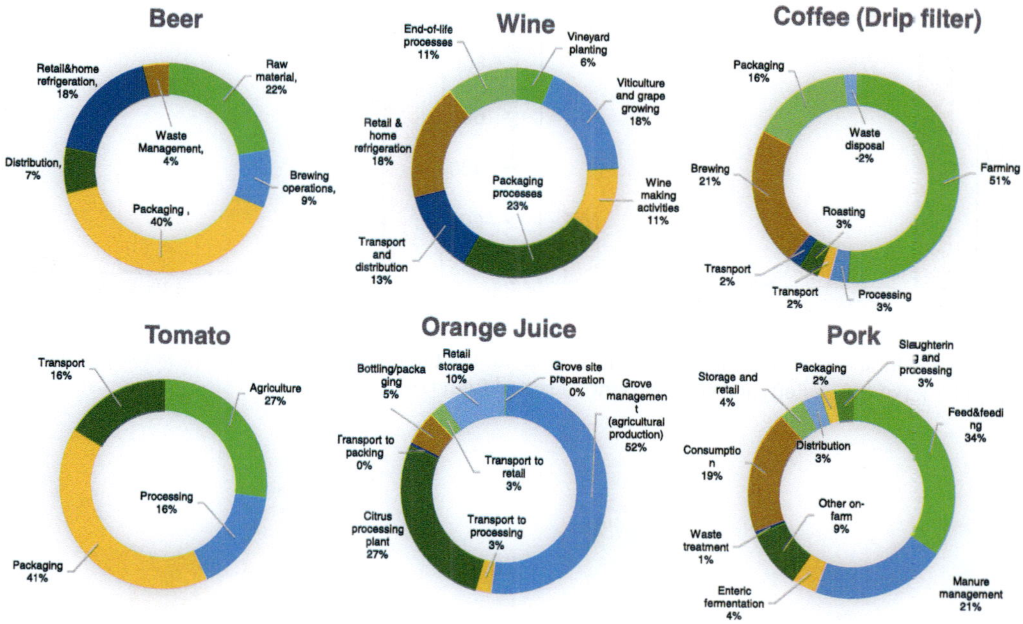

Figure 18.4 Greenhouse gas emission distribution for selected products across their supply chain; data for beer [6], wine [7], coffee [8], tomato [9], orange juice [10], and pork [11].

of the packaging systems should be a good effort. In the case of pork, any reduction of the EFP during the pig-raising stage will have a larger contribution to a reduction of the total EFP, a systems approach.

18.5 The Role of Packaging in Waste Creation

Another significant consideration when evaluating packaging systems is the impact of waste creation due to the use of different packaging systems, and the role of packaging in saving food. In 2018, in the US, food waste generation (FWG) in the MSW stream represented around 22% [12] and around the world it was 44% in 2016 [13].

Figure 18.5 describes the environmental impact in terms of total GHG emissions in the life cycle of 1 kg of coffee (left) and cheese (right). Food manufacturing and food waste were the largest contributors to the system GHG emissions, while packaging, distribution, and packaging waste were only small contributors.

Similarly, Figure 18.6 shows the GHG profile of an aggregated number of food items – meat, fish and eggs, dairy, and fruits, vegetables, and nuts, consumed by a family of four people (three adults and one child) in Australia over 1 week. Packaging represents between 2% and 12% of the different categories. Wasted food for each category represents between 13% and 22%, with fruits, vegetables, and nuts being the highest. We can envision that there are packaging design options that simultaneously decrease waste and the environmental impact of the packaging itself. However, sometimes we may need to increase the EFP of the packaging to achieve overall food waste reduction and the lowest EFP of the entire product/package system [15].

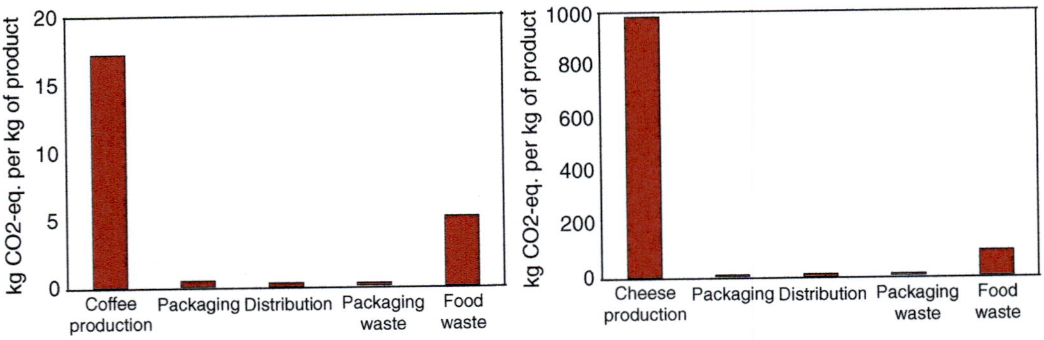

Figure 18.5 Total greenhouse gas emissions throughout the life cycle of manufacturing, packaging, and supply of 1 kg of coffee (left) and cheese (right); adapted from [14].

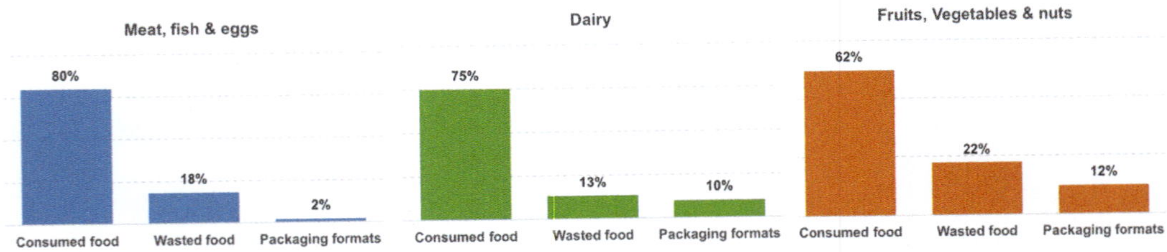

Figure 18.6 Greenhouse gas distribution between food consumed, food wasted, and packaging materials for meat, fish and eggs, dairy, and fruits and vegetables for a four-person household over 1 week. The climate impact is larger for the food wasted than for the packaging in all cases; reproduced from [16] under CC BY 4.0 license.

18.6 Impact of Transportation on the Environmental Footprint of the Product/Package

Products consumed around the world such as coffee, bananas, cocoa, sugar, and salt are grown in different parts of the world and transported to distant locations for consumption. However, the method of transportation of these products varies, such as planes, trucks, trains, and ships. All these modes of transportation have different EFP impacts, and they will translate to the product/package that we acquire. So, the type of impact that transportation has on each stage of the supply chain of a product must be differentiated and understood.

A closer look at the US food system suggests that transportation accounts for about 14% of the total energy used, about 5% from personal grocery shopping trips, and about 9% from distributing raw and processed food (see Figure 18.2).

It is natural to think that the farther a product travels to market from the production site, the greater its environmental contribution to global warming. This is the right perception if the mode of transportation is always the same. However, not all transportation modes emit the same emissions and not all modes have the same efficiency. For example, let us consider transporting a passenger using different transportation methods such as air, road, rail, or ocean freight. Figure 18.7 shows that transportation by plane and car has larger GHG impact than rail and ferry. The same analogy applies to the distribution of products and packages. Generally, air or road transport result in more emissions than rail and ocean freight transportation. For example, transporting a product by ship from Shanghai, China to Los Angeles, CA, for 19,720 nautical miles (22,693 miles),

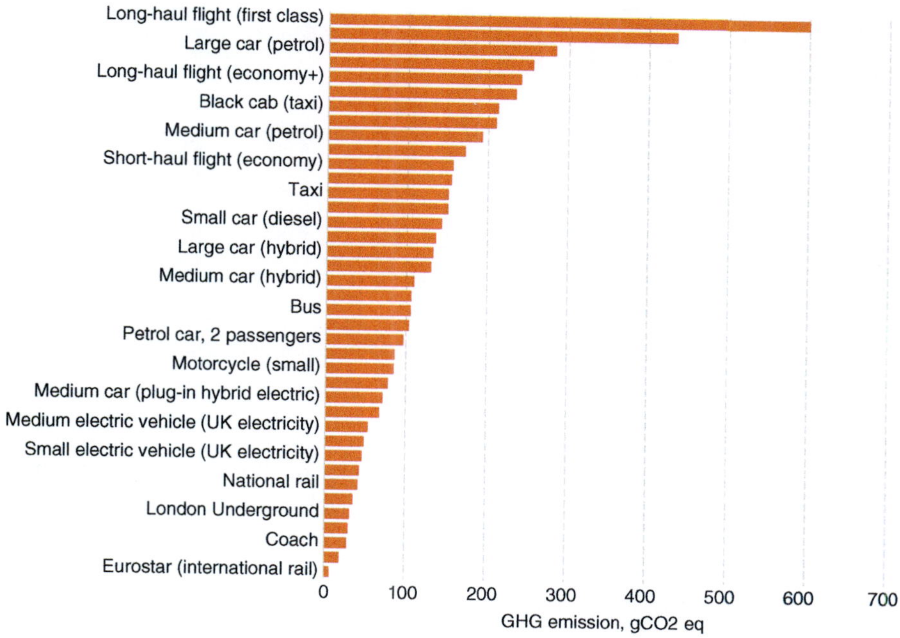

Figure 18.7 Carbon footprint in grams of CO_2-eq of different transportation modes for a passenger traveling for 1 km; data obtained from [17] and reproduced under CC BY 4.0 license.

may have a lower EFP than shipping by truck from Los Angeles, CA to Roselle Park, NJ, for 2,764 miles. But, again, this will depend on the cube efficiency of the shipping container and the type of truck or freight considered.

So, it will make sense to select always the lowest EFP mode of transporting goods as long as factors such as shelf life, urgency, and time constraints are not a determining factor. However, the lowest EFP mode of transportation may not always be an available choice since we may be limited by space and regulations, such as in transporting goods in suburban or urban areas. We can select between railroad and trucks to transport between some states and achieve the lowest EFP of transported goods. However, dynamic and modern supply chains are constrained not only by distance but also by time; factors that we must consider optimizing during the distribution of goods.

18.7 Impact of Consumer Behavior on Waste Creation and the Environmental Footprint of the Product/Package System

Consumption of products varies among consumers, and their habits have a large impact on their use and the creation of waste. Studies have been conducted to understand the role of packaging in saving food waste in households and the supply chain and the consumer role in the generation of waste. According to Hebrok and Boks, the three dominating solutions for reducing consumer food waste are: (1) technology that helps people plan, share, and keep an overview of stock, (2) packaging and storing solutions that extend shelf life, and (3) information and awareness campaigns [18].

Figure 18.8 shows the main reasons for food waste in supermarkets and in homes in Norway as reported in 2010. For consumers, the main reasons given for wasting food were either an exceeded

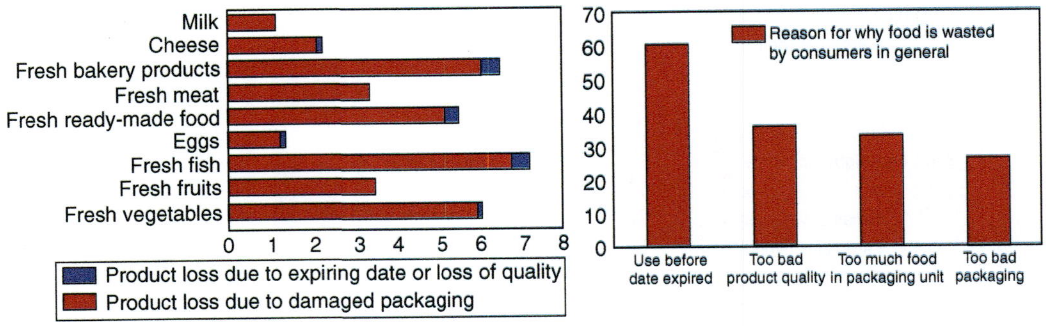

Figure 18.8 Principal reasons for food product waste in the supermarket sector (left) and at home (right) in Norway; adapted from [19].

expiration date or that the food had lost its freshness or appearance quality. On the other hand, consumers are less likely to report faulty or unsuitable packaging (such as an oversized unit) as the cause of food waste. In contrast, in the retail sector, the utmost important cause of food waste was product loss due to damaged packaging.

In a recent study, Williams et al. stated that although the significance and relationship of food waste packaging and the role of packaging differ for each food category as suggested in Figure 18.5, it is particularly important to better understand this relationship for products having high EFP, such as meat and dairy. They further emphasized that for these product categories, packaging designers must ensure that the package contains the right amount of product, provides relevant information about the product, and is supportive of consumer uses such as being easy to use, reseal, and empty [1].

We modeled the FWG of milk at the consumer level considering factors such as the amount and type of consumption, type and number of containers bought, buying behavior, and shelf life of milk using discrete-event simulation [3]. We recommended that the GHG emissions from a one-person household can be decreased with increased top-up shopping trips (i.e., shopping at the corner store) – Figure 18.9. For households of four, similar recommendations were provided to those for one-person households. When purchasing 2 gallons or 4 half gallons of milk in weekly shopping, spoilage was approximately the same. However, when four-person households purchased 3 half gallon containers, their total spoilage decreased significantly, thus decreasing their total GHG emissions without increasing the amount of unavailable milk. So, we concluded that for one-person households and four-person households, spoilage reductions were large enough to justify the purchase of less milk in smaller containers [3]. We further demonstrated that decreasing the amount and increasing the frequency of the amount of milk purchased significantly reduced spoilage without a considerable increase in the amount of milk unavailable.

18.8 Impact of End-of-life Scenarios on the Environmental Footprint of Packaging Systems

In Chapters 11 to 16, different waste management methods to reduce and manage waste were covered. However, the right packaging strategy does not only depend on the product, but also on the waste management methods at EoL. To bring this into perspective, we should evaluate the impact of the EFP of different packaging reduction and EoL scenario options. Let us consider a few scenarios, such as reduction, reuse, recycling, incineration with energy recovery, and landfilling.

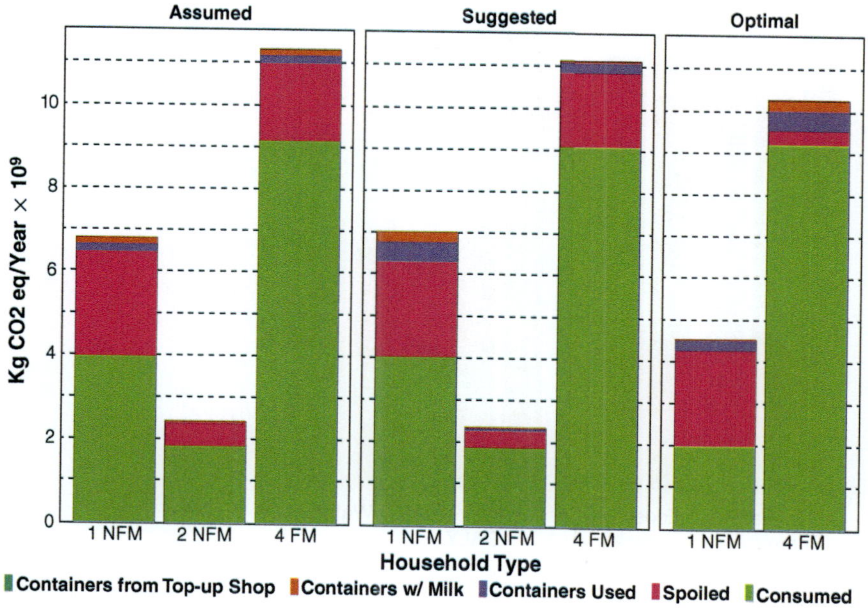

Figure 18.9 "Total GHG emissions of 1 (32 million, 1NFM), 2 (6 million, 2NFM), and 4-person (15 million, 4FM) US households. 'FM' stands for family, while 'NFM' stands the non-family household. Columns are separated by the container size purchased. 'Assumed' containers represent what is currently assumed to be purchased by households, new containers were the initially 'Suggested' purchasing changes, and 'Optimal' represents the optimized container purchase. 1NFM, 2NFM, and 4FM households were assumed to purchase 1HG, 1 G, and 2 G containers, respectively. 1NFM, 2NFM, and 4FM households were suggested to purchase 2Qs, 2HGs, and 4HGs, respectively. Lastly, 1NFM and 4FM households were optimized for consumption and spoilage when purchasing 1Q and 3HGs, respectively"; reproduced from [3] under CC BY 4.0 license.

Figure 18.10 shows the EFP of reducing the weight of a 100 12 fL oz (355 mL) glass bottles by 10% and 20% as discussed in Chapter 11. We can observe a direct correlation of the EFP reduction and weight reduction. However, if a bottle weight reduction of 20% translates to 5% bottle breakage, we must realize that we will need 5% more bottles to deliver a set volume of the product to the consumer (e.g., 105 × 12 fl oz). So, our EFP will be increased by the five extra bottles that are required as well as the EFP of the product contained in these bottles. If the bottle contains water locally packaged, the EFP may be low; however, the EFP will be much more significant if the bottle contains milk or wine shipped from overseas.

Now, let us consider another EoL scenario. We want to reuse the glass bottle 100 times. For this example, we will neglect the EFP of cleaning the bottle for every use, which will increase each EFP, in particular water emissions [20]. We can observe a significant reduction of the EFP of the glass bottle when reusing is considered. This makes sense since the initial high EFP of the heavy glass bottle is distributed over 100 uses (see Figure 18.11). In the case of a plastic bottle, the initial EFP of the bottle is much lower. Thus, the benefit of reuse may not be as significant. Figure 18.11 also shows a preliminary estimation that increasing the distribution distance of the reuse network has a direct impact on the final EFP of the reused glass bottle. If the return trip of the bottle increases from 100 to 1,000 miles, the GHG emissions increase by 53%. However, it is still lower than the single-use glass bottle due to the large initial EFP of the glass bottle. The same absolute EFP gain benefit done by reuse may not be obtained for lighter and less environmentally intense products such as plastics and aluminum, as suggested in Table 18.1. However, reuse cycles are complex and further calculations should be performed before implementation.

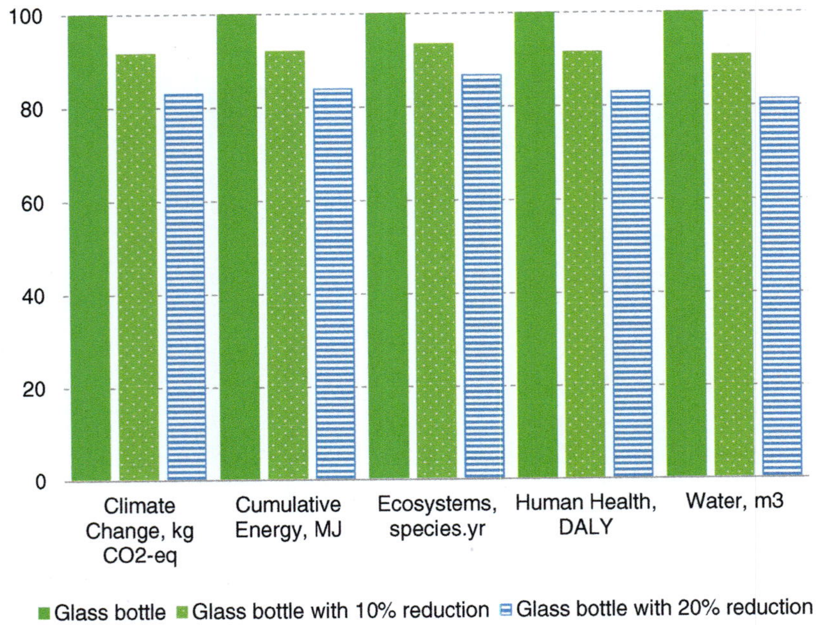

Figure 18.10 Environmental footprint indicators of a glass bottle and 10% and 20% reduction in its weight. The EFPs were estimated with PackageSmart.

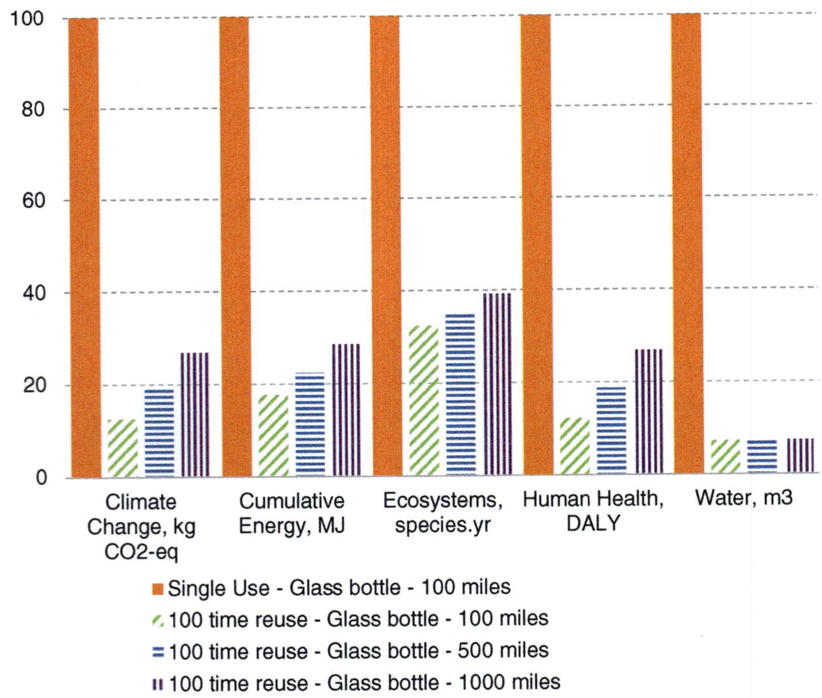

Figure 18.11 Environmental footprint of single-use glass bottle and 100 time-reused glass bottles distributed 100, 500, 1,000 miles return cycle. The EFP were estimated with PackageSmart. The y-axis indicates the EFP of a single bottle as the norm.

Further analysis of the EoL scenarios can be conducted considering incineration with energy recovery and landfill. However, in the case of glass or aluminum, no meaningful energy is recovered in incineration and glass and metal are entombed in landfill for centuries, so there are no benefits or reduction of their overall EFP obtained through these waste management options. In the case of plastic, their energy can be recovered in MSW incinerators as discussed in Chapter 15.

18.9 Case Study 18.1. Environmental Footprint of Milk Package Containers in the US

Although milk is less impactful than meat, it is consumed in large quantities and accounts for 13% – 7.7 billion kg – of the total 60 billion kg of US food waste [21]. So, preservation and proper packaging of milk is crucial for sustainable development as milk is a food staple around the world. Of the total milk produced, 32% is wasted, with consumers accounting for the largest portion of fluid milk waste generation, 20%, compared to retailers, which account for 12% [21]. For every kg of fluid milk consumed, approximately 2.05 kg of CO_2-eq are emitted, including package EoL [22]. Consumption of milk accounts for 50 billion kg of CO_2-eq each year. If 20% of milk is wasted, then consumer milk waste amounts to 10 billion kg of CO_2-eq released annually [3].

Bureck et al. evaluated the EFP of 12 in-home and 6 on-the-go fluid milk delivery systems from cradle-to-grave. The fluid milk delivery systems modeled included a farm gate-to-grave system boundary described by seven main stages: raw milk transport from the farm, container manufacturing and filling, processing plant, distribution center, retail center, milk consumption for in-home and on-the-go environments, EoL and necessary transport between stages (Figure 18.12). The functional unit (FU) of the study was one thousand kilograms of milk consumed.

The authors reported the complete model from farm gate-to-grave of consumed milk in the US and reported EFP indicators such as climate change impact, marine eutrophication, particulate matter formation, freshwater eutrophication, human toxicity (cancer and noncancer), photochemical oxidant formation, and ecotoxicity. Figure 18.13 shows the EFP of farm gate-to-grave of consumed milk in the US in the classic monolayer HDPE chilled 1 gal container. Of the nine reported EFP indicators, the eight highest contributions correspond to raw milk, followed by consumption and the processing plant. In addition, the container impacts are about 15% for cumulative energy demand and below 5% for all other EFP indicators.

Figure 18.14 shows the GHG emission/FU in relation to the weight of each container. Although the study was not conducted as a comparative assertion due to the lack of sufficiently robust data to make comparisons, a few conclusions can be drawn. Lightweight packages showed lower GHG than heavy ones (e.g., pillow pouch vs. gable-top carton chilled 16 oz.), and a large variation of the GHG can be seen for each container.

18.10 Case Study 18.2. The Perceived and Actual Environmental Footprint of Glass, Plastic, and Aluminum Beverage Packaging

It may be clear by now that consumers' perceived EFP and the actual EFP of packaging are often not the same. We tend to have an impression and a preconceived preference for what we believe is a more sustainable package (e.g., glass bottle over aluminum can or plastic bottle). However, we have demonstrated using LCA that the optimal EFP can be quite the opposite of what we perceive.

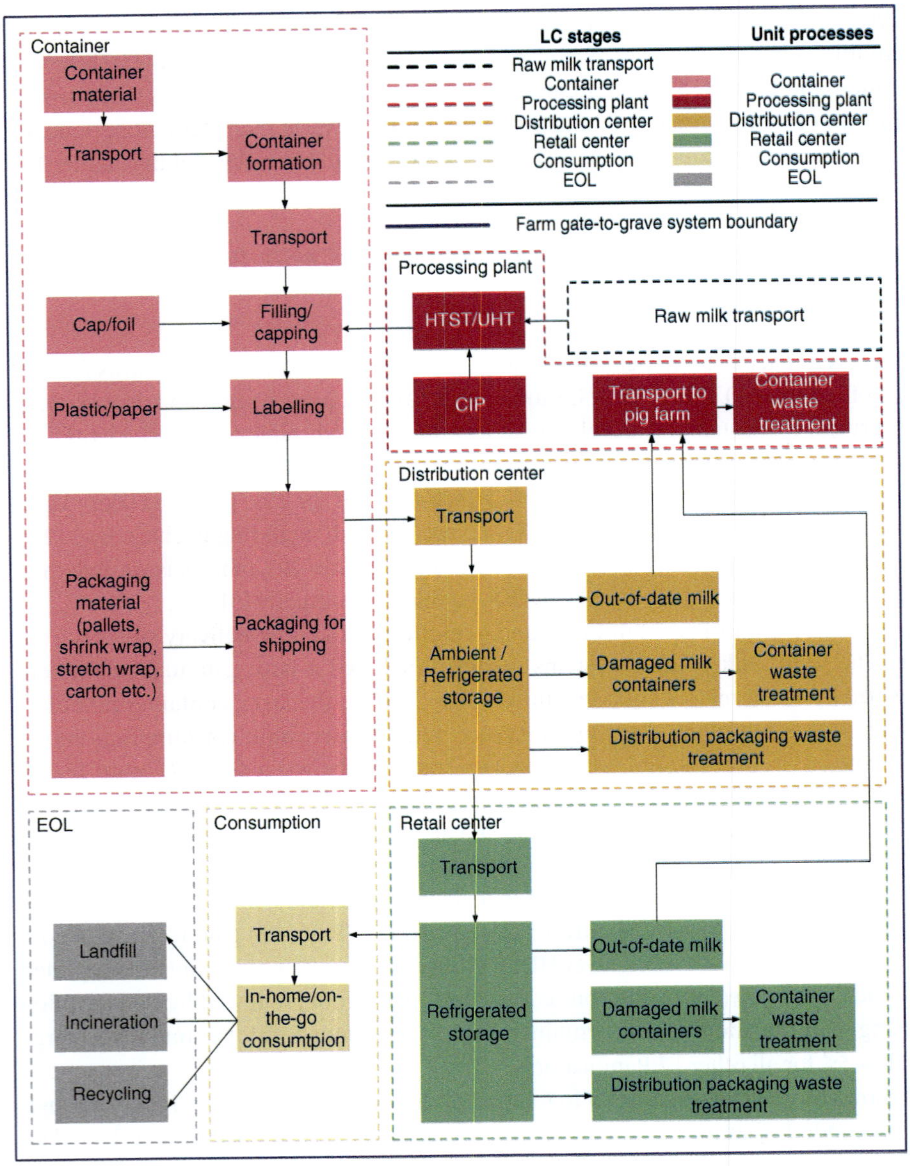

Figure 18.12 System boundaries for modeling the farm gate-to-grave of milk and its different seventh stages; reproduced from [4] with permission from John Wiley & Sons.

This perception is essential to understand and correct since nature and global climate change do not care where the emissions are coming from (e.g., kg CO_2-eq) as long as we minimize our EFP. Consumers should have the correct information to make the right choices about their product consumption and selection of packaging systems.

Recently, De Feo et al. (2022) [23] evaluated the perceived environmental attributes of a glass bottle, plastic bottle, and aluminum can for beverages (see Figure 18.15). They asked young students from an Italian university questions about their perception of these beverage containers. The survey was intended to understand what factors influence the students' purchase decisions and preferences and evaluate whether environmental sustainability of packaging

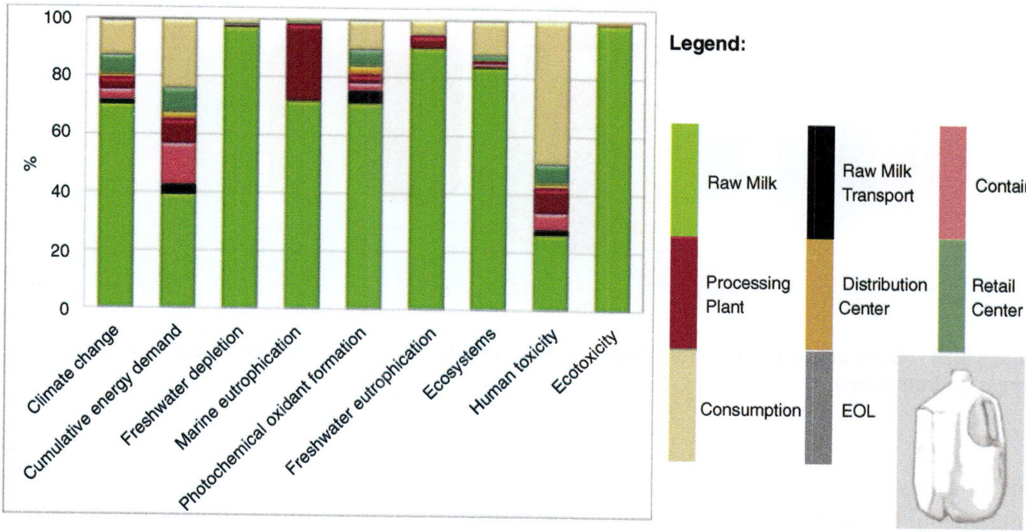

Figure 18.13 Environmental footprint from farm gate-to-grave of consumed milk in the US in the classic monolayer HDPE chilled 1 gal container.

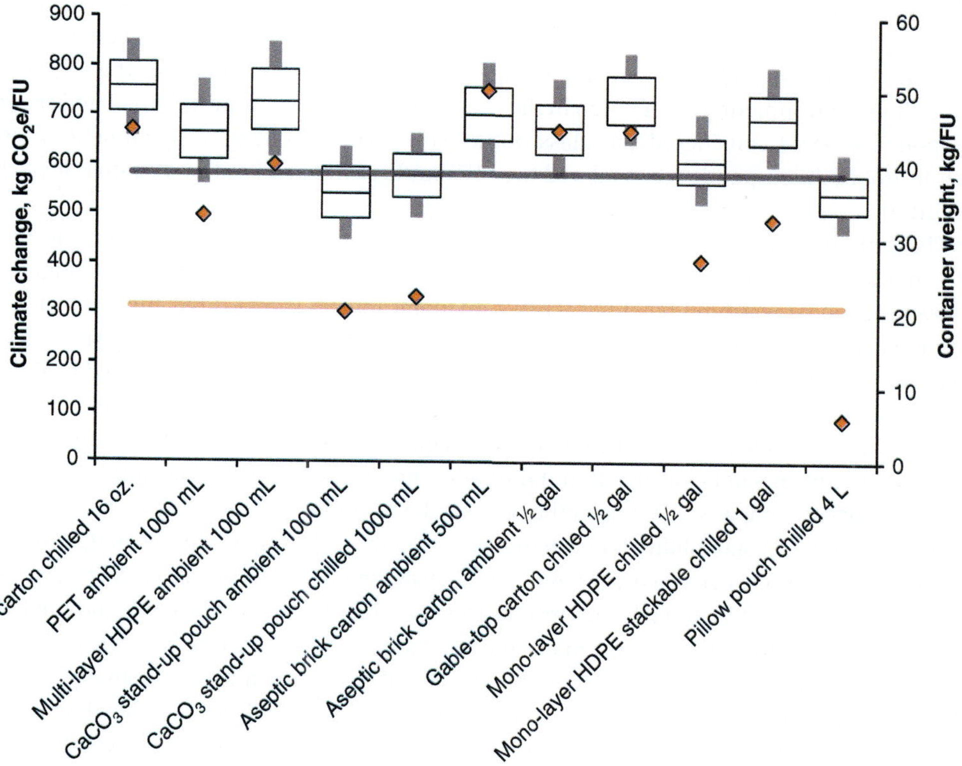

Figure 18.14 Greenhouse gas emissions comparison of in-home milk containers against the classic monolayer HDPE chilled 1 gal container for delivery of thousand kilograms of milk consumed. The gray line is the baseline impact for the HDPE chilled 1 gal container. The weight of the containers is plotted in color on the right-hand axis; reproduced from [4] with permission from John Wiley & Sons.

(a) (b) (c)

Figure 18.15 Glass bottle, plastic bottle, and aluminum can used for the study, reproduced from [23], with permission from Elsevier.

materials was one of them. Additionally, they assessed the actual EFP of the beverage containers using LCA. Then, they correlated the information obtained through the survey with an LCA study of the containers.

Figure 18.16a shows the perceived sustainability for the different packaging alternatives. Young students perceived that glass bottles were 86% very or extremely sustainable and PET bottles 73% not at all or low sustainable. The aggregate data from the survey (Figure 18.16b) show that 75% of the students selected the glass bottle as the most sustainable. The aluminum can was selected by 20% of the students. The results obtained from the University of Salerno students were compared with an adapted questionnaire sent to members of the *Legambiente* – an Italian environmental association in the Campania region of Southern Italy. The *Legambiente* responders (89%) considered the glass bottle as a very or extremely sustainable alternative while 76% of the responders considered the PET bottle as not at all or low sustainable packaging solution.

The authors also conducted an LCA to quantify the EFP of a glass bottle, aluminum can, and a PET bottle (all 33 cL), PET – 50 cL, as well as aseptic cartons with sizes 33 and 20 cL. Figure 18.16c shows the global warming potential (GWP) midpoint indicator and three endpoint indicators – human health, ecosystems, and resources, as reported by Recipe (2016) (H). They showed that the glass bottle had the largest EFP for the GWP and three endpoint indicators followed by the PET 33 cL, and then there was a trade-off among the indicators for the other containers. The highest impact of the glass bottle was attributed to the required glass bottle weight to contain the 33 cL. The aseptic carton – not considered in the survey – had the lowest EFP. The authors then defined an indicator for comparison between the perceived and actual sustainability as a ratio of the perceived and actual sustainability for each indicator. Figure 18.16d shows the 18-midpoint indicator comparison between perceived (PS) and actual sustainability (AS) of the packaging systems as reported by ReCiPe (2016) (H). The results show a large discrepancy for the glass bottle, with response values larger than 1 and many greater than 2, indicating that the perceived environmental sustainability was more than double the actual environmental sustainability. In contrast, the PET bottle and aluminum can were much more environmentally sustainable than consumers perceived, with the largest gap for the PET bottle, maybe due to an inaccurate perception of plastic. This study shows the importance of communication between the stakeholders, the scientific community, and consumers. So, better communication techniques and education, as recommended by the authors of the study are needed, to eradicate misinformation, misconception, and prejudices about packaging systems.

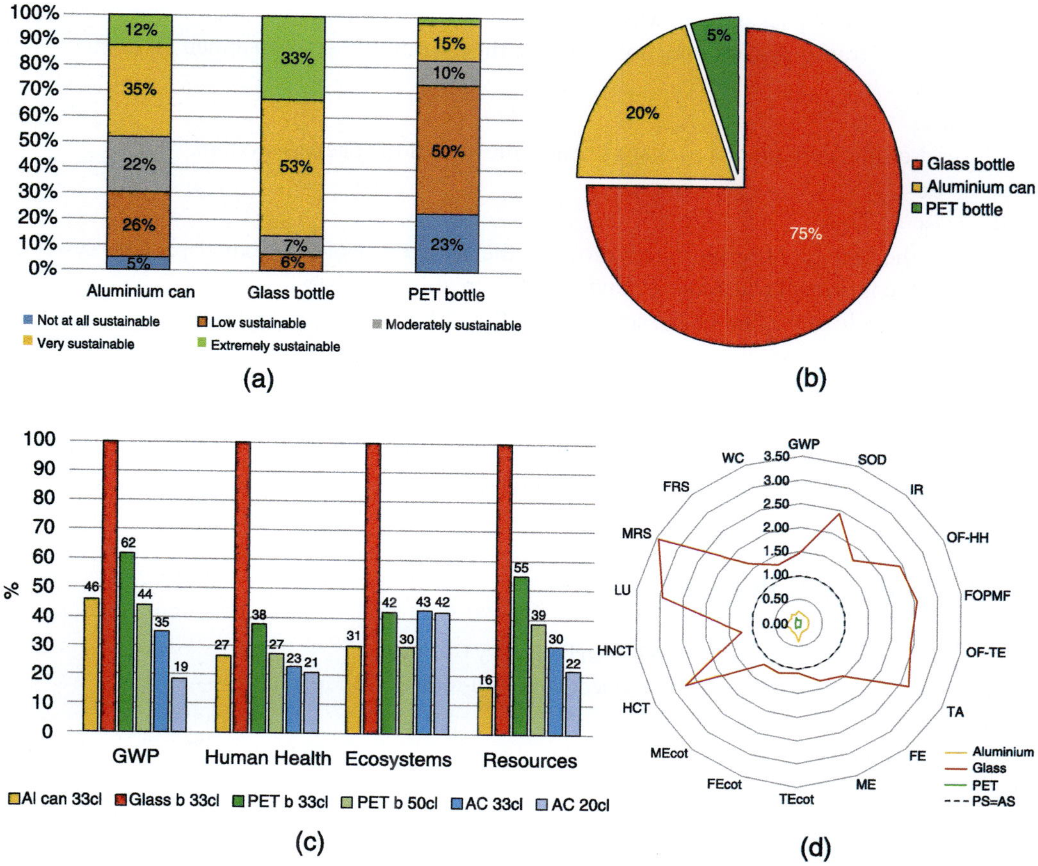

Figure 18.16 Results of the study: (a) perceived sustainability by consumers for different packaging alternatives; (b) consumers' choice of most sustainable packaging solution; (c) percentage comparison of the environmental impact of packaging alternatives estimated as global warming potential midpoint indicator and three endpoint indicators, human health, ecosystems, and resources of ReCiPe (2016) (H); (d) 18 indicators comparison between PS and AS of the packaging systems referred to the midpoint impact categories of ReCiPe (2016) (H); reproduced from [23] with permission from Elsevier.

18.11 Study Questions

1 Discuss the importance of understanding the entire environmental footprint of a product/package system.

2 For which types of products are the packaging systems likely to have large EFP contributions?

3 Packaging has the lowest EFP contribution in the supply chain of:

A Spinach

B Milk

C Mushrooms

D Cheese

E Beef

F None of the above

4 Discuss the role of packaging in waste creation.
5 The mode of transportation does not influence the overall EFP of a consumed product.
  A True
  B False
6 The end-of-life scenario of a package impacts the EFP of the package system.
  A True
  B False
7 Explain the impact of transportation mode on the overall EFP of the product/package system.
8 Discuss the impact of consumer behavior on the creation of waste.
9 Discuss the role of the end-of-life scenarios on the EFP of packaging systems.
10 Packaging represents:
  A Less than 10% of the energy of the US food supply
  B Between 10% and 30% of the energy of the US food supply
  C Between 30% and 50% of the energy of the US food supply
  D More than 50% of the energy of the US food supply

18.12 Additional Resources

1 EPA – Sustainable Management of Food – Food Recovery Hierarchy
https://www.epa.gov/sustainable-management-food/food-recovery-hierarchy
2 Food waste
https://www.nrdc.org/food-waste
3 Food Print
https://foodprint.org/issues/the-problem-of-food-waste
4 Food Waste in America: Facts and Statistics
https://www.rubicon.com/blog/food-waste-facts
5 FDA – Food Waste Animations
https://www.fda.gov/food/consumers/food-waste-animations
6 Food Loss and Food Waste
http://www.fao.org/food-loss-and-food-waste/flw-data

References

1 Williams, H., Lindström, A., Trischler, J., Wikström, F., and Rowe, Z. (2020). Avoiding food becoming waste in households – The role of packaging in consumers' practices across different food categories. *J Clean Prod* 265. https://doi.org/10.1016/j.jclepro.2020.121775.
2 Heller, M. (2017). *Food product environmental footprint literature summary: Food transportation*, 15. https://www.oregon.gov/deq/FilterDocs/PEF-FoodTransportation-FullReport.pdf (accessed November 15, 2020).
3 Stankiewicz, S.K., Auras, R., and Selke, S. (2019). Modeling American household fluid milk consumption and their resulting greenhouse gas emissions. *Sustainability* 11: 2152. https://doi.org/10.3390/su11072152.

4 Burek, J., Kim, D., Nutter, D., Selke, S., Auras, R., Cashman, S. et al. (2018). Environmental sustainability of fluid milk delivery systems in the United States. *J Ind Ecol* 22: 180–195. https://doi.org/10.1111/jiec.12531.

5 Heller, M.C., Selke, S.E.M., and Keoleian, G.A. (2019). Mapping the influence of food waste in food packaging environmental performance assessments. *J Ind Ecol* 23: 480–495. https://doi.org/10.1111/jiec.12743.

6 Heller, M. (2017). *Food product environmental footprint literature summary: Beer*. http://www.oregon.gov/deq/FilterDocs/PEF-Beer-FullReport.pdf (accessed November 15, 2020).

7 Heller, M. (2017). *Food product environmental footprint literature summary: Wine*, 1–5. https://www.oregon.gov/deq/FilterDocs/PEF-Wine-FullReport.pdf (accessed November 15, 2020).

8 Hassard, H.A., Couch, M.H., Techa-Erawan, T., and Mclellan, B.C. (2014). Product carbon footprint and energy analysis of alternative coffee products in Japan. *J Clean Prod* 73: 310–321. https://doi.org/10.1016/j.jclepro.2014.02.006.

9 Manfredi, M. and Vignali, G. (2014). Life cycle assessment of a packaged tomato puree: A comparison of environmental impacts produced by different life cycle phases. *J Clean Prod* 73: 275–284. https://doi.org/10.1016/j.jclepro.2013.10.010.

10 Heller, M. (2012). *Food product environmental footprint literature summary: Citrus*, 1–5. https://www.oregon.gov/deq/FilterDocs/PEF-Citrus-FullReport.pdf (accessed November 15, 2020).

11 Heller, M. (2012). *Food product environmental footprint literature summary: Pork*, 1–5. https://www.oregon.gov/deq/FilterDocs/PEF-Pork-FullReport.pdf (accessed November 15, 2020).

12 US Environmental Protection Agency. (2020). *Advancing Sustainable Materials Management: 2018 Tables and Figures Assessing Trends in Materials Generation and Management in the United States*. U.S. EPA.

13 Kaza, S., Yao, L., Bhada-Tata, P., and Van Woerden, F. (2018). What a waste 2.0: A global snapshot of solid waste management to 2050. *The World Bank*. https://doi.org/10.1596/978-1-4648-1329-0 (accessed November 15, 2020).

14 Svanes, E., Vold, M., Møller, H., Pettersen, M.K., Larsen, H., and Hanssen, O.J. (2010). Sustainable packaging design: A holistic methodology for packaging design. *Packag Technol Sci* 23: 161–175. https://doi.org/10.1002/pts.887.

15 Williams, H. and Wikström, F. (2011). Environmental impact of packaging and food losses in a life cycle perspective: A comparative analysis of five food items. *J Clean Prod* 19: 43–48. https://doi.org/10.1016/j.jclepro.2010.08.008.

16 Wikström, F., Verghese, K., Auras, R., Olsson, A., Williams, H., Wever, R. et al. (2019). Packaging Strategies That Save Food: A Research Agenda for 2030. *J Ind Ecol* 23: 532–540. https://doi.org/10.1111/jiec.12769.

17 Which form of transport has the smallest carbon footprint? (2021). *Our World in Data*. https://ourworldindata.org/travel-carbon-footprint (accessed June 2, 2021).

18 Hebrok, M. and Boks, C. (2017). Household food waste: Drivers and potential intervention points for design – An extensive review. *J Clean Prod* 151: 380–392. https://doi.org/10.1016/j.jclepro.2017.03.069.

19 Hanssen, O.J., Meiler, H., Svanes, E., and Schakenda, V. (2012). Life cycle assessment as a tool in food waste reduction and packaging optimization – packaging innovation and optimization in a life cycle perspective. In: *Life Cycle Assessment Handbook: A Guide for Environmentally Sustainable Products* (ed. M.A. Curran), 345–367. https://doi.org/10.1002/9781118528372.ch16.

20 Chariyachotilert, C., Joshi, S., Selke, S.E., and Auras, R. (2012). Assessment of the properties of poly(L-lactic acid) sheets produced with differing amounts of postconsumer recycled poly(L-lactic acid). *J Plast Film Sheeting* 28: 314–335. https://doi.org/10.1177/8756087911434337.

21 Buzby, J.C. and Hyman, J. (2012). Total and per capita value of food loss in the United States. *Food Policy* 37: 561–570. https://doi.org/10.1016/j.foodpol.2012.06.002.

22 Thoma, G., Popp, J., Nutter, D., Shonnard, D., Ulrich, R., Matlock, M. et al. (2013). Greenhouse gas emissions from milk production and consumption in the United States: A cradle-to-grave life cycle assessment circa 2008. *Int Dairy J* 31: S3–14. https://doi.org/10.1016/j.idairyj.2012.08.013.

23 De Feo, G., Ferrara, C., and Minichini, F. (2022). Comparison between the perceived and actual environmental sustainability of beverage packagings in glass, plastic, and aluminum. *J Clean Prod* 333: 130158. https://doi.org/10.1016/j.jclepro.2021.130158.

Index

Life Cycle of Sustainable Packaging: From Design to End of Life, First Edition. Rafael A. Auras and Susan E. M. Selke.

Companion Website: www.wiley.com/go/Auras/lifecycleofsustainablepackaging

d

m

Printed and bound by CPI Group (UK) Ltd, Croydon, CR0 4YY
17/09/2026
14964258-0004